CHEMICAL PROCESS INDUSTRIES

VOLUME II
Second Edition

CHEMICAL PROCESS INDUSTRIES

VOLUME II

Second Edition

W.V. MARK
M.Sc., Ph.D., C.Eng.

Edited by
S.C. BHATIA
B.E. (Chem), M.B.A.

CBS PUBLISHERS & DISTRIBUTORS PVT. LTD
New Delhi • Bengaluru • Chennai • Kochi • Mumbai • Pune

ISBN : 81-239-1142-4

First Edition: 2001
Second Edition: 2007
Reprint: 2008, 2015

Published by:
Satish Kumar Jain for CBS Publishers & Distributors Pvt. Ltd.,
CBS Plaza, 4819/XI Prahlad Street, 24 Ansari Road,
Daryaganj, New Delhi – 110002, India
delhi@cbspd.com, cbspubs@airtelmail.in • www.cbspd.com
Ph.: 23289259, 23266861, 23266867 • Fax: 011-23243014

Branches:

- ***Bengaluru:*** Seema House, 2975, 17th Cross, K.R. Road,
Bansankari 2nd Stage, Bengaluru - 560070
Ph: +91-80-26771678/79 • Fax: +91-80-26771680
E-mail: cbsbng@gmail.com, bangalore@cbspd.com
- ***Chennai:*** 20, West Park Road, Shenoy Nagar, Chennai - 600030
Ph: +91-44-26260666, 26208620 • Fax: +91-44-42032115
E-mail: chennai@cbspd.com
- ***Kochi:*** 36/14, Kalluvilakam, Lissie Hospital Road, Kochi - 682018
Ph: +91-484-4059061-65 • Fax: +91-484-4059065
E-mail: cochin@cbspd.com
- ***Mumbai:*** 83-C, Dr. E. Moses Road, Worli, Mumbai - 400018
Ph: +91-9833017933, 022-24902340/41 • E-mail: mumbai@cbspd.com
- ***Pune:*** Bhuruk Prestige, Sr. No. 52/12/2+1+3/2,
Narhe, Haveli (Near Katraj-Dehu Road Bypass), Pune - 411041
Ph: +91-20-64704058/59, 32342277 • E-mail: pune@cbspd.com

Printed at :
. S. Offset Printers, Delhi

Preface to the Second Edition

The new second edition strengthens the text book, which has now been thoroughly revised and edited by incorporating the suggestions made by various professors and students. The salient features of this revised edition are—considerable changes and addition have been made in the entire text to clear the points that were not apparent or needed additional information for better understanding. Some chapter have been completely overhauled and rewritten with the inclusion of additional and latest technology. The concept of 'Green chemisty' includes—waste minimisation, solvent selection and selecting alternative synthetic routes from sustainable resources. Keeping this in mind an additional chapter on 'Green Technologies through Chemical Engineering' has been added.

This volume, dealing with organic aspects is the second volume of the comprehensive textbook on chemical process industries. This invaluable model textbook is specifically intended for B.E/B.Tech institutions which offer courses in Chemical Engineering in India, including the Indian Institutes of Technology and its primary aim is to provide an exhaustive coverage of principles and concepts which form the basis of chemical process engineering. As engineering education becomes more complex, it is essential for a chemical engineer to be familiar with the realisation of scientific principles in chemical process industries. Even the most sophisticated engineering curriculum must give the student some cultural background of the process industries such as, for example, how sulphuric acid is produced or how petroleum refining and the caustic soda industries have been affected by increasing demands. In addition to academic knowledge, the chemical engineer must have an appreciation of the professional role in the chemical process industries.

The encyclopaedic range of the subject of chemical process industries has been condensed into a handy, easy-to-use reference book, presented in a concise, compact and lucid manner. Stressing the essential facts for the student, it forms the necessary criteria for a systematic study of chemical process industries, the textbook informs, clarifies and explains how chemical processing transforms raw materials into usable and profitable products by chemicals.

Each chapter covers an important aspects of chemical processing industries with an accurate, up-to-date account of each topic. Chapter 1 highlights the role of chemical industry in this millennium. No organic industry is more important to technical civilisation than petroleum. Therefore, Chapter 2 outlines origin, exploration, constituents and contain details on petroleum refinery. Chapter 3 is devoted to mechanism of polymerisation and various methods of polymerisation are also discussed. Chapter 4 familiarises students with synthetic fibres and film industries based on polyfibres, polyamides, polyester, acrylics etc.

Chapter 5 describes in detail natural and synthetic rubbers. The preparation, properties, compounding and applications of various synthetic rubbers are the highlight of this chapter. Plastic industries, their classifications and manufacturing aspects are discussed in chapter 6. Chapter 7 contain information about various vegetable oils, animal oils, fats and waxes. Chapter 8 is devoted to soaps and detergents, emphasising the significant differences between them in the manufacturing process and chemical combination, resulting in differences in action. Essential oils, perfume, flavour and fragrances are dealt in Chapter 9. Chapter 10 is devoted to surface coating industry and its various products like pigments, paints, varnishes and lacquers are also discussed in detail.

Pulp and paper industries and their manufacturing process are discussed in chapter 11. Chapter 12 deals with sugar and starch industries, sugar derivatives like sucrose, fructose, maltose and lactose are also discussed in detail. Chapter 13 describes fermentation and distillery emphasising on manufacturing and uses of industrial ethyl and butyl alcohol. Acetone, acetic acid, vinegar and citric acid also form the part of this chapter. Chapter 14 is devoted to food processing industries, food additives, milk and milk products, poultry, and meat processing methods are the subject of this chapter. Chapter 15 explores leather and tannery, their raw materials, manufacturing, uses and also health and safety aspects. Gelatin and various adhesives like hot melt, rubber based and high temperature adhesives also form the part of this chapter. Chapter 16 deals with dye and dye intermediates and various methods used for manufacturing dye intermediates and organic dyes are also discussed. Chapter 17 discusses agrochemical industries and the utilisation of chemicals as pesticides and insecticides. Chapter 18 is devoted to coal and coal chemicals, utilisation of coal and coal cleaning and air pollution in coal manufacturing are also dealt in this chapter. Chapter 19 focuses on pollution control. Various environmental pollutants, air, water, solid wastes, radioactive wastes and noise pollution, their causes and control are discussed in detail in this chapter. Green chemistry includes concepts such as waste minimisation, solvent selection, atom utilisation, E factor, intensive processing and alternative synthetic routes from sustainable resources. Keeping this in mind chapter 20 is devoted to green technologies through chemical engineering.

The simple, lucid and readable style of text will help the students to comprehend the concepts and information presented, and ensures an easy grasp of key-points which are illustrated by diagrams and tables. A carefully presented index allows the student easy access to each topic covered in the book. Additional information on modern technologies and a special emphasis on environmental issues are the highlights of this well-graded textbook.

While painstacking care has gone into producing a useful and exhaustive textbook, the author would welcome constant constructive criticism and a creative feed back from students and professors, and interaction that will certainly be fruitful.

The text could not have been completed without the help of many colleagues who provided figures, photographs and offered important advice during the course of its preparation.

The Chemical process industries text book first edition printed in 2001 received very well and it is expected that the revised second edition will also be received with same interest.

Appreciation is also extended to Dr. J. L. Mayo who spent many hours preparing high-quality computer graphics when he had much more interesting things to do.

W. V. Mark

Author

Contents at Glance

Contents in Detail

Volume II : Chemical Process Industries

CHAPTER 2

CHAPTER 3

CHAPTER 4

CHAPTER 5

CHAPTER 6

CHAPTER 7

CHAPTER 8

CHAPTER 9

CHAPTER 10

CHAPTER 11

CHAPTER 13

CHAPTER 14

CHAPTER 17

CHAPTER 19

Chapter 1

Chemical Industry in the New Millennium : An Overview

INTRODUCTION

Technology is the application of science to develop and control processes, methods or techniques for the manufacture of industrial and consumer products in diverse fields and industries. The chemical process industry is an industry whose product(s) results from one or more chemical or physico-chemical changes; extraction, separation, or purification of a natural product, with or without the aid of chemical reactions; the preparation of specifically formulated mixtures of materials, either natural or synthetic. Examples are: plastics, rubber, leather, food, dye and synthetic organic industries; petroleum, paper, textile, and perfume industries; glass, cement, fertiliser, soap, and paint industries. Many of these involve one or more unit operations of chemical engineering, as well as such basic processes as polymerisation, oxidation, reduction, hydrogenation, etc., usually with the aid of a catalyst. This definition could be interpreted to include ore processing, separation and refinement, as well as the manufacture of metal products; however, these are usually considered to comprise the metal and metallurgical industries.

The discipline of chemical engineering has continued to contribute in numerous and important ways to worldwide industrial progress over the past decade. Significant strides have been made in process and equipment design methods. This progress has been due in part to factors such as increased computer utilisation and availability of better material of construction. The focus on the increased value of energy and the need for improved waste management have also affected design decisions. Basically, however good chemical engineering practice is still founded on the sound application of the two aspects of empiricism and theory.

Chemical engineering is concerned with the development and operation of processes for production purposes, engineering aspects, such as design, construction, installation and operation of plants and equipment for manufacturing chemical products.

A chemical engineer's work is to develop and work out processes for bulk manufacture of chemicals and useful industrial products. He ensures that raw materials and equipment are purchased to the prescribed specifications and that manufacturing processes are controlled properly for obtaining the maximum yield with the minimum cost. Quite often, he is called upon to discover new raw materials, find out indigenous substitutes for the imported raw materials, and test and standardise raw materials and finished products.

To achieve the above, he employs either the established or the new methods, using different equipment and machines, involving many processes or operations. Thus the responsibility of a Chemical Technologist ranges from research and production to general administration. With higher assignments, the chemical engineer finds himself increasingly engaged in administrative and managerial affairs of the industry and may finally become responsible for the whole organisation, looking after its various aspects: production, research, development, sales, labour welfare, etc.

In his work, the chemical engineer is assisted by a team of junior technologists, research workers, operators, mechanics and laboratory technicians. These workers mainly carry out day-to-day analysis of raw materials and finished products, to control the processes at various stages for obtaining finished products of a standard quality, to operate repair and maintain the machinery in working order and to take precautionary measures to avoid contamination.

The work of a chemical engineer is essentially indoors, under varying conditions. It may be the congenial atmosphere of a laboratory or that of a hot, humid, dusty and smoky factory with all sorts of obnoxious smells. The work may also require at times handling of bulk quantities of corrosive and injurious acids, gases and other chemicals. However, many chemical engineers are engaged in products sales—market research, product development, technical sales, and customer technical service etc.

CHEMICAL INDUSTRY IN THIS MILLENNIUM

The world is presently experiencing changes of momentous magnitude. The chemical industry is no exception. Information revolution has swept all aspects of our lives. Physical and financial boundaries have virtually collapsed. Skills, capital and market have become global; transcending earlier barriers. Customers' demands are becoming increasingly exacting and investors acutely conscious of their power, are unsparing towards management and uncompromising in their expectations of return. Both customers and shareholders arc ready to punish inefficiency mercilessly and unhesitatingly. To this volatile atmosphere, scientific and technological innovations add exciting opportunities and inevitably incredible unpredictability too.

Against such a daunting backdrop, it will be an audacious adventure in futurology to predict where the chemical industry is heading towards, say, in the next fifty years. We will, therefore, focus on the major global trends, their effect on chemical industry in the coming decades, crucial technological drivers, operational trends and perspectives for India's chemical industry.

Global Chemical Industry

Size, structure and growth

Let us take a look at the global chemical industry in terms of its size, structure, transition and future directions. With more than 70,000 commercial products, few other industries approach this degree of complexity and diversity. The global chemical industry is valued at one and a half trillion US dollars today, about 6% of the global GDP of 28 trillion dollars. In terms of size it is five times the size of the Indian economy. The total world trade in chemicals is valued at US$ 400 billion 10% of the value of global trade. Geographically, W. Europe accounts for 35%, USA for 26%, Japan for 15% and the rest of Asia for 12% of its output. In terms of sectors, petrochemicals dominate the global chemical industry with a share of 25%. Pharmaceutical is the next largest sector with a share of 16.5%, closely followed by Performance chemicals at 16%, both valued at around US$ 250 billion. Global chemical industry growth has consistently been outpacing GDP growth by a factor of 1.5%.

Global dominance and consolidation

Nineties have seen a lot of activities in consolidation of businesses globally along product lines, departing from the diversification era of the eighties. Companies are choosing to stay in a field whether it is petrochemicals or speciality or life science and aiming for global dominance. For example, Hoechst, the largest chemical company in the world has slimmed from 120 businesses to only 30. Montell, formed out of the merger of Montedison's Himont and Shell's Polypropylene business created overnight at global player with 20% of market across the globe. The frantic merger activity in the oil industry is unprecedented, the latest one being 75 billion USD take-over of Mobil by Exxon to make it the largest oil company in the world. This was preceded by another mega merger between BP of UK and Amoco Corporation of US. What is prompting these mergers? Hard economics seem to be the driver behind these moves. Exxon-Mobil is talking about pre tax synergy of 3.0 billion USD, BP-Amoco is hopeful of saving 2.0 billion USD by global integration of assets and management, besides complementary operation and capacity utilisation. This also means higher shareholder value, higher risk taking ability for multi-billion dollar, large gestation projects, in new geographical areas.

Concept of outdated anti-trust and monopoly

If we look at life sciences field, similar trends are visible. Pharmaceutical and agrochemical companies are joining forces and recreating themselves as enterprise based on biological innovation. Only recently, France's Rhone Poulenc and German giant Hoechst merged to form what they called as the world's largest life science company called 'Aventis', joining the ranks of 'Novartis', a company formed by the merger of Ciba-Geigy and Sandoz of Switzerland, arguably the founder of life science concept. All these mergers aim to take advantage of the common technology platform at the level of DNA. So the main justification seems to be one of technology, as biological research becomes more complex and expensive with cost of developing a drug running into hundred of million of dollars. These companies are sharing the high risk of failure in this field and of course the huge rewards which success brings.

These trends for future indicate a move away from creating new capacities: spinning off to focus on core business/segment; merging for synergy, re-grouping for segment leadership. We may see mitigation of the boom and bust cycle, which bedeviled the commodity chemicals business in the past. Chemical industry in the developed world is adopting strategies, which are centred around a global market place. The biggest challenge it faces is value creation. In its quest for value creation, they will bring their full force into play in terms of organisation, technology, finance, synergy and specialisation. This will result in fewer larger companies—we may have half the companies we have today in the next decades.

Global Forces of Change and Evolution of a New Era

Emergence of Asia

In such an industrial milieu, as we enter the third millenium, Asia is poised to emerge as the world's centre of economic activities. Notwithstanding the current Asian turmoil, which will retard the growth for three or four years, two of the largest countries in the region, China and India, together accounting for 25-30% of the world's population, would continue to achieve GDP growth of over 5% over the next few decades. Consequently, developing countries aggregate GDP should triple by 2010. It is likely that China will overtake the USA as the world's largest economy in less than 40 years in GDP terms or even sooner in terms of purchasing power. This shift in the market concentration would have a profound impact on the structure of the global chemical industry.

While new players in the Asian region are emerging to play a global role (including players in the Middle East enjoying feedstock advantages), major global players faced with saturated markets in their home turf would seriously be muscling for dominance in the region for next level of growth. With deeper pockets and technological strength they would contest every additional ounce of this new market. Therefore, the ability to attain supremacy in the Asian markets will determine the winner in the chemical/ commodity industry.

Growth and sustainable development

As global population crosses 8 billion by 2025; the majority of the growth occurring in the developing countries; with people living in this region aspiring for a better quality of life; with sweeping industrialisation waves engulfing two-thirds of the world in Asia and Latin America; finite resources of this planet would be under severe pressure. Forests, soil, water and ocean resources would be pushed beyond limits. The present mode of development with ever increasing uses of energy and associated pollution is not sustainable. The biosphere cannot be tested for its resilience. A new development paradigm is called for. The implication for the chemical industry would be enormous, as technology and processes would be challenged to meet these demands. Responsible care approach and mutual cooperation on the part of developed and developing world and vision of sustainability would drive the operational and envrionmental strategy. There would be three elements on environmental strategy.

Pollution prevention

Emphasis would be on reducing waste and energy use. Emerging global standards for environmental management ISO 14,000 should create strong incentives for companies.

Product stewardship

To reduce environmental effect during the entire life cycle of the product (from cradle to the grave).

Clean technology

Companies need to invest in tomorrow's clean technology. Companies can no longer bury their heads in the sand. Chemical industry needs to lead the way in helping shape public opinion and driving changes in consumer behaviour. After all it makes good business sense to pursue strategies for a sustainable world. The chemical industry also has a major role to play in meeting challenges of society by going beyond environment care with a broader objective of ensuring food security in a large number of developing nation by partnering with the farming community in increasing the yield, and reducing post harvest wastage during transportation, storage and distribution. It can do a remarkable service in ensuring healthcare for all by creating cost-effective medical care and preventive care. In the long term these initiatives would lead to a healthier society and eventually expanding the consumer base for the industry.

India Strategy

We all need to put our creative forces together and develop an India strategy for the chemical industry. The following priority areas need our attention.

New mindset to match dynamic external environment

(1) We need to plan for world class initiatives—in manufacturing, businesses and support functions. We need to benchmark ourselves with the best in the world and leave our cottage industry mindset behind, if we are to succeed in our mission to take on the global chemical companies.

(2) We need to focus on the niche areas in the Indian market that give competitive strengths. We must first seize the immense Indian opportunity before we spread ourselves thin to capture markets in remote corners of this world. Indian chemical industry has invested

heavily in capacity creation in areas like petrochemicals, which would far exceed the current demand in the country and lead to supply led growth in the coming years.

(3) International competiitveness is a must in all ventures. Unless the venture can stand the test of times in a down cycle with nil import duties, it would not make sense to invest. We must focus and consolidate our strengths through mergers and acquisition as free trade and capital is merciless with under-performers.

(4) Chemical plants are capital intensive. We must leverage India's significant capital cost advantages over the developed world. We need to integrate infrastructure needs of the plant for ensuring a cost effective operation. We need to invest much more in research and development and do it strategically.

(5) We have to be friends of the nature and community as a whole, with sustainable environmental policy and responsible care.

(6) We must leverage and utilise people skills effectively. India is blessed with enormous managerial, scientific and technological talents and a large base of computer software operators. These skills are so very appropriate in the changing world order.

(7) In our view, the real challenge is in harnessing people and organisational skills. With similar access to capital, technology and markets what could differentiate an organisation's competitive skills is it's people and organisational skills. We must value this intellectual capital justly.

Future of Chemical Industry in India

India has the opportunity to create history in the chemical industry in the next few decades. The large domestic opportunity, it's inherent strength in intellectual capital, access to global financial markets, strong knowledge of chemistry, strength in software engineering and it's rich bio diversity provide us this great opportunity to make India a global force in chemical industry. Indians are talented as evidenced by thier success in competitive foreign environment. They are hardworking and can be inspired to a "Will do" mind frame to compete for the industry, compete for themselves and for the country for realising the great Indian potential which is rightfully theirs. Industry has to nurture organisations, raise the bar higher to inspire, achieve global competitiveness in whatever they do, partner with institutions and government to implement a India strategy for the chemical industry with the ultimate objective of improving the quality of life of one and quarter billion people who can hold their head high in tomorrow's new world.

Let chemical industry make it's contribution with utmost concern for its community and a sustainable world. The telented force of Indian scientists working in various chemical research laboratories have already done India proud. Catalysts are the heart of a chemical process and of chemical technology. They have also developed new process technologies. The contribution in commercialisation of advanced material of various research laboratories has been laudable. We have to be happy at the success we have achieved in the synthesis of drugs such as Vitamin B6, eco-friendly pesticides, etc. The scientists have earned acclodes for developing disease resistant and high yielding varieties of sugarcane, cardamom, turmeric, ginger and eucalyptus. Tissue culture based bamboo was a path breaking work. The work done by our scientists in the area of biotechnology and plant molecular biology is of far reaching significance.

India looks toward this outstanding centre of excellence with great hopes. Specially, chemical industry looks forward to working closely, creatively and in a co-operative environment. Although the path ahead is not lined with rose petals, we have to make many hard choices along the way. It would be a choice between self-reliance and dependence, a choice between globalisation and isolation, between consolidation and marginalisation, between leadership and surrender and a choice between do or die. So, come let us all put our minds together and dedicate ourselves to create a strong chemical industry on the global map. Let us resolve with determination and far sight to place Indian chemical industry on the global map. We owe it to our future generations.

Chapter 2

Petroleum Refinery and Petrochemicals

NOMENCLATURE IN THE PETROLEUM INDUSTRY

Most petroleum types are complex mixtures that are difficult to characterise in detail, and therefore, many definitions used by the exploration, production, and refining sectors of the industry to describe petroleum and its products lack precision. Traditionally, the unit of crude-oil production has been the barrel, equal to 0.159 m^3, or 5.61 ft^3. Today, however, petroleum reserves are often given in metric tons. Density may be reported in metric units but is more often given in degrees API. The relationship between specific gravity and ÅPI, also called API gravity, is defined by

$$\text{Sp gr (at } 16°\text{C)} = \frac{141.5}{(\text{ÅPI} + 131.5)}$$

or

$$\text{ÅPI} = -\frac{141.5}{\text{sp gr (at } 16°\text{C)}} - 131.5$$

Other terms relating to physical properties are determined under defined conditions established by ASTM.

Gas production is generally given in cubic metres or cubic feet (28.3 m^3 = 1000 ft^3) and reserves are often quoted in trillions of cubic feet (10^{12} ft^3 = 28.3 × 10^9 m^3). Dry gas contains mostly methane; it is called wet gas if it contains more than 4 litres per 100 m^3 of natural gas.

Cycloalkanes (or cycloparaffins) are commonly called naphthenes. Naphtha, however, is a distillate boiling below 204–260°C. Gas oil boils slightly higher than kerosene (232–426°C). Cylinder oil is a viscous lubricating oil. The term distillate is sometimes used to denote distillate fuel oil as opposed to residual fuel oil.

ORIGIN OF PETROLEUM

Petroleum is a naturally occurring complex mixture consisting predominantly of hydrocarbons, frequently with other hydrogen and carbon compounds containing significant amounts of nitrogen, sulphur and oxygen as well as small amounts of nickel, vanadium, and other elements. It can occur in solid, liquid, or gaseous form as asphalt, crude oil, or gas, respectively.

Organic matter is incorporated into sediments as they are deposited and some biogenic gas may be generated; with increasing depth of burial, temperature converts the organic matter to petroleum-like material; part of this material migrates from the source rock through carrier beds to the reservoir; here, compositional changes may be produced by increasing temperature, water washing, and bacterial degradation.

Organisms produce a wide range of organic compounds, and the types incorporated into source rocks influence the type of petroleum generated. Much of the algal-derived amorphous matter (kerogen) is rich in hydrogen and poor in oxygen and generates oil, whereas the woody type of kerogen is rich in oxygen and poor in hydrogen and generates mainly gas. The gener tion of petroleum is non-biological, induced by temperature and influenced by time. It follows essentially first-order kinetics and an increase of 10°C approximately doubles the rate. An important exception to thermal generation is the bacterial formation of methane. The bacteria are anaerobic, effective in sulphate-free anoxic conditions, and give shallow fields with gas that is isotopically light.

Movement of a separate crude-oil phase occurs when source rocks develop high internal pressures which squeeze oil droplets out and into adjacent permeable carrier beds. Here, buoyancy is the main driving force and migration distances can be in excess of 100 km. Oil may be remobilised after its initial accumulation in the reservoir which can lead to considerable compositional changes if both gas and oil are involved.

Petroleum composition can be changed by rising temperature or by contact with flowing water which may remove the more soluble components. If water brings bacteria and oxygen into contact with oil below 75°C, substantial degradation can result giving a higher-density, more sulphur-rich crude oil.

COMPOSITION OF PETROLEUM

The term petroleum, literally, rock oil, is applied to the deposits of oily material found in the upper strata of the earth's crust, where they were formed by a complex series of chemical reactions from organic materials laid down in previous geological eras.

The elemental compositions of petroleum vary greatly from one deposit to another. Carbon and hydrogen make up the bulk, but other elements present include sulphur (trace to 8%), nitrogen (trace to 6%), oxygen (trace to 1.8%), and nickel and vanadium.

Modern techniques, such as gas chromatography (gc) and mass spectrometry (ms) and their combination (gc-ms), permit complete molecular characterisation of the lower boiling range (through C_7) without prior separation. However, isolation and identification of individual compounds is not practical for the higher distillate range. The crude is usually distilled, followed by liquid-liquid chromatography to separate saturated, aromatic, and polar compounds. The vacuum residuum includes the heaviest compound types that do not distill under commercial operating conditions. Special techniques are required for characterisation.

Saturated Hydrocarbons

n-Alkanes are found in all crudes and throughout the entire boiling range. *n*-Alkanes are present through *n*-$C_{44}H_{90}$ in the distillate range and have been identified as high as *n*-$C_{78}H_{158}$. Properties and concentrations vary greatly; carbon number and concentrations have a strong effect upon the pour point, i.e., the temperature at which crude or its products turn from a liquid to a solid. This is an important physical property often defining the necessary degree of processing.

All possible isoalkanes through C_8 have been identified in petroleum, together with many of the C_9 and some of the C_{10} compounds and a number of isoprenoids.

Cycloalkanes isolated from petroleum contain five-, six-, or seven-membered rings. Alkyl derivatives of cyclopentane and cyclohexane are commonly found; dicycloalkanes have also been reported, e.g., *cis*- and *trans*-decahydronaphthalene and hexahydroindan. Cycloalkanes containing up to seven rings have been detected by mass spectrometry.

Analysis

All possible isomers with carbon number 7 or less and most of the C_8s can be separated by capillary gas chromatography. Hydrocarbon mixtures are analysed by mass spectrometry according to the number of rings present in the molecule. These methods have been standardised and published by ASTM. However, the saturated compounds need to be separated from the aromatic compounds before analysis, usually by liquid chromatography over silica gel. Alkenes interfere with this analysis.

Unsaturated Hydrocarbons

Both alkenes (olefins) and aromatic compounds occur in petroleum. Alkenes are rare, whereas aromatic compounds of almost every known type have been found, e.g., benzene and alkylbenzene derivatives (4.6–12.6%) through the 22 isomeric C_{10}-alkyl isomers, indan and tetra-hydronaphthalenes and their methyl isomers, naphthalene and its 1- and 2-alkyl derivatives, biphenyl and derivatives, and many multiring condensed aromatic compounds. Instrumental ASTM methods permits rapid compound-type analysis of aromatic compounds.

Sulphur, Nitrogen and Oxygen Compounds

Sulphur is found in all crudes in concentrations ranging from traces to 8.0%, mostly in the 0.5–1.5% range. It is present in aliphatic and aromatic compounds and also as elemental sulphur, hydrogen sulphide, and carbonyl sulphide (COS). Sulphur-specific flame photometric detectors in gas chromatography permit the rapid identification and measurement of sulphur compounds in the lower-boiling fractions. It is important to know the distribution of sulphur compounds within a crude. As the reserves of high quality, low sulphur crude diminish, crudes of poorer quality with high sulphur contents must be converted catalytically into low sulphur products.

Nitrogen concentrations are usually below 0.2%; but may be as high as 1.6%. Nitrogen compounds are concentrated in the high boiling fractions and in the residuum; very little is present in the fraction boiling below 300°C. Compounds containing oxygen as well as nitrogen are also found, e.g., hydroxypyridines and hydroxyquinolines. The nitrogen compounds affect processing, particularly of the higher-boiling fractions. Aromatic nitrogen-containing molecules are refractory and some are basic and may interact and adsorb onto acid catalysts and promote coke formation.

The oxygen content of most crudes is low. Oxygen is present mainly in the form of carboxylic acids, e.g., fatty acids through stearic, branched chain acids, alicyclic acids, and dicarboxylic acids.

Residuum

Specifications of a residuum fraction must always include the boiling range taken at the end of the distillate. Compound types are separated by chromatographic techniques. Unique to the residuum is solvent precipitation or deasphalting. The portion of the residuum insoluble in a large excess of *n*-heptane is known as asphaltenes (up to 25%), and the heptane-soluble fraction as maltenes. Saturated compounds may constitute 40 wt% of the residuum, usually much less.

Metals

In trace quantities, all metals through atomic number 42 (molybdenum) have been found, except rubidium and niobium; a few heavier elements have also been detected. Present in all crudes, nickel and vanadium are the most important. Analytical methods can detect metal concentrations in crudes at 10 ppb (10 ng/g).

A number of compounds, identified in petroleum, are referred to as biological markers or geochemical fossils. These are compounds whose basic carbon skeleton has survived chemical changes that occurred during petroleum genesis and whose structures are frequently found in terrestrial and marine life.

RESOURCES

Petroleum resources are distributed widely in the earth's crust as gases, liquids, and solids. The products derived from these naturally occurring resources are used mainly as energy sources, although substantial volumes serve as feedstocks in the plastics, chemical, and other industries.

Petroleum resources are classified as follows: The *resource base* is the total amount present in a specific volume of the earth's crust (also called in-place petroleum); *resources* represent the total amount of petroleum, including reserves, that is expected to be produced in the future; *reserves* constitute the petroleum that has been discovered and can be produced at the prices and technology that existed when the estimate was made; *proved reserves* are estimates of reserves contained primarily in the drilled portion of the fields; *indicated reserves* constitute known petroleum that is currently producible but cannot be estimated accurately enough to qualify as proved; *inferred reserves* are producible but the assumption of their presence is based upon limited physical evidence and considerable geological extrapolation; sub-economic resources constitute the petroleum in the ground that cannot be produced at present prices and technology but may become producible; *undiscovered resources* are estimated totally by geological speculation with no evidence by drilling available.

DRILLING FLUIDS

Drilling fluids and muds perform a variety of functions that influence the drilling rate, cost, efficiency, and safety of the drilling operation. They are pumped down a hollow drill string through nozzles in the bit at the bottom, and up the annulus formed by the hole or casing and the drill string to the surface. The bit is turned by rotating the entire string from the surface or by using a down-hole motor. After reaching the surface, the drilling fluid is passed through a series of screens, settling tanks or pits, hydroclones, and centrifuges to remove formation material brought to the surface. It is then treated with additives to obtain the desired properties. Once treated, the fluid is pumped back into the well and the cycle repeated.

Gas-based muds, ranging from compressed dry air or natural gas to mist to stable foams, are used mostly for the drilling of hard rock. About 90% of all drilling fluids is water-based systems, ranging from clear water without additives to high density muds containing clays and various organic additives. Offshore wells are frequently drilled with a seawater system, which needs more additives to achieve the desired flow and filtration properties.

Oil-based drilling fluids have diesel, a low-aromatic-content oil, or occasionally crude oil as a continuous phase with both internal-water and solid phases. Fluids with more than 5–10 vol% water are called invert oil-based muds or inverts. Oil muds are employed for high temperature wells where water-based systems may be unstable and where problems may arise from water-sensitive shale formations.

Properties

The density is adjusted with dissolved salts or powdered high density solids to provide a hydrostatic pressure against exposed formations in excess of the pressure of the formation fluids. In addition, the hydrostatic pressure of the mud column prevents collapse of weak formations into the borehole. Fluid densities may range from that of air to > 2500 kg/m^3 (20.8 lb/gal). Most drilling fluids have densities > 1000 kg/m^3 (8.33 lb/gal), the density of water. The hydrostatic pressure imposed by a column of drilling fluid is expressed as follows :

$$P = 0.098 L\rho_m \quad (0.052\, L\rho_m)$$

where P = hydrostatic pressure in kPa (psi), ρ_m = drilling-fluid density in kg/m^3 (lb/gal), and L = column length or well depth in m (ft). Wells with bottom-hole pressure of 100 MPa (14,500 psi) at 5000 m are not unusual.

Viscosity and velocity must be high enough to remove drill cuttings and other formation material that may fall into the well bore. Low viscosity fluids are circulated at high rates and vice versa.

Density control

Mud density is controlled accurately by suitable weighting materials that do not adversely affect the other properties. Most important is the specific gravity of the weighting agent as well as its chemical inertness. Barite (barium sulphate, $BaSO_4$) meets these requirements; it is virtually insoluble in water and does not react with other mud constituents. The barite content in drilling mud can be as high as 2000 kg/m^3 (700 lb/ bbl). Other weighting materials include galena, haematite, magnetite, ilmenite, siderite, celesite, dolomite, and calcite.

The drilling rate is improved by solids-free fluids, which are designed for work-over and completion operations to avoid formation damage. Weighted fluids without solids are provided by solutions of various water-soluble salts (Table 2.1) for maximum densities. The incorporation of a limited amount of drilled solids is an economical way to increase mud density, but it also reduced penetration rates.

Table 2.1. Soluble salts that increase the density of drilling muds and work-over and completion fluids.

			Saturated brine, 20-25°C	
Salt	*Formula*	*Specific gravity*	*Wt%*	*Density, g/cm³*
Potassium chloride	KCl	1.984	26.5	1.18
Sodium chloride	NaCl	2.165	28.5	1.20
Sodium carbonate	Na_2CO_3	2.532	22.5	1.24
Calcium chloride	$CaCl_2$	2.150	46.1	1.46
Potassium carbonate	K_2CO_3	2.428	52.9	1.56
Sodium bromide	NaBr	3.203	53.7	1.65
Calcium bromide	$CaBr_2$	3.353	58.7	1.70
Zinc chloride	$ZnCl_2$	2.910	81.2	2.14
Zinc bromide	$ZnBr_2$	4.201	81.7	2.65

Viscosity

The type of solids and the solids content of water-based muds control the viscosity. The reactive solids are composed of commercial clays, of which bentonite (montmorillonite), attapulgite, and sepiolite are the most important; organophilic clays, prepared from bentonite or attapulgite and aliphatic amine salts, are used for oil-based muds.

Organic polymers are used to increase viscosity and also to control filtration rates. Water-soluble polymers include starch, guar gum, sodium carboxymethyl cellulose, hydroxyethyl cellulose, xantham gum, polyacrylates, and polyacrylamides.

Viscosity is decreased by a reduction of solids, or deflocculation of clay-water suspensions. Thinning is obtained by reducing plastic viscosity, yield point, or gel strength, or a combination of these properties. Typical mud-thinning chemicals are polyanionic materials that are absorbed on the positive-edge sites of the clay particles, thereby reducing the attractive forces between the particles without affecting clay hydration.

Filtration control

Filtration control is particularly important in permeable formations where the mud hydrostatic pressure exceeds the formation pressure. It reduces drill-string sticking, torque, and drag due to filter-cake development; minimises damage to protective formations; and, in some areas, improves boreholes stability. Materials for this purpose include clays, organic polymers, and lignite derivatives. The bentonite present in freshwater systems acts as the primary filtration agent; concentration vary widely, but may range up to 100 kg/m^3 (35 lb/bbl).

Although a combination of bentonite clay and lignosulphonate provides filtration control in many water-based muds, organic polymers are generally needed as filtration aids, especially starches in concentrations of 6–29 kg/m^3 (2–10 lb/bbl). Fermentation by micro-organisms is prevented by saturating with salt or adding a preservative, e.g., paraformaldehyde or another biocide (ca 0.6–1.4 kg/m^3 or 0.2–0.5 bbl). Carboxymethyl cellulose is most commonly used in fresh- to seawater fluids (1.4–5.7 kg/m^3 or 0.5–2.0 lb/bbl).

Alkalinity control and removal of contaminants

Water-based drilling fluids are generally maintained at an alkaline pH by the addition of sodium hydroxide (3–14 kg/m^3 or 1–5 lb/bbl), lime (6–57 kg/m or 2–20 lb/bbl), or magnesium oxide.

Drilling-fluid contaminants are defined as any material or condition that adversely affects fluid performance. Drill solids are removed mechanically by various combinations of screens, hydroclones, and centrifuges. Electrolytes are removed by dilution and chemical treatment.

Shale stabilisation

Shales can be stabilised with an oil mud that prevents direct contact between the shale and the emulsified water. The most recent shale-protective agents are the polymer/potassium muds. The reaction of potassium ions with clay results in potassium fixation and formation of a less water-sensitive clay. Potassium chloride is generally used in combination with a polymer. Ammonium chloride, ammonium sulphate, and diammonium phosphate have also been used, along with a number of anionic polymers.

Surfactants perform various functions in drilling fluids. Lignites and lignosulphonates act as emulsifiers, whereas aluminium stearate is used as a defoamer.

To function properly a drilling fluid must be circulated through the well and back to the surface. Loss of fluid, owing to openings in the formation, can result in a loss of hydrostatic pressure at the bottom of the hole, influx of formation fluids, and possibly loss of well control.

Lost-circulation zones are sealed off by the addition of flakes (shredded cellophane or paper, laminated plastic), fibrous materials (wood fibres, saw-dust straw), or granules (ground walnut shells, rubber, plastics, limestones); concentrations of these materials range from 14 to 143 kg/m^3 (5–50 lb/bbl).

Lubricants and spotting fluids are added to overcome frictional resistance of the drilling string or sticking of the drill pipe. Blends of anionic and non-ionic surfactants and natural products are used as well as gilsonite and air-blown asphalts. String and casing corrosion is generally caused by oxygen dissolved or entrained in the mud, and can be prevented by excluding or scavenging corrosive gases and by maintaining a high pH.

ENHANCED OIL RECOVERY

Crude oil exists in contact with gases and water in sub-terranean rock formations called reservoirs. Wells serve as pressure sinks that allow the oil to be brought to the surface. Normal depletion, called primary recovery, results from the driving forces provided by dissolved gases of an associated aquifer. Following primary production, fluid may be injected into a portion of the wells to provide a secondary energy source. This is called secondary oil recovery, e.g., waterflooding or immiscible-gas flooding. Continued fluid injection, in time, results in ever-increasing production of injected fluid and eventually only injected fluid (i.e., water) is produced, leaving unrecovered oil behind as a residual oil saturation.

Enhanced oil recovery (EOR), often referred to as tertiary oil recovery, is recovery of oil over and above that which can be recovered by primary and secondary methods (Fig. 2.1). Estimates of the potential of EOR are highly speculative, but can be 5–10% of the oil discovered to date. Residual oil remains in a reservoir because of the high interfacial tension (IFT) between oil and water. Of importance is the interrelationship between capillary and viscous forces given by the capillary number:

$$\text{capillary number} = (\text{velocity} \times \text{viscosity})/\text{IFT}$$

where the velocity refers to the injected fluid and the IFT is measured between the oil and the injected fluid. High capillary numbers are required to get low residual oil saturations. For miscible fluids, the capillary number is infinite.

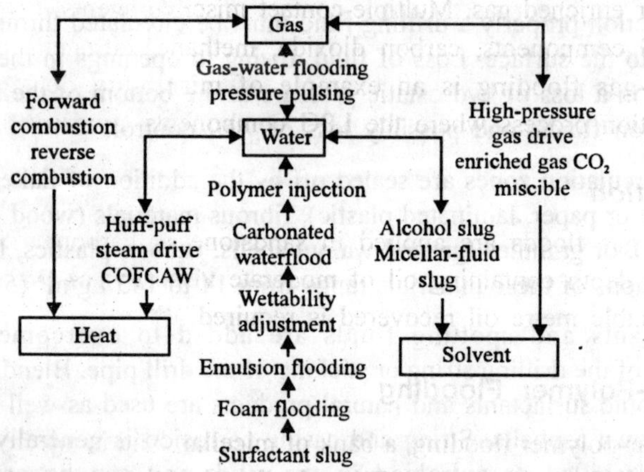

Figure 2.1. Improved methods for oil recovery (connecting lines indicate several combinations commonly in use), COFCAW is the combination of forward combustion and waterflooding.

Because of the density differences between gas, oil, water, and injected fluids, oil may be trapped high in the reservoir and not contacted by an injected water-base fluid. Gravitational forces are expressed in terms of densities. The economics of EOR methods are governed by the amount of oil that can be recovered and the cost of the injected fluids. A high initial investment is needed and oil production is delayed after initial injection of the fluid.

Miscible-Gas Drives

A gas can either miscibly or immiscibly displace oil, depending on the physical properties of the gas and the oil and the reservoir temperature and pressure. In a miscible-gas process, the residual oil is mobilised by an injected solvent (miscible gas) and the mobilised oil is displaced to a production well. A thorough understanding of the physical/chemical phenomena (displacement mechanisms) occurring within the oil-moving and solvent-moving zones is needed for a project design.

Miscible displacement may be controlled by either a first-contact or multiple-contact mechanism. In the former, the reservoir oil and injected solvent are mixed in all proportions at reservoir conditions and no phase separation occurs. In the latter, oil and solvent cannot be mixed in all proportions without phase separation upon equilibration. Miscibility is developed by either a condensation or a vapourisation (extraction) process.

First-contact miscible solvents include liquefied petroleum gas fractions (LPGs) or enriched gas. Multiple-contact miscible agents include methane plus LPG components, carbon dioxide, methane, nitrogen, and flue gas. Enriched-gas flooding is an example of miscibility developed by the condensation process where the LPG components condense into the oil.

Application

Miscible gas floods are applied in sandstone or carbonate reservoirs of sufficient depth containing oil of moderate viscosity; ca 900–5400 m^3 of gas per cubic metre oil recovered is required.

Micellar-Polymer Flooding

In micellar-polymer flooding, a bank of micellar fluid is injected to displace residual oil from a reservoir. The micellar fluid is pushed by a polymer-thickened water bank. The mechanism is based on reducing the IFT between crude oil and water. The fluid consists of a surfactant in a solvent in which the surfactant ions are arranged in oriented aggregates or micelles. As the fluid moves through the porous medium, mass transfer occurs between the injected fluid and the in-place fluids, and an oil-moving zone creates an ultra-low IFT condition which results in residual oil being displaced. The desired mobility of the various banks can be adjusted, and the proper mobility keeps the integrity of the bank and maximises sweep efficiency. This technique is still in the developmental stage.

Thermal Processes

Thermal processes utilise heat generated above ground or *in situ* to enhance oil recovery; they are applied primarily to heavy oils. The simplest process is steam injection. Steam is used as a driving fluid to displace oil from an injector to a producer, or in a cyclic-injection production sequence using the same well as both injector and producer with a soak period to heat the well, the so-called steam-soak or steam huff-and-puff process. Underground or *in situ* combustion processes utilise injected air or oxygen and hydrocarbons or coke within the reservoir as fuel.

REFINERY PROCESSES

Petroleum products derived from crude oil are a convenient source of energy. Since petroleum liquids are easy to handle and store, they are well suited for transportation fuels, e.g., for cars and airplanes. Other refinery products include lubricants, waxes, asphalt, solvents, and specialities such as liquefied petroleum gas (LPG), hydraulic fluids, and others. Petroleum is the principal raw material source for petrochemicals such as plastics, synthetic elastomers, certain alcohols, and other important products. The various fuel and chemical markets have their own product-quality requirements and it is the function of the refining operation to separate crude and other raw materials into fractions that are then processed to meet product specifications.

Crude petroleum contains a wide range of hydrocarbons from light gases to residuum that is too heavy to distill even under vacuum. Crude oil is primarily made up of paraffins, cycloparaffins (naphthenes), and aromatic compounds in varying proportions, some sulphur compounds, a small amount of nitrogen, but no appreciable amount of oxygen or olefins.

Refining processes can be grouped into three classes; separation, usually distillation to give the desired type of compounds; conversion, usually cracking, to change molecular weight and boiling point; and upgrading, e.g., hydrotreating, to meet product-quality specifications.

In general, refineries are located near a large body of water, partly to supply cooling water but also for transportation. Large refineries process about 80,000 m^3 (5×10^5 bbl) of crude per day, which corresponds to one supertanker load. Rail transportation would require 1600 tank cars per day to carry the same amount of crude.

Since the 1970s, the great increase in crude cost has been accompanied by greater emphasis on high value products at the expense of fuel products.

For example, fuel oil previously used in large power plants is displaced by coal or nuclear fuel. Increased attention to environmental aspects has led to demand for low sulphur products despite the trend toward higher sulphur crudes. Unleaded gasoline is another example of environmental concern.

Petroleum refining has shown a rapid growth and is now the largest manufacturing industry in India. Efforts to decrease the gasoline consumption (km/L) of vehicles by reducing their size and weight have been successful. More efficient engines are in wide use, particularly diesel engines, in which the fuel is injected directly into the combustion chamber. Diesel engines have efficiencies of about 35% versus ca 25% efficiency for gasoline engines.

Refineries range in size from 1600 m^3/d (10,000 bbl/d) to over 64,000 m^3/d (4×10^5 bbl/d). Small refineries make only gasoline, diesel, and domestic heating oils. Large refineries include the manufacture of lubricating oils and greases. Refining is also the main source of raw materials for petrochemical manufacture. A large steam-cracking unit for 500,000 t/yr of ethylene may consume ca 2×10^6 t/yr of oil feed (40,000 bbl/d), which may be 10% of the crude used in very large refinery. Clearly, chemical and fuel-refining operations must be carefully coordinated.

Processing Steps

Desalting

Salt and clay or other suspended solids are removed by washing with water at 65–90°C to reduce viscosity. Typical salt content of crude may be 280 g/100 m^3; desalting may remove over 90% without the loss of oil.

Distillation

The crude is separated in continuous-fractionation plate towers, as shown in Fig. 2.2. Primary distillation takes place at atmospheric pressure and the bottom temperature is limited to 370–400°C to prevents thermal cracking.

Naphtha, the fraction taken from the top, is mainly used for motor gasoline and processed further for octane improvement by catalytic reforming. The middle distillate includes diesel, and jet fuels, and heating oil. Kerosene and certain speciality solvents distill between these two fractions. The bottom fraction can be used as fuel oil but is usually vacuum distilled in order to increase the yield of high value distillate oil for catalytic cracking.

Vacuum distillation provides low sulphur oil by hydrosulphurising vacuum gas oil, which is then blended back into untreated vacuum bottoms. In addition, various speciality materials are obtained, such as wax and lube fractions.

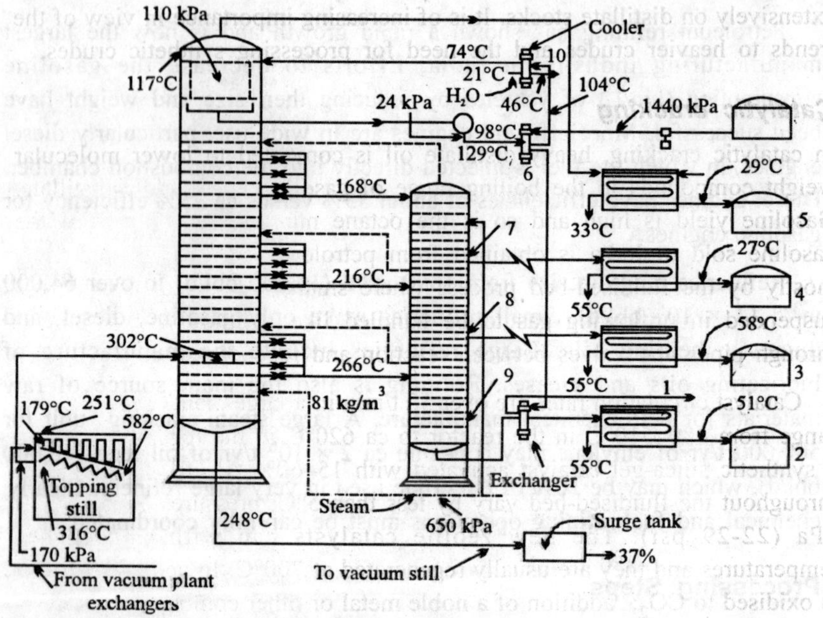

Figure. 2.2. Atmospheric distillation of crude. Multidraw crude oil topping plant. To convert kPa to psi, multiply by 0.145. 1. To vacuum-plant exchangers; 2. Gas oil 10.6%; 3. Kerosene 19.8%; 4. Naphtha 5.8%; 5. Light gasoline 26.8%; 6. Exchanger; 7. Steam; 8. 26 kg/m^3; 9. 66 kg/m^3; 10. 46°C.

Hydroprocessing

Hydroprocessing improves the quality of various products or cracks heavy carbonaceous materials to lower-boiling, more valuable products. Mild hydrotreating removes sulphur, nitrogen, oxygen, and metals, and hydrogenates olefins. A fixed-bed may be employed at 1.5–2.2 MPa (200–300 psig) and 350–400°C, without catalyst regeneration. Severe conditions are 7–21 MPa (1000–3000 psig) and 350–500°C with catalyst regeneration.

Hydrogen consumption increases with severity and depends on the amount of sulphur removed and the feed content of aromatic materials and olefins, which also consume hydrogen. Net consumption can range from 18 m^3/m^3 (100 ft^3/bbl) feed for hydrofinishing to well over 180 m^3/m^3 (1000 ft^3/bbl) feed in hydrocracking operations.

Hydrocracking

In hydrocracking, high molecular weight compounds are cracked to lower boiling materials. Severity is increased by operating at higher temperatures and longer contact time than in hydrotreating. Hydrocracking is used extensively on distillate stocks. It is of increasing importance in view of the trends to heavier crudes and the need for processing synthetic crudes.

Catalytic cracking

In catalytic cracking, heavy distillate oil is converted to lower molecular weight compounds in the boiling range of gasoline and middle distillate. Gasoline yield is high and so is the octane number. About half of the gasoline sold in India is obtained from petroleum by catalytic cracking, mostly by the fluidised-bed process where small particles of catalyst are suspended in upflowing gas to be handled like a liquid and circulated through pipes and valves between reaction and regeneration vessels.

Catalyst circulation rates are over 50 t/min in a large plant. Temperatures range from 480–510°C in the reactor to ca 620°C in the regenerator using a synthetic silica-gel catalyst activated with 15–60% Al_2O_3. Temperatures throughout the fluidised-bed vary by less than 5°C; pressures are 150–200 kPa (22-29 psi). The new zeolite catalysts can withstand higher temperatures and they are usually regenerated at 700°C. In addition, all CO is oxidised to CO_2; addition of a noble metal or other combustion catalysts in ppm concentrations assures complete combustion. With zeolite-type catalysts, 80-90% conversions are obtained. A feed rate of 16,000 m^3/d (10^5 bbl/d) is not uncommon for large units having a 15-m dia regeneration vessel and a 6.5-m dia reactor; the structure may be 65 m high. Configurations of typical cracking units are shown in Fig 2.3, and typical yields in Table 2.2.

Coking

Both delayed and fluid coking are used on very heavy low value residuum feeds to obtain lower-boiling cracked products. Coking can be considered as high severity thermal cracking or destructive distillation. It is generally used on vacuum residuum (which is not a suitable feed for catalytic cracking) to generate lighter components. In delayed coking, a cyclic batch operation, the oil is held at 450°C and 35–70 kPa (5–10 psi) to deposit coke while cracked vapours are taken overhead. Fluid coking is a continuous process similar in design to fluid catalytic cracking.

Table 2.2. Catalytic-cracking operation.

	Operation	
	Maximum gasoline	*Maximum middle distillate*
Feed		
Boiling range, °C	260–566	343–566
Conversion, vol %	85	60
Yield[a]		
Methane and C_2, wt%	2.8	2.4
Propane, vol%	3.0	2.0
Propylene, vol%	7.0	6.0
Butane, vol%	8.8	5.2
Butene, vol%	7.6	7.9
C_5 naphtha up to 221°C, vol%	69.5	47.1
Heating oil, vol%	10.0	35.0
Slurry, vol%	5.0	5.0
Coke, wt%	5.3	5.4
RON[b] clear	92	92
Carbon on regenerated zeolite catalyst, wt%	< 0.20	0.35

[a] Because of expansion, vol% adds up to more than 100%.
[b] Research of octane number.

The products include gas, naphtha, gas oils, and coke. Hyderotreated coker gas oil augments supplies of residential heating oil; heavier fractions are cracked catalytically.

The sulphur content of the coke may be as much as 50% higher than that of the feedstock, but the coke can be used as fuel if SO_x control is provided. Needle coke, a premium product especially suited for graphite production, is made by delayed coking of low sulphur, highly aromatic heavy oil obtained from catalytic or thermal cracking. It is the world's main source of graphite for carbon electrodes.

In flexicoking, high sulphur coke is gasified to clean fuel gas in a fluidised-solids circulating system. The coke is gasified with steam and either air or oxygen. The gas is scrubbed to remove sulphur compounds and impurities and thereby provides a cleaner fuel. Gasification is carried almost to completion while ash and metals build up to a high concentration in the purge solids. Yield pattern is given in Fig. 2.4.

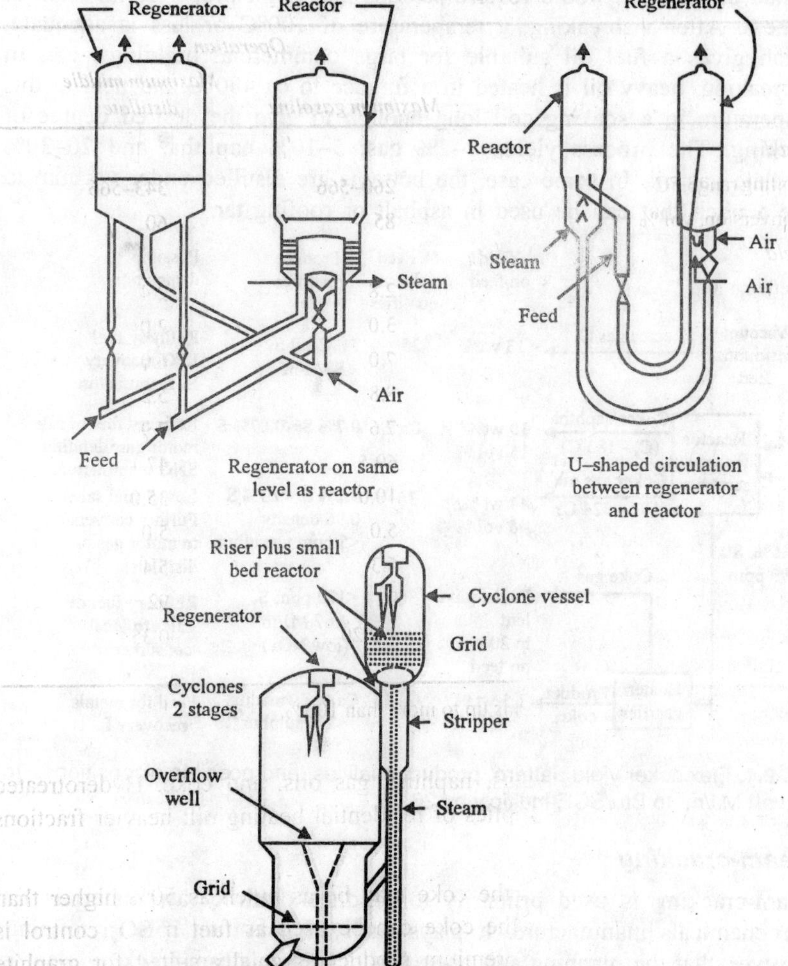

Fig. 2.3. Typical fluid-catalytic-cracking-unit configurations.

Visbreaking

Viscosity breaking is one of the few thermal-cracking processes still used in refining other than coking. It reduces the viscosity of heavy fuel oil and permits handling at lower temperature. Vacuum residuum can usually be burned directly but would require perhaps 200°C for proper atomisation in burners. After visbreaking, a temperature of 100°C or less is adequate, which gives a fuel oil suitable for large commercial buildings, etc. In visbreaking, heavy oil is heated in a furnace to ca 480°C and held at this temperature in a soaking coil long enough to give the desired degree of cracking. The process yields 1–2% gas, 5–10% naphtha, and 20–30% distillate gas oil. In some case, the bottoms are distilled under vacuum to give a pitch that can be used in asphalt or roofing tar.

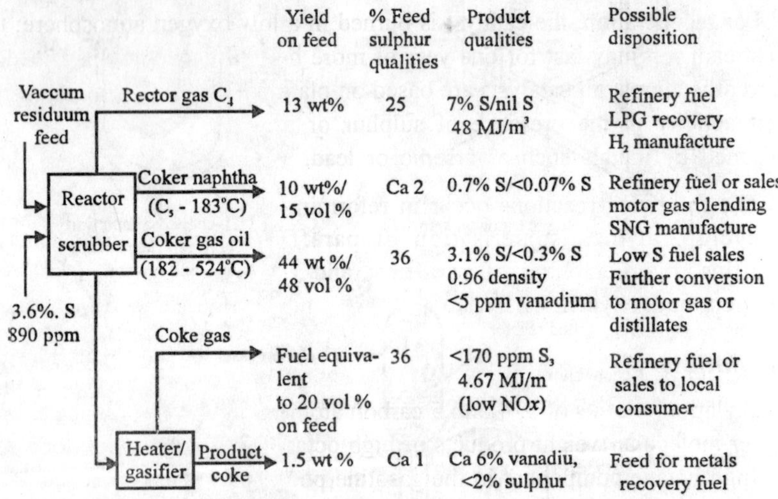

Fig. 2.4. Flexicoker yield pattern, product qualities, and possible dispositions. To convert MJ/m^3 to Btu/SCF, multiply by 26.85.

Steam-cracking

Steam-cracking is used primarily to produce olefinic raw materials for petrochemicals manufacture. It plays such an important role in refining, however, that the planning and operation of a large steam-cracking plant is usually closely integrated with a large refinery. Steam-cracking is carried out at ca 800–850°C and slightly above atmospheric pressure. Feedstocks range from ethane to vacuum gas oil; heavier feeds give higher yields of by-products such as naphtha, etc. Steam is added with the feed to reduce the hydrocarbon partial pressure, thereby giving more olefins. The amount of steam may be equal in weight to the oil feed.

Catalytic reforming

Naphtha recovered directly from crude by distillation is called straight-run or virgin naphtha. It is too low in octane (30–50 clear octane) to meet quality requirements for motor gasoline and is upgraded by catalytic reforming. The feed is a naphtha fraction boiling in the 80–230°C range. Lighter fractions are not amenable to reforming, whereas heavier fractions would give products that remain outside of the gasoline boiling range since reforming gives only a small reduction in boiling point. Typical operating conditions are 430–520°C and 1–6 MPa (145–870 psi). By-product hydrogen amounts typically to 36–45 m^3/m^3 of feed (202–253 ft^3 /bbl) with a yield of C_5 and higher of 80–85 vol%.

For regeneration, the catalyst is burned in a low oxygen atmosphere; the on-stream run may last for one year or more before regeneration is needed. Currently, almost all catalysts are based on platinum or palladium. They lose their activity in the presence of sulphur or nitrogen compounds and are poisoned by metals such as arsenic or lead.

The following reactions occur in reforming (Fig. 2.5), dehydrogenation of cycloparaffins; isomerisation of paraffins; dehydrocyclisation of paraffins, hydrocracking of paraffins; and removal of sulphur, olefins, nitrogen, and oxygen.

Alkylation

In alkylation, olefins of 3, 4 and 5 carbon atoms react with isobutane to give higher molecular weight products of high octane. Thus, alkylation increases the high octane component in the gasoline pool. Having a 93–96 octane on either the ASTM scale (MON = motor octane number) or research scale (RON = research octane number), the product is especially valuable for unleaded gasoline. The octane number quoted in retail sales is an average of MON and RON.

Sulphuric or hydrofluoric acids are employed as catalysts. Sulphuric acid alkylation operates at 4–15°C and a pressure slightly higher than atmospheric. Feed consisting of olefins and isobutane is injected into the reaction emulsion through high velocity nozzles, and the mixture is agitated with mechanical stirrers. A large amount of sulphuric acid is consumed owing to the solubility of by-products and accumulation of water. With a hydrofluoric acid catalyst, the operating temperature is 25–45°C; the pressure of 0.7 MPa (101.5–145 psi) maintains the HF as a liquid phase.

Extraction

Extraction separates by chemical type rather than by boiling point and is much more expensive than distillation. It is used primarily for removal of aromatic material from naphtha fractions, upgrading middle distillates, and preparation of high quality lubricating oils.

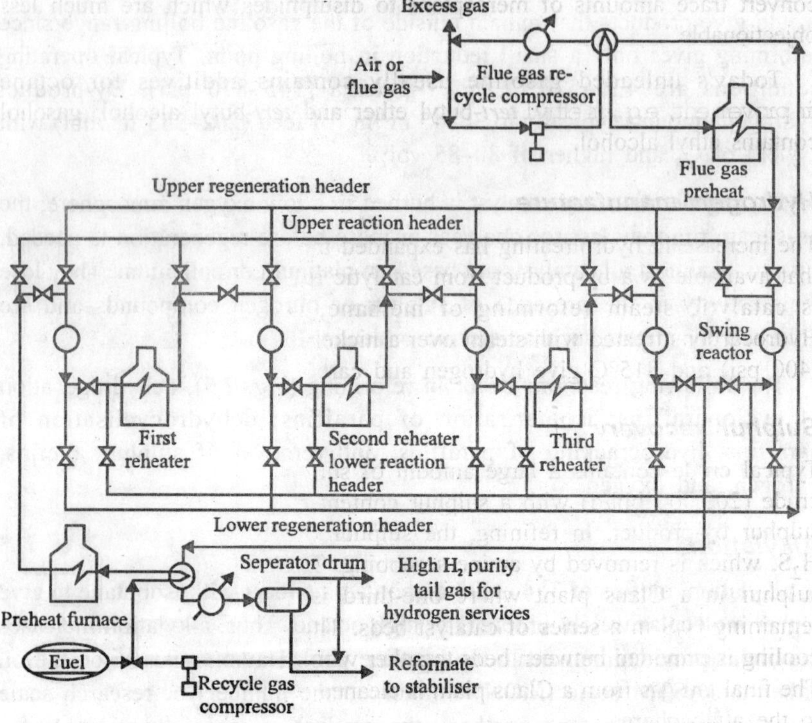

Fig. 2.5. Fixed-bed cyclic catalytic reforming.

Lubricants

Petroleum is the main source of lubricants for a wide range of uses. Lubricants are derived from distillates boiling above 370°C and they are separated by vacuum distillation and refined to meet specifications. Aromatic material is removed from the raw distillate by extraction with *N*-methylpyrrolidine-2-one, phenol, or furfural. Wax is crystallised by chilling and filtered. Various additives play an important role in meeting quality requirements.

Gasoline Blending and Treatment

Separate components are blended to give various grades. Elaborate instrumentation controls in-line blending and offers large savings compared to blending in tanks. Leaded gasoline of reduced lead content is still being made, but may be phased out by about 1990. Naphtha-treatment processes, such as the Merox process, are used to extract sulphur compounds and to convert trace amounts of mercaptans to disulphides which are much less objectionable.

Today's unleaded gasoline usually contains additives for octane improvement, e.g., methyl *tert*-butyl ether and *tert*-butyl alcohol; gasohol contains ethyl alcohol.

Hydrogen manufacture

The increase in hydrotreating has expanded the need for hydrogen beyond that available as a by-product from catalytic reforming. A common source is catalytic steam reforming of methane, refinery gas, or naphtha. Hydrocarbons treated with steam over a nickel-base catalyst at ca 2.8 MPa (400 psi) and 815°C give hydrogen and carbon oxides.

Sulphur recovery

Typical crude contains a large amount of sulphur. Thus, 10×10^6 t/yr of crude (200,000 bbl/d) with a sulphur content of 3% would yield 870 t/yr sulphur by-product. In refining, the sulphur compounds are converted to H_2S, which is removed by amine scrubbing. The H_2S is converted to pure sulphur in a Claus plant where one-third is allowed to react with the remaining H_2S in a series of catalyst beds. The reaction is exothermic and cooling is provided between beds together with separation of liquid sulphur. The final tail gas from a Claus plant is cleaned further before being released to the atmosphere.

Other refinery facilities

A complete refinery is a highly complex and integrated installation requiring many auxiliary facilities (so-called off-sites), such as large storage capacities for feed and products, extensive port and dock facilities, and utilities including steam, electricity, cooling water and waste-water cleanup. Some utility requirements can be supplied from coal.

Conservation

Recently, conservation in refineries has produced energy savings of 15–25%. Efficiency of existing equipment has been increased by using less air in furnaces, and pumps with less excess capacity operate at more nearly optimum efficiency. Pressure energy in the flue gas from regeneration can be recovered, and heat pumps save distillation energy. Steam leaks, losses by evaporation, and leakage from valves have been greatly reduced. The overall efficiency of a refinery is ca 90%, individual processes ranging from

95%; for distillation and catalytic cracking to about 80% for hydrocracking residuum (Table 2.3).

Table 2.3. Thermal efficiencies of various processes.

Operation	*Thermal efficiency, %*
Methanol from coal	50
Hydrogen manufacture	
from coal	55
from methane	60
SNG from coal	65
Coal liquefaction	70
Oil from shale	80
Fuel gas from coal	80–85

Environmental Aspects

Environmental aspects of refining concern air, water, and solid-waste problems (EPA classification). Petroleum refining was one of the first industries to be covered by Federal regulations mandating ambient air-quality standards and allowable emissions (Table 2.4). State standards are at least as strict; Japan has very strict standards.

Table 2.4. National primary and secondary ambient air-quality standards.

Pollutant	*Type of standard*	*Averaging time*	*Concentration* $\mu g/m^3$	*Concentration* *ppm*
Carbon monoxide	Primary and secondary	1 h	40,000	35
		8 h	10,000	9
Hydrocarbons[a]	Primary and secondary	3 h	160[b]	0.24[b]
Nitrogen dioxide	Primary and secondary	1 yr	100	0.05
Photochemical oxidants[c]	Primary and secondary	1 h	160	0.08
Particulate matter	Primary	24 h	260	
		24 h	75	
	Secondary	24 h	150	
		24 h	60[d]	
Sulphur dioxide	Primary	24 h	365	0.14
		1 yr	80	0.03
	Secondary	3 h	1.300	0.5

(Contd...)

Pollutant	Type of standard	Averaging time	Concentration µg/m³	ppm
Lead	Primary	90 d	1.5	
Ozone	Primary and secondary	1 h	235	0.12

[a] Non-methane.

[b] As a guide in devising implementation plans for achieving oxidant standards.

[c] Expressed as ozone by the federal reference method.

[d] As a guide to be used in assessing implementation plans for achieving the annual maximum 24-h standard.

Air-pollution control is directed toward the principal polutants; SO_x, NO_x particulates, CO, and hydrocarbons. Control measures on furnaces include the use of low sulphur fuels. The EPA standards regulate emission of flue gases from furnaces and catalyst regenerators as well as hydrocarbon vapours generated by leaks, evaporation, and handling.

Water pollutants include oil, H_2S or other sulphur compounds, and ammonia, phenols, chromium compounds, and chlorine used to control corrosion and biological fouling. The main sources of waste-water are cooling systems that may contain additives to inhibit corrosion or fouling; low toxicity inhibitors are now available. Process water containing H_2S, ammonia, and oil is termed sour water. It is stripped with steam, and treated in an API-type separator where the oil rises to the surface and is removed with the aid of air flotation.

Solid wastes consist mainly of spent catalysts, and ash if coal is the refinery fuel. Catalytic cracking units are subject to the following regulations: emission of particulates directly from regenerators should not exceed 1 kg/kg of coke burned; a stack opacity of 30% maximum; and a maximum of 0.05 vol% of CO in gases discharged to the atmosphere.

Plant Economics

Utilities and general off-site expenditures include utilities production (equipment to generate steam and electricity), tie-in lines for power and water; roads and motor-vehicle access; port and dock facilities; tanks for feed and products; product loading and shipping; pipelines and railroads; repair shops, spare parts, and maintenance; technical service and engineering support; analytical and testing laboratories, an instrument and controls shop; business and accounting offices; and housing and facilities for construction crews. Investment for these facilities may amount to 40%

of the costs of a new plant in a large existing refinery and often doubles the costs for a new installation.

Operating costs are strongly related to investment and include maintenance and repair; utilities; chemicals; catalysts and additives; and labour. Recently the costs have increased drastically both for the increased investment and the need for a higher return in order to make additiional investment attractive. Capital charges include taxes and insurance.

Natural petroleum sources are still the lowest in cost of liquid fuels, but liquid from oil shale, fuel gas by coal gasification, and methanol plus substitute natural gas (SNG) from coal are now nearly competitive in some situations.

PETROLEUM PRODUCTS

The importance of petroleum products is reflected in the great consumption of crude oil, the starting material for these products.

Petroleum products include power fuels for passenger cars, trucks, aircraft engines, and gas turbines (gasoline, jet fuel, diesel fuel and LPG); heating oils; illuminating oils (kerosene); solvents (petroleum spirits, aromatic solvents); lubricants (engine oils, greases); building materials (asphalts); insulating and waterproofing materials; and other products, such as cutting oils, heat-treating oils, hydraulic oils, petroleum jelly, etc.

The amounts produced are influenced by several factors, especially feed costs, processing costs and market demands. Distillate fuels usually denote burner fuels, diesel fuels, and similar intermediate-volatility products taken as condensed vapours from a distillation process. Gasoline is also a distillate product but is identified separately. Residual fuels are the unvapourised products from distillation and include asphalt, petroleum coke, and heavy industrial and bunker fuels.

Because of Government regulations for the control of automobile emissions, new automobiles are equipped with catalytic mufflers for the reduction of unburned hydrocarbons emanating from engine exhausts. Since catalytic mufflers are rendered ineffective by leaded fuels; the production of unleaded gasoline has increased. High compression engines tend to develop knock, so that refiners developed processes to improve anti-knock properties, even though they may reduce gasoline yields.

Aviation gasoline for commercial aircraft has been largely replaced by jet fuels, also called aviation turbine fuels.

Many petroleum products of commerce are blended, compounded, or otherwise modified from primary stock. Trade names are not necessarily those given by the refiner, and a product may have several trade names, depending on uses. Specifications for automotive gasoline are given in Table 2.5, and for diesel fuel in Table 2.6.

Table 2.5. Specifications for automotive gasoline.

	Volatility class				
Property	*A*	*B*	*C*	*D*	*E*
Octane number[a]	no limit specified				
Distillation temperature, °C					
10% evaporated, max	70	65	60	55	50
50% evaporated, min	77	77	77	77	77
50% evaporated, max	121	118	116	113	110
90% evaporated, max	190	190	185	185	185
end point, max	225	225	225	225	225
Temp. for vapour-liquid ratio of 20, min, °C	60	56	51	47	41
Vapour pressure, max, kPa[b]	62	69	79	93	103
Lead content, max, g/L					
Unleaded grade	0.013	0.013	0.013	0.013	0.013
Conventional grade	1.1	1.1	1.1	1.1	1.1
Corrosion, copper strip, max, no.	1	1	1	1	1
Gum, existent, max, mg/100 mL	5	5	5	5	5
Sulphur, max, wt%					
Unleaded grade	0.10	0.10	0.10	0.10	0.10
Conventional grade	0.15	0.15	0.15	0.15	0.15
Oxidation stability, min	240	240	240	240	240

[a] Measure of anti-knock for spark ignition.

[b] To convert kPa to mm Hg, multiply by 7.5.

Table. 2.6. Specifications for diesel fuel oils (ASTM D 975-78).

	Grade		
Property	*1-D*	*2-D*	*4-D*
Distillation (90%) point, °C	288 max	282-338	
Flash point, min, °C	38	52	55
Water and sediment, max, vol%	0.05	0.05	0.05
Carbon residue on 10% bottom, max, %	0.15	0.35	

(Contd...)

Property	*Grade* *1-D*	*2-D*	*4-D*
Ash, max, wt%	0.01	0.01	0.01
Viscosity at 40°C, kinematic, mm^2/s (= cSt)	1.3–2.4	1.9–4.1	5.5–24.0
Sulphur, max, wt%	0.50	0.50	2.0
Corrosion, copper strip, max, no.	3	3	
Cetane number[a], min	40	40	30

[a] Measure of ignition quality for autoignition.

PETROCHEMICALS

Petrochemicals are essentially the chemical compounds derived from petroleum sources. Thus cracking of petroleum produces ethylene which is converted to ethylene glycol, the latter being a typical petrochemical. The term is also applied to substances such as ammonia, because hydrogen used to form the ammonia is derived from natural gas. Thus synthetic fertilisers are considered to be petrochemicals. A number of chemical products are recovered from the abundantly available raw materials which include crude oil, natural gas, refined petroleum fractions and refinery gases. The bulk of petrochemicals obtained from common hydrocarbons are broadly grouped as under.

(1) *Paraffins (saturated):* Methane, ethane, propane, butanes (iso and normal), etc.

(2) *Olefins (unsaturated):* Ethylene, propylene, butylenes (butene 1 and 2 and isobutylene), etc.

(3) *Aromatics:* Benzene, toluenes, xylenes, etc.

The peculiarity of petrochemical industry is that most of the products can be made from more than one hydrocarbon raw material, depending on the availability of feedstock and economic feasibility of the processes.

The common techniques employed in petrochemical manufacture are the thermal and catalytic cracking, oxidation, hydrogenation, alkylation, chlorination, polymerisation, dehydrogenation, isomerisation, solvent extraction, crystallisation and other related processes. Apart from these, the manufacture of petrochemicals involves the application of a number of continuous physical operations, such as distillation, extraction, filteration, mixing, pumping, etc. To illustrate the scope of manufacture and possible lines of development, various petrochemical products obtained from paraffins, olefins, aromatics, etc. are given below.

Petrochemicals from Methane

Ammonia, nitric acid, ammonium nitrate, ammonium sulphate, urea, carbon dioxide, methyl alcohol, formaldehyde, isobutyl alcohol, acetylene, vinyl chloride, acrylonitrile, neoprene rubbers, acetaldehyde, acetic acid, perchloroethylene, trichloroethylene, carbon tetrachloride, chloroform, methyl chloride, methylene chloride, carbon black, carbon disulphide, etc. (Fig. 2.7).

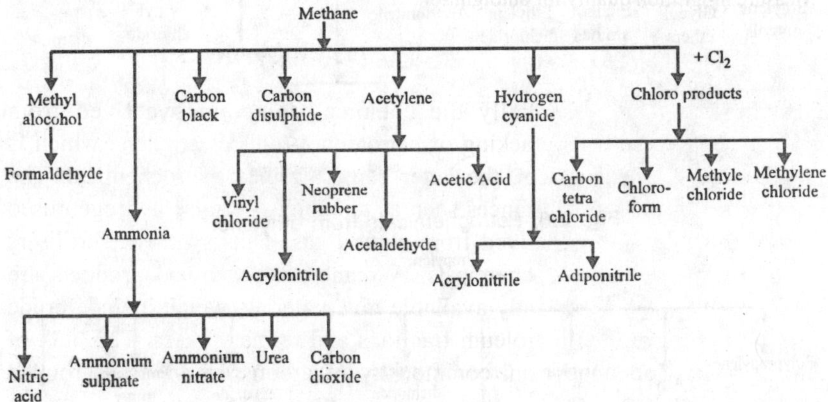

Fig. 2.7. Petrochemicals from methane.

Petrochemicals from Ethylene

Ethylene oxide, ethylene glycol, monoethanolamine, diethanolamine, triethanolamine, diethylene glycol, triethylene glycol, acrylonitrile, ethyl alcohol, acetaldehyde, acetic acid, acetic anhydride, polyethylene, ethyl benzene, styrene, polystryrene, ethyl chloride, ethylene dichloride, vinyl chloride, ethylene diamine, ethylene dibromide, etc. (Fig. 2.8).

Petrochemicals from Propane and Propylene

(1) *Propane:* Ethylene, methyl alcohol, formaldehyde, acetaldehyde, acetone (Fig. 2.9).

(2) *Propylene:* Isopropyl alcohol, acetone, propylene trimer, nonyl phenol, propylene tetramer, dodecyl benzene, propylene oxide, propylene glycol, dipropylene glycol, allyl chloride, allyl alcohol, glycerine, epichlorohydrine, epoxy resins, propylene dichloride, acrolein, cumene, phenol, methyl styrene, polypropylene, etc.

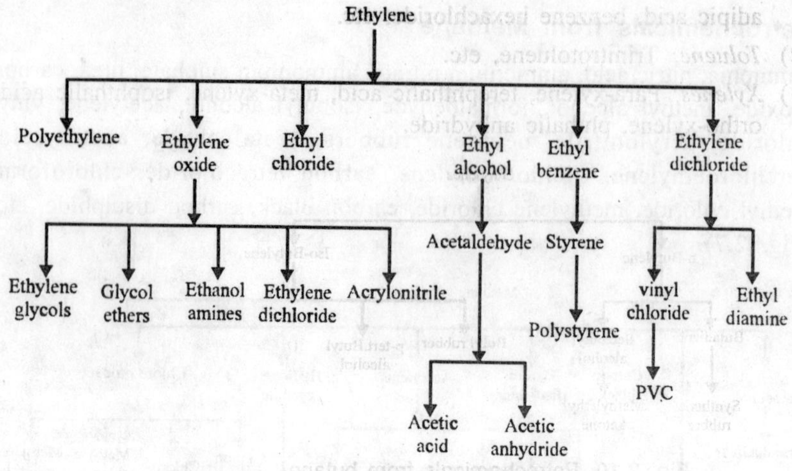

Fig. 2.8. Petrochemicals from ethylene.

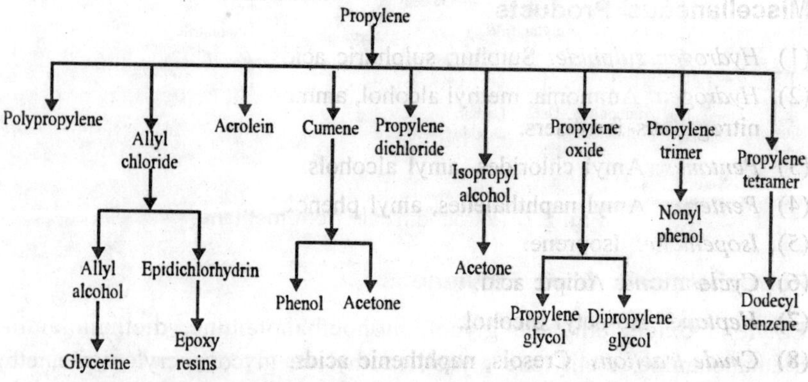

Fig. 2.9. Petrochemicals from propane and propylene.

Petrochemicals from Butanes and Butylenes

(1) *Butanes:* Acetic acid, acetaldehyde, acetic anhydride, methyl alcohol, formaldehyde, propionic acid, n-proply alcohol, butyric acid, n-butyl alcohol, isobutly alcohol (Fig. 2.10).

(2) *Butylenes:* N-butylenes, butadiene, synthetic rubber, sec-butyl alcohol, methyl ethyl ketone, isobutylene, butyl rubber.

Petrochemicals from Aromatics

(1) *Benzene:* Ethyl benzene, styrene, synthetic rubber, polystyrene, cumene, methyl styrene, dodecyl benzene, phenol, maleic anhydride,

adipic acid, benzene hexachloride, etc.

(2) *Toluene:* Trinitrotoluene, etc.

(3) *Xylenes:* Para-xylene, terephthalic acid, meta-xylene, isophthalic acid, ortho-xylene, phthalic anhydride.

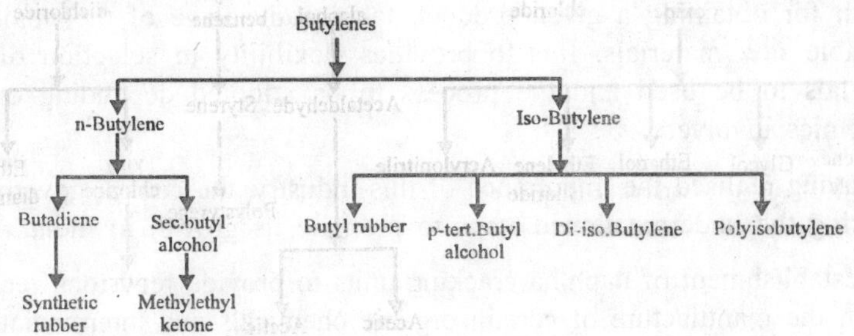

Fig. 2.10. Petrochemicals from butanes and butylenes.

Miscellaneous Products

(1) *Hydrogen sulphide:* Sulphur, sulphuric acids, carbon disulphide.

(2) *Hydrogen:* Ammonia, methyl alcohol, ammonium compounds, urea and nitrogenous fertilisers.

(3) *Pentanes:* Amyl chlorides, amyl alcohols.

(4) *Pentenes:* Amyl naphthalenes, amyl phenol.

(5) *Isopentane:* Isoprene.

(6) *Cyclohexane:* Adipic acid.

(7) *Heptane:* Iso-octyl alcohol.

(8) *Crude fractions:* Cresols, naphthenic acids, naphthalenes, sulphonates.

As a result of rapid development in the field of petrochemicals, a net work of industries, such as plastics, synthetic rubber, dyestuffs, intermediates, nitrogenous fertilisers, synthetic detergents, synthetic fibres, automotive chemicals, printing inks, etc. have come into existence and have experienced an enormous expansion in almost all industrial countries of the world.

The popularity of petrochemicals is attributed to the fact that it is possible to derive products of high purity from oil and natural gas. The range of products obtained from petroleum refining provide an easy and abundant supply of chemicals which are difficult to produce otherwise. In some cases the industries are entirely dependent on petroleum sources for supply of basic raw materials.

The promotion and rapid growth of petrochemical industry is ascribed to the cheap substitutes if provides in place of products hitherto obtained from conventional sources. The industry has made considerable progress because of the recent technological advances. In this field the focus has always been on the adoption of a suitable method from the multiple routes known for obtaining a given product, taking advantage of the numerous available raw materials. It also provides flexibility in selection of raw materials to be used and the process to be adopted depending on the economics involved.

Having realised the importance of this industry, the central government is taking the undermentioned steps to promote its growth in India.

(1) Establishment of naphtha cracking units to provide feedstock required in the manufacture of certain organic chemicals and intermediates;
(2) Utilisation of hydrocarbons and gases for the production of polymers and plastics, synthetic rubber, detergents, insecticides and pesticides, synthetic fibres, carbon black and other products; and
(3) Manufacture of nitrogenous fertilisers from naphtha and associated gases.

LUBRICATION AND LUBRICANTS

The process of reducing friction between two surfaces moving tangentially with respect to one another by interposing a substance of low shear strength between them is called lubrication and the interposed substance is known as a lubricant. Thus primary purpose of lubrication is suspension of moving surfaces to minimise friction and wear. The loss of energy due to friction is considerably reduced in this manner. The lubricant, in fact, acts in a number of manners. For example, it acts as a coolant or heat transfer medium by removing the heat of friction generated as a result of rubbing of surfaces. In lubrication, a thin film of lubricants is formed, between two rubbing surfaces, which does not allow a direct contact between the bubbling surfaces. In internal combustion engines, it acts as a seal by sealing the piston and cylinder wall at the compression rings. Thus there is no leakage of gases at high pressure in the combustion chamber.

Functions of Lubricants

The main functions of the lubricants are as follows :

(1) It reduces wear and tear by avoiding direct contact between the rubbing surfaces.

(2) It reduces loss of energy in the form of heat by acting as coolant.

(3) It increases the efficiency of the machine.

(4) It minimises the liberation of heat.

(5) It reduces the maintenance as well as running cost of machine.

(6) It also acts as a seal to internal combustion engines.

Lubrication Principles

Several distinct regimes are commonly described to systematise the fundamental principles of lubricants. These range from complete separation of moving surfaces by a fluid lubricant, through partial separation in boundary lubrication to dry sliding where solid material properties and surface chemistry dominate. Presently, a change from sliding motion to rolling contact with elasto hydrodynamic lubrication also possible.

Tests of Lubricants

The most important tests of lubricants which are of prime importance in lubrication are copper strip corrosion, oxidation stability, inorganic sulphates and chlorides, steam emission number, viscosity, viscosity-temperature relation, viscosity index, volatility etc. In order to get efficient lubrication, there should be no change in these properties during lubrication. Moreover, there should be no effect of any chemical change such as decomposition, oxidation, reduction at higher temperature and emulsification during lubrication.

Viscosity. The behaviour of a lubricating oil or lubricant is determined by its viscosity. Viscosity is defined as the force in dynes necessary for the movement of 1 sq. cm. layer of a fluid with a velocity of 1cm. per sec., past another parallel layer 1 cm away. The unit usually employed is centipoise. The measurement of viscosity is usually done in C.G.S. units and the unit is stoke. In case of thin oil which have very low viscosity, it is measured in centi-stokes. The viscosity measured by this method is called kinematic viscosity. In case of lubricants, viscosity depends upon the temperature. The viscosity has been found to decrease with increase in temperature.

Specific gravity. The specific gravity of an oil is defined as the ratio of the weight of a given volume of oil to the weight of the same volume of water. The test can be conducted either by using a hydrometer or a density bottle if greater accuracy is desired. The temperature of the test is important and always specified by the standard specification. It has been calculated that for insulating oils specific gravity varies by 0.0005 per every degree centigrade, the value decreasing with increasing temperature. The specific gravity of the oil is considered to have no important bearing on the utility of the oil. However, an upper limit of 0.89 to 0.90 is specified so that any water present in the oil may remain at the bottom only and not tend to float on the oil. A value lower than 0.85 adversely affects other properties, particularly viscosity so that normal insulating oils have specific gravity ranging from 0.85 to 0.90. The specific gravity is used extensively to facilitate volume to weight conversions.

Copper strip corrosion. Crude petroleum usually contains sulphur compounds, most of which are removed during the refining processes (sweetening). This test is designed to detect any traces of free or combined sulphur that may be present in an oil. A clean polished strip of pure copper metal of the specified dimensions is heated in a fixed quantity of the oil at a fixed temperature for the specified period of time. A black, brown, or grey tarnish indicates the presence of corrosive sulphur. The presence of sulphur promotes oxidation of oil in service and causes excessive formation of sludge. The sulphur also corrodes the copper and silver metal parts of the equipment.

Oxidation stability. The oxidation test is perhaps the most important test an oil must pass before it is accepted for service. An oxidation test submits the oil sample to accelerated oxidising conditions in a relatively short time than it is likely to meet in practice. The aim is to assess the behaviour of the oil during its long service from the results of the severe laboratory oxidation.

Though the conditions laid down in various oxidation tests of the standards are different, the accelerated conditions are brought by ne or more of the following methods :

(1) Increase in temperature to the order of 100°–170°C.

(2) Inclusion of metals or soluble catalysts.

(3) Bubbling of air or oxygen in the oil during oxidation.

The oxidation stability test as laid down by the Indian Standard consists in ageing 100 grams of the oil at 150°C for a period of 45 hours during which purified air at the rate of two litres per hour is bubbled into the oil. A copper foil 51 × 32 × 0.1 mm after being cleaned and rolled into a cylinder is placed in the oil to act as a catalyst. The oxidation is carried out in a round bottomed flask of fixed dimensions fitted with a condenser through which cold water is circulated. This ensures that the volatile products of oxidation do not escape. At the end of the oxidation period the oil is diluted to four times its volume with normal-heptane and the amount of sludge and acids formed determined by analysis. The maximum limits fixed for new oils are acidity – 2.5 mg/KOH/g and sludges 1.2% by weight. Formation of deterioration products above this level has been correlated by experience, to a faster rate of deterioration in service. Following are some of the other variation of the oxidation test :

(1) The oil is heated in a metal bomb for a fixed period under specified conditions of temperature and pressure, in an atmosphere of oxygen. The sludge and acids formed are subsequently determined (American method).

(2) The oil is aged in the presence of an electrical field and oxygen and the deterioration products determined after a fixed period of time (Swedish method).

(3) The tensile strength of cotton fibres immersed in the oil, which is being aged continuously, is tested periodically. The sludge formed is also determined (Swiss method).

The repeatability of the oxidation tests is poor and standards often fix the limits of difference between duplicate tests. The rate at which an oil will deteriorate during service can be assessed by the oxidation absorption tests. Such tests though not specified by any standard are widely used to assess the quality of oils.

Inorganic sulphates and chlorides. The inorganic salts such as sulphates and chlorides may be present in the oil as left-overs from the refining processes. These salts ad to corrosivity of the oil. Their presence is determined by making an extract of the oil with hot water. The aqueous extract is treated with Barium chloride solution for sulphates or silver nitrate solution for chlorides. Any presence of a white haziness or precipitate indicates sulphates or chlorides respectively. A new oil must not contain any quantity of these salts.

Steam emulsion number. Water normally cannot remain in suspension in the oil. However the presence of soaps, polar contaminants etc., tends to increase its water retention capacity. The steam emulsion number is a measure of this capacity. In this test a known amount of steam is passed into a fixed quantity of the oil under prescribed conditions. The mixture is allowed to stand and the time taken for the condensed steam to separate out is noted. This time in seconds is the steam emulsion number.

This test confirms the absence of soluble impurities if the oil has a low steam emulsion number.

Viscosity index. It may also be regarded as the average decrease in viscosity per degree rise of temperature between 100°F and 210°F. In the case of most lubricants, the viscosity index (V.I.) ranges from –50 to +110. Lubricants having below or above viscosity index ranges than this are also known.

Pour point. It is an important property of a sample of lubricating oil and is defined as the temperature at which the oil ceases to flow when cooled under standard conditions. This property indicates the dissolved wax concentration in a given sample of lubricating oil.

Cloud point. It is another index of dissolved wax concentration, which may be defined as the temperature at which the cloudiness appears in a sample of oil, under standard conditions, because of wax separation. Lubricants also undergo oxidation, hydrolysis and pyrolysis. The property of lubricants to undergo oxidation is different for different lubricants and mainly depends upon the composition and purity of the lubricant. As a result of oxidation, the lubricant undergoes decomposition and its lubricating property is also affected. Hence certain anti-oxidants are employed in the lubricants to avoid oxidation.

Flash point. It is very important to test the *flash point* of the oil to be used as lubricant. It is not advisable to use an oil as lubricant, which flashes or reaches combustion temperature under the conditions of the lubrication. This is due to the fact that spontaneous combustion of oil vapours occurs if the oxidation is so rapid that heat of oxidation is not dissipated so quickly as to avoid the reaching of flash point or combustion temperature. When lubricants such as esters are exposed to moisture at high temperatures and in the presence of acids and alkalies, they undergo decomposition as a result of hydrolysis. Polyethers and halides are hydrolysed only under drastic conditions of hydrolysis and poly siloxanes are hydrolysed easily by alkaline solutions. It is, therefore, necessary to use those lubricants, which are not capable of hydrolysing under lubricating conditions.

Interfacial tension

The interfacial tension between the oil and water is measure of the molecular attractive force between their unlike molecules at the interface, It is expressed in dynes-centimetre. When an oil sample is poured in water the hydrocarbons comprising the oil arrange themselves at the interface and form a bond with the water molecules. This interface possesses a high tension. When the oil contains dissolved impurities which are polar in nature these tend to diffuse into the water through the interface consequently lowering the interfacial tension.

The interfacial tension is determined by means of suitable tensiometers (Torsion balance) in which the force required to lift a planar ring of platinum or a thin glass plate from the oil/water interface into the oil is measured. This value varies with temperature and ageing of the interface, and consequently the prescribed test conditions must be rigidly adhered to. A minimum interfacial tension value of 40 dynes/cm specified in some standards ensures freedom from all types of dissolved impurities.

Neutralisation number

Neutralisation number or the acid content of an oil is defined as the number of milligrams of potassium hydroxide required to neutralise completely the acids present in one gram of the oil. Neutralisation number is the sum of the organic and inorganic acids in the oil. To determine this a weighed quantity of the oil sample is boiled with a previously neutralised mixture of Alcohol and water (60:40). The resulting mixture is titrated while hot against a standard solution of potassium hydroxide using a suitable indicator. Organic acidity is the difference between the total and inorganic acidity. The maximum prescribed limit for total acidity in new oils is 0.05 mg/KOH/g.

Saponification value

The saponification number is defined as the number of milligrams of potassium hydroxide required to saponify completely the combined and free acids present in one gram of the oil.

It is determined by refluxing a weighed quantity of the oil dissolved in ethyl methyl ketone with an alcoholic solution of potassium hydroxide. The unconsumed alkali is measured by titration with standard hydrochloric acid. The amount of potassium hydroxide thus consumed is calculated. Under these conditions many of the soaps, metals, sulphur and its compounds, halogens and nitrogen compounds consume the alkali, in addition to the substances indicated in the acidity test. Also these are potential contaminates and if present in large quantities lead to a faster rate of oil deterioration.

Mechanism of Lubricants

Depending on the operating conditions three mechanisms may occur :

(1) Complete fluid or hydrodynamic lubrication.

(2) Boundary lubrication.

(3) Extreme pressure lubrication.

Complete fluid or hydrodynamic lubrication

In such type of lubrication, one of the rubbing surfaces is slightly displaced in relation to the other and a wedge shaped film of the lubricant is formed between the two surfaces rubbing against each other. The pressure developed in the wedge shaped film supports the load and prevents the direct contact between the rubbing surfaces. Since a solid-solid interface has been replaced by two solid-liquid interfaces (Fig 2.6), viscosity of the lubricant becomes a crucial factor. An example of this type of lubrication is the rotation of a shaft with respect to a stationary bearing. In Fig 2.6, the lubricant occupies the annular space between the shaft and the bearing and forms a hydrodynamic wedge. So long the shaft rotates, the wedge will remain and prevent contact between two solids. When the load becomes very high, the lubricant is squeezed out of the wedge and friction occurs.

Complete fluid lubrication is provided to delicate instruments such as watches, clocks, guns, sewing machines, scientific instruments etc. The most satisfactory and widely used lubricants for this type of lubrication are hydrocarbon oils, which are blended with selected long chain polymers in order to maintain suitable viscosity of the oil. Some anti-oxidants such as aminophenols are also added to hydrocarbon petroleum fraction because these usually contain minute amounts of unsaturated hydrocarbon petroleum fraction because these usually contain minute amounts of unsaturated hydrocarbons which are susceptible to oxidation under operating conditions.

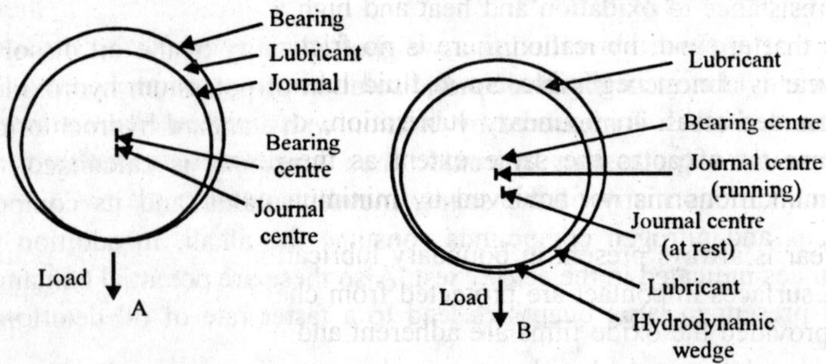

Fig 2.6. Fluid film formation in bearing "A" at rest. "B' rubbing.

Boundary lubrication

Boundary lubrication takes place under any of the following conditions :

(1) When the relative speed of the lubricated surface is insufficient to produce pressure in the lubricant film to carry the load. Such

conditions are reached when a heavily loaded bearing is stopped or started.

(2) When the load on bearing exceeds the pressure generated in the lubricant film.

(3) When the surfaces are not perfect and the load is carried on high spots.

Boundary lubrication is therefore suitable when a continuous film of lubricant cannot persist and direct metal to metal contact is possible because of various reasons, for example, when a shaft starts moving from rest of speed is very low or load is very high or the viscosity of the oil is extremely low. Under all these circumstances friction between the lubricated surfaces lies on absorbed or boundary layer of lubricant. The load is carried by the layer of the absorbed lubricant or both the metal surfaces. The coefficient of friction in such a case is generally 0.05 to 0.15. This friction is generally dependent upon the nature of the material of the surfaces and also on some characteristics that can be determined by the chemical properties of the lubricant.

The effectiveness of boundary lubrication primarily depends upon the structure and chemical properties of the oil to be used. The lubricant used in such type of lubrication should have long chain and active atoms or groups capable of forming linkages with the metals or other surfaces. The other properties of boundary lubricants are good oiliness, low pour point, high resistance to oxidation and heat and high viscosity index. It should be noted that in fluid lubrication there is no friction between the surfaces and the wear is almost negligible. So in fluid lubrication wear decreases as the friction decreases. In boundary lubrication, decrease in friction does not decrease the wear to the same extent as the decrease in friction and so minimum friction is not achieved by minimum wear.

Wear is always present in boundary lubrication and it can be reduced if metal surfaces in contact are protected from chemical attack by oxide films etc., provided the oxide films are adherent and quite hard. Certain additives such as sulphur and phosphorus are also good in preventing the surfaces from chemical attack. Compounds which contain strongly polar groups and capable of absorbing on the metal surfaces are very effective and wearing agents. Graphite and molybdenum disulphide are the two most usual solid lubricants, which either alone or as stable suspension in oil are also capable of producing boundary lubrication. These lubricants form films on metal surfaces, which possess low internal friction and can withstand compression and high temperatures.

Extreme pressure lubrication

This mechanism is prevalent under conditions of high load and high speed and occurs at high temperatures. Under these conditions, the lubricant film may decompose and vapourise. This necessitates the addition of some chemicals to lubricant with a view to increase the chemical stability.of the film. The additives contains chlorinated esters, sulphated oils and aryl(tricresyl) phosphates. These react with metal surfaces to form high melting chlorides, sulphides and phosphides. Their action is chemical, they do not reduce friction but prevent formation of welded joints.

Substances Used as Lubricants

Vegetable oils and fatty acids are most important substances which are used as lubricants. Tallow, lard, whale oils, neat foot oil, rosin oil, castor oil, olive oil, coconut oil, palm oil, oleic acid etc. are the various substances used as lubricants. Mineral oils compounded with fatty acids are known as compounded oils and these are also suitable as lubricants.

Lubricating oil not only reduce wear and friction between two moving or sliding metallic surfaces by providing a continuous fluid film in between them, but they also serve as coolant, sealing agent, corrosion preventors etc.

Characteristics of good lubricating oil

The characteristics of good lubricating oil are :

(1) High boiling point or low vapour pressure.

(2) Suitable viscosity for particular service condition.

(3) Low freezing point.

(4) High thermal stability.

(5) High oxidation resistance.

(6) Non-corrosive nature.

(6) Resistance to decomposition at higher or elevated temperatures.

Additives for Lubricating Oils

Although the bulk of additives is used in automotive lubricants, they also are used in oils for turbines, diesel and aircraft engines, two-cycle engines, hydraulic equipment, gears, and metal working. The common types of additives in the approximate order of their frequency of application are oxidation inhibitors, rust inhibitors, anti-wear agents, detergents-dispersants, pour-point depressants, viscosity index (VI) improvers and foam inhibitors.

Additives containing polar groupings such as fatty acids and their derivatives (e.g. multivalent metal soaps, known as metallic soaps), long chain alcohols and organic phosphides are used to ease boundary lubrication.

Certain additives such as long chain alkylated ring compounds, alkyl naphthalenes, polymers of long chin alkyl acrylate, metallic soaps of fatty acids etc. are used to reduce the temperature at which a liquid lubricant may become solid. This type of additives are known as pour depressants. Their function is colloidal and they prevent the wax crystals to grow.

Phenol and certain condensation products of chlorinated wax with naphthalene are also used as pour point depressing additives. They also prevent the separation of wax from the oil.

Additives such as phenols, their derivatives, amines, organic phosphides etc. are added to the lubricants to prevent the oxidation of lubricant and known as anti-oxidants.

The anti-oxidants or inhibitors retard the oxidation of oil by getting themselves preferentially oxidised. They are generally added to lubricant used in internal combustion engines, turbines etc., where oxidation causes a serious problem.

Additives containing strongly polar groups are added to increase the oiliness and film strength of the lubricants. Examples of such additives are amyl phenyl phosphate, methyl di chloro stearate, dibenzyl disulphide, tricresyl phosphate, sulphurised sperm oil etc.

Oiliness of a lubricant can also be increased by adding oiliness carriers such as vegetable oils (coconut oil, castor oil etc.) and fatty acids (palmitic acid, oleic acid, stearic acid etc.)

Additives such as high molecular weight compounds like hexanol are added as viscosity index improvers. Additives are also added as detergents of deflocculents to the lubricants. Examples of such additives are salts of phenol, carboxylic acids, sulphonic acids and alkyl phosphoric acids etc. Additives added as wear inhibitors prevent wearing of the lubricating surface. Alkaline earth phenolates are generally used as anti-corrosive wear agent and tricresyl phosphate is generally used as anti-abrasive wear agent.

Corrosion preventatives are organic compounds of phosphorus or antimony. Their function is to protect the metal from corrosion by preventing contact between the metal surfaces and the corrosive substances.

The gummy and acidic constituents of lubricants can be removed by a suitable refining processes. The tendency of emulsification of the lubrication oil is checked by certain additives to the lubricant. The additives are also added or used for number of other purposes. For example, polymeric thickners are used as additives for improving the viscosity index of the lubricant.

Extreme pressure additives such as compounds containing phosphorus, lead sulphur and chlorine (e.g. fatty acids, esters ketones, chlorinated wax, sulphurised wax etc.) are added to make lubricating oil to bear extreme pressure. Under extreme pressure a thick film of oil is not capable of maintaining the film and it needs to have high oiliness. High pressure additives are absorbed on the metal surface or chemically react with metal producing a surface layer of low shear strength on the metal surface. Hence tearing up of metal is prevented. High pressure additives also react on metal surface at high temperatures forming surface alloys and prevent welding to gather of the rubbing parts under severe operating conditions.

Additives such as fatty acids, oil soluble alkaline earth salts of fatty acids etc. are added to the lubricant as rust preventives. Additives such as alkali metal salts of carboxylic and sulphonic acids, non-ionic emulsifiers such as mono-esters of polyhydric alcohols, ether alcohols etc. are added to the lubricant which help the formation of emulsion with water. These additives are known as emulsifiers. It should be noted that additives are added to the lubricant taking all the factors including properties, application and effects of individual additives into consideration.

No single oil serves as the most satisfactory lubricant for various modern machineries. The properties of petroleum oils are greatly improved by adding various additives given above but the exact adjustment of the additives to a lubricating oil is one of the most complex problem. The so-called blended oils, are capable of giving desired lubricating properties, required for a particular machinery.

LUBRICANTS OF MINERAL ORIGIN

Lubricants are either of mineral origin or vegetable origin but lubricants of mineral origin are generally used as lubricants. Lubricants of vegetable origin such as vegetable oils and fatty acids are rarely used as lubricants, but they are generally used as additives to the mineral lubricants to improve their properties and qualities.

Large number of lubricants or lubricating oils are obtained from crude petroleum distillation as residue. These residues are either paraffins or asphaltic. Paraffinic residues are steam distilled to get an oil distillate and steam refined cylinder oil as the main fractions. Asphaltic residues are distilled to get various fractions with different viscosities. The lubricating oils obtained from mineral residues have a large number of impurities such as paraffin waxes, resins, asphalts etc. For a good lubricant these properties are undesirable, so it is necessary to remove them. For example, waxes increase the pour point and render the lubricant unfit for use at low temperatures. Asphaltic compounds undergo decomposition at higher temperature, causing carbon deposition and sludge formation. Resinous and asphaltic compounds are easily oxidisable and form sludge when the lubricant is used on the surfaces.

Methods for Removing Impurities

Sharpley's method

Sharpley's method is generally used for removing wax from lubricating oil. This method is based on the difference of density between precipitated wax and the solvent lubricating oil solution from which wax is separated by cooling or chilling. In this method, the lubricating oil containing wax and the solvent for dilution are mixed together and then heated to get a complete solution. The clear solution so formed is chilled down slowly to –20°F and then charged into a centrifugal rotating machine. The wax is separated from the solution of the solvent and lubricating oil. It should be noted the lubricating oil should be completely soluble in the solvent at moderately high temperature and wax should be so precipitated that it can be easily removed.

Acetone Benzene Solvent or Propane Dewaxing Method

Wax can also be removed by using acetone–benzene solvent method or propane dewaxing method. In acetone–benzene solvent method acetone and benzene (1:20) are mixed and then added to the oil at 120°F. The oil solvent mixture is then chilled to –30°F. The precipitated wax is removed by using continuous filtration drum type filter. In propane dewaxing method, lubricating oil is mixed with liquid propane solvent (propane remains liquid) at a pressure of about 200 lbs. per square inch in the ratio of 1:2. Asphalt is thus precipitated which is removed by cooling is brought about by releasing the pressure and obtaining autogenous refrigeration. As a result, wax is thrown out.

Solvent Extraction Method

Aromatic compounds are removed by solvent extraction using different methods such as sulphur dioxide benzene process, nitrobenzene process, furfural process, phenol extraction process, chlorex process etc. In sulphur dioxide benzene process, aromatic compounds are removed by using a mixture of benzene (30:70) and sulphur dioxide.

In nitrobenzene method, the solvent and lubricating oil are mixed and extraction temperature is brought to 40°F. Two layers are separated when the mixture is allowed to stand. The lower layer contains naphthenic hydrocarbons in naphthalene and upper layer contains paraffin compounds with some dissolved nitrogen. The two layers are continuously separated and the solvent is removed by evaporation in a vacuum evaporator.

In furfural method, furfural is used as a solvent for the extraction of lubricating oil. It removes olefinic and aromatic compounds but does not remove paraffinic compounds. The lubricating oil is then treated with 200% by volume of furfural for extraction.

In phenol extraction method, phenol is used as an intermediate solvent between selective solvents like nitrobenzene and furfural and the high potency solvents. Phenol is cheap, stable and can be removed easily from the oil. It also reduces the viscosity sulphur content, oxidation number and carbon content as residue.

Chlorex process

In chlorex process, chlorex (β-β' dichloroethyl ether) is used for the extraction in the chlorex; oil ratio of 0.7:1.15. In this process viscosity and carbon residue of the lubricating oil are reduced and pour point is improved. Moreover, extraction also causes improvement in colour and oxidation stability.

Advantages

(1) It is less expensive.

(2) It provides satisfactory removal of asphaltic impurities and hence minimises the chances of carbon deposit formation during the use of refined lubricating oil.

(3) It gives high yield of refined oil and provides a very refined product which does not show much variation in viscosity due to increase in temperature.

Disadvantages

(1) It yields oil which is less resistant to oxidation.

(2) The refined oil also posses lower oiliness and hence its oiliness is improved by adding suitable additives.

Acid Treatment

Unsaturated hydrocarbons from the oil are removed by acid treatment. The lubricating oil is treated with 10% by volume of the oil of concentrated sulphuric acid. The principle of counter-current is used in the treatment. In order to avoid side reactions, generally the temperature is kept low. The oil is then separated from sulphuric acid sludge and is treated with dilute alkali and then with clay.

The acid refining can be followed after dewaxing process. If the oil still contains undesirable constituents like naphthenic and asphaltic impurities. These are eliminated by treating the oil with concentrated sulphuric acid and then agitated. Some of the impurities dissolved in the acid while others are converted into sludge. The later is removed by filtration and the filtrate is neutralised with requisite amount of NaOH so as to remove excess acid. The oil is finally decolourised by passing through fullers' earth at 100°–140°C. It should be noted that during above treatment of lubricating oils, not only the objectionable material and oxidisable impurities are removed, but polar bodies in lubricants are also removed. These polar bodies attach themselves to lubricated surfaces and provide better lubricant because of their chemical affinity. Hence the additives are now-a-days added to ultra purified and non-reactive oily substances and these additives are capable of giving polar properties to lubricating oils.

Classification of Lubricants

Lubricants are classified on the basis of their states of aggregation or consistency. They are :

(1) Liquid lubricants

(2) Semi-solid lubricants

(3) Solid lubricants.

Liquid lubricants (lubricating oils)

Most of the lubricating oils are of petroleum origin. When petroleum is distilled, a residual fraction collects at the bottom of the fractionating column. This oil on refractionation under reduce pressure gives three fractions—light, medium and heavy oils. These have to be purified before use as lubricants. The oil contains wax, asphalt (tar), coloured substances and oxidisable impurities. Wax prevents fluidity of the lubricant at low temperatures; asphalt deposits on the engine parts during use of the lubricant; and oxidisable impurities cause sludge formation.

Semi-solid lubricants or greases

A grease is a semi-solid lubricant obtained by thickening a lubricating oil through the addition of a metallic soap. The thickener also called a gelling agent, is usually a sodium, calcium or lithium soap. The long chain fatty acids forming the soap may be stearic acid, oleic acid or palmitic acid. If sodium soap is added to the lubricating oil, the grease is known as soda base grease; if calcium soap is added, lime base grease or cup grease results. The functions of the soaps are optimising the operating temperature, improving the resistance to oxidation and hydrolysis as well as adjusting the consistency. The different types of greases, their properties and uses are listed in Table. 2.10.

Though soaps are the usual thickening agents in the greases, graphite, asphalt (tar) or bentonite (clay) may also be used. Among synthetic greases, the silicones are the foremost.

Table 2.10. Properties and uses of various types of greases.

Greases	*Properties*	*Uses*
Soda base grease	Low water resistance; hence can be used only in dry conditions	Used in high temperature lubrication upto 175°C in ball bearings
Lime base or cup grease	General purpose greases. Can be used only upto 70°C. More resistant to water than soda base grease	Lubrication of pumps and tractors
Lithium-soap grease	Resistant to both heat and water.	For aircraft lubrication; from –55 to 150°C

Conditions in which Greases can be Used

(1) Oil is squeezed out due to heavier load or low speed, or the oil does not remain in place due to vertical flow or jerks.

(2) The design of the machine does not allow the lubricant to be retained.

(3) Lubricant should not contaminate the manufactured product, such as food, paper, textile etc.

(4) Lubricant has to act as a screen to prevent entry of dirt or moisture.

(5) Electrical motor bearings are to be fabricated.

(6) Frequent application of lubricant is inconvenient, as in automobile wheel bearings.

The important properties of greases are consistency, penetration index and drop point, of which, dropping point determination helps in finding the maximum temperature of use for grease. The dropping point of grease is the lowest temperature at which the grease, when heated under specified conditions, becomes sufficiently fluid to drop from a cup through a hole of standard dimensions.

Solid Lubricants

The reasons for which greases are preferred over lubricating oils hold preference of solid lubricant as well. Only those solids which have laminar micro structure can be used as lubricants. Graphite and molybdenum disulphide are ideal lubricants on this basis.

Graphite may be used as powder or as colloidal dispersion in water or in oil. Colloidal dispersion of graphite is denoted by the dag (deflocculated Acheason's Graphite). Dispersions in water is known as aquadag and oily dispersion is called oildag. Graphite is not effective as a lubricatnt in vacuum and gets oxidised above 370°C. Graphite is used as a lubricant in I.C. engines, air compressors, lathes etc.

SYNTHETIC LUBRICANTS

Many lubricants even with additives can be used under various abnormal conditions, for example when the temperature is extremely high or the atmosphere is chemically reactive. Under such conditions synthetic lubricants having appropriate properties to be used under abnormal conditions are generally employed. Such synthetic lubricants are prepared by hydrocarbons, polyglycols, esters, polycarbonates, amines, silicones and fluorine derivatives etc.

In fact synthetic lubricants are developed to meet the most severe and drastic operations such as those existing in aircraft engines in which oil may have to be pumped at –50°C. During take off or landing the same oil may get heated to as high a temperature as 250°C. Hence it is necessary for the oil to maintain satisfactory properties. The oil to be used at high temperatures and high operating pressures is expected to have high resistance to oxidation and high viscosity, and at the same time it must also possess sufficiently low freezing point and it should also be non-inflammable.

Modern Synthetic Lubricants

Modern synthetic lubricants are prepared by taking all these factors into consideration and have number of distinguishing characteristic features. They are inflammable and have high flash points, high viscosity index and high thermal stability at higher temperatures. For most common purposes blended petroleum oils are good and synthetic lubricants are needed only in special circumstances, e.g. in certain specialised military and industrial uses.

Polymerised Hydrocarbons

Polymerised hydrocarbons such as polymerised ethylene, propylene, butylene etc. having molecular weight ranging from 250 to 50,000 are used as synthetic lubricants. Although these hydrocarbons have chemical and physical properties similar to the lubricants obtained from petroleum fractions, but they have so distinct disadvantages over equivalent petroleum products. For example they can be used at high temperatures as residue free high temperature lubricants, they are light in colour and free from non-hydrocarbons, having better electrical properties.

Polyglycols and Related Compounds

Polyglycols and related compounds such as polyethylene glycol, polypropylene glycol, their ethers and esters, higher polyalkylene oxides, polythioglycols and polyglycidyl ethers can be used as water soluble and water insoluble lubricants.

Water Diluted Lubricants

Water diluted lubricants can be prepared by polyethylene glycol or polypropylene glycol to be used in rubber bearings and joints. After applying to metal surfaces they can be easily removed by water flushing. Higher polyalkylene oxides and polyglycidyl ether absorb a considerable amount of water, although they are not soluble in water. Temperature has no effect on their viscosity. Polyalkylene oxides undergo decomposition into the volatile parts under high temperature and these volatile parts undergo oxidation. Hence these lubricants are very useful at high temperatures and used in roller bearings of sheet glass machinery.

Natural Fats and Oils as Esters of Glycerides

They have high freezing point and can be easily oxidised. On account of this reason they cannot be used as lubricants. Diesters of glycol and aliphatic dibasic acid as well as that of aromatic dibasic acids and triesters of phosphoric acid etc. have been found to be good synthetic lubricants because their viscosity is not much affected by increase in temperature. Their volatility is very low, they can be easily oxidised by this property, this can be reduced by adding additives. They also act as good lubricant even at low temperatures. Esters as lubricants can however be decomposed by steam, acids and alkalies.

Organic Amines and Amides

Organic amines and amides are good synthetic lubricants having high viscosity index and low pour point. Silicones and the products obtained by hydrolysis and decomposition of organic chloro silicones are also employed as synthetic lubricants, because they have low viscosity temperature coefficients and are not readily oxidised upto about 200°C. They however oxidise quickly above 225°C. Silicones are very useful for low temperature lubrication of small parts.

Fluorocarbon and Other Compounds

Fluorocarbon and other compounds, e.g. tertiary amines, ethers, esters which can be completely fluorinated are also used as synthetic lubricants. They are chemically inert and resist to all chemicals except sodium. They are not decomposed easily be heat. They are also not easily oxidisable.

Preparation of Greases

By saponification of fats

Greases are generally prepared by saponification of fat such as tallow or fatty acid with alkali such as lime caustic soda, followed by adding hot lubricating oil with constant agitation soaps are also added which act as gelling agents and give an interconnected structure. At high temperature, the interconnected structure which is held by intermolecular forces ceases to exist, because soap (gelling agent) dissolves in the oil and hence the grease undergoes liquefaction.

From fatty acids and lubricating oil

Greases can also be prepared by dissolving fatty acid in part of lubricating oil. Metal hydroxide is then added to it in form of dry oil slurry or as aqueous solution. The mixture of soap and oil is dehydrated by stirring and the balance of lubricating oil is added and homogenised. Since the gelling agents in soap based grease are crystallities of fibrous shape. Their gelling property is due to their ability to form a tangle in the interstices as a result of which the oil is held, the soap dissolved in oil. As a result, the fibrous structure no longer exists and liquefaction of grease takes place.

Greases can be made stable to oxidation and to provide lubrication under extreme conditions of load and sliding speed by adding additives to lubrication oils. The consistency of the greases considerably changes when the temperature is suddenly decrease, because the intersection of gelling agent and viscosity of oil changes with increase in temperature. The temperature at which the consistency of greases changes, determines the upper temperature limit of the applicability of the greases. Under certain conditions it is not possible to use liquid lubricant, so in such cases solid lubricants such as talcum, graphite, boron nitride, molybdenum disulphide are used.

Chapter 3

Mechanism of Polymerisation

INTRODUCTION

There are several useful materials which have high molecular weight. Examples include alkaloids, dyes, etc., and a group of materials known as polymer. The difference between the two types lies in the fact that the former has a very complex structure whereas the polymer has rather a very simple chemical formula. A polymer is a large molecule built up by successive repetition of small chemical unit (mer) which, in turn, is derived from simple compound(s) called monomer(s). The number of mers present in a polymer molecule is called *degree of polymerisation (DP).* Hence,

$$\text{Molecular weight of a polymer} = \text{DP} \times \text{mer weight}$$

The molecular weight of polymers may vary between 10^2 and 10^6 range. However, as long as the molecular weight is low and within few thousands, the polymer is known as oligomer. When the molecular weight is 5000 or above they are called high polymer, super-molecules, giant molecules, macromolecules.

$$n\text{CH}_2=\underset{\text{CN}}{\underset{|}{\text{CH}}} \longrightarrow n\text{CH}_2\text{—}\underset{\text{CN}}{\underset{|}{\text{CH}}}\text{—} \longrightarrow (\text{—CH}_2\text{—}\underset{\text{CN}}{\underset{|}{\text{CH}}})_n\text{—}$$

Acrylonitrile monomer — Acrylonitrile-mer — Polyacrylonitrile

n = Degree of polymerisation

Though derived from simple chemical compound (monomer), polymers exhibit properties that are very much different from those of the monomer of any other low molecular weight compound. The differences are a matter of degree rather than of kind, i.e., they are mainly physical and mechanical rather than chemical. The unique properties of high polymers are due to unusual shape and large size of the molecules. The large size and surface result in a large magnitude of total force of intermolecular attraction. Also due to great length of the molecules, polymers are capable of undergoing enormous changes in dimension in passing from a straightened to a coiled or folded condition. In simple molecules, both these possibilities are absent.

CLASSIFICATION OF POLYMERS

Polymers may be classified as natural (such as cellulose, wool, protein, natural rubber, etc.) and synthetic (e.g., nylon, bakelites, styrene, butadiene rubber etc.). They may also be of organic origin like wool or nucleic acid and inorganic, e.g., diamond, ceramics, etc. More scientifically, high polymers are usually classified on the basis of the outline of the principal chains of atoms. When the polymers contain isomers joined together in the form of a rope or chain, it is called linear.

$$-CH_2-\underset{}{\overset{CH_3}{\overset{|}{C}}}=CH-CH_2-CH_2-\overset{CH_3}{\overset{|}{C}}=CH-CH_2-$$

Linear polyisoprene

When the linear structure bears side chains composed of the same linear structure as the main chain, it is called branched polymer.

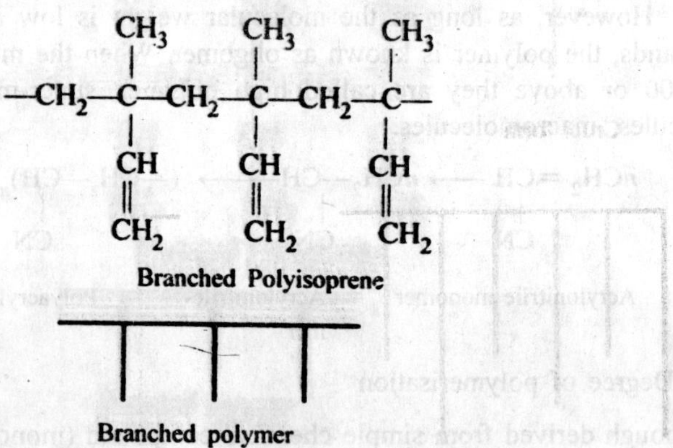

Branched Polyisoprene

Branched polymer

If the branches are of a different structure than the linear polymer chain (backbone polymer) it is called a graft polymer.

$$
\begin{array}{c}
\qquad CH_3 \qquad\qquad\qquad\qquad\qquad CH_3 \\
—CH—C{=}CH—CH_2{=}CH—C{=}CH—CH_2— \\
CH_2 \qquad\qquad CH_2 \\
C_6H_5—CH \qquad C_6H_{11}—CH \\
CH_2 \\
C_6H_5—CH
\end{array}
$$

Graft polymer of isoprene and styrene

Graft polymer

There are several other types of branched polymers which are shown below :

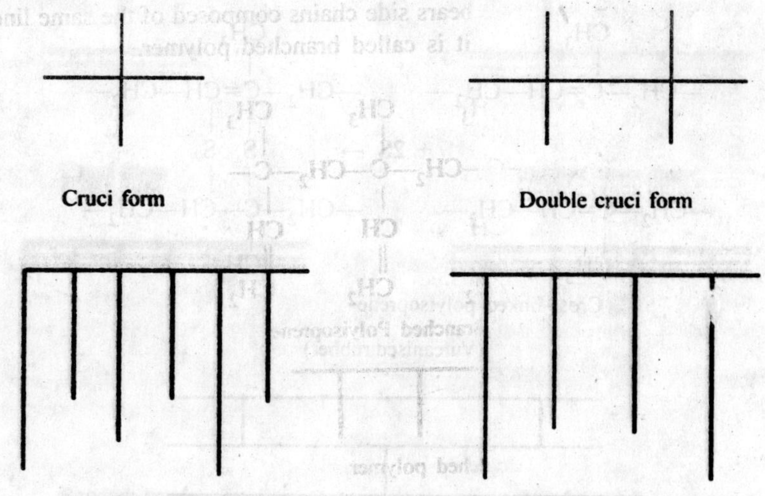

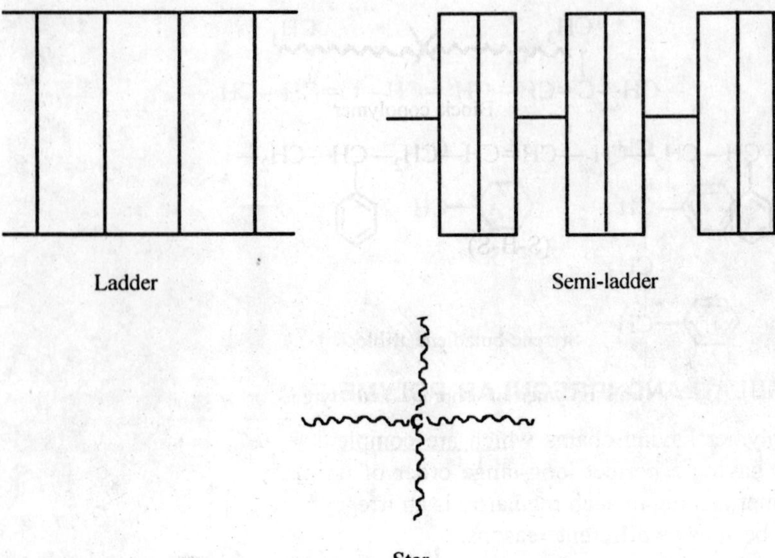

Ladder Semi-ladder

Star

When the branches connect one principal chain to another so that a three dimensional network develops then the polymer is called as cross-linked (space) polymer.

$$\begin{array}{c} CH_3 \\ | \\ -CH_2-C{=}CH-CH_2- \\ \\ -CH_2-C{=}CH-CH_2- \\ | \\ CH_3 \end{array} \;+\; 2S \rightarrow \begin{array}{c} CH_3 \\ | \\ -CH_2-C{=}CH-CH_2- \\ |\quad\; | \\ S\quad\; S \\ |\quad\; | \\ -CH_2-C-CH-CH_2- \\ | \\ CH_3 \end{array}$$

Cross-linked polyisoprene

(Vulcanised rubber)

Cross-linked polymer

When a polymer is made up of two or more distinct but different blocks of ner then it is called a block copolymer.

Block copolymer

$$-CH(C_6H_5)-CH_2-CH_2-CH{=}CH-CH_2-CH(C_6H_5)-CH_2-$$

(S-B-S)

Styrene-butadiene triblock polymer

REGULAR AND IRREGULAR POLYMERS

A polymer having chains which are completely identified to one another, thus, having a perfect long-range order of units is a regular polymer. Any polymer lacking in such regularity is an irregular polymer. Chain irregularity may be due to different reasons.

Thus, during formation of polymers the repeating units may combine in two ways:

$$-CH_2-CH(Cl)-CH_2-CH(Cl)-CH_2-CH(Cl)-$$

Head-to-tail (or 1, 3 addition)

Polyvinyl chloride

$$-CH_2-CH(Cl)-CH(Cl)-CH_2-CH_2-CH(Cl)-CH(Cl)-CH_2-$$

Head-to-head or tail-to-tail (or 1, 2 addition)

In the case of dienes (rubber forming monomers) depending upon the nature of addition either a linear or branched polymer may form.

$$\overset{1}{-CH_2}-\overset{2}{CH}=\overset{3}{CH}-\overset{4}{CH_2}-\overset{1}{CH_2}-\overset{2}{CH}=\overset{3}{CH}-\overset{4}{CH_2}-$$

Linear polybutadiene (or 1, 4 addition)

$$\overset{1}{-CH_2}-\overset{2}{CH}=\overset{3}{CH}-\overset{4}{CH_2}-\overset{1}{CH_2}-\overset{2}{CH}=\overset{3}{CH}-\overset{4}{CH_2}-\overset{1}{CH_2}-\overset{2}{CH}(-CH={}^{4}CH_2)-$$

Branched polybutadiene (or 1, 2 addition)

In the isolated state the macromolecule may change its shape, and each polymer may turn differently relative to one another. When polymers crystallise, the units of certain sections of the polymer becomes fixed. Thus the molecular polymer chain is made up of repeating sections of identical spatial structure called repeat units. The distance from the beginning of one repeat unit to beginning of the next is called repeat distance or identity period. Polymers having same chemical composition but differing in spatial arrangement (and thus in identity period) are called *cis*- and *trans*-isomer. Thus:

$$\sim CH_2-\overset{H_3C}{C}=CH-CH_2-CH_2-\underset{CH_3}{C}=CH-CH_2-CH_2-\overset{CH_3}{C}=CH-CH_2\sim$$

$$\longleftarrow 8.16\ \text{Å} \longrightarrow$$

cis-isomer (Natural rubber)

$$\sim CH_2-\overset{H_3C}{C}=CH-CH_2-CH_2-\overset{CH_3}{C}=CH-CH_2-CH_2-\overset{CH_3}{C}=CH-CH_2\sim$$

$$\longleftarrow 4.8\ \text{Å} \longrightarrow \longleftarrow 4.8\ \text{Å} \longrightarrow$$

trans-isomer (Gutta percha)

The repeat distance of crystalline rubber is 8.16 Å, while that of gutta-percha (a plastic) is 4.8 Å. Going from gutta-percha to rubber the repeat distance should have been 9.6 Å (i.e. 4.8 × 2), but it is less due to changes in valency angles. Crystalline polyethylene is a plane zig-zag carbon chain with identity period equal to the length of one zig-zag:

CH_2 CH_2 CH_2
CH_2 CH_2
← 2.53 Å →

Another type of configurational isomerism is D-isomerism, which is due to the presence of an asymmetric carbon atom in the polymer chain.

$$R-\underset{B}{\overset{A}{\underset{|}{\overset{|}{C}}}}-R'$$

The asymmetry is determined by unequal length and possible differences in space configuration of the two parts of the molecular chains connected to each carbon atom (R≠R'), in the presence of two different substituents A and B. Polymers in which all adjacent asymmetric carbon atoms over the length of at least one macromolecular chain have the same space configuration are called *isotactic polymers.* Polymers whose macromolecules are made up of units with opposite space configuration of each consequent carbon atom (asymmetric) in the chain are called syndisotactic polymers. A simpler (though crude) way of looking at these is that if the substituent (say A) appears on the same side of the chain in space then the structure is called isotactic and if they occur alternately on both sides of chain then it is in syndisotactic configuration.

A A A A

(Isotactic)

A A

A A

(Syndisotactic)

Polymers with an irregular arrangement in space are called *atactic.*

POLYMERISATION

It is a process by which simple chemical compounds are reacted to form a macromolecule. Essentially the reaction takes place intermolecularly and is capable of proceeding indefinitely (to grow to an infinite length) functionally. Basically, they are of two types—condensation polymerisation and addition polymerisation.. According to the mechanism of the reaction they are called as step polymerisation and chain polymerisation. The essential distinguishing features of the two are listed below:

	Chain polymerisation	*Step polymerisation*
1.	Only growth reaction adds repeating units one at a time to the chain.	Any two molecular species present at a time can react.
2.	Number of units decreases steadily throughout the reaction.	Monomer disappears early in the reaction at DP 10, 1% monomer remains.
3.	High polymer is formed at once; polymer molecular weight changes little throughout the reaction.	Polymer molecular weight rises steadily throughout the reaction.
4.	Long reaction time gives high yields but affect molecular weight little.	Long reaction times are essential to obtain high molecular weight.
5.	Reaction mixture contains only monomer, polymer, any other added agent and about 10^{-8} part of growing chains.	At any stage all molecular species are present in a calculable distribution.

STEP POLYMERISATION

This type of polymerisation can take place in two ways. In the step growth system, monomers react intermolecularly (condensation) with the formation of the polymer accompanied by a small, volatile molecule.

$$nA \longrightarrow P + X$$

$$n\text{Cl—CH}_2\text{—CH}_2\text{—Cl} + \text{Na}_2\text{S}_x \longrightarrow \left[\text{—CH}_2\text{—CH}_2\text{—}\underset{\text{S}}{\underset{|}{\text{S}}}\text{—}\underset{\text{S}}{\underset{|}{\text{S}}}\text{—}\right]_n + 2\text{NaCl} + (n\text{—}x)\,\text{S}$$

Ethylene dichloride — Sodium polysulphide — Thiokol rubber

In the step polymerisation the reaction takes place by addition of mers but the product is not accompanied by the formation of any small molecule.

$$nB \longrightarrow (B)n$$

$$n\text{OC—(CH}_2)_5\text{—NH} \longrightarrow \left[\text{—OC—(CH}_2)_5\text{—NH}\right]_n$$

Caprolactam — Polycaprolactam (Nylon-6)

For a step polymerisation, the monomer(s) must have certain characteristics which include the following:

(1) The monomer structure should be such that it is incapable of forming 5- or 6- membered ring.

(2) The average functionality of the monomer or monomers should be at least two or more, if it is two a linear polymer will be formed and if it is more than two a cross-linked polymer will be formed. The functionality (f), DP (n) and extent of polymerisation (P) are related by Carother's equation:

$$P = \frac{2}{f} - \frac{2}{nf}$$

ADDITION POLYMERISATION

Addition (chain) polymerisation is the polymerisation of unsaturated (ethylenic) monomers typically involving in a chain reaction. A chain reaction is a series of consecutive reactions in which an activated species converts reactants to products, another activated molecule being formed in the process. The most common chain reaction proceeds via a free radical mechanism, though it may also take place through a cationic, anionic or coordination complex mechanism.

The monomers which are most susceptible to addition polymerisation are derivative of ethylene (vinyl compounds, diene, etc.) producing an unsymmetrical double bond.

$H_2C{=}CH_2$ — Ethylene

$H_2C{=}CHR$ — Vinyl compound

$H_2C{=}CR_1R_2$ — Vinylidene compound

$H_2C{=}CH{-}CH{=}CH_2$ — Diene

Depending upon the nature of attacking agent (free radical or ionic) the double bond may undergo either homolytic (free radical) or heterolytic fission.

$$>C{=}C< \; + \; R \longrightarrow >C(R){-}\dot{C}<$$

Free radical

$$\overset{\delta+}{C}=\overset{\delta-}{C} + R^+ \longrightarrow \underset{R}{C}-\overset{+}{C}$$

Carbonium ion

$$\overset{\delta+}{C}=\overset{\delta-}{C} + R^- \longrightarrow \underset{R}{C}-\overset{-}{C}$$

Carbonium

COPOLYMERISATION

This is a process by which two or more monomers are polymerised together to produce a polymer having a mer derived from contributions by each of the monomers:

$$\underset{\text{Acrylonitrile}}{CH_2{=}CH(CN)} + \underset{\text{Butadiene}}{CH_2{=}CH{-}CH{=}CH_2} + \underset{\text{Styrene}}{CH{=}CH_2(C_6H_5)}$$

$$\downarrow$$

$$-CH_2-\underset{CN}{CH}-CH_2-CH{=}CH-CH_2-\underset{C_6H_5}{CH}-CH_2-$$

Copolymer of acrylonitrile-butadiene-styrene or ABS

Another type of copolymerisation is done by using a polymer substrate on which a reactive site is developed followed by reaction with a monomer activated by the reactive site. This process is known as *graft copolymerisation.*

$$\underset{\text{Natural rubber}}{-CH_2-\overset{CH_3}{C}=C-CH_2-} + \underset{\text{Styrene}}{CH{=}CH_2(C_6H_5)} \longrightarrow$$

$$\downarrow$$

$$\begin{array}{l} \quad\quad\;\; CH_3 \\ \quad\quad\;\;\; | \\ -CH-C=HC-CH_2- \\ \quad\quad\;\;\; | \\ -CH-CH_2- \\ \quad | \\ \;\; C_6H_5 \end{array}$$

Styrene grafted natural rubber

Block copolymerisation may be prepared when two growing polymer chains react only through end groups.

$$\underset{C_6H_5}{CH}-CH_2-\underset{C_6H_5}{CH}-CH_2-\underset{C_6H_5}{CH}-CH_2 \;+\; -CH_2-\overset{CH_3}{C}=CH-CH_2-CH_2-\overset{CH_3}{C}=CH-CH_2- \longrightarrow$$

$$-\underset{C_6H_5}{CH}-CH_2-\underset{C_6H_5}{CH}-CH_2-\underset{C_6H_5}{CH}-CH_2-CH_2-\overset{CH_3}{C}=CH-CH_2-CH_2-\overset{CH_3}{C}\;CH-CH_2$$

Styrene - isoprene block copolymer

Copolymerisations are usually carried out by the same type of initiators used for making homopolymers. However, anionic polymerisation and Ziegler Natta catalysts have been found most suitable for producing block and stereoregular polymers.

Structures of some important polymers are given as under:

$$nCH_2=CH_2 \longrightarrow \left[-CH_2-CH_2-\right]_n \quad \text{Polyethylene}$$

$$nCH_2=\overset{CH_3}{CH} \longrightarrow \left[-CH_2-\overset{CH_3}{CH}-\right]_n \quad \text{Polypropylene}$$

$$nCH_2=\overset{CH_3}{\underset{CH_3}{C}} \longrightarrow \left[-CH_2-\overset{CH_3}{\underset{CH_3}{C}}-\right]_n \quad \text{Polyisobutylene}$$

$$nCH_2{=}\underset{\substack{|\\ Cl}}{CH} \longrightarrow \left[-CH_2-\underset{\substack{|\\ Cl}}{CH}-\right]_n \qquad \text{Polyvinyl chloride}$$

$$nCH_2{=}\underset{\substack{|\\ OR}}{CH} \longrightarrow \left[-CH_2-\underset{\substack{|\\ OR}}{CH}-\right]_n \qquad \text{Polyvinyl ether}$$

$$nCH_2{=}\underset{\substack{|\\ CN}}{CH} \longrightarrow \left[-CH_2-\underset{\substack{|\\ CN}}{CH}\right]_n \qquad \text{Polyacrylonitrile}$$

$$nCH_2{=}CH-CH{=}CH_2 \longrightarrow \left[-CH_2-CH{=}CH-CH_2-\right]_n \qquad \text{Polybutadiene}$$

$$nCH_2{=}\overset{\substack{CH_3\\ |}}{C}-CH{=}CH_2 \longrightarrow \left[-CH_2-\overset{\substack{CH_3\\ |}}{C}{=}CH-CH_2-\right]_n \qquad \text{Polyisoprene}$$

$$nCH_2{=}\overset{\substack{Cl\\ |}}{C}-CH{=}CH_2 \longrightarrow \left[-CH_2-\underset{\substack{|\\ Cl}}{C}{=}\overset{\substack{Cl\\ |}}{CH}-CH_2-\right]_n \qquad \text{Polychloroprene}$$

$$nNH_2-(CH_2)_6-NH_2 \rightarrow \left[H-NH-(CH_2)_6-NH-OC-(CH_2)_4-CO\right]_{n-1}-OH$$

Hexamethylene diamine → Nylon 66

$$+ \qquad\qquad\qquad\qquad +$$

$$nHOOC-(CH_2)_4-COOH \qquad\qquad H_2O$$

Adipic acid

$$nNH_2-(CH_2)_5-CO \longrightarrow -\left[NH-(CH_2)_5-CO-\right]_n$$

(ring closed: NH_2 linked to CO)

Nylon 6

$$nHO\,(CH_2)_4-OH$$

Butanediol

$$+ \longrightarrow H-\left[O(CH_2)_4-O-\overset{\substack{O\\ \|}}{C}-NH-(CH_2)_6-NH-CO-\right]_{n-1}OH$$

$$nOCN-(CH_2)_6-NCO$$

Hexamethylene diisocyanate — Perlon

$$n\text{Cl}(\text{CH}_2)_x\text{Cl} + n\text{Na}_2\text{S} \longrightarrow \left[\text{—(CH}_2)_x\text{—S—}\right]_n + n\text{NaCl}$$

Polysulphide
(Thiokol A)

$$n\text{HO—Si(CH}_3)_2\text{—OH} \longrightarrow \text{HO}\left[\text{—Si(CH}_3)_2\text{—O—}\right]_{n-1}\text{H} + (n-1)\,\text{H}_2\text{O}$$

Polydimethylsiloxane
(Silicone)

$$\text{CH(C}_6\text{H}_5)\text{=CH}_2 + \text{CH}_2\text{=CH—CH=CH}_2 \longrightarrow \left[\text{—CH(C}_6\text{H}_5)\text{—CH}_2\text{—CH}_2\text{—CH=CH—CH}_2\text{—}\right]$$

Styrene butadiene rubber

$$\text{CH}_2\text{=C(CH}_3)\text{—CH}_3 + \text{CH}_2\text{=C(CH}_3)\text{—CH=CH}_2 \longrightarrow \left[\text{—CH}_2\text{—C(CH}_3)_2\text{—CH}_2\text{—C(CH}_3)\text{=CH—CH}_2\text{—}\right]$$

Butyl rubber

$$\text{CH}_2\text{=CH—CN} + \text{CH}_2\text{=CH—CH=CH}_2 \longrightarrow \left[\text{—CH}_2\text{—CH(CN)—CH}_2\text{—CH=CH—CH}_2\text{—}\right]$$

Nitrile rubber

METHODS OF POLYMERISATION

There are several methods of conducting a vinyl polymerisation. They only vary in the physical state of dispersions. But numerous theoretical and practical considerations dictate the use of one method in preference to the other for a particular monomer. The most important of these procedures are only four :

(1) Bulk polymerisation

(2) Solution polymerisation

(3) Emulsion polymerisation

(4) Suspension polymerisation

Apart from the nature of chemical reaction involved and the type of polymer to be produced, the selection of a polymerisation technique for production of polymers on an industrial scale is based on several technical and economic factors.

Technical Factors

(1) Type of raw materials to be used and accuracy of charging them into the reactor

(2) Reactor design

(3) Agitation of the reaction mass

(4) Removal of exothermic heat of the reaction and effective temperature control

(5) Choice of heat transfer medium or media

(6) Control of operating conditions to control the rate and degree of polymerisation

(7) Movement and handling of the reaction mass and of the polymer

(8) Separation of polymer from the reaction mass

(9) Type, number and quality of polymers to be produced

(10) Choice of materials of construction

(11) Prevention of contamination from external sources

(12) Provision for cleaning of pipes and equipment

(13) Flexibility of operation

(15) Safety hazards

Economic Factors

(1) Capital investment and operating costs
(2) Cost of raw materials and yield
(3) Rate of production per unit reactor volume
(4) Scale of operation
(5) Cost of recovery and purification of unreacted monomer, solvent, etc.
(6) Cost of unit operations like filtration, washing, drying and pelletising of polymer to get it into a processable form

BULK POLYMERISATION

This is the simplest method and involves the use of monomer and initiator only as the major components. So the polymer could be obtained in a high state of purity and the maximum utilisation of the production capacity can be made. However, the process is not without disadvantages, some of the more important ones being the problems connected with heat transfer, agitation, control of molecular weight, removal of residual monomer, initiator, etc. The molecular weight distribution is also broad.

High temperature polycondensation may also be considered as a type of bulk polymerisation. Most of the points discussed below are equally applicable to high temperature polycondensation also (Fig. 3.1).

The heats of polymerisation of vinyl monomers are high and fall in the range of 20 kcal/g mole. Heat dissipation becomes a problem with increasing viscosity. This may lead to local superheating leading to large variations in size distribution, lower molecular weight, discolouration and added danger of cross-linking leading to useless product formation which may stick to the walls of the reaction kettle and can only be removed with great difficulty. Also, there are chances of explosion and fire hazards. Sometimes, at low conversion percentages auto acceleration effect sets in and the reaction be better stopped. Due to these reasons, this method was not previously recommended for general industrial practice as a continuous method. However, recent advances in mitigating the heat of polymerisation has made it possible to use this method for continuous operation, and the advantages of this method are becoming more and more evident. Due to these advantages, the recent tendency is to polymerise vinyl polymers by this method.

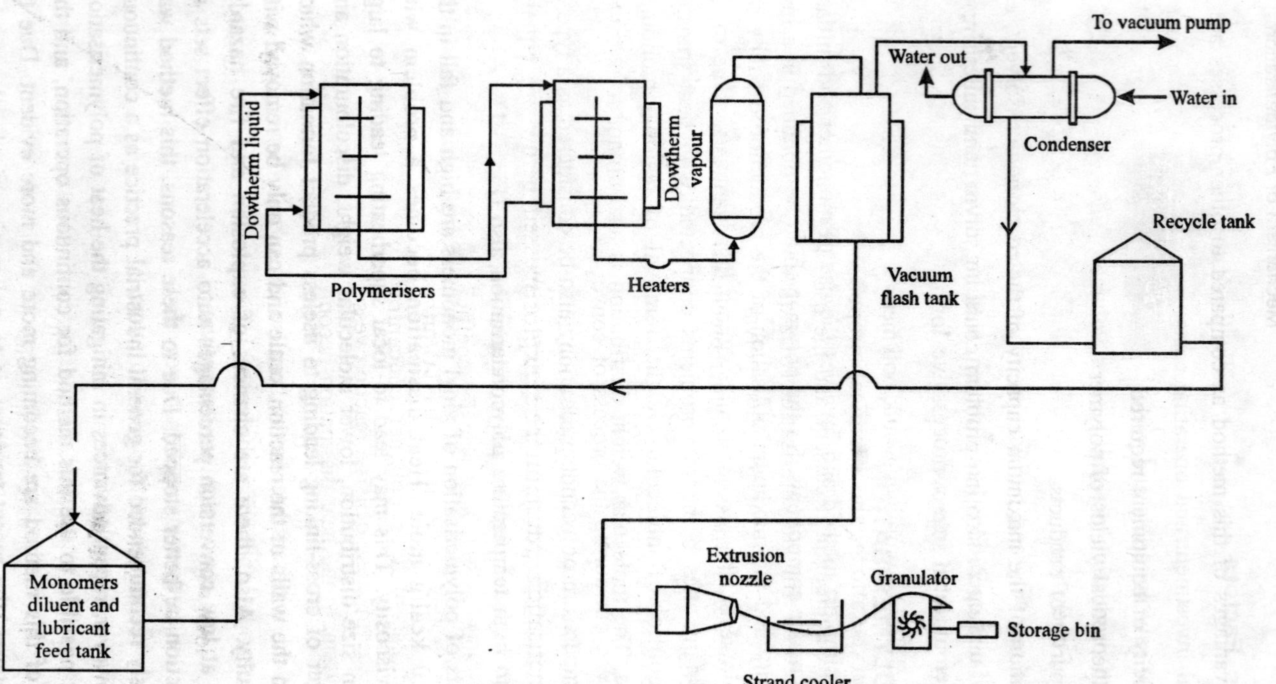

Fig. 3.1. Schematic flowsheet—mass polymerisation process.

The advantages of this method as compared to other processes are as follows:

(1) Simplicity of equipment required

(2) Near theoretical yields of polymer

(3) High purity of products

(4) Utilisation of the maximum capacity of the reactor.

Also if the heat dissipation problem could be solved then the polymer properties would also improve.

The feasibility of employing bulk polymerisation depends upon, among several considerations, the chemical properties of the monomer and polymers, the equipment necessary and the use of the polymer. It is most obvious method for polymerisation of liquid vinyl monomers. Bulk polymerisation is homogeneous when polymer dissolves in the monomer and is heterogeneous when the polymer precipitates out from the mixture during reaction.

In case of vinyl polymerisation, large amount of heat (10-30 kcal/g. mole) is given out during the process of conversion of double bonds into single bond. This heat of polymerisation must be carefully and quickly dissipated in order to minimise explosion hazard, uncontrolled polymerisation, charring, broad molecular weight distribution of the product, etc. The overall heat transfer coefficients in bulk polymerisation systems are in the range of 5-30 BTU/hr. -sq. ft.-°F, depending upon various factors, such as polymer concentration, viscosity, amount of turbulence, etc. In case of large reaction vessels, where surface to volume ratio is very small, the heat dissipation becomes really difficult.

Free radical initiators are better because of their safe decomposition temperature ranges and wide choice of initiators over a wide range of temperatures. With ionic initiators the activation energy is low (4-10 kcal/g. mole) and rate of polymerisation is very fast. So control of the reaction is difficult and it is always advisable to avoid their use if a suitable free radical initiator is available.

A polymerisation reactor would normally have facilities for agitation, supply and removal of heat, connections for measurement of temperatures, level and pressure release arrangement in addition to inlet and outlet connections. In modern plants, elaborate control, alarm and automatic cut-off arrangements are also provided.

The rate of heat of dissipation is the most crucial consideration in the reactor design. Power requirements for effective mixing of the viscous reaction mass (essential for heat transfer) are generally high. Reactors of special designs have been developed to give satisfactory heat transfer. These reactors are quite complex in design and hence expensive. Most of the reactor designs employ high ratios of heat transfer area to the reactor volume by providing jacket and large number of internal cooling coils or plate coolers which also serve as baffles. Some processes use evaporative cooling and condensation of vapours with or without compression.

Agitation is done mostly by mechanical agitators. In view of high viscosities encountered special agitator design is necessary. Paddle, anchor, screw or ribbon agitators are generally used. A double-ribbon mixer suitable for service in a vertical, cylindrical reactor at 100 r.p.m. for a 200,000 cp viscosity reaction mass has been developed. The power requirement for agitation is usually high and depends upon the geometry of the vessel as well as the agitator and physical characteristics of the reaction mass at operating temperatures. This factor has to be taken into account before selecting the type and size of agitator for a specific service as most of the power input to the agitator drive is converted into heat in the viscous reaction mass which is sometimes as high as 30-40% of the total heat generated in the process.

In most of the cases, cold water (35°–65°F) is normally used as the *heat-transfer medium*. Other agents are methanol, hexane or glycols. Direct expansion of the refrigerant in the reactor jacket may also sometimes be used. Dowtherm is normally used for high temperature processes like those for the manufacture of nylon 6, polyester, etc.

For the highly viscous polymerisation mix, *movement of the reaction mass* is preferably carried out with positive displacement screw or jacketed gear, pumps equipped with mechanical seals.

Methods

Bulk polymerisation may be carried out either by batch process or by continuous process.

Batch operation is mostly employed when it is desired to produce a number of small formulations in small quantities in the same equipment. Only about 70% of the cycle time is used for carrying out the reaction and hence this process is uneconomic for large-scale preparation. But this process is advantageous to produce the polymer in the desired form and shape from the monomer directly like perspex sheets, cast phenolic resins, etc. In a batch reactor the monomer and catalyst are added simultaneously.

From time to time little amount of fresh monomer is to be added. The polymer which forms a particular shape is taken out after proper residence time in the reactor.

In *continuous process* the conversion of monomer to polymer may be carried out to certain per cent and thereafter the concentration of monomer-polymer is maintained and fed into another reactor for better heat-transfer. The process is generally carried out in two stages.

The pre-polymerisation (first stage) is carried out in nitrogen atmosphere in the reaction vessels, which are heated by hot jacket or hot water coils and the monomer is added continuously with stirring at certain rate into the vessel. The monomer is polymerised to about 30%.

The polymerisation mix, which is highly viscous and in a clear state, is fed into the upper end of a vertical reactor, which in turn, is further subdivided into a number of zones, for post-polymerisation (second stage) operation. Reactor zones are kept at desired temperatures by means of superheated steam, except the lowest one, which is electrically heated to get maximum heat necessary for complete conversion of monomer. The average residence time into the respective zones is the deciding factor for the success of the process because viscosity increases with polymerisation and agitation problems appear. The highest possible conversion of monomer is carried out at the lowest zone and the resulting molten polymer is discharged into the extruder unit. Polymer is conveyed through conveyor belts over an air or water cooling systems, followed by chopping and grinding of the solid polymer.

SOLUTION POLYMERISATION

In this method, the monomer is dissolved in a suitable solvent and initiator is then added to start the polymerisation reaction. The heat of polymerisation is mostly mitigated by evaporation (followed by condensation and recycling) of the solvent. After the required extent of polymerisation, the polymer is precipitated by adding a non-solvent to the mix. Sometimes the polymer itself is insoluble in the polymerisation. The solid polymer is then filtered off, washed free of any adhered monomer or solvent and dried before use. The solvent-unreacted monomer-non-solvent mixture is separated by distillation or any other suitable methods and reused.

The disadvantages of this polymerisation technique are bulk handling, tiresome mechanical separation processes, purification of the solvents for reuse and difficulty to remove the solvent from the monomer completely. However, the most serious problem is due to chain transfer (Table 3.1) to solvent restricting the polymer molecular weight value, and/or reducing the rate of polymerisation. Solvents also may determine the stereoregularity of the polymer. Sometimes the induction period of the polymerisation may be quite long increasing uselessly the time of production. For these reasons, this method is not normally used for industrial production of polymer.

There are certain systems in which water can be used as solvent for polymerisation (N-vinylpyrolidone, acrylonitrile, acrylic acid, etc.) thus providing a cheap solvent and also permitting the use of inexpensive inorganic redox initiator.

Table 3.1. Recipe for emulsion polymer styrene-butadiene rubber (SBR).

Ingredients	*Parts by wt.*
Butadiene	50-85
Styrene	50-15
Emulsifier	1-5
Buffer	1-5
Catalyst	0.1-1
Orientation catalyst	01-1
Water	100-400

(Temperature 35-60°C, pressure 3-5 atmosphere, time 5-20 hrs.)

EMULSION POLYMERISATION

This method combines the advantages of bulk and solution methods, namely, ease of temperature control, rapid production of polymer and high average molecular weight. This method is applicable to almost all vinyl compounds and dienes except those which have a good water solubility like acrylonitrile and vinyl acetate (recent techniques had made it possible to polymerise these monomers). Capital investment is low. A number of products like elastomers, plastics and latices are prepared industrially by this method.

Bulk polymerisation differs from suspension polymerisation in initiation, which in this case occurs in the aqueous phase.

In this method, the monomers are dispersed as fine droplets in a large amount of water and then emulsified with common soap or special emulsifying agent. To stabilise the emulsion, protective colloids like casein, gum, gelatin or dextrine are used. The initiator is usually a redox type which may be either water-soluble or organo-soluble. The reaction mass remains in the form of an emulsion after the polymerisation is over. It may be used as such in a number of applications like adhesive, surface-coating and textile finishing. For other uses, the emulsion is broken by adding an electrolyte or a solvent, followed by washing and drying of the coagulated polymer (Fig. 3.2).

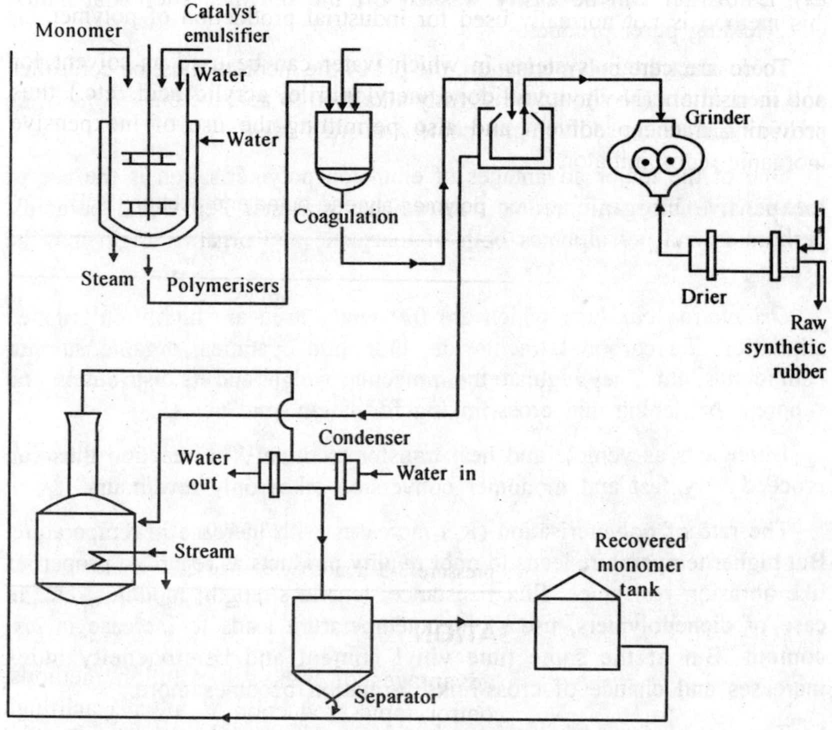

Fig. 3.2. Schematic flowsheet—emulsion polymerisation process.

A *disadvantage* of this method is that impurities, e.g., emulsifier, stabiliser, etc. left in the product may damage electrical or other properties.

Emulsifiers derived from natural products, such as, fatty acids and resins dominate the field. As unsaturated compounds hinder polymerisation so saturated compounds should be used as emulsifier. The rate of polymerisation is dependent on number of carbon atoms in the soap upto C_{12} and then becomes independent of it. Industry normally uses a soap containing 16–18 carbon atoms. Rosin acid soap is better to form higher tacky products. But they impart an yellowish tinge in the polymer. Synthetic detergents, which are available now economically from petrochemical sources, are finding increasing use as emulsifier. The advantages are :

(1) They are stable at neutral and slightly acidic media.

(2) Monomers soluble in water and affected at high pH like acrylonitrile and acrylic acid can be polymerised.

(3) Emulsifier can be easily washed off the polymer after coagulation yielding purer product.

To conduct the polymerisation, the pH of the medium must be controlled and in normal case should be slightly alkaline. To achieve this, *buffers* like sodium acetate or sodium phosphate are often used.

One of the major advantages of emulsion polymerisation is the use of inexpensive inorganic redox free-radical catalysts. Peroxides, peracids, perborates and persulphates both of inorganic and organic origin may be used.

Orientation catalysts which are frequently used are highly chlorinated aliphatics like carbon tetrachloride, inorganic cyanides, organic sulphur compounds, etc. They regulate the molecular weight and its distribution and suppress branching and cross-linking formation tendencies.

Water acts as vehicle and heat transfer medium. The reaction thus can proceed very fast and monomer conversion takes only few hours.

The rate of polymerisation (R_p) increases with increase in temperature. But higher temperature leads to poor quality products as regard to properties like abrasion resistance, flux resistance, tensile strength, modulus, etc. In case of dienepolymers, use of high temperature leads to increase in *cis*-content. But at the same time vinyl content and heterogeneity index increases and chance of cross-links formation becomes more.

The nature of emulsion polymerisation is quite complex because it is the result not only of the interplay of the various polymerisation reactions but also of the various colloidal phenomena governing reactions which occur at the interface. In bulk, solution and suspension polymerisations, R_p varies inversely with DP whereas in emulsion polymerisation R_p varies directly with DP. This observation indicates towards a different mechanism.

SUSPENSION POLYMERISATION

This method was developed with a view to improve and simplify the emulsion polymerisation method. The principle is to add no emulsifying agent but to keep relatively large drops (particle size 0.1-1 mm) of the monomer dispersed in the non-solvent by mechanical agitation. The monomer and water are placed in the reactor and vigorously agitated. The catalyst is then added and the initiation takes place in the aqueous phase. As the polymerisation proceeds, the dispersed drop increase in viscosity and becomes sticky. At this point if the suspension is not stabilised by any agent then upon collision the droplets agglomerate and drop away, thus, breaking the suspension. This is prevented by the addition of a small amount of suspension stabiliser. If the stabilising agent works properly, polymerisation continues and the drops pass beyond the sticky phase and become hard without agglomerating upon collision. When the reaction is complete, the reaction product is in the form of pearls (hence the name pearl or bead polymerisation) and does not have to be recovered by coagulation. The stabiliser is removed from the surface of the polymer simply by washing. The material is dried and used (Fig. 3.3).

The suspension stabilisers used may be either water-soluble organic polymers (like methyl cellulose and its derivatives, polyvinyl alcohol, salts of acrylic acid polymer and its copolymers, etc.) or water-insoluble inorganic compounds (like kaolin, magnesium silicates, aluminium hydroxide, etc.). With these inorganic stabilisers often a small amount of surface active agent is used. Inorganic stabilisers can be more easily removed by washing from the surface of the polymer than the organic stabilisers. The type and quantity of the suspending agent influences particle size, shape, distribution, porosity and bulk density of the polymer.

Water-monomer ratio is an important factor also, since water not only acts as suspension medium but also as the heat transfer medium. In general, from the consideration of polymer property and economy the ratio is maintained at 1.5–2.5.

Oxygen retards the rate of polymerisation. The particle size of the final polymer may be affected by varying the vacuum in the reactor prior to charging the monomer.

pH has got good effect on the polymerisation. Lower the pH the material is coarser. For polyvinyl chloride production, a pH of 3–4 is most suitable. Also, lower the pH, poorer is the heat stability of the polymer.

Agitation is very critical particularly for the particle size distribution. Paddle type agitator produces a sharper particle size distribution compared to propeller type agitator. Temperature should be accurately maintained. A wide range gives a polymer having poor mechanical properties and lower heat stability.

This method is very suitable for non-polar monomers and is largely used in industry for the polymerisation of styrene, methylmethacrylate, vinyl chloride, vinyl acetate and vinylidene chloride. Monomers having significant water solubility, such as, acrylonitrile and acrylic acid, may be polymerised efficiently by this method by adding electrolytes to sort out the monomer from solution.

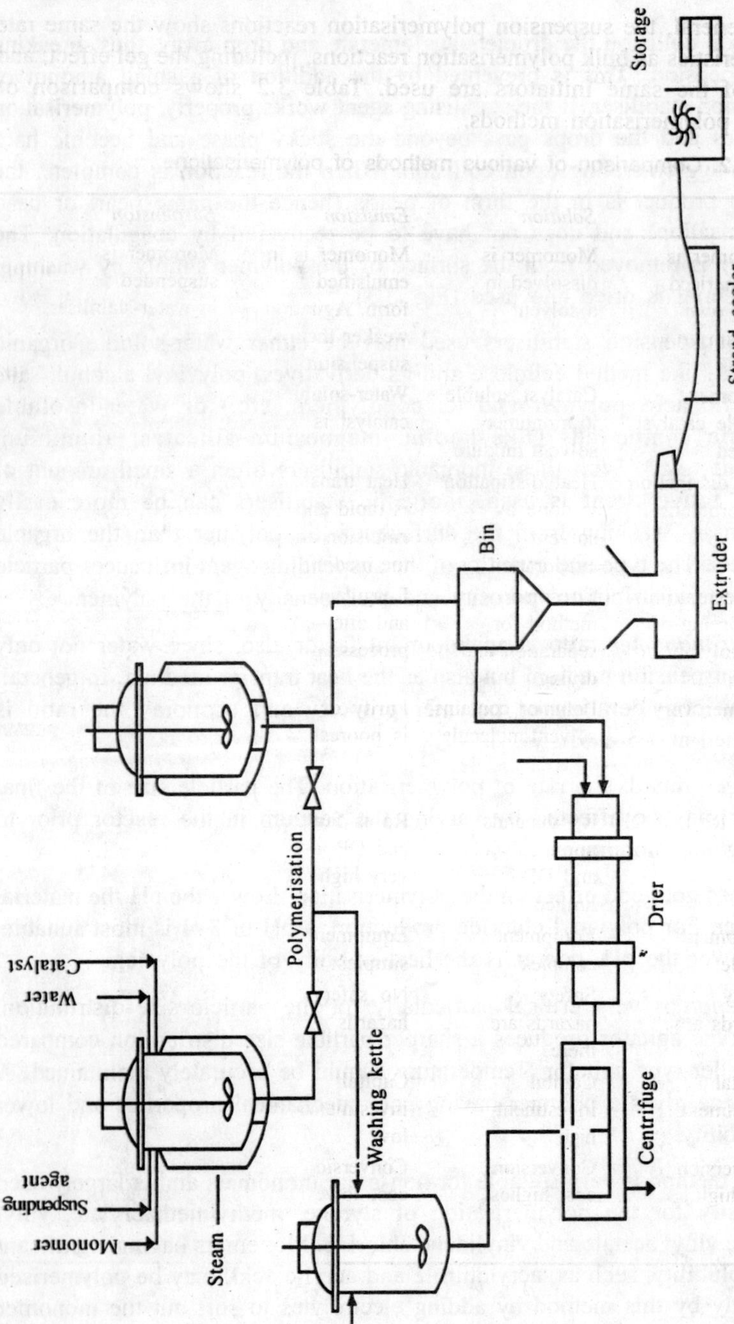

Fig. 3.3. Schematic flowsheet—suspension polymerisation.

In general, the suspension polymerisation reactions show the same rate characteristics as bulk polymerisation reactions, including the gel effect, and many of the same initiators are used. Table 3.2 shows comparison of various polymerisation methods.

Table 3.2. Comparison of various methods of polymerisation.

	Bulk	*Solution*	*Emulsion*	*Suspension*
1.	Monomer is polymerised in its own medium only	Monomer is dissolved in a solvent	Monomer is in emulsified form. Agitation weaker than suspension	Monomer is suspended in water-stabiliser, very rapid agitation
2.	Monomer soluble catalyst is used	Catalyst soluble in monomer-solvent mixture	Water-soluble catalyst is used	Monomer soluble catalyst used
3.	Heat dissipation is a problem	Heat dissipation is done by solvent evaporation	Heat transfer is rapid and emulsion acts as the medium	Heat transfer rapid and takes place through water
4.	No separation and drying of the polymer	Tiresome method for separation and drying	Coagulation and after-processing costly	Separation and drying procedures are simple
5.	Polymer very pure	Polymer contains solvent molecule	Purity of polymer is poorest	Polymer pure but has lower stability and clarity than in bulk method
6.	High Rp and DP	Rp depends upon solvents and DP limited	Rp is fastest and DP also very high	High Rp and DP
7.	Equipment simple	Equipment complex	Equipment simplest	Equipment simple
8.	Safety hazards are more	Safety hazards are there	No safety hazards	No safety hazards.
9.	Capital investment low	Capital investment high	Capital investment low	Capital investment low
10.	Conversion cost high.	Conversion cost highest.	Conversion cost lowest.	Conversion cost higher than emulsion but lower than bulk.

Chapter 4

Synthetic Fibres and Film Industries

INTRODUCTION

Synthetic or man-made fibres constitutes one of the largest chemical industries at the present. The achievement of the chemists and chemical engineers is one of the landmarks in the history of science. Fibres are now classified into three classes: natural fibres, semi-synthetics and wholly or true synthetics. Wool, cotton, etc., are natural fibres; semi-synthetic fibres are viscose, acetate fibres, etc. (these are prepared starting with natural fibres to impart special properties) while synthetic or true synthetics are nylon, dacron or terylene, orlon, saran, tefflon, etc.

The performances of the synthetic fibres are remarkable. Natural fibres have been in use to clad human beings since the dawn of history and all their achievements are directed towards that semi-synthetics filled the shortcomings in the properties of the natural fibres and also diversified and brightened the fibres. But the synthetic fibres, besides doing all the work of the natural and synthetic fibres, are being applied to do tough jobs, e.g., cords, conveyor belts, parachutes, filter clothes, etc. Their qualities are also varied and power of dyeing remarkable.

The advantages of synthetic fibres are numerous. Besides their unlimited availability these differ from natural fibres in most of their physical properties. Taking wool and nylon as representatives of natural and synthetic fibres it is found that nylon is superior in tensile strength, elastic recovery, strength, stiffness, water absorbency and heat resistance. Nylon is also affected less by time, sunlight, alkalies, or organic solvents. It resists the action of moths and mildew to which wool is susceptible.

Of the disadvantages, one principal defect is that synthetics may cause some allergic effects on individuals.

In spite of everything natural fibres are the first choice of man, as the product of nature has akinness to the human system.

DISTINCTION BETWEEN NATURAL AND SYNTHETIC FIBRES

Synthetic fibres may be defined as man-made textile yarn, whose chemical base is cellulose or other organic substances. More accurately defined, 'synthetic fibres' is the name given to filaments made from various solutions of modified cellulose or organic substances by passing the solution through an orifice and solidifying it in the form of filament.

Synthetic fibres are made of either natural fibres, or organic substances by chemical means. The composition of a synthetic fibre derived from a natural fibre changes through chemical reactions, e.g., cotton is a natural fibre containing cellulose of structure $(C_6H_9O_4OH)_n$ and rayon made from cotton has the structure $(C_6H_9O_4OH)3_n$. Distinction between natural and synthetic fibres is given as under :

Natural	*Synthetic*
1. These are present in nature, e.g., wool, etc.	These are to be made from natural fibres like cotton or from organic substances, with the composition changed by chemical reaction.
2. Natural fibre (natural silk) is a protein fibre secreted by insects.	Artificial fibres are prepared chemically by condensation method, e.g., rayon.
3. The formula of natural fibre is $(C_6H_9O_4OH)_n$.	The formula is $(C_6H_9O_4OH)3_n$ and these are used under the same artificial silk.
4. They are inferior to synthetic fibres.	These are superior to natural fibres in quality.
5. They may be of ununiform length.	These are of uniform length.
6. They have low fixing quality.	These are of high fixing quality.
7. They have high affinity for sulphur and vat dyes.	These have low affinity for sulphur and vat dyes.
8. They give effective odour on burning.	Synthetic fibres do not give such odour.

Cotton and wool possess natural crimp. Crimp is also introduced in some cases in continuous-filament yarns to alter their appearance and feel, e.g., in nylon carpet yarn. *Denier* is a measure of weight per unit length.

Each fibre has a cross-section, and special characteristics are imparted by manufacturing polyfibres with different cross-sections. A fibre has a cross section corresponding to 1 denier if 9,000 m of it weighs 1 g.

Films, as well as fibres, are changing. Nitrocellulose, once largely used especially for photographic film, has given way to acetate and polyester. Many other films are suited to the demand for their special properties, e.g., cellophane, saran, teflon and polyethylene.

POLYFIBRES AND FILMS

The rapidly growing and versatile non-cellulosic fibres are classified chemically and by the method of spinning is given in Table 4.1. The manufacture of all true synthetic fibres first begins with the preparation of a polymer consisting of extremely long, chainlike molecules. The polymer is spun in one of three ways resulting in most cases in a weak, practically useless fibre until it is stretched further to orient the molecules and set up crystalline lattices. Although the range of any one polymer is always limited by controlling the degree of orientation, crystallinity, and average chain length, a single polymer can be used to make a number of fibres with different mechanical properties; i.e., some can be weak and stretchy, others strong and stiff. The two elements important in determining the range of the polymer's mechanical properties are the attractive forces between the molecules and the flexibility and length of the molecular chains.

Table 4.1. Representative synthetic fibres and films.

Classification[a]	*Spinning*[b]
Polyfibres and film	
Polyamides, or nylon, fibres	
Nylon 66, nylon 6, Qiana	Melt
Polyesters	
Fibres: Dacron, trevira, kodel, fortrel	Melt
Films: Mylar, cronar, kodar, estar	Melt
Acrylics and modacrylics	
Orlon fibre	Dry
Acrilan fibre	Wet
Creslan fibre	Wet
Dynel fibre (vinyl-acrylic)	Dry
Verel fibre	Dry
Vinyls and vinylidines	
Saran fibre and film	Melt
Vinyon N fibre[b]	Dry
Spandex	
Lycra	Dry

(Contd...)

Classification[a]	*Spinning*[b]
Numa	Wet
Glospan	Wet
Olefins	
Polyethylene films	Melt
Polypropylene fibres and films : Avisun, herculon	Melt
Glass fibres	
Fibreglass	Melt
Cellulosic fibres and films	
Regenerated cellulose	
Fibres : Rayon (viscose), cuprammonium	Wet
Film : Cellophane	Wet
Cellulose esters	
Acetate fibres and films: Acele, estron	Dry
Triacetate fibre : Arnel	Dry

[a] Classified as melt, dry or wet.
[b] 60% vinyl and 40% acrylonitrile.

The three spinning procedures are melt, dry, and wet. Melt spinning, developed for nylon and also used for polyester, polyvinyl, polypropylene, and others, involves pumping molten polymer through spinneret jets. The polymer solidifies into filaments as it strikes the cool air. In dry spinning, as described for acetate, the polymer is dissolved in a suitable organic solvent. The solution is forced through spinnerets, and dry filaments result upon the evaporation of the solvent in warm air. Orlon and vinyon are spun in this manner. Viscose rayon, acrilan, and dynel staple are examples of wet-spun fibres. Here, as the solution emerges from the spinneret, it is coagulated in a chemical bath. Comparison of various selected synthetic fibres is shown in Table 4.2.

POLYAMIDES

Polyamides are condensation products that contain recurring amide groups as integral parts of the main polymer chains. Linear polyamides are formed by condensation of bifunctional monomers. If the monomers are amino acids, e.g., 6-aminohexanoic acid, the polymers are called AB types (A representing amine groups and B representing carboxyl groups). If the polymers are formed from condensation of diamines and dibasic acids, they are called AABB types. Polyamides are frequently referred to as nylons. Although they generally are considered as condensation polymers, polyamides also are formed by addition polymerisation. This method of preparation is especially important for some AB polymers, in which the monomers are cyclic lactams, e.g., ε-caprolactam, hexahydro-2*H*-azepin-2-one or 2-pyrrolidinone, γ-aminobutyrolactam.

Table 4.2. Comparison of selected synthetic fibres.

Fibre	*Wool*	*Acrylic*	*Modacrylic*	*Polyesters*	*Nylon*
Tensile strength, g/denier	1.0–1.7	4.0–5.0	3.0	4.4–6.6	4.7–5.6
Elongation, %	25–35	16–21	16	18–22	25–28
Elastic recovery	0.99 at 2%	0.97 at 2%	0.80 at 2%	90 –100 at 4%	100 at 8%
Strength, psi	20,000–29,000	59,000–74,000	44,000–66,000	78,000–116,000	68,000–81,000
Stiffness, g/denier	3.9	24	30	23–63	20
Abrasion resistance	90	330		1,570	2,520
Water absorbency	21.9% at 90 RH[a]	2% at 95 RH[a]	. . .	0.5% at 95 RH[a]	8% at 95 RH[a]
Effect of heat	Becomes harsh 100°C, decomposes at 130°C	Sticking point 235°C	Sticking point 235°C	Sticking point 240°C	Melts 263°C
Effect of age	Little	Little	Little	Little	Slight
Effect of sun	Weakened	Very resistant	Slight	Little	Weakened
Effect of acids (concentrated, room temp.)	Resistant	Resistant	Resistant	Resistant	Weakened
Effect of alkalies	Susceptible	Partly resistant	Resistant	Resistant	Resistant
Effect of organic solvents	Resistant	Resistant	Resistant	Resistant	Resistant
Dyeability	Good	Good	Good	Difficult	Good
Resistance to moths	None	Wholly	Wholly	Wholly	Wholly
Resistance to mildew	Good	Wholly	Wholly	Wholly	Wholly

[a]RH: Relative humidity.

POLYAMIDE FIBRES

Polyamide fibres have monomer units joined by amide groups $[CONHRNHCOR']_n$ and are prepared from diamines and dicarboxylic acids, or in the case of $[RCONH]_n$, from lactams. If R and R' are aliphatic, alicyclic, or mixtures containing less than 85 wt% aromatic moieties, the polyamides usually are referred to as nylon. If more than 85 wt% of the repeating units are aromatic in structure, the fibres are called aramids.

Nylon-6 and Nylon-6,6

Properties

The main properties of representative nylon-6 and -6,6 yarns are listed in Table 4.3.

Table 4.3. Physical properties of Nylon-6 and -6,6 continuous-filament yarns[a].

Property	*Normal tenacity*	*High tenacity*
Tenacity, N/tex[b]	0.4–0.6	0.75–0.84
wet	0.36–0.51	0.64–0.7
loop	0.40–0.49	0.57–0.69
knot	0.35–0.42	0.52–0.72
Tensile strength, MPa[c]	580–635	855–952
Elongation at break (conditioned), %	23–43	12–17
wet	28-41	14–21
Tensile modulus (conditioned), N/tex[b]	2.2–3.1	2.9–4.1
Average toughness, N/tex[b]	0.07–0.14	0.06–0.14
Moisture regain at 21°C, %		
65% rh	4.5	3.3–4.5
95% rh	7.1–7.8	7.8

[a] Conditioned at 65% rh and 21°C.
[b] To convert N/tex to gf/den, multiply by 11.33.
[c] To convert MPa to psi, multiply by 145.

Tensile strength

The stress-strain properties of nylons depend to a large extent on spinning speed and draw ratio. The properties are controlled by the intimate morphology of the fibres, especially the crystalline orientation and the amorphous orientation.

Crepe and recovery

Crepe and recovery with respect to time for nylon yarns under different initial loads for different total elongations have been measured. Initially, recovery is rapid, and most of it occurs within a few minutes after release of the load. The outstanding elastic recovery of the nylons is responsible for the cling or fit in women's hosiery and for their extensive market acceptance.

Staple

The production of staple is simpler than that of filament yarn, since the breaking of a filament in the spinning step usually is not a problem, and uniformity of filaments is not as critical because the staple is blended. Several hundred to several thousand filaments can be spun from one spinneret, and many of the yarns can be combined into a large tow for drawing, crimping, and cutting. Polymerisation is usually coupled with spinning to lower processing cost. As for filament, the polymer is metered to spinnerets, the filaments are quenched, and finish is applied. Yarns then are combined to form a tow, which is drawn by being passed over hot rolls. More finish can be applied at this step. The tow may be crimped, then cut to staple.

The use of nylon staple rather than nylon continuous filament stems from a need to satisfy the demand for a soft and warm staple hand or sensation when fabric is touched with a hand, and for ease of styling in carpets. Best performance in many types of fabrics usually requires blends of fibres. Since intimate blending only can be achieved in staple form, the combining of the blend components takes place early in the textile spinning process, i.e., either before or immediately after carding. Therefore, the properties of nylon staple must be compatible with each of the other blend components, and the fibre must also be adaptable to processing on the respective spinning systems. The outstanding contributions of nylon staple to carpet, apparel, and upholstery fabrics are its abrasion resistance and its low moisture absorption. The former property increases the wear life of the textile structure, the latter contributing to faster drying.

Thermal and moisture

The thermal behaviour of nylon fibre and the interrelated effects of moisture have a pronounced influence on the physical properties of the yarn and its products. They are of basic importance in fibre manufacture, in converting the yarn into fabric, and in ultimate use.

Thermal properties of fibres have not been extensively investigated, because much of the practical thermal behaviour of textiles is strongly influenced by other factors. Thermal and light-resistant properties are to a large extent related to the specific stabilising substances used and to certain aspects of the processes by which the yarns are made. These vary from producer to producer and are changed from time to time by each producer.

Light and heat stabilisation

Light. In the presence of light, titanium dioxide, which normally is used as a delusterant, is considered to react with oxygen to form peroxide autocatalytically, and the peroxide degenerates polyamides in the absence of stabilisers, e.g., manganese salts. Additives, e.g., hypophosphorus acids, phosphites, phosphates, etc., also are used.

Heat and light

Water, carbon dioxide, and ammonia are the three principal gaseous products of nylon pyrolysis. The preferred stabilisers in non-delustered fibres from degradation by heat and light are copper salts.

Stabilisation against high temperature strength loss at high loads

One of the significant factors that has made it possible to produce a nylon-6 tyre yarn is the ability to increase the breaking temperature under high loads to essentially equal nylon-6,6-9,9. Dialkyldihydroacridine compounds are employed in polyamide stabilisation at high temperature and high loads.

Stabilisation against heat disorientation in liquids

The disorientation of polyamides as vulcanisation temperatures of tyres are raised becomes an increasingly significant problem. The problem can be alleviated partially by maintaining some degree of tension on the tyre cord at the end of the pressure release after vulcanisation. It can also be avoided partially by minimising the moisture content of tyre yarns prior to vulcanisation.

Electrical conductance

Electrical conductance of nylon is very low. Conductivity increases as moisture content rises.

Manufacture and processing

Essentially all polyamides other than the aromatic (aramid) types are melt-spun.

Texturing

In general, textured yarns are yarns that have greater apparent volume or are made more extensible by mechanical distortion of the filaments. The distortion may be produced by buckling the filaments under endload compression, by either bending them over an edge of small radius or twisting the strands as a whole. Bulked nylon yarns are either fine tex [1.7–22.2 tex (15–200 den)] for woven, knitted stretch, and textured fabrics in apparel applications or heavy tex [110–400 tex (100–3600 den)] for carpet.

Modified cross-sections

Conventional spinneret orifices are circular. Improved machining methods, advances in fibre-production technology, and the study of the effect of profiled fibres, i.e., with non-circular cross-sections, on fabric characteristics, e.g., lustre, sparkle, opacity, air permeability, resistance to showing soil, and heat insulation, have given greater importance to the production of modified cross-section filaments.

Finishes

The main functions of a fibre finish are to provide surface lubricity and yarn cohesion. Spin finishes are applied to the fibre during spinning. Yarn lubricant or coning oil defines all lubricants added subsequent to application of spin finish. Yarn lubricants are used in textile milling and processing.

The finish generally is an emulsion or a water-soluble mixture of one or more lubricants and an antistatic agent to avoid static charge, which would cause the filaments to repel each other, giving a yarn with poor cohesion and so flaring the filaments, which causes uneven processing of the yarn. Wetting agents usually are added to aid the spreading of the finish on the yarn. The concentration of finish on the yarn and after evaporation of water usually is 0.3–0.8 wt%.

Additives

A variety of substances usually are added to the reactive mixture before or during polymerisation to change the properties and nature of the polymer.

Additives include delustering agents, spin-dying agents, antistatic agents, and flame retardants.

Dyeability (fibre characteristics)

Chemical structure, crystallinity, molecular orientation, and fabric preparation affect the rate and the extent of nylon dyeability.

Dyebath characteristics

In general, uniform dyeing is most easily achieved when dyeing conditions are selected so that dye, diffusing into and out of the fibre, establishes an equilibrium with an appreciable amount of dyestuff remaining in the bath. Under this condition, non-uniform fabrics eventually achieve a level shade. Disperse dyes are unexcelled for this purpose.

Modified Nylon Fibres

Modified polyamides include bicomponent and biconstituent fibres, block copolymers, and random copolymers.

Other Nylons

Other nylons include Qiana and dimensionally stable nylons and miscellaneous polyamides.

Uses

Principal commercial uses of nylon-6 and -6,6 are in carpets, apparel, tyre reinforcement, and other industrial applications. Qiana is used primarily in apparel.

CAPROLACTAM

Caprolactam, 2-oxohexamethylenimine, hexahydro-2*H*-azepin-2-one, is one of the most widely used chemical intermediates. However, almost all of the annual production of 2.3×10^6 metric tons is consumed as the monomer for nylon-6 fibres and plastics. Cyclohexanone, which is the most common organic precursor of caprolactam, is made from benzene by either phenol hydrogenation or cyclohexane oxidation. Reaction with ammonia-derived hydroxylamine forms cyclohexanone oxime which undergoes molecular rearrangement to the seven-membered ring ε-caprolactam.

Physical Properties

Caprolactam (mol wt 113.16) is a white, hygroscopic, crystalline solid at ambient temperature, with a characteristic odour. It is very soluble in water and in most common organic solvents and is sparingly soluble in high molecular weight aliphatic hydrocarbons. Molten caprolactam is a powerful solvent for polar and non-polar organic chemicals. Selected physical properties and solubilities of caprolactam are listed in Table 4.4.

Reactions

Caprolactam is a cyclic amide and therefore undergoes the reactions of this class of compounds. It can be hydrolysed. *N*-alkylated, *O*-alkylated, nitrosated, halogenated and subjected to many other reactions.

Health and Safety Factors

Caprolactam has a low order of toxicity and presents no appreciable health hazard if it is handled properly. Threshold limit values for caprolactam dust and vapour are 1 and 20 mg/m^3, respectively (time-weighted averages).

Table 4.4. Physical properties of caprolactam $CH_2(CH_2)_4CONH$.

Properties	*Values*
Melting point, °C	69.3
Density (at 77°C), g/cm^3	1.02
Bulk density, kg/m^3[a]	600–700
Boiling point at 101.3 kPa[b], °C	266.9
Refractive index	
at 40°C	1.4935
at 31°C	1.4965
Viscosity, mPa·s (= cP)	
at 70°C	12.3
at 80°C	8.5
Specific heat, J/(kg·K)[c]	
solid	
at 25°C	1380
at 35°C	1420
liquid	
at 70°C	2117
at 110°C	2412
at 178°C	2608
vapour	
at 100°C	1640
Thermal conductivity, W/(m·K)	0.169
Heat of fusion, J/g[c]	108.86
Heat of vapourisation at 80°C, J/g[c]	511
Heat of combustion (liquid at 25°C), J/g[c]	–31,900
Heat of formation (liquid at 25°C), J/g[c]	–2,840
Flash point (closed cup), °C	125
Fire point, °C	140

[a] To convert kg/m^3 to lb/ft^3, divide by 16.02.
[b] To convert kPa to mm Hg, multiply by 7.5.
[c] To convert J to cal, divide by 4.184.

POLYESTER FIBRES

A polyester fibre is a manufactured fibre in which the fibre-forming substance is any long-chain synthetic polymer composed of at least 85 wt% of an ester of a dihydric alcohol (HOROH) and terephthalic acid (*p*-$HOOCC_6H_4COOH$). The most widely used polyester fibre is made from linear poly(ethylene terephthalate) (PET).

$$H\left[OCH_2CH_2O\overset{O}{\overset{\|}{C}}-C_6H_4-\overset{O}{\overset{\|}{C}}\right]_n OH$$

Properties

Structural

Drawn polyester fibres may be considered to be composed of crystalline and non-crystalline regions. The structural unit or unit cell for PET has been deduced by means of x-ray diffraction techniques and is considered triclinic, with one repeating unit.

Mechanical

Typical physical and mechanical properties of polyester fibres are summarised in Table 4.5.

Table 4.5. Physical properties of poly(ethylene terephthalate).

	Filament yarn		*Staple and tow*	
Property	*Regular tenacity*[a]	*High tenacity*[b]	*Regular tenacity*[c]	*High tenacity*[d]
Breaking tenacity[e], N/tex[f]	0.25–0.50	0.5–0.86	0.2–0.5	0.52–0.63
Breaking elongation, %	19–40	10–34	25–65	18–40
Elastic recovery, %	88–93 at 5%	90 at 5%	75–85 at 5%	75–85 at 5%
Initial modulus, N/tex[f]	6.6–8.8	10.2–10.6	2.2–3.5	4.0–4.9
Specific gravity	1.38	1.39	1.38	1.38
Moisture regain[g], %	0.4	0.4	0.4	0.4
Melting temperature, °C	258–263	258–263	258–263	258–263

[a] Textile filament yarns for woven and knit fabrics.

[b] Tyre cord and high strength, high modulus industrial yarns.

[c] Regular staple for 100 wt % polyester fabrics, carpet yarn, fibrefill, and blending with cellulosics or wool.

[d] High strength, high modulus staple for industrial sewing thread and blending with cellulosics.

[e] Standard measurements are conducted in air at 65% rh and 22°C.

[f] To convert N/tex to g/den, multiply by 11.33.

[g] The equilibrium moisture content of the fibres at 21°C and 65% rh.

Chemical

Polyesters show good resistance to most mineral acids, but dissolve with partial decomposition in concentrated sulphuric acid. Hydrolysis is highly dependent on temperature. Basic substances attack the fibre in two ways. Strong alkalies, e.g., caustic soda, etch the surface of the fibre and reduce its strength. Ammonia and other organic bases, e.g., methylamine, penetrate the structure initially at the non-crystalline regions and cause degradation of the ester linkages and a general loss in physical properties. The polyesters display excellent resistance to conventional textile bleaching agents and are resistant to cleaning solvents and surfactants. Most of these effects are conditioned by the structural morphology of the specimens.

Other

Poly(ethylene terephthalate) fibres display good resistance toward sunlight, and addition of modern light stabilisers and antioxidants enhances the polymer's performance. The resistance to mildew, aging and abrasion generally is excellent.

Manufacture and Processing

Terephthalic acid or its dimethyl ester reacts with ethylene glycol to form a diester monomer, which polymerises to the homopolymer PET. The molten polymer is extruded or spun through a spinneret, forming filaments which are solidified by cooling in a current of turbulence-free air. The spun fibre is drawn by heating and stretching the filaments to several times their original length to form a somewhat oriented crystalline structure with the desired physical properties.

Waste recovery

Glycol and antimony catalysts are recycled into the polymerisation system. Polymer and fibre waste can be separated into two chemical components, usually by methanolysis; the DMT that forms is purified and repolymerised. The waste also can be melted directly and converted to a lower quality fibre.

Texturing

Continuous-filament, 5.6–33-tex (50–300-den) yarn can be textured easily by several techniques, e.g., false-twist, stuffer-box, knitdeknit, and air-jet methods. In texturing, filaments are crimped or looped at random to give the yarn greater volume or bulk; then they are heat-set for dimensional stability. The most common textured yarns are the false-twist textured yarns made with aggregates of tiny, high speed spindles, stacked friction disks, or cross-belt nips, all of which impart a high degree of twist.

Dyeing

Polyester fibres are dyed almost exclusively with disperse dyes because of the lack of reactive dyes sites in the fibre.

Finishing

Poly(ethylene terephthalate) is a thermoplastic and, therefore, heat is applied to set the structural memory desired for maximum performance of PET fibre either alone or in blends. This important heat-setting technique known as no-cure-permanent-press, or durable press, provides the wrinkleproof, ease-of-care, and low laundering shrinkage properties for which polyester fabrics are noted.

Uses

Polyester fibres are used in staple, filament, and spun-bonded textile fabrics.

POLYESTERS (UNSATURATED)

Unsaturated polyesters are macromolecules with polyester backbones derived from the interaction of unsaturated acids or anhydrides and polyhydric alcohols. The reaction normally proceeds at 190–220°C until a predetermined acid value viscosity relationship has been achieved. Solutions of these polymers in vinyl monomers, e.g., styrene, often are called polyester resins. They are compounded with fillers or fibres, or both, in the liquid stage and then are cured with the aid of free-radical initiators to yield thermoset articles. Market penetration, especially in the area of fibre-glass reinforcement, is greatly enhanced as a result of greater latitudes in compounding and processing than is possible with other polymeric systems. Unsaturated polyesters can be mass-cast, laminated, moulded, pultruded, and made into gel coats in a variety of colours. Depending on the application, the physical and chemical properties of the product often can be met by judicious choice of polyester backbone ingredients and the type and amount of the diluent vinyl monomer.

Polyester resins are considered one of the three principal thermoset systems of economic importance. The other two are phenolics and epoxy resins.

Properties

As the molecular weight of unsaturated polyesters increases, the cured products exhibit improvements in certain properties, e.g., heat-deflection temperature (HDT), hardness, tensile strength, and flexural strength, until a constant value is reached. With an increase in molecular weight, the effect of the end groups becomes less pronounced. However, polyesters that terminate with carboxyl groups generally are characterised by higher viscosities and improved physical properties compared to those with hydroxyl terminations.

The structure of the backbone of the polyester characterises the physical properties of the cured resin. For most cured polyester resins, optimum physical properties are obtained when the degree of unsaturation, as represented by mol% fumarate, is 50–70%. Retention of physical properties at higher temperatures is not necessarily aided by higher cross-link density, although an increase in HDT may be observed.

Unsaturated polyester resins rarely are used alone without some type of filler or fibrous reinforcement. Fillers tend to increase the flexural modulus, HDT, and penetrating hardness and to degrade other physical properties. Fibre glass, however, transfers its strength to the matrix resin and improves most physical properties, depending on the length of the fibre and the amount of glass used.

Raw Materials

Raw materials include ethylenically unsaturated acids, aromatic and saturated aliphatic acids, glycols, monomers, inhibitors, accelerators, and initiators.

Manufacture

The manufacture of polyester is carried out in large stainless-steel lined kettles, which are provided with overhead fractionating columns to reduce the loss of components that are less volatile than water. The columns normally are packed and steam-jacketed. The most widely used manufacturing method is the direct fusion process (Fig. 4.1). The oxide method also is used.

Health and Safety Factors

Most of the concerns for safety and health associated with polyester resins pertain to worker exposure to styrene and its emission into the environment.

ACRYLIC AND MODACRYLIC FIBRES

Acrylic fibre is the generic name for a manufactured fibre in which the fibre-forming substance is any long-chain synthetic polymer composed of at least 85 wt% acrylonitrile units. Modacrylic fibre is the generic name for a manufactured fibre composed of at least 35 and less than 85 wt%

acrylonitrile units. Commercially, the only modacrylics marketed have 25–60% of monomers such as vinyl chloride or vinylidene chloride and consequently possess a high degree of flame resistance.

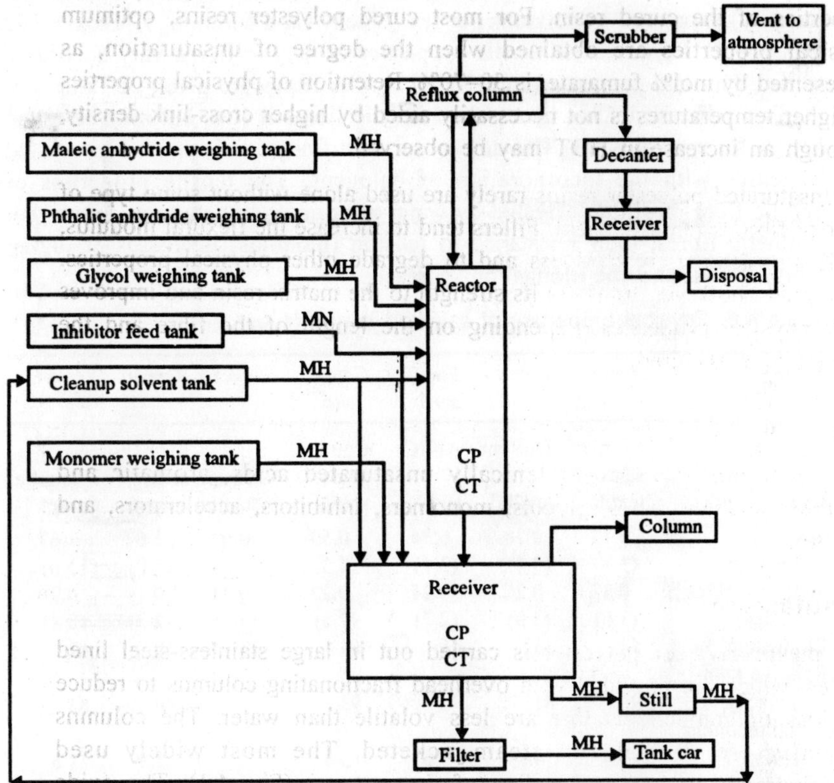

Fig. 4.1. Typical flow diagram of unsaturated polyester by fusion method. MH = mechanical transfer, MN = manual transfer, CP = closed process, CT = cleaning operation.

Acrylic Fibres

Many of the properties of acrylic fibres are determined by the inherent chemical nature or physical behaviour of the long-chain polyacrylonitrile (PAN) molecules in oriented structures.

Acrylic fibres are sold in the form of staple, tow, "tops" (in which the tow has been stretch-broken or cut to form a sliver), and a small amount of continuous filament. Staple lengths vary from 25 to 150 mm depending on the spinning system used to convert the fibre into spun yarn. Fibre size is expressed in denier (the weight in grams of 9000 metres) or tex (the weight in grams of 1000 metres). Specific stress on the fibre during extension or at break has been written as g/den or gf/tex and may be converted to the International System unit for specific stress (modulus and tenacity) expressed in newtons per tex (N/tex) via a formula: 1 N/tex = 102 gf/tex = 11.3 g/den. The tensile properties of acrylic fibres vary according to use. Physical properties of selected acrylic fibres are given in Table 4.6. Boiling-water shrinkage of standard acrylic fibres is low, but may be much higher for special types. The density of most fibres is about 1.17 g/cm^3 and increases with the incorporation of halogens. The tensile properties of acrylic fibres are little affected by water at normal temperatures. Like other synthetic fibres, their resistance to tension (modulus) decreases at higher temperatures and the elongation increases.

Table 4.6. Physical properties of selected acrylic fibres[a].

Fibre	*Acrilan*	*Orlon*	*Euro-acril*	*Cour-telle*	*Badische*	*Acrilan*	*Zefran*
Producer	Mon-santo	DuPont	ANIC	Court-aulds	Badische	Mon-santo	Badische
Use	textile	textile	textile	textile	industrial	carpet	carpet
Tex	0.13	0.38	0.34	0.50	0.67	1.67	1.67
(denier)	(1.2)	(3.4)	(3.1)	(4.5)	(6.0)	(15.0)	(15.0)
Tenacity, N/tex	0.32	0.26	0.24	0.22	0.31	0.19	0.20
(g/den)	(3.6)	(3.0)	(2.7)	(2.5)	(3.5)	(2.2)	(2.3)
Elongation, %	42	33	35	64	33	56	62
Relative knot tenacity, %	90	81	91	93		80	82
Initial modulus, N/tex	3.9	4.0	3.6	3.3		1.9	1.8
(g/den)	(44)	(45)	(41)	(37)		(21)	(20)
Boiling water shrinkage, %	1.0	0.7	0.7	1.2	< 1.0	1.0	1.0

[a] Tensile properties measured at 65% rh and 21°C. Extension rate, 100% per minute.

Acrylic fibres are used in some uses requiring flame-retardance. The general flame-resistance test for carpets specifies a maximum burn distance after ignition from a methanamine (hexamethylenetetramine) pill. Most acrylic fibres produced specifically for carpet use contain sufficient amounts of halogen to pass this test. Where higher degrees of flame retardancy are required, blends of acrylic and modacrylic fibres are used. Acrylic fibres have outstanding resistance to sunlight and to attack by insects or micro-organisms in the soil, properties which make them particularly suitable for outdoor uses. They are resistant to most ordinary chemicals such as inorganic acids, alkalies, organic solvents, oxidising agents, and dry-cleaning solvents.

Fibre analysis and identification

Acrylics can be distinguished from other chemical fibres by their solubility in hot dimethylformamide and their density of 1.12–1.20 g/cm^3.

Manufacture

In addition to acrylonitrile, most polymers for acrylic fibres contain minor amounts of various monomers. Neutral monomers such as methyl acrylate, methyl methacrylate, and vinyl acetate are co-polymerised with acrylonitrile to increase solubility of the polymer in various solvents and the diffusion rate of dyes into the fibre. Small amounts of ionic monomers are often also included to enhance dyeability. Halogen-containing monomers are added to improve flame resistance.

Although acrylonitrile is moderately soluble in water, the polymer is insoluble in both acrylonitrile and water. The copolymers are generally made by aqueous heterogenous or solution polymerisation, using both batch and continuous processes. In a typical continuous process, monomer, water, and initiator are fed to a continuously stirred, overflow reactor at atmospheric pressure and a temperature of 30–70°C. The slurry of polymer, water and unreacted monomer is filtered; the polymer is washed and dried; and the monomer is recovered from the filtrate and returned to the reactor. Advantages of this process are simplicity, control, and applicability to a variety of polymer compositions. Initiators are usually water-soluble redox combinations such as persulphate-bisulphite, activated by a small amount of iron, or chlorate-sulphite.

Solution polymerisation is used to prepare acrylic polymers directly in a form suitable for wet or dry spinning. Solvents include dimethyl sulphoxide, dimethylformamide, and aqueous solutions of zinc chloride or various thiocyanates. Preferred initiators are various organic peroxides and azo compounds. Ultraviolet light may also be used as an initiator. The principal advantage of solution polymerisation is reduced manufacturing cost through elimination of process steps.

Mass or bulk polymerisation can also be used to prepare acrylic polymers when precautions are taken to prevent autocatalytic polymerisation, which can become explosive. A key requirement described in recent patents is an initiator system that produces an easily stirrable monomer-polymer slurry from which heat can be readily removed. Typical conditions listed for continuous polymerisation using a cumene hydroperoxide-sodium methyl sulphite redox initiator are a reaction temperature of 30–60°C and a one hour residence time.

Acrylonitrile copolymers used for acrylic fibres are white, easy-flowing powders of viscosity average molecular weight commonly in the range of 100,000–150,000. The polymers decompose before melting is above 180°C, exothermic decomposition becomes rapid. This characteristic, undesirable for acrylic fibre because it causes discolouration, is used advantageously for the manufacture of graphite fibres.

Seven solvents for polyacrylonitrile are used commercially. Dimethylformamide is the principal solvent since it is used in both dry- and wet-spinning processes, followed in importance by dimethylacetamide. Aqueous solutions of sodium thiocyanate and nitric acid are the most important inorganic solvents, along with $ZnCl_2$, dimethyl sulphoxide, and ethylene carbonate. Fibre production from sulphuric acid solution has been reported, but no commercial processes are known. Effective solvents for polyacrylonitrile are highly polar, aprotic compounds which break the intermolecular nitrile-nitrile attraction. Solubility and solution properties are affected by the polymer composition and molecular weight. The presence of 3–10% neutral comonomers increases acrylic polymer solubility and decreases solution viscosity at a given concentration. Molecular weight, composition, spinning concentration, and coagulation conditions are adjusted to give acceptable fibre formation and good tensile properties.

Fibre Production

Acrylic fibres are produced either by wet spinning (78%) or dry spinning (22%) to optimise fibre structure and properties; increase production rate as determined by linear speed, filament tex, and tow size; and improve process stability and fibre uniformity. The combination of process conditions to achieve all these features simultaneously is strongly influenced by the desired filament size or tex. In all spinning systems, fibre tex is controlled by the material balance:

$$\text{tex/filament} = 1.075 \times 10^4 \; (pW_2Q/SV_1)$$

where tex = g/km; p = solution density, g/L; W_2 = weight fraction of polymer in the spinning solution; Q = flow rate per spinnerette hole in L/s; V_1= first godet (take-out roll) velocity (after coagulation bath or below dry tower) in m/s; and S = overall stretch ratio after first godet, including relaxation shrinkage. These parameters also affect fibre structure and properties.

The main difference between wet- and dry-spinning occurs in the initial fibre-formation stage. In wet spinning, fine streams of viscous polymer solution are extruded through spinnerette holes into a liquid bath containing a coagulant that is miscible with the solvent. Fibre formation occurs rapidly as the polymer comes out of solution and is drawn onto the first godet or take-out roll. Wet spinning is a three-component system (polymer-solvent-non-solvent) in which two transitions may occur; gelation (the gradual and continuous transition of the fluid solution into a homogeneous elastic gel) and phase separation. Three modified wet-spinning processes have also been described for the production of acrylic fibres. In dry spinning, fibre formation occurs as solvent is evaporated from streams of polymer solution drawn downward through heated gas. Fibre formation in the polymer-solvent systems used for dry spinning results only from gelation. Other transitions do not occur because the systems are miscible, temperatures are high, and the solvent in the fibres lowers their transition temperatures. Gelation occurs gradually as the solvent evaporates and the concentration of polymer in the fibre increases. Tows from the different wet- and dry-spinning systems all undergo the essential steps of washing, orientation or stretching, finish application, and drying accompanied by collapse, relaxation, and crimping, but the sequence of these steps varies.

Bicomponent acrylic fibres are composed of two polymers in separate areas of the cross-section. Each polymer is continuous along the fibre length and the two are permanently joined at an interface. The polymers are selected to shrink or swell differently in response to heat or moisture. As a result, when the fibres are properly treated, differential shrinkage will occur and a spiral or helical crimp will form. Bicomponent fibres may be produced in any of the spinning processes described earlier. Special equipment is required prior to the spinnerette to channel two separate solution streams into each hole.

Acrylic fibres are dyed mainly with cationic dyes which achieve bright, fast colours. A small quantity of acrylic fibre spun from special polymers containing basic groups is dyed with anionic dyes. An over-simplified but useful view of the mechanism of dyeing with cationic dyes has three steps: surface adsorption, diffusion into the fibre, and dye-fibre bonding. Dyeing in non-aqueous liquids (solvent dyeing) is a useful process for colouring acrylics. This process requires less energy than aqueous dyeing and greatly reduces water pollution.

Uses

Acrylic fibres are used extensively in the apparel market for sweaters, single- and double-knit jersey fabrics, hosiery and craft yarns, and pile and fleece goods. They are used in the home furnishing market for carpets, blankets, curtains, and drapes. A small amount goes into industrial uses.

Modacrylic Fibres

Since modacrylic fibres usually contain more than 50 mol% acrylonitrile, their properties are related to those of acrylic fibres. They differ importantly in the lower maximum temperatures at which the modacrylics are dimensionally stable, a property resulting from the presence of the halogen-containing monomers. Halogen also affects other fibre properties, such as moisture regain, light stability, and resistance to degradation on heating.

The most important property of modacrylic fibres is their flame retardance. The commercial significance of this property has increased rapidly with consumer concern and government regulations. However, conclusions about the flame retardancy of the modacrylic fibres, and any new fibre, can be drawn only in relation to specific products and tests. Results are dependent upon several factors in addition to fibre composition.

Like the acrylics, modacrylics are highly resistant to biological agents. They are also stable to aqueous solutions of most salts, acids, and bases, and are generally resistant to dry-cleaning solvents although some will be dissolved by acetone and other ketones. Their lightfastness is good but not equal to that of the acrylics. Physical properties of modacrylic fibres are shown in Table 4.7. Most modacrylic fibres are quite lustrous.

Table 4.7. Physical properties of modacrylic fibres.[a]

Fibre	*Properties*
Denier	3.0
Tex	(0.33)
Tenacity, N/tex	0.24
(g/den)	(2.7)
Elongation, %	35
Boiling water shrinkage, %	34.3
Dry heat shrinkage temperature for 10%, °C	187
Moisture regain, %	2.7
Density, g/cm^3	1.36

[a] Tensile properties measured at 65% rh and 21°C. Extension rate, 100% per minute.

Manufacture

All modacrylic fibres of commercial importance are made from copolymers of acrylonitrile and a halogen-containing monomer, with or without additional monomers to enhance dyeability and other physical properties. The two most commonly used monomers are vinyl chloride and vinylidene chloride; the more expensive vinyl bromide, vinylidene bromide, and 2,3-dibromopropyl acrylate are also used. As with acrylic fibres, monomers incorporated to increase ease of dyeing with basic dyes usually contain sulphonic acid groups. Polymers for modacrylic fibres can be made by emulsion, heterogeneous, and solution polymerisation, both batch and continuous. Initiation is usually with the persulphate-bisulphite-iron redox system. Polymer separation, washing, and drying follow standard procedures. In solution polymerisation, with dimethylformamide or dimethyl sulphoxide as solvents, azo and peroxy initiations can be used, as well as light. After removing unreacted monomers, these solutions are spun directly into fibres by conventional wet-and dry-spinning techniques. The equipment, factors involved in process development and product optimisation, and the sequence of process steps are generally the same as those discussed under spinning of acrylic fibres.

Uses

As a result of flammability legislation and the improvement in textile performance, modacrylics are now used in a wide range of fabrics. The largest volume sales are in the apparel market, particularly in children's sleepwear and pile fabrics, and home furnishings, notably for draperies and curtains that must pass flame-retardant standards. They are also used in a number of industrial applications where their flame retardancy, chemical resistance, and other properties are desirable, such as for paint rollers, battery separators, and protective uniforms. In addition, they are used to manufacture wigs.

DIETARY FIBRE

Dietary fibre is defined at present, as the plant-derived food polysaccharides and lignin that are not digested by enzymes of the human gastrointestinal tract. These polysaccharides include cellulose, hemicelluloses (xylan, arabinoxylan, xyloglucan, glucomannan), pectic substances and associated polysaccharides (arabinan, arabinogalactan, galactan), as well as fructant, galactomannan, and b-glucan. Dietary fibre is a complex mixture composed of cellulose with lesser amounts of other polyasaccharides and lignin, the precise composition being related to the plant source and also to the stage of maturity as well as to food processing effects. A resurgence of interest in dietary fibre was stimulated by epidemiological evidence of differences in colonic disease patterns between certain cultures whose diets contain larger quantities of fibre and western people who consume more highly refined diets.

The primary sources of fibre in the diet are vegetables, cereals, and to a lesser extent, fruits. Currently, certain processed foods and breads are supplemented in their fibre content by incorporation of purified cellulose or cereal bran. In addition, water-soluble cellulose derivatives, seaweed polysaccharides (alginate, cerrageenan), seed mucilaginous polysaccharides, highly complex plant-exudate polysaccharides (gum arabic, tragacanth, etc.), and microbially-synthesised polysaccharides such as xanthan gum are added to foods and contribute to the dietary-fibre content.

Physiological Properties

Dietary fibre has a profound effect on the quality and appearance of the faces and on the passage time of digesta through the gastrointestinal tract. The particle size and shape, density, and water-holding capacity influence the flow rates. Increased levels of fibre increase the flow rates and frequency of defecation and the fecal weight. High fibre diets play a role in excretion of bile acids and cholesterol. The composition and level of dietary fibre may affect the generation and action of tumorigenic substances in the colon; however, evidence is not yet available to prove this hypothesis.

The colon bacteria possess inducible enzymes capable of degrading and fermenting many of the fibre polysaccharides. Some possess cellulolytic activity. The volatile fatty acids generated by bacterial action are partially absorbed through the colon wall and are available as a supplementary energy source.

Physio-chemical Properties

Water-holding capacity, ion-exchange capacity, solution viscosity, density, and molecular interactions are characteristics determined by the chemical structure, the crystallinity, and surface area of the component polysaccharides.

Uses

Food use: used to control consistency of various food products; to improve texture and appearance; to enhance stability of food formulations; as an additive. Non-food use: filter aid and binders.

CHEMICAL FIBRES

The generally used term man-made, referred to here as chemical, as applied to fibres differentiates between the natural fibres and those produced by chemical and physical means in manufacturing processes. The advent of nylon, and the subsequent proliferation of fibres from synthetic polymers, has brought about the subdivision of chemical fibres into two broad classes: those made from natural polymers such as the cellulosics, and those derived from synthesised polymers. The latter, which are called synthetics (or sometimes non-cellulosics), now represent the majority of chemical fibres. In this category fall the nylons, the polyesters, the acrylics, the polyolefins, and many others.

Chemical Fibres Based on Natural Polymers

Rayon

The predominant method of formation is the viscose process based on the finding of Cross, Bevan, and Beadle that cellulose is soluble in a mixture of caustic soda and carbon disulphide. Relatively purified wood-pulp cellulose, called dissolving pulp or chemical cellulose, is treated first with caustic soda, then with carbon disulphide, to form a xanthate which is, in turn, soluble in aqueous alkali solution. Control of conditions in the spinning solution, called viscose, and of the bath into which the viscose is extruded, importantly affect final fibre properties.

Cellulose acetate

Cellulose acetate is dry-spun, predominantly in continuous filament form. The fibre-forming base polymer is made by acetylating dissolving pulp with acetic anhydride usually in acetic acid solvent with sulphuric acid as catalyst. The fully acetylated cellulose initially produced is partially hydrolysed to produce secondary acetate. Following catalyst removal, precipitation of the acetate flake, washing, and drying, the flake is dissolved in acetone containing a small amount of water. This spinning "dope" is subjected to extensive filtration and then dry-spun to evaporate the acetone solvent.

Cellulose triacetate

Cellulose triacetate production is similar to that of acetate, differing in that the initial reaction product is not hydrolysed to a product with less than 92% of the hydroxyl groups esterified and the spinning solution for the triacetate flake is made with a mixture of methylene chloride and a small amount of methanol or ethanol.

Other fibres

Other fibres based on natural polymers include: alginate fibres, based on alginic acid from seaweed, and protein fibres, several of which have been based on casein from milk, the fibres being produced by wet spinning.

Synthetic Fibres

Nylon-6,6

Nylon-6,6 is so designated because both the adipic acid and the diamine from which the polymer is synthesised contain six carbon atoms. The fibre-forming polymer is formed by heating nylon-6,6 salt under pressure to remove water as the adipic acid and hexamethylene diamine condense to form a long-chain linear polymer in what is called a melt polymerisation. When the desired average molecular weight is reached, pressure is reduced, and the polymer is usually extruded in spaghetti-like strands, cooled, and cut into chips. In some cases, the molten polymer is conveyed directly to melt-spinning equipment.

Nylon-6

Nylon-6 manufacture is based on the use of caprolactam, the internal cyclic amide of aminocaproic acid, derived petrochemically. Nylon-6 polymer is also melt-spun, and the manufacture, fibre properties, and uses are very similar to those of nylon-6,6.

Acrylic fibres

The fibre-forming acrylic polymers are high in molecular weight and are produced primarily in aqueous medium by free-radical-initiated addition polymerisation. The regular acrylics are usually copolymers having minor amounts of one or more comonomers such as methyl acrylate, having to provide accessibility or sites for dyestuffs.

Like other synthetics, acrylic fibres are stretched to develop orientation and fibre strength.

Polypropylene

Polypropylene constitutes the majority of commercial polyolefin fibres. In stereoregular form, polypropylene has a crystalline melting point of about 165°C, compared to ca 125°C for low pressure (stereoregular) polyethylene and ca 110–120°C for high pressure polyethylene.

These fibres are melt-spun and produced in monofilament, multifilament, and staple forms. For uses where colour is requisite, either the base polymer is modified to provide dyeability or the fibre is spun-coloured that is, produced with pigments incorporated in the polymer and fibre during melting. Other synthetic fibres are summarised in Table. 4.8.

Table 4.8. Relevant F.T.C. generic names for manufactured textile fibres.

Generic name	*Definition of fibre-forming substance*[a,b]
Acetate	cellulose acetate; triacetate where not less than 92% of the cellulose is acetylated
Acrylic	at least 85% acrylonitrile units
Aramid	polyamide in which at least 85% of the amide linkages are directly attached to two aromatic rings
Azlon	regenerated naturally occurring proteins
Glass	glass
Modacrylic	less than 85% but at least 35% acrylonitrile units
Novoloid	at least 85 % cross-linked novolac
Nylon	polyamide in which less than 85% of the amide linkages are directly attached to two aromatic rings
Nytril	at least 85% long chain polymer of vinylindene dinitrile where the latter represents not less than every other unit in the chain
Olefin	at leat 85% ethylene, propylene, or other olefin units
Polyester	at least 85% ester of a substituted aromatic carboxylic acid, including but not restricted to substituted terephthalate units and para-substituted hydroxybenzoate units
Rayon	regenerated cellulose with less than 15% chemically combined substituents
Saran	at least 80% vinylidene chloride
Spandex	elastomer of at least 85% of a segmented polyurethane
Vinal	at least 50% vinyl alcohol units and at least 85% total vinyl alcohol and acetal units
Vinyon	at least 85% vinyl chloride units

[a] All percentages are by weight.

[b] Except for acetate, azlon, glass, novoloid, and rayon, the fibre-forming substance is described as a long chain synthetic polymer of the composition noted.

NYLON

Nylon yarn is a new synthetic fibre now available in commercial quantities. It comes closer to silk in constitution and properties than any other synthetic fibre manufactured previously. Nylon is popular because of its use in preparing nylon hose and other articles of wearing apparel.

Nylon is a generic term applied to all superpolymers made from long chain dibasic acids combined with long chain diamines. "Nylon 66" is the specific nylon now made on a large scale. It is the superpolymer formed by the combination of adipic acid and hexamethylene diamine. The two substances, both solids at the ordinary temperature, are mixed in a solution to form nylon salt. Stabilising agents are added to control molecular weight and viscosity, the solution is evaporated, and the liquid residue transferred to an autoclave in which it is heated for many hours. The product is a melt consisting of the final superpolymer. The melt is extruded in ribbon form on to a chilled wheel; the ribbon now becomes a tough horny material, and is cut into small chips convenient for storage. To make the fibre, the chips are melted by passing them between heated grids, the liquid is fed into a pump and is supplied under pressure of the pump through the spinnerets which form the filaments. The latter are congealed at once by a gentle blast of air and wound on bobbins.

Next, the primary filaments, rather useless in their existing form, are cold drawn to about four times their original lengths with a decrease in diameter. By this operation, the properties are changed completely, for they now form elastic fibres, transparent, lustrous and of high tensile strength. The explanation for the change is that the random arrangement of the superpolymer molecules gives place to an oriented one. The cold drawn filaments are twisted, sized and re-wound on spools or cones, ready for the textile mill.

Nylon may be made to give up its lustrous appearance like rayon, by the addition of finely powdered colloidal titanium dioxide, or similar pigments. Nylon may be coloured in the molten mass, or the yarn or the finish fabrics may be dyed with dyes for acetate rayons.

Properties of Nylon

Nylon is not soluble in water or most of the ordinary organic solvents. It, however, dissolves in phenol, cresol and formic acid. Nylon 66 melts at 263°C. The tensile strength is 4.5 to 7 gm per denier, dry and 4.0 to 6 gm, wet, depending on the degree of molecular orientation and chain length. For comparison, a good silk has a tensile strength of 4.5 gm, dry, and 3.8 gm, wet. Nylon is stronger than any natural fibre. It has good flexing qualities and excellent stretch recovery power and high tenacity. It can be dyed by all classes of dyes. The raw materials for nylon manufacture are adipic acid and hexamethylene diamine. These raw materials are made on a large scale.

Health and Safety Factors

Ecological and pollution considerations

The cuprammonium and the viscose processes consume huge amounts of water for each kilogram of product; ca 420–750 L (110–200 gal) of water per kilogram of rayon is needed directly for processing and 8–10 times that amount of water must be handled to provide the plant supplementary-service facilities.

In the cases of the solvent spinning process, all recovery and recycle stages are considered to be closed loops with essentially none of the solvent being lost. This must be achieved because the solvent systems are too expensive to permit anything less than almost complete recycling.

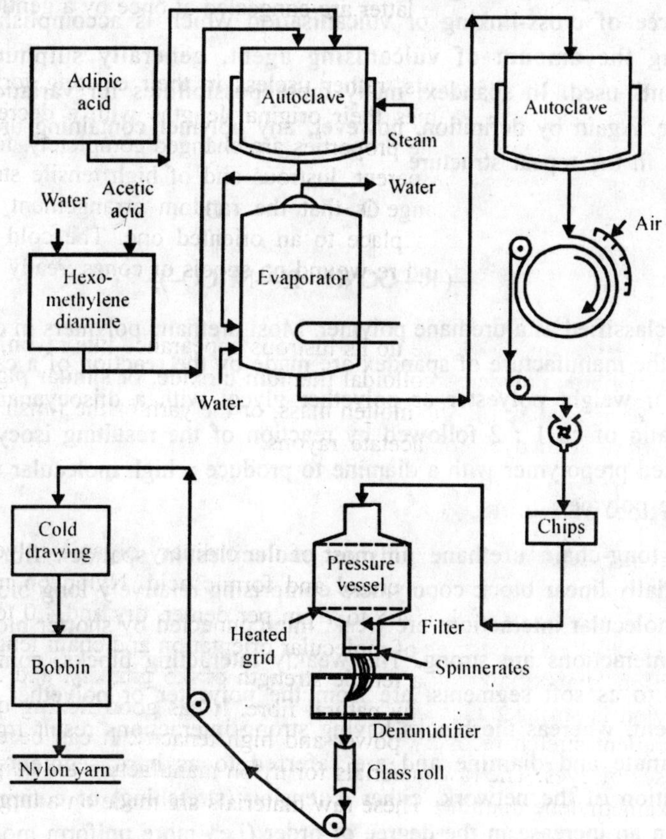

Fig. 4.2. Flowchart for manufacture of nylon yarn.

ELASTOMERIC FIBRES

An elastomeric fibre can be made from any natural or synthetic polymeric material that has high elongation and good recovery properties. At present, only natural rubber and urethane polymers are raw materials for commercially successful elastomeric fibres.

The term spandex has come into common trade usage throughout the world in referring to fibres based on elastomeric urethane polymers.

Properties

In both rubber thread and spandex, mechanical properties may be varied over a relatively broad range. In rubber, variations are made by changes in the degree of cross-linking or vulcanisation which is accomplished by changing the amount of vulcanising agent, generally sulphur, and accelerants used. In spandex, many more possibilities for variation are available. Again by definition, however, any polymer containing urethane linkages in the repeat structure

$$-\!\!\left(R-O\overset{\overset{\displaystyle O}{\|}}{C}NH-R'-NH\overset{\overset{\displaystyle O}{\|}}{C}O-\right)_{n}$$

may be classified as a urethane polymer. Most urethane polymers in current use for the manufacture of spandex are made by the reaction of a ca 2000 molecular weight polyester or polyether glycol with a diisocyanate at a molar ratio of ca 1 : 2 followed by reaction of the resulting isocyanate-terminated prepolymer with a diamine to produce a high molecular weight urethane polymer.

The long-chain urethane polymer molecules in spandex fibres are substantially linear block copolymers comprising relatively long blocks in which molecular interactions are weak, interconnected by shorter blocks in which interactions are strong. The weakly interacting blocks, commonly referred to as soft segments, are from the polyester or polyether glycol component, whereas the blocks having strong interactions result from the diisocyanate and diamine and are referred to as hard segments. Any deformation of the network, either extension (stretching) or compression results in an increase in the degree of order (i.e., more uniform molecular alignment) and is opposed by entropy forces within the soft segments.

Manufacture

Cut rubber

In producing cut-rubber thread, smoked rubber sheet or crepe rubber is milled with vulcanising agents along with stabilisers and pigments. The milled stock is calendered into sheets of thicknesses of 0.3–1.3 mm, depending on the final size of rubber thread desired. Multiple sheets (usually 20 to 40) are layered, heat-treated to vulcanise, then slit to form ribbons containing the same number of threads as layered sheets.

Extruded latex thread

In the manufacture of extruded latex thread, a concentrated natural-rubber latex is blended with aqueous dispersions of vulcanising agents, pigments, and stabilising agents. The compounded latex is held under carefully controlled temperature conditions for 8–36 hour in order to allow some initial vulcanisation to occur.

The matured latex is extruded at constant pressure through precision-bore glass capillaries into an acetic acid bath where coagulation into thread form occurs. Threads are removed from the coagulation bath, washed, dried, then a finish is applied and the threads are wound on individual bobbins or formed into multi-end ribbons and wound on ribbon spools or boxed in the free-ribbon form. The thread may be finally vulcanised by heating the ribbon spools or bobbins in vulcanising ovens for up to 18 hour; however, most modern manufactures vulcanise continuously in ribbon form before winding up on boxing.

Spandex

Spandex is produced commercially by four very different processes—melt extrusion, reaction spinning, solution dry spinning, and solution wet spinning. These are illustrated in Fig. 4.3.

Use

Both spandex and natural-rubber fibres may be used in any area where extendability with good recovery is required. Spandex has replaced most rubber in foundation garments, swimwear, and other critical performance-use areas, owing to better aging properties, higher modulus, availability of finer deniers, and superior durability.

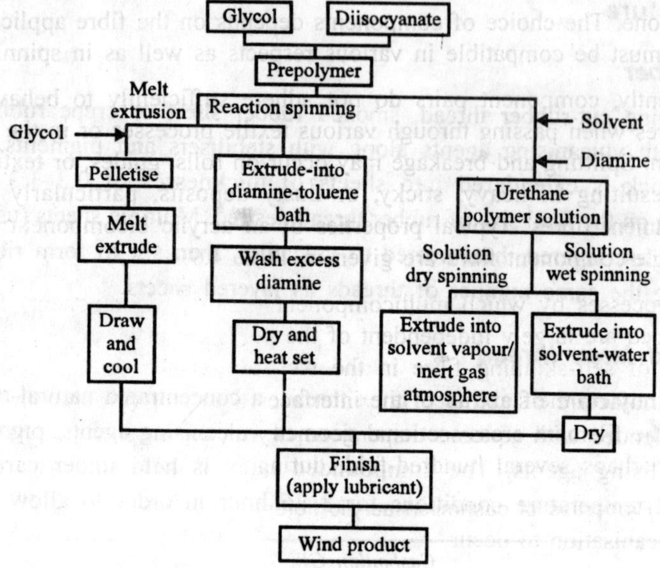

Fig. 4.3. Spandex fibre production methods.

MULTICOMPONENT FIBRES

The designation for fibres composed of two polymers is bicomponent fibres (bico fibres). Since there are few examples of three or more fibre-forming polymer components, bicomponent tends to be used as a general term. It is important to distinguish between composite fibres (where every individual fibre or filament has two or more components) and composite yarns, also called heteroyarns.

Bicomponent fibres are classified on the basis of the cross-sectional and longitudinal arrangement of the components. The simplest class is known as side–side fibres irrespective of the cross-sectional shape, which is not necessarily circular. This class includes fibres with a single interface and with both components possessing an external boundary. The second class is composed of fibres with a single interface but in which only one component has an external boundary. These fibres are known as sheath-core fibres.

Materials and Processes

The manufacture of bicomponent fibres differs little from that of single-component fibres. Many pairs of fibre-forming polymers may be employed, and it is not necessary for both components to be capable of forming fibres is spun alone. The choice of components depends on the fibre application and they must be compatible in various respects as well as in spinning.

Frequently, component pairs do not adhere sufficiently to behave as single fibres when passing through various textile processes or in the final application. Splitting and breakage may occur on rolls, guides, or texturing devices resulting in heavy, sticky, or dusty deposits, particularly with matrix-filament fibres. Typical properties of an acrylic bicomponent fibre and a single-component fibre are given in Table 4.9.

The processes by which multicomponent (mainly bicomponent) fibres are produced are largely independent of the polymers used, except for the formation of self-skinning fibre in the wet-spinning process. Production processes rely on the stability of the interfaces between the components to maintain the filament cross-sectional geometry as the cross-section area is reduced, perhaps several hundred-fold, during extrusion and drawdown.

Table 4.9. Properties of cashmilon acrylic fibres.

Property	*Cashmilon GW, bicomponent*	*Cashmilon FW, monocomponent*
Strength, N/tex[a]		
dry	0.25	0.25–0.32
wet	0.21	0.23–0.28
Elongation, %		
dry	40–45	28–40
wet	45–50	30–45
Loop strength, N/tex[a]	0.31	0.18–0.27
Loop elongation, %	20–25	5–20
Modulus of tensile elasticity, %	99 ± 2	(90–92) ± 3
Specific gravity	1.18	1.188
Moisture regain at 20°C, %	1.8–1.6	1.6
Shrinkage in hot water, %	4	2–4

[a] To convert N/tex to gf/den, multiply by 11.33.

A form of multicomponent fibres is produced by spinning a blend of immiscible polymers, e.g., nylon and polyester. A matrix-fibril fibre is produced in which the minor component forms short individual fibrils dispersed in a fibre composed of the principal component. True multicomponent fibres are prepared by the controlled combination of liquid polymer streams after the individual streams have entered the spinning pack. If only side-side fibres are required and a degree of uncertainty regarding the position of the interface can be tolerated, a conventional spinnerette may be fed with a polymer stream containing alternating layers of the two components. Other methods of preparing multicomponent fibres require the controlled flow of the components to the spinnerette holes or beyond. Most of these methods require the components to be combined and formed into the required cross-sectional configuration just above the spinnerette holes.

In another method, both components must be of the same material but with different orientation. Each filament is formed by the combination of two polymer streams below the spinnerette hole. Side-side bicomponent fibres are also produced by feeding a mixed polymer stream to a simple spinnerette with the holes aligned with the stream interfaces. The mixed polymer streams are generated as concentric rings or parallel laminae.

The pipe-in-pipe method is the simplest for the formation of mixed streams (Fig. 4.4). When it is employed for the production of matrix-filament fibres, a number of core-sheath filaments are combined and narrowed in a conical chamber below the spinnerette holes before extrusion (combining-chamber method). One form of the septum method is similar to the pipe-in-pipe technique. The core stream is inserted directly into the sheath stream via a fine tube. The wall of this tube may be regarded as a circular septum.

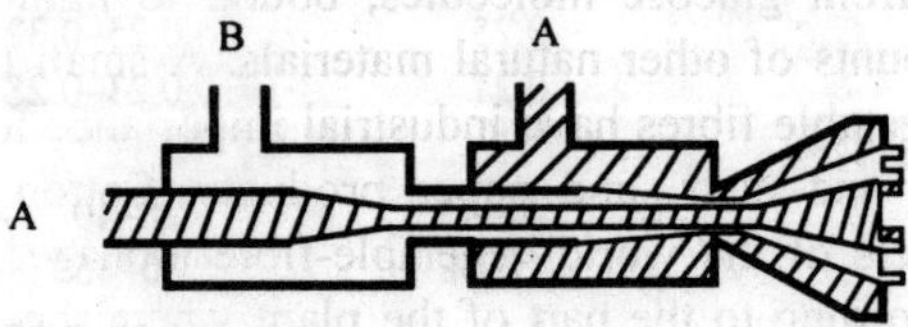

Fig. 4.4. Pipe-in-pipe method for mixed-stream formation.

The need to control the sheath polymer has led to the castellation method, in which a flow of polymer sheath is controlled by passage through a fine gap between the upper (distribution plate) and lower spinnerettes. The gap is ca 100 μm and extends evenly over the spinnerette, resulting in the production of uniform fibres.

Matrix-filament fibres are also manufactured by a side-injection technique with a septum device in which a tube with fine side holes passes through a chamber with one component to connect another chamber containing the second component to the spinnerette hole.

Uses

Where the components are inert, non-toxic, and non-allergenic, multi-component fibres behave similarly. By far the most important application is the creation of desirable textile properties or aesthetic effects. For example, although the initial aim was to achieve a stearic crimp similar to that of wool, much more sophisticated effects are now obtained.

Microdenier (< 0.1 tex per filament) fibres formed from matrix-filament fibres are used in the production of knitted or woven fabrics with appearance different from those made of fibre in the normal filament range (0.1–0.5 tex per filament). In other applications, one component serves as an adhesive. Core-sheath fibres are generally used in which the sheath melts ca 20°C lower than the core. Multicomponent fibres are used to modify physical properties of textiles, e.g., for antistatic applications.

The inclusion of more than one component in a fibre may confer various optical properties. For example, a bichromic shot effect is obtained with a fibre of flat cross-section and a second component located at the rounded-end portions. Multicomponent fibres have been proposed at containers or diffusion barriers.

VEGETABLE FIBRES

Natural fibres of vegetable origin are constituted of cellulose, a polymeric substance made from glucose molecules, bound to lignin and associated with varying amounts of other natural materials. A small number of a vast array of these vegetable fibres have industrial importance for use in textiles, cordage, brushes and mats, and paper products. Cotton fibre makes up about three-quarters of the world vegetable-fibre tonnage. Vegetable fibres are classified according to the part of the plant where they occur and from which they are extracted.

Physical Properties

The vegetable fibres are stronger but less extensible than cotton, i.e., they have a higher breaking length and elasticity modulus with a lower strain (extensibility) and work modulus. They approach glass in stiffness (resistance to deformation) and are considerably stiffer than man-made fibres, but have lower toughness (ability to absorb work). Kapok and other seed-hair fibres are relatively low in strength but have great buoyancy.

Chemical Composition

The chemical compositions (Table 4.10) vary greatly between plants and within specific fibres depending on genetic characteristics, part of the plant, and growth harvesting and preparation conditions.

Table 4.10. Chemical compositions of various vegetable fibres, wt%.

Type of fibre	*Cellulose*	*Hemi-cellulose*	*Pectins*	*Lignin*	*Water-soluble compounds*	*Fats and waxes*
Cotton	91.8		6.4		1.1	0.7
Flax (retted)	71.2	18.6	2.0	2.2	4.3	1.7
Flax (non-retted)	62.8	17.1	4.2	2.8	11.6	1.5
Italian hemp	74.4	17.9	0.9	3.7	2.3	0.8
Jute	71.5	13.4	0.2	13.1	1.2	0.6
Manila	70.2	21.8	0.6	5.7	1.5	0.2
New Zealand cotton	50.1	33.4	0.8	12.4	2.4	0.1
Ramie	76.2	14.6	2.1	0.7	6.1	0.3
Sisal	73.1	13.3	0.9	11.0	1.4	0.3

Economic data

Vegetable fibres, excluding cotton and flux, have been relatively unimportant in the total fibre supply and have become less important in relation to converted cellulosic and non-cellulosic fibres.

OLEFIN FIBRES

Olefin fibres, which also are called polyolefin fibres, are manufactured fibres in which the fibre-forming substance is any long-chain, synthetic polymer of at least 85 wt% ethylene, propylene, or other olefin units. The olefin fibres of commercial importance are polypropylene and, to a lesser extent, polyethylene. Fibres have been produced on a laboratory scale from several other polyolefins, e.g., poly(1-butene), poly(3-methylbutene), and poly(4-methyl-pentene).

Properties

Physical

Some physical properties of commercial polyethylene and polypropylene fibres are listed in Table 4.11. Polyethylene and polypropylene fibres differ from other synthetic fibres in two important respects; their nearly total lack of water absorption ensures that wet properties are identical with their properties at standard conditions (65% rh, 21°C) and their low specific gravity leads to a higher covering power, i.e., one kilogram of polypropylene can produce a fabric, carpet, etc., with as much as 45% more fibre per unit area than one kilogram of polyester.

Table 4.11. Synthetic fibre properties.

Polymer	Fibre Type	Tenacity, MPa (N tex[a]) Standard	Wet	Modulus, GPa (N tex)[a]	Breaking elongation, % Standard	Wet	Elastic recovery, % [b]	Density, g cm^3
Low density Polyethylene	monofilament	92–280 (0.1–0.3)	92–280 (0.1–0.3	0.18–1.0 (0.2–1.1)	20–80	20–80	95 (at 5%)	0.92
High density polyethylene	monofilament	290–570 (0.3–0.6)	290–570 (0.3–0.6)	1.7–4.2 (1.8–4.4)	10–45	10–45	100 (at 5%)	0.95–0.96
Polypropylene	staple and tow	280–560 (0.3–0.6)	280–560 (0.3–0.6)	0.28–3.3 (0.3–3.5)	20–120	20–120	70–100 (at 5%)	0.90–0.96
	monofilament	270–540 (0.3–0.6)	270–540 (0.3–0.6)	1.6–4.8 (1.8–5.3)	14–30	14–30	98 (at 5%)	0.90–0.91
	multifilament	180–630 (0.2–0.7)	180–630 (0.2–0.7)	1.2–3.2 (1.3–3.5)	20–100	20–100	94–98 (at 5%)	0.90–0.91
Polyester	staple and tow	280–830 (0.2–0.6)	280–830 (0.2–0.6)	1.2–2.1 (0.9–1.5)	12–55	12–55	81 (at 3%)	1.38
	multifilament	280–690 (0.2–0.5)	280–690 (0.2–0.5)	1.2–3.6 (0.9–2.6)	24–42	24–42	76 (at 3%)	1.38
Nylon-6,6	staple and tow	230–680 (0.2–0.6)	230–570 (0.2–0.5)	1.0–4.5 (1.0–4.5)	16–75	18–28	82 (at 3%)	1.13–1.14
	multifilament	230–570 (0.2–0.5)	230–570 (0.2–0.5)	0.45–2.4 (0.4–2.1)	26–65	30–70	88 (at 3%)	1.13–1.14
Acrylic	staple and tow	230 (0.2)	120–130 (0.1–0.2)	1.0 (0.9)	20–28	26–34	73 (at 3%)	1.16
Rayon	staple, regular tenacity	150–450 (0.1–0.3)	150–300 (0.1–0.2)	0.75–2.3 (0.5–0.15)	15–30	20–40	82 (at 2%)	1.46–1.54
	staple, high tenacity	450–750 (0.3–0.5)	300–600 (0.2–0.4)	1.7–6.6 (1.1–4.4)	9–26	14–34	70–100 (at 2%)	1.46–1.54
Cellulose acetate	staple, multi-filament	130 (0.1)	130 (0.1)	0.53–0.70 (0.4–0.53)	25–45	35–50	48–65 (at 4%)	1.32

[a] To convert N/tex to gf/den, multiply by 11.33; N/tex = kJ/g; N/tex × density (g/cm^3) = GPa. To convert GPa to psi, multiply by 145,000.

[b] Parenthetical number denotes the amount of initial extension.

Chemical

The general chemical characteristics of olefin fibres are extreme hydrophobicity and inertness to a wide variety of inorganic acids and bases and to organic solvents at room temperature. These properties derive from the hydrocarbon character of the fundamental unit and from the great molecular weights of commercial polymers.

Thermal

Comparatively, polyethylene and polypropylene undergo all thermal transitions at substantially lower temperatures than either polyesters or nylons. The thermal degradation of olefin fibres is oxygen-sensitive, and may be problematic in fibre formation and in use. In general, the absorption of oxygen by the polymer results in chain scission and, therefore, molecular weight degradation as a result of high temperature formation of hydroperoxides.

Ultraviolet degradation

Sunlight affects olefin fibres, and linear polyethylene is somewhat less susceptible than other polyolefins. The effect is similar to that for thermal oxidation, i.e., chain scission and molecular weight degradation, although the mechanism of photo-oxidation is different.

Flammability

One fundamental measure of the flammability of polymers is the oxygen index which defines the minimum oxygen concentration necessary to support combustion. Polyolefins are more combustible on the basis of this index than many other common polymers.

Dyeing

Olefin fibres are inherently difficult to dye, because there are no sites for the specific attraction of dye molecules, i.e., no hydrogen-bonding or ionic groups, and dyeing only can take place by virtue of weak dye-fibre van der Waals forces. As a result, the most common method of colouring olefin fibres is to add pigments to the melt before extrusion.

Stress-strain relationships

As is common for all polymeric fibres, the stress-strain curves for olefin fibres depend on strain rate, temperature, molecular weight, and fibre morphology, especially molecular orientation and crystallinity.

Crepe and stress relaxation

Olefin fibres exhibit crepe, i.e., time-dependent deformation, of which only a portion is recoverable under load; and the inverse phenomenon, i.e., stress relaxation, which is the spontaneous relief of internal stress and occasionally is accompanied by small spontaneous length changes. Both of these behaviours depend on fibre molecular weight, molecular orientation, and crystallinity.

Dynamic mechanical properties

The response of olefin fibres to cyclic deformation is determined by the temperature and frequency of the deformation and by the structural characteristics of the molecule and fibre morphology.

Manufacture and Processing

Olefin fibres are fabricated commercially by one of several modifications of the melt-extrusion technique, of which the fundamental elements are the continuous expulsion of molten polymer through a die, i.e., spinneret, the solidification of the extrudate by heat transfer to the surrounding fluid medium, and the winding of the solid extrudate onto packages. Further processing may include drawing the fibre to as much as six times its original length and a variety of heat treatments to relieve thermal stresses within the fibre. Texturising processes, which are combinations of deformations and heat treatments, also may be applied.

Novel olefin fibres

Novel olefin fibres include hard elastic fibres, high modulus fibres, and bicomponent and biconstituent fibres.

Uses

Olefin-fibre use patterns are concentrated in home furnishing and industrial areas, where good fibre-mechanical properties, relative chemical inertness, low moisture absorption and, sometimes, low density contribute to desirable product properties. Olefin fibre use in apparel has been restricted by low melting temperatures which make ironing of polyethylene and polypropylene fabrics impossible.

CELLULOSE

Cellulose is the principal fibre cell-wall material of green terrestial and marine plants, produced also by a few bacteria, animals and fungi, and thus the most abundant natural material (– 40% in wood, over 70% in bast and leaf fibres, 95% in cotton, 70% in the cell wall of the green alga *Valonia ventricosa;* ca 5 × 10^{11} metric tons biosynthesised yearly). Cellulose is a long linear polymer of anhydroglucose units and this is reflected in the thread-like structures of cellulose found in plant cell walls (elementary fibrils, approximately 3.5 nm in width and indefinite length) which are further laterally associated to provide strength (as microfibrils, generally 10–30 nm in breadth). Although the biochemical pathway for the biosynthesis of cellulose from glucose in plants is relatively well understood, the mechanisms of the glucose in plants is relatively well understood, the mechanisms of the formation of long chains, of association as microfibrils and of deposition in plant cell walls are still being researched.

Cellulose is partly ordered (crystalline) and partly disordered (amorphous), presumably the result of regions of regularity and non-regularity within the elementary and microfibrils. Accessibility of cellulose is the relative ease by which the hydroxyl groups can be accessed by reactants. The amorphous regions are highly accessible and react readily, whereas the crystalline regions with close packing and hydrogen bonding can be completely inaccessible.

CELLULOSE ACETATE AND TRIACETATE FIBRES

Acetate fibre, also known as secondary acetate, is the generic name for a cellulose acetate fibre which is a partially acetylated cellulose. Triacetate is the generic name for cellulose triacetate fibre, also known as primary acetate, an almost completely acetylated cellulose. Acetate and triacetate differ only moderately in the degree of acetyl substitution on cellulose, yet they have different chemical and physical properties.

Cellulose acetate is the reaction product of cellulose and acetic anhydride. The degree of polymerisation (DP) of the cellulose used for esterification is ca 1000-1500, whereas the esterification product DP is ca 300. Secondary cellulose acetate is obtained by acid catalysed hydrolysis of the triacetate to an average degree of substitution of 2.4 acetyl groups per glucose unit. The degree of acetylation is specified by the acetyl value (%) and combined acetic acid (%); the two are always in the ratio of 43:60. The combined acetic content of commercial cellulose triacetate is 61.5%, and of cellulose acetate is 55%. Commercial cellulose acetates contain small amounts of residual-free carboxyl and sulphate groups as well as hydroxyl and acetyl groups. The salt form is ordinarily more thermally stable. Cross-linking between residual sulphate and carboxyl groups can cause an artificially high solution viscosity which limits the concentration of cellulose acetate used in yarns.

Properties

Physical properties

Cellulose acetate fibres have a low degree of crystallinity and orientation even after heat treatment; cellulose triacetate develops considerable crystallinity after heat treatment. Fibres of both have a bright lustrous appearance. Duller, whiter yarns are produced by the addition of pigments. The appropriate azo, anthraquinone, or diphenylamine disperse dye is selected to ensure colourfastness. Fading inhibitors are used to resist the effects of nitrogen oxides and ozone.

The specific gravity values of 1.32 for acetate and 1.30 for triacetate are accepted for fibres of combined acetic acid contents of 55 and 61.5%, respectively.

Triacetate fibre is hydrophobic. An hysteresis effect is noted for both acetate and triacetate fibres over the entire range of relative humidities. Heat-treated triacetate fibre has a lower regain than nonheat-treated fibre, values are in the range of 2.5–3.2%. The average percentages of water-imbibition are acetate, 24%; non-heat-treated triacetate, 16%; heat-treated triacetate, 10%.

Like other thermoplastic fibres, acetate sticks, softens, and even melts when ironed at high temperatures; between 190–205°C acetate softens and sticks, but fuses at ca 260°C. The sticking and glazing temperatures of nonheat-treated triacetate fibre are in the same range, those of heat-treated triacetate fibres considerably higher (triacetate mp ca 300°C). Mechanical properties are not altered significantly; however, as the temperature is raised the modulus of the acetate and triacetate fibres is reduced, and they extend more readily under stress. Both fibres are weakened by prolonged exposure to elevated temperatures in air; their resistance is lowered upon exposure to direct weathering.

Acetate and triacetate yarns readily develop static charges, and often have antistatic finishes added. The resistivity of cellulose acetate is 1.27 MΩ-cm; of non-heat-treated cellulose triacetate is 3.81; of heat-treated cellulose triacetate is 15.2; and of specially scoured and heat-treated cellulose triacetate 10.16.

The physical forms of the fibres and the geometrical construction of the fabric determine the mechanical and aesthetic properties of textile fabrics.

Performance criteria include hand, drape, wrinkle resistance and recovery, strength, and flexibility. Fibre mechanical properties are described by stress-strain and recovery behaviour under conditions of tensile, torsional, bending, and shear loading. Under standard conditions, acetate and triacetate tenacity ranges from 0.10–0.12 N/tex (1.3 gf/den); loop ranges from 0.09–0.10 N/tex (65% rh, 21°C); and elongation at break is 25–45%.

The ability of a material to resist deformation under applied tensile stress is measured by the modulus of elasticity (Young's modulus). In visco-elastic materials, the apparent modulus of a textile fibre is defined as the ratio of stress to strain in the initial, linear portion of the stress-strain curve. Most stress-strain data are reported under standard conditions of 21°C and 65% rh. Strains of more than 10% are avoided in textile processing so that the dimensional or shape stability of the resultant fabric will be acceptable. The modulus of elasticity can be affected by drawing (elongating) the fibre and by other manufacturing procedures. Values for commercial acetate and triacetate fibres are generally 2.2–4.0 N/tex (25–45 gf/den). The wet modulus of fibres at various temperatures influences the degree of creasing the mussiness caused by laundering. The ability of a fibre to absorb energy during straining is known as toughness, or work of rupture. The ratio of the work recovered to the total work absorbed is known as resilience.

The visco-elastic behaviour of a fibre is the elongation of a stretched fibre described as a combination of instantaneous extension and a time-dependent extension or creep. Conversely, recovery of visco-elastic fibres is typically described as a combination of immediate elastic recovery, delayed recovery, and permanent set or secondary creep, which is residual extension which is not recoverable. The bending properties of a fibre generally depend upon the visco-elastic behaviour of the material, measured in flex rigidity of $g.m^2$.

Chemical properties

Under slightly acidic or basic conditions at room temperature, acetate and triacetate fibres are very resistant to chlorine bleach at the concentrations normally encountered in laundering. Triacetate fibre is significantly more resistant than acetate to alkalies encountered in normal textile operations. Acetate should be dyed at temperatures no higher than 85°C or in a pH above 9.5; triacetate may be dyed with alkali up to pH 9.5 and temperatures up to 96°C. Acetate and triacetate are unaffected by dilute solutions of weak acids, or by perchloroethylene dry-cleaning solutions. Strong mineral acids cause serious degradation. Triacetate is very resistant to micro-organisms.

Cellulose Acetate and Triacetate Flakes

Acetate and triacetate flakes are white amorphous solids produced in granular, flake, powder, or fibre form which are used as raw materials in the preparation of fibres, films, and plastics. Density varies with physical form, ranging from 100–500 kg/m^3.

Manufacture of flakes

Acetate and triacetate flakes are prepared by the esterification of high purity chemical cellulose with acetic anhydride. The solution properties of the resultant acetate or triacetate flake may be adversely affected by impurities even at low concentrations; extrusion processes require the polymer to flow freely through small diameter capillaries (e.g., 30–80 μm). Wood pulp, which has replaced cotton linters as a source of chemical cellulose, must have the non-cellulose impurities (e.g., lignin, hemicellulose) removed.

Secondary acetate processes

Three main processes used are: (i) A solution process, the most common, in which acetylation is obtained with acetic anhydride using glacial acetic acid as the solvent, and suphuric acid as catalyst. Two major variants of this process are high catalyst (10–15% sulphuric acid based on cellulose) and low catalyst (< 7% sulphuric acid) procedures. (ii) A solvent process in which methylene chloride is substituted for all or part of the acetic acid and acts as a solvent for the triacetate as it is formed, perchloric acid is used as catalyst. (iii) A heterogeneous process (non-solvent or Schering process) in which an inert organic liquid, such as benzene or ligroin is used as a non-solvent to prevent the acetylated cellulose from dissolving as it is formed.

The cellulose ester produced is never solubilised and its physical form is similar to the original cellulose fibre. To produce a secondary acetate soluble in acetone, it is necessary to acetylate the cellulose during the dissolution step and then hydrolyse while still dissolved to the required acetyl value. This four-step process involves: (i) preparation of the cellulose for acetylation; (ii) acetylation at temperatures less than 50°C using a 5–15 wt% excess of acetic anhydride and yielding a 15–25% cellulose acetate concentration; (iii) hydrolysis, normally at 50–100°C using steam injection for 1–24 hour to reduce the number of acetyl groups present in each anhydroglucose unit at the end of acetylation from slightly less than 3.0 to ca 2.4; (4) recovery of cellulose acetate (via precipitation, washing, and drying) and of solvents.

Batch preparation of triacetate

The batch triacetate process differs from the preparation of secondary acetate in that there is little hydrolysis and the temperature is in the range of 50–100°C. The triacetate hydrolysis step (or desulphation) removes only the sulphate groups from the polymer by slow addition of a dilute acetic acid solution containing sodium or magnesium acetate or triethanolamine to neutralise the liberated sulphuric acid. Cellulose triacetate product has a combined acetic acid content of 61.5%.

Continuous flake—manufacturing process

The basic processing stages are identical to those of the batch process but some equipment and materials handling techniques differ. For example, the acetyliser is a specialised, high energy input reactor with a range of materials-handling capabilities which receives matted pulp, acetic acid, catalyst, and acetic anhydride and produces a highly viscous, homogeneous solution of cellulose triacetate. The hydrolysis step consists of a series of retention tanks which provide a controlled time-temperature exposure to reach the desired combined acetic acid and residual sulphate level.

Manufacture of Cellulose Acetate and Triacetate Fibres

A four part dry-extrusion process is primarily used to convert a solution of cellulose acetate or triacetate polymer into fibre form: (i) the cellulose acetate is dissolved in a volatile solvent, e.g., acetone for secondary acetate (ca 95% acetone and 5% water) and chlorinated hydrocarbons (i.e., methylene chloride containing 5–15% methanol) for triacetates; (ii) the solution is filtered to remove insoluble matter; (iii) the solution is extruded to form fibres by forcing the solvent through the spinnerettes (which has numerous holes 30–80 mm in diameter) into a cabinet of warm air; (iv) the yarn is lubricated and taken up on a suitable package, finish or lubricant giving the extruded yarn the frictional and anti-static properties required for further processing. The solvent used to form the extrusion dope is recovered by adsorption, or condensation.

Acetate and triacetate continuous-filament yarns, staples, and tows are manufactured with varying properties, e.g., tex (wt (g) of a 1000-m filament) or denier (wt (g) of a 9000-m filament), cross-sectional shape, and number of filaments. The same basic extrusion technology that produces continuous-filament yarn also produces staple and tow.

Uses

Principal applications are women's apparel and home furnishing fabrics; and cigarette filters.

RAYON

Today the most important synthetic textile fibre is rayon or artificial silk. Rayon may be defined as a man-made yarn or fibre whose chemical base is cellulose. 'Rayon is defined as the genetic term for filament made from various solutions of modified cellulose by pressing or drawing the cellulose solution through an orifice and solidifying it in the form of a filament.' Rayon, therefore, means those fibres composed of a chemical compound of cellulose, viz., cellulose acetate, as well as those formed by the regeneration of cellulose from a chemical solution, viz., viscose cuprammonium or nitrocellulose. The term 'rayon' has now replaced the term 'artificial silk' as a name for the above cellulose-based synthetic fibres.

Two primary products are made by the rayon industry—rayon filament yarn made of a number of fine continuous rayon filaments, grouped and tightly twisted together, and rayon staple fibre by cutting rayon filaments into uniform lengths which are subsequently spun into yarn like cotton or wool.

Manufacture of Rayon

Rayon filament yarn is produced by three processes.

(1) Viscose process

(2) Cellulose acetate process

(3) Cuprammonium process.

The nitrocellulose process, from which rayon was first commercially produced, is of little use now.

Viscose Rayon (Viscose Process)

Most of the world's rayon is made by the viscose process. The term rayon designates a range of products that have widely varying properties, for the most part owing to the versatility of the viscose process. This versatility which is the result of the many process steps and spinning changes that can be made, is a mixed blessing, since each steps of processing and spinning requires close attention to guarantee the desired product properties. The viscose process is a demanding process that requires continuous, year-long-operation to prevent gelling of the system and to yield high quality products.

Process

The stage in the preparation of a satisfactory cellulose xanthate solution are given in Fig. 4.5. In order to have cellulose react with carbon disulphide to form the corresponding cellulose xanthate, cellulose must be converted to alkali cellulose. Normally this is done by placing cellulose-pulp sheets in a steeping press and filling the press with a closely controlled concentration of sodium hydroxide at a desired level of 18–20 wt%, depending on the type of cellulose used.

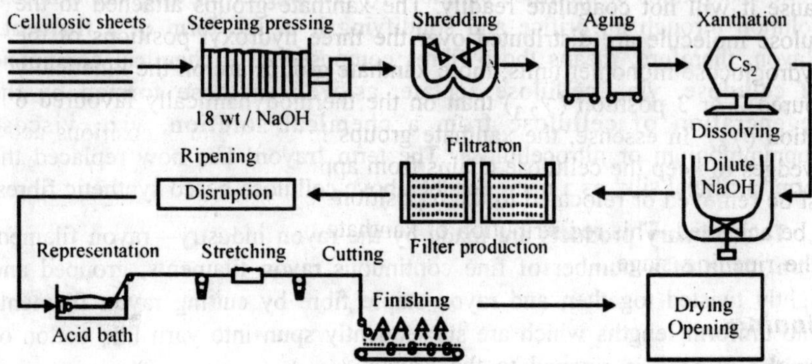

Fig. 4.5. Viscose process.

In addition to converting the alkali cellulose to a crumblike material, shredding serves additional functions, e.g., the squeezing action of the shredder blades distributes the caustic more uniformly in the cellulose. Most alkali cellulose sheets are wetter at the edges than in the centres and, when the caustic solution drains, the bottoms of the sheets have more alkali. Thus, shredding gives a more uniform distribution of caustic in the alkali-cellulose crumb.

Aging is used to decrease and control the cellulose DP, i.e., the alkali cellulose stands in covered containers in a temperature-controlled room. The alkali crumb should never become dry. In the continuous process, the alkali cellulose may be dried while conveyed in an air stream; therefore, humidified air must be recirculated to the conveying section. In the batch process, drying can occur during aging if the cans are not covered properly.

Xanthation is a critical part of the viscose process; many reactions proceed simultaneously during the step. Pure cellulose xanthate is white; any yellow colour is owing to by-product formation, usually in the form of trithiocarbonate. The lower the xanthation temperature and the better the alkali-cellulose preparation, the lighter the colour of the xanthate crumb and the less the hydrogen sulphide (H_2S) evolution during spinning.

Xanthate crumb is dumped into large, stirred tanks containing dilute caustic solution to dissolve the cellulose xanthate into a clear, honeylike viscous dope known as viscose. The caustic and water in the dissolver are measured to give the desired cellulose: alkali ratios. Many dissolvers have standpipes that force the mixtures upwards in a cascading fashion while the moving blades propel the mixtures in a circular manner.

When the cellulose xanthate is first dissolved, it is not ready for spinning because it will not coagulate readily. The xanthate groups attached to the cellulose molecule are distributed over the three hydroxyl positions of the anhydroglucose monomer units; more xanthate groups are on the kinetically favoured 2 or 3 position ($\gamma_{2,3}$) than on the thermodynamically favoured 6 position (γ_6). In essense, the xanthate groups in the 2 and 3 positions act as wedges to keep the cellulose chains from approaching one another; these must be removed or relocated to the 6 positions before closer chain packing can be achieved. This redistribution of xanthate groups is the main function of the ripening stage.

Spinning

Once the viscose is ripened to the proper xanthate level, it is ready for spinning. Chief goals in spinning are control of coagulation is regeneration rates and the maximal use of the differences in these rates to obtain maximum responses to stretching.

Stretching

An advantage of retarding regeneration is to make use of the densification of the extruded dope so as to obtain better molecular alignment on stretching. A fixed amount of stretching can occur; this is balanced between what is called jet stretch and godet stretch. Jet stretch is the ratio of the linear speed at the first godet roll to the average linear flow rate in the spinnerette. Godet stretch is the stretching between the first and the last godet rolls. The higher the godet stretch, the less residual elongation is left in the final fibre. Therefore, the manufacturer balances fibre strength and residual elongation.

Crimped fibres

The crimp is produced by selective control of the conditions in the primary and secondary regeneration baths wherein a gel-core fibre is allowed to leave the primary bath and is exposed to hot-acid conditions in a secondary bath. Under these conditions, significant tensions are established through the fibre, which causes permanent chemical crimping. These fibres also have high wet modulus and give cover, hand, bulk, and working performance in finished fabrics superior to that of ordinary rayon. When produced in higher deniers, Prima blends with polyester closely resemble wool fabrics.

Hollow fibres

This fibre is much lighter than regular rayon and therefore gives significantly more coverage per unit weight. Moisture absorbing capacity is about 59% greater than ordinary rayons; this makes Viloft a strong candidate for use in towels, sheets, pillow cases, and garments requiring improved absorbency. It is produced by adding sodium carbonate to the viscose and then controlling the regeneration conditions to capture the released CO_2 gases to give the hollow fibre.

Solvent-Spun Rayon

Several companies are exploring the possibility of using organic solvents with closed recycle-recovery loops to produce rayon type fibres through solvent-spun processes because of the continued projected need for a fibre such as rayon. The Newell process (recently developed) which appears to be based on spinning solutions of up to 20 wt% cellulose dissolved in hot aqueous *N*-methyl-morpholine *N*-oxide. This system appears to be amenable to closed-loop recovery-recycle and relatively rapid spinning speeds.

A new solvent system for cellulose has been developed that can give up to 16% solutions of 500 DP cellulose. This system, which does not cause cellulose degradation even on extended heating or air exposure, utilizes lithium chloride in dimethylacetamide or in *N*-methyl-2-pyrrolidinone to form solutions of cellulose that are stable in extended storage. This system permits total recycle and recovery of all components to give a non-polluting process. The economics of the process appear to be competitive with a new viscose plant installation.

Cuprammonium Rayon

Essentially all of the rayon made today is produced by the viscose process; nevertheless, there are some uses where the cuprammonium process has retained an advantage because of specific performance factors. Fibres from cuprammonium rayon are significantly more supple than viscose fibres and are used where a very soft hand is desired. The use of films and hollow fibres from the cuprammonium process for making artificial kidneys is critically important in this branch of medicine. At present, almost all artificial-kidney units use membranes prepared from such films and fibres. These films and fibres exhibit superior clearance performance for urea, creatinine, and metabolites, possess better dewatering characteristics, and cause less blood clotting as compared to any synthetics or corresponding products from the viscose process.

Dissolving

The ability of selected cupric ion, ammonia and alkali mixtures to dissolve cellulose was originally reported in 1847. Such solutions, known as Schweitzer's reagent, are used not only for making cuprammonium rayon, but also for measuring dilute-solution viscosities that can be related to degrees of polymerisation of cellulose.

Spinning

The spinning of cuprammonium yarns depends on the use of a funnel into which the coagulating-regenerating fluid flows and into which the dope is extruded.

As the following liquid travels down the funnel, its velocity increases and the entrained gelatinous cellulose fibres are stretched up to 400%. Subsequently, the fibres are washed as free as possible of occluded material, which is recycled, and the fibres enter a 5% H_2SO_4 bath where final removal of copper is achieved and where any remaining alkali or ammonia is converted to the corresponding sulphate.

Cellulose-Acetate Process (Acetate Rayon)

The accurate process of rayon yarn manufacture ranks second only to the viscous process in volume of production. The principal raw materials used by this process are refined cotton linters, special high alpha-dissolving wood pulp, glacial acetic acid, acetic anhydride, sulphuric acid, acetone, titanium dioxide and water. The following are the general principles and stages of manufacture of acetate yarn.

Acetylation

The first operation is acetylation of the cellulose. Batches of cotton linters or wood pulp which have been pre-heated with acetic acid are placed in an acetylator, previously charged with a solution containing a mixture of acetic anhydride and glacial acetic acid, together with a small quantity of sulphuric acid for a catalyst. During this operation, the temperature is rigidly controlled and the reaction carried to completion. In the final stage of acetylation, the solution is a heavy, viscous, semi-transparent, acrid-smelling fluid, commonly called the 'primary acetate'.

Hydrolysis

The second stage is the hydrolysis of the primary acetate into secondary acetate. In this operation, a solution of primary acetate in strong acetic acid is partially diluted with water, and then allowed to stand for sometime at increased temperatures, during which time the primary acetate is partly saponified (or hydrolysed) to the secondary or acetone soluble state whose acetyl-content approximates the diacetate. The secondary acetate is precipitated in the form of white flakes by pouring the solution into a large tank of cold water. The cellulose acetate flakes are washed in warm water until free of acid, then dried and stored in bins.

Spinning solutions preparation

The spinning solution is prepared by dissolving these acetate flakes in acetone. The preparation of this spinning solution is done in a closed mixture machine into which the cellulose acetate flakes and the acetone are added in the proper proportions and gradually mixed together. About 12-24 hours are required to dissolve the cellulose acetate completely. In the final spinning solution, the cellulose acetate completely. In the final spinning solution, the cellulose acetate content may range from 15–20 per cent. From the mixture, the solution is taken into tanks where several batches of solution are blended. The contents of these blending tanks are then filtered several times and finally the solution is drawn into storage tanks where all the air bubbles are removed. The solution is then ready for spinning.

Spinning acetate yarn

Acetate yarn is spun by the dry spinning system. In the dry method of spinning, the spinning solution is delivered from storage tanks to the spinnerette on the spinning machine and is forced downward through the solution under controlled pressures into an enclosed spinning chamber. The spinning solution emerging from the spinnerette is met by a rising current of warm air which evaporates the acetone. By the time the acetate filaments have reached the bottom of the spinning chamber, coagulation has been completed and the filaments dried.

Cuprammonium Process

In this process of manufacture, the purified cotton linters are commonly used as the principal source of cellulose. First the purified cotton linters are cleaned by cooking in a mild caustic solution under pressure and then bleached. These linters, then essentially 100 per cent pure alpha-cellulose, are mixed with basic copper sulphate and 28 per cent aqueous ammonia. This solution must be prepared under rigid chemical control to obtain a solution of proper viscosity and cellulose content. The solution is then filtered and stored in tanks.

The solution, when used, is carried into spinning machines and spun by the stretch spinning device. By this method, the spinning solution is pumped through spinnerettes into a glass funnel through which de-aerated water flows. On the way through this funnel, the single filaments are stretched and slowly solidified, the water meanwhile extracting most of the ammonia and part of the copper. On leaving the funnel, the filaments pass through a bath of mild H_2SO_4 which fully coagulates the filaments and at the same time extracts more of the copper. The filaments are then wound on reels. These reels are sprinkled with plain water, removing any of the remaining acid and $CuSO_4$, after which the yarn is washed in a soap bath, dried and then treated with a second soap bath to soften up the yarn, and dried again. The yarn is now ready for spinning into the desired form.

Chapter 5

Rubber Industries

INTRODUCTION

Rubber is the name given to elastic, high-molecular compounds capable of changing their shape (deforming) when acted upon by an external force and of rapidly resuming their original shape when the force is removed. The elastic properties of rubber are due to the fact that the linear macromolecules of the material which normally are of a bent, curled or curved shape, easily stretch out of contract under the effect of an applied force; when the force is removed, the molecules return to their initial shape and length.

If sulphur is added to natural rubber which is then heated, the sulphur will form links between the linear macromolecules of the rubber, i.e., form bridges between them; by this larger macromolecules of a web structure will be produced. This process is called vulcanisation of the rubber, and the product of reaction between the rubber and sulphur is vulcanised rubber. Vulcanisation improves the mechanical properties of the rubber and increases the range of temperatures within which the elasticity of the material is retained; it also increases the resistance of the rubber to organic solvents, etc.

Natural and synthetic rubber is the basic material in the fabrication of rubber-textile and rubber-metal articles which are so extensively used in various branches of the economy. These are motor car and aircraft tyres, drive belts and conveyer belts, flexible hoses and sleeves, machine parts and various gaskets, electric insulating materials, rubberised fabrics and rubber footwear, articles of personal hygiene and many other rubber products. The consumption of rubber has increased year by year. This explains the attention devoted to developing the rubber industry in India and in many other countries.

Rubbers are classified as natural rubber and synthetic rubber. Until the nineteen thirties only natural rubber, obtained from the latex of rubber plants which grow in countries with a damp tropical climate (mainly in Malaya, Indonesia, etc.), was used.

NATURAL RUBBER

Natural rubber can be produced from many plants, however actually the latex is obtained from a small number of so-called rubber plants. The most important of these is a tree-like plant *Hevea brasilliensis*, whose latex contains 20 to 35% of a rubber constituent, formed by biochemical synthesis. The latex is drawn from notches in the tree bark in the same way as turpentine. This is called tapping.

Latex contains the rubber in the form of tiny pear-like particles or globules, which are suspended in water and bear a negative electric charge; this prevents them from coagulating. Latex also contains a small amount of protein substances (2%), resins (up to 2%), sugary substances (0.4%), and ash (0.5%). To remove the charge from the rubber particles formic or acetic acid diluted with water is added to the latex; this causes the latex to coagulate and the rubber is separated in the form of doughy, bulky mass. The rubber is filtered off, thoroughly washed with water and passed through corrugated rollers. The rubber sheets produced are dried and smoked in a smoke chamber so that they will keep without spoiling. The sheets are packed in bales. Natural rubber is a polymer of isoprene, $CH_2{=}\underset{\displaystyle CH_3}{\underset{|}{C}}{-}CH{=}CH_2$. The macromolecules of natural rubber are made up of

$$-CH_2-\underset{\displaystyle CH_3}{\underset{|}{C}}=CH-CH_2-$$

structural units.

The degree of polymerisation exceeds 2,000 and the molecular weight is from 136,000 to 340,000. Natural rubber possesses high strength and elasticity within a wide range of temperatures.

Natural latex is also used directly, without separating the rubber from it. It is used as a glue, for impregnating textiles and making microporous rubbers.

SYNTHETIC RUBBER

Mainly synthetic rubber is used in the rubber industry. At the present time the chemical industry turns out a wide range of synthetic rubbers, both made from one individual monomer by addition polymerization or from a blend of two or three monomers by copolymerisation. The synthetic rubbers can be divided into the two groups of general-purpose and special-purpose rubbers. This division is purely nominal. General-purpose rubbers are used for making tyres and other rubber articles of mass consumption. Special rubbers have certain improved properties. Such rubbers can operate under severe conditions, for instance at low or at high temperatures resist highly corrosive media (acids, alkalies, oxidants, solvents, etc.)

STYRENE-BUTADIENE RUBBER

Styrene-butadiene rubber (SBR) comprises more than half of world's production of synthetic rubber and exceeds natural rubber supply. In fact, tyre industry depends more on SBR than on natural rubber because it is available in plenty and serves the purpose better after the incorporation of fillers.

Raw Materials

The principal raw materials for SBR manufacture are monomers, styrene and butadiene. Other raw materials are oils used for extending soaps, antioxidants, short stopping agents, coagulating agents, catalysts, modifiers, water and hydrocarbon solvents.

Styrene

Styrene can be obtained as follows:

$$\underset{\text{Benzene}}{C_6H_6} + \underset{\text{Ethylene}}{CH_2=CH_2} \xrightarrow[95°C]{AlCl_3} \underset{\text{Ethyl benzene}}{C_6H_5CH_2CH_3} \xrightarrow[\text{bauxite } 600° \, -H_2]{\text{Metallic oxides or}} \underset{\text{Styrene}}{C_6H_5CH=CH_2}$$

Butadiene

Butadiene is chiefly obtained by thermal cracking of petroleum. It can also be obtained by oxidation of alcohol.

$$C_2H_5OH + \text{Air} \xrightarrow{\text{Catalyst}} CH_3CHO + H_2O$$

$$CH_3CHO + C_2H_5OH \xrightarrow{\text{Catalyst}} \underset{\text{Butadiene}}{CH_2 = CH - CH = CH_2} + 2H_2O$$

Manufacture

Different ratios of styrene and butadiene can be used but the ratio mostly used is styrene 25% and butadiene 75%. If the percentage of styrene is increased to 50% or above, a tough product—a reinforcing resin is obtained. SBR is obtained mostly by emulsion polymerisation using free radical catalyst (Fig 5.1). Thus:

$$\underset{\text{(25\%)}}{\underset{\text{Styrene}}{C_6H_5-CH=CH_2}} + \underset{\text{(75\%)}}{\underset{\text{Butadiene}}{CH_2=CH-CH=CH_2}} \rightarrow$$

$$\left[CH_2-CH=CH-CH_2-CH_2-CH(C_6H_5) \right]_n$$

Polymerisation at 5–10°C is called cold process and the product is called cold SBR while the one made at 50°C is called hot SBR. The cold initiators are redox combinations such as *p*-methane hydroperoxide/ferrous sulphate which result in higher molecular weight at lower temperatures. Hot process uses initiators such as potassium persulphate. There is higher degree of branching. Cold SBR gives better abrasion resistance and dynamic properties. Higher molecular weights obtained also allow higher levels of oil extension.

Structure

SBR is a random copolymer. The butadiene units are present about 20% in 1,2 configuration, 20% in *cis* 1,4 and 60% in *trans* 1,4 if polymerisation is brought about at 50°C which increases if made at low temperature because of irregular structure, it does not crystalline.

Compounding and Applications

The compounding of SBR is similar to that of natural rubber and makes use of same fillers, vulcanising systems, antioxidants and anti-ozonates. Emulsion SBR is a general purpose rubber and can be blended with any other such rubber and possesses cure compatibility. It has certain weaknesses as compared to natural rubber such as less strength and resilience and its heat build up is greater. A reinforcing filler is essential to

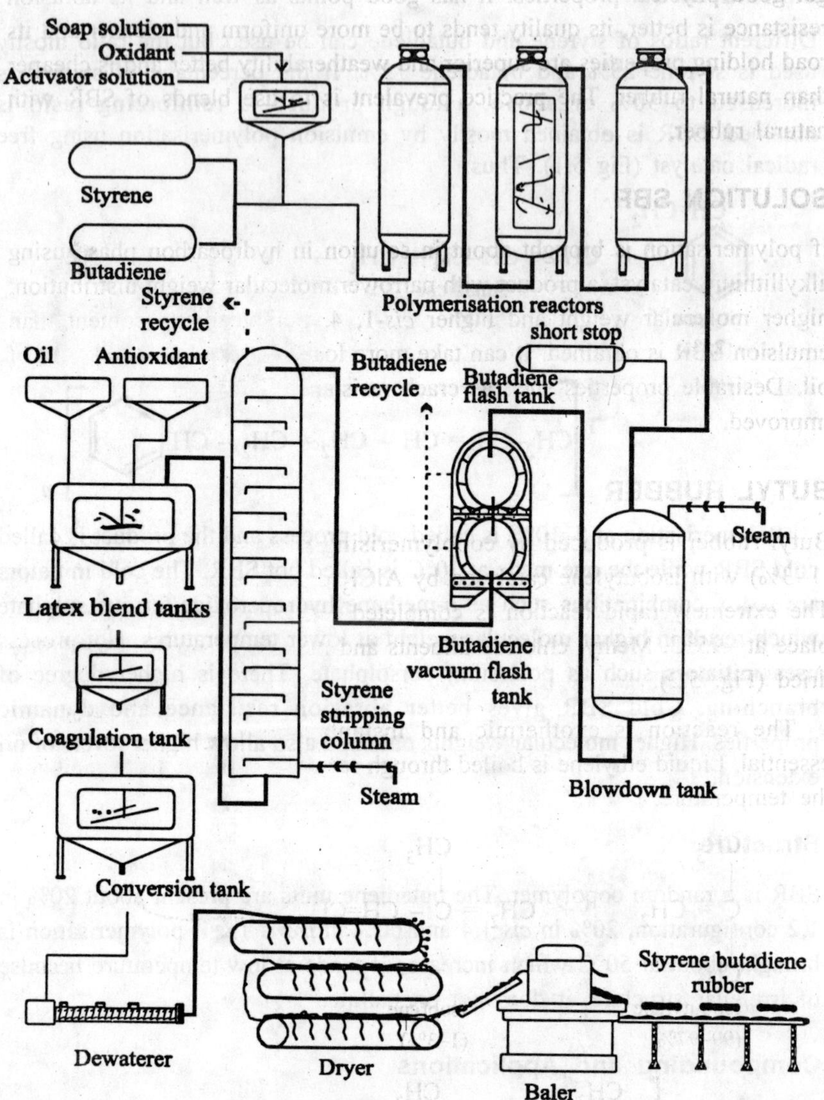

Fig. 5.1 Manufacture of emulsion styrene-butadiene rubber (SBR).

get good physical properties. It has good points as well and its abrasion resistance is better, its quality tends to be more uniform and cleaner and its road holding properties are superior and weatherability better and is cheaper than natural rubber. The practice prevalent is to use blends of SBR with natural rubber.

SOLUTION SBR

If polymerisation is brought about in solution in hydrocarbon phase using alkyllithium catalyst, a product with narrower molecular weight distribution, higher molecular weight and higher *cis*-1, 4, polybutadiene content than emulsion SBR is obtained. It can take more loading of carbon black and of oil. Desirable properties such as crack resistance and tread wear are also improved.

BUTYL RUBBER

Butyl rubber is produced by copolymerising small amounts of isoprene (1–3%) with isobutylene catalysed by $AlCl_3$ dissolved in methyl chloride. The extremely rapid reaction is completed less than a second and takes place at –95°C. Methyl chloride diluents and monomer should be carefully dried (Fig. 5.2).

The reaction is exothermic and instantaneous, so cooling is very essential. Liquid ethylene is boiled through reactor cooling coils to maintain the temperature.

$$\underset{\substack{\text{Isobutylene}\\(99\text{–}97\%)}}{CH_3-\overset{\displaystyle CH_3}{\overset{|}{C}}=CH_2} \;+\; \underset{\substack{\text{Isoprene}\\(1\text{–}3\%)}}{CH_2=\overset{\displaystyle CH_3}{\overset{|}{C}}-CH=CH_2} \xrightarrow[\text{Catalyst } AlCl_3 \text{ in methylchloride or } BF_3 \text{ gas}]{\text{Temp. }-95^\circ C}$$

$$\left[-\underset{\underset{\displaystyle CH_3}{|}}{\overset{\overset{\displaystyle CH_3}{|}}{C}} - CH_2 - CH_2 - \overset{\overset{\displaystyle CH_3}{|}}{C} = CH - CH_2 \right]_n -$$

Structure

Butyl rubber is amorphous under normal conditions but crystallises on stretching. The isoprene units are mostly present in *trans* 1,4 structure and the polymer is linear. The molecular weight is controlled at 2,00,000–3,00,000 level by addition of chain transfer agents during its manufacture.

Properties

Butyl polymers are among the most widely used synthetic elastomers in the world and rank third among synthetic rubber production. Because of very low residual unsaturation, they posses unique age-resistant properties.

The special property of the rubber is its being impermeable to air and because of its low unsaturation being very resistant to attack by oxygen, ozone chemicals and moisture.

Applications

About three-fourth of butyl rubber produced is used for inner tubes for tyres. Other uses are mainly in the field of mechanical goods. The ozone resistance of butyl rubber together with moisture resistance of its saturated structure finds use as a high quality electrical insulation.

Compounding

Butyl rubber can be vulcanised by three methods: (i) sulphur vulcanisation; (ii)cross-linking with dioxime or denitiose compounds; and (iii) polymethylol-phenol resin.

Sulphur vulcanisation

Thiouram $(R_2\text{–N–}\overset{\overset{\displaystyle S}{\|}}{C}\text{–S})_2$ or dithiocarbamate $(R_2\text{–N–}\overset{\overset{\displaystyle S}{\|}}{C}\text{–S})_x$ where N is metal, accelerators which are very potent are used along with sulphur and zinc oxide. Thiozoles such as mercaptobenzothiazole, if used reduce scorching during processing. Vulcanisation temperature range from 135–190°C.

Dioxime cure

The cure can be brought about with *p*-quinone dioxime in the presence of PbO_2 which acts as an oxidising agent.

Polymethylol-phenol resin cure

Phenolmethylol groups of phenol-formaldehyde resin bring about curing by reacting with the isoprenoid units. Thus

$$HOH_2C-\underset{}{\overset{OH}{\bigcirc}}-\left[CH_2-\underset{R}{\overset{OH}{\bigcirc}}\right]_n-CH_2-\underset{R}{\overset{OH}{\bigcirc}}-CH_2OH$$

If hydroxyl group of the methylol group is replaced by bromine, a more reactive cure is obtained.

SYNTHETIC POLY-ISOPRENE RUBBER

Manufacture

Isoprene is derived from petroleum. It is mixed with a hydrocarbon solvent such as *n*-pentane. Two types of catalysts can be used. One is based on a co-ordination catalyst of titanium tetrachloride and an aluminium alkyl such as triisobutyl aluminium. (Al–Ti) while the other anionic catalyst is butyl lithium. Reaction takes place at 50°C and moderate pressure. The process makes use of solution polymerisation.

Structure

The polymer made by co-ordination catalyst has a molecular weight distribution similar to that of natural rubber whereas one made using butyl lithium is higher in molecular weight and requires longer mastication than natural rubber.

The synthetic *cis*-1, 4-polyisoprene prepared with Ziegler type Al–Ti catalyst is almost similar to natural rubber in its crystallisation behaviour and has identical crystal structure.

Properties and applications

The properties of *cis*-1,4-polyisoprene are nearly identical with those of natural rubber and this polymer is preferred because of greater clealiness and uniformity. It is virtually colourless due to high purity and freedom from contamination and is used for medicinal purposes. It has high elongation break which leads to its use in rubber threads for golf balls. Tensile strength and tear resistance are somewhat low as compared to natural rubber, other properties viz., aging, hysterisis are similar to natural rubber.

POLYBUTADIENE RUBBER

Structure

The presence of two double bonds in butadiene can give rise to a number of configuration of the resulting polymer. However, it is *cis*-1,4-polybutadiene which is most important as a rubber.

$$\begin{array}{c} \quad HC{=}CH \qquad\qquad HC{=}CH \\ -CH_2 \diagup \qquad \diagdown CH_2{-}CH_2 \diagup \qquad \diagdown CH_2{-} \end{array}$$

***cis*-1,4-polybutadiene**

Manufacture

It is manufactured similar to polyisoprene employing solution polymerisation using pure dry monomer and hydrocarbon solvents. Organic lithium compounds or co-ordination catalysts containing metals in reduced valence states are employed. Care has to be taken to eliminate completely air, moisture and impurities which can have detrimental effect on the properties of the resultant polymer.

Properties

The high *cis*-1,4-polybutadiene exhibits excellent dynamic properties, low hysteresis and good abrasion resistance. A glass-transition temperature of 102°C is reported for this polymer which accounts for its excellent low temperature, performance, high resilience and abrasion resistance. It also processes well in blends with other polymers e.g., natural rubber and SBR. It should not however be used more than 40-50%, above which it imparts poor skid resistance.

ETHYLENE-PROPYLENE RUBBER

There are two types of ethylene-propylene rubbers, EPM and EPDM. EPM stands for simple copolymer of ethylene and propylene ['E' stands for ethylene and 'P' for propylene and 'M' for polymethylene ($-(CH_2)_x-$) type backbone]. In case of EPDM, 'D' denotes a third co-monomer, a diene which is introduced to bring about unsaturation in the molecule.

They are the fastest growing elastomer, because of its outstanding properties such as good weather resistant in ozone environment.

Structure

Ethylene and Propylene are by-products of petroleum cracking and are quite cheap,

$$H_2C{=}CH_2$$

Ethylene

$$CH_3-\overset{\overset{\displaystyle H}{|}}{C}=CH_2$$

Propylene

$$-(CH_2-CH_2-\overset{\overset{\displaystyle CH_3}{|}}{CH}-CH_2)_n$$

Ethylene propylene copolymer

Because of saturated nature of the polymer molecule (EPM), the sulphur vulcanisation cannot be employed. Peroxide cure is used instead which, however, is expensive.

Unsaturation can be introduced by copolymerising ethylene and propylene with a non-conjugated diene (third co-monomer). There are two double bonds in a diene. Only one of these polymerises and the unreacted double bond can be utilised for sulphur vulcanisation.

The three dienes most commonly used are:

1. 1,4 hexadiene

   $CH_2 = CH-CH_2-CH = CH-CH_3$

2. Dicyclopentadiene

   CH_2

3. Ethylidene Norbornene

   H H_3

   = C–C

These are used 2–5% and the resultant rubbers are known as EPDM rubbers (ethylene-propylene-diene monomer elastomers).

Manufacture

The polymerisation is brought about by solution or slurry process, the solvents being hexane, heptane or octane. The catalyst system used for polymerising consists of vanadium halide along with a halogenated aluminium alkyl.

The ethylene content in EP elastomer varies between 75–45%. The monomers are randomly distributed resulting in amorphous type copolymer. Higher ethylene containing polymers possess crystallinity and consequently higher green strength but poorer mill processing behaviour.

Applications

The outstanding properties of ethylene propylene rubber is due to its complete saturation which leads to air-aging properties and ozone resistance.

It is used in tyre sidewalls, tank linings, wire and cables, mechanical goods, agricultural equipments and automotive applications. It was considered as a special purpose rubber. However, it has found use in wide variety of applications and has replaced natural rubber, styrene-butadiene rubber and butyl rubber where resistance to ozone and weathering is most desired. Table 5.1 shows resistance to various chemical agents on EPM and EPDM elastomers.

Table 5.1. Resistance to chemical agents on EPM and EPDM elastomers.

Chemical agent	*Resistance*
Ozone	Excellent
Oxygen	Excellent
Water	Excellent
Alkalies	Excellent
Acids, dilute	Excellent
Acids, concentrated	Fair
Hydraulic fluids	Good
Animal oil	Fair
Vegetable oil	Fair
Oxygenated solvents	Good

Processing

EPDM rubber requires reinforcement to be of practical value since the mechanical properties of the unfilled rubber are poor. Carbon black is the most useful filler, although silica clay and talc may also be used. Naphthenic oils provide the best compatibility. The rubber is inherently not tacky. If tack is desired, a tackifier has to be added. Sulphur vulcanisation is possible. Common accelerators may be used. Table 5.2. shows comparison of the performance of EPDM and Hypalon rubbers.

Table 5.2. Comparison of sulphur cured EPDM and HYPALON.

Property	*EPDM*	*Hypalon*
Oxygen and ozone resistance	Excellent	Excellent
Colour stability	Excellent	Excellent
Gum strength (psi)	500	5000
Oil resistance	Poor	Fair
Heat resistance	Very good	Very good
Flame resistance	Poor	Good
Low temp. properties	Excellent	Fair-excellent

MODIFIED POLYETHYLENE OR CHLOROSULPHONATED POLYETHYLENE

Manufacture

Chlorosulphonated polyethylene (Hypalon) is unique among elastomers as it is derived from a plastic. It is prepared by dissolving polyethylene in a suitable solvent and passing simultaneously SO_2 and Cl_2 in the presence of a free radical catalyst to bring about chlorination and chlorosulphonation.

Structure

The chlorine atom breaks up the regularity of the polyethylene chain so that crystallisation is no longer possible and an elastomeric character is imparted to the polymer and sulphonyl chloride groups provide site for cross-linking. It contains nearly 1.5% sulphur and 27% chlorine.

$$\left[\left\{ (CH_2)_3 - \underset{\displaystyle Cl}{\underset{|}{(CH)}} - (CH_2)_3 \right\}_{12} - \underset{\displaystyle SO_2Cl}{\underset{|}{CH}} \right]_n$$

The elastomer can be cross-linked by reacting with lead or magnesium oxide in presence of water.

Properties

The rubber is completely saturated and so has excellent resistance to chemicals, weather and ozone (Table 5.3). It has flame resistance due to the presence of chlorine.

Applications

It is used for making white tyre sidewalls after blending with Neoprene and natural rubber and gives a beautiful white colour and a surface free from cracking and ozone attack. It finds use in hose, electrical applications and in reservoir liner construction.

Table 5.3. Chemical resistance of hypalon.

Chemical	*Rating*
Acids, mineral	A
Acids, organic	A
Alcohols	A
Aldehydes	C
40% Formaldehyde	A
Alkalies, caustic	A
Aniline	B
Aromatic solvents	C
ASTM Oil No. 1	A
ASTM Oil No. 3	B
ASTM Reference Fuel A	A
ASTM Reference Fuel C	C
Asphalt	B
Chlorinated solvents	C
Esters	C
Ethers	B–C
Gasoline	B
Hydrogen peroxide	A
Inorganic salts	A
Ketones	B–C
Triethanolamine	A
Water	A

For room temperature exposure, using vulcanisates based on Hypalon 40: (Chlorine contact 35%, sulphur 1%).

A–little or no effect

B–minor to moderate effect

C–severe effect

POLYCHLOROPRENE (NEOPRENE)

Neoprenes are among the earliest synthetic rubbers produced from chloroprene (2-chloro-1, 3-butadiene) by emulsion polymerisation using free radical catalyst.

$$CH_2 = \underset{\displaystyle |}{\overset{Cl}{C}}\!\!\!\! - CH = CH_2 \longrightarrow \left[CH_2 - \overset{Cl}{\overset{|}{C}} = CH - CH_2 \right]_n$$

Structure

The ozonolysis and X-ray examination show the molecular structure to be linear with addition at 1,4 position and *trans*, configuration almost entirely. Thus

$$\begin{array}{ccccccc} -CH_2 & & H & & CH_2{-}CH_2 & & H \\ & \searrow C = C \swarrow & & C{=}CH & & C{=}C & \\ Cl & & CH_2{-}CH_2 & & Cl & & CH_2{-} \end{array}$$

Chloroprene, the starting material is obtained as follows:

$$\underset{\text{Acetylene}}{2\ CH \equiv CH} \xrightarrow[\text{CuCl}_2]{\text{Catalyst}} \underset{\text{Vinyl acetylene}}{CH \equiv C - CH - CH_2} \xrightarrow[(\text{CuCl}_2)]{+\ HCl} \underset{\text{Chloroprene}}{CH_2 = CCl - CH = CH_2}$$

The process is expensive. It can also be made by chlorination of butadiene to 3,4-dichlorobutene-1 followed by dehydrochlorination.

Vulcanisation

Vulcanisation of neoprene is different from that of other rubbers and it can be vulcanised by heat alone. Zinc oxide and magnesium oxide are good vulcanising agents. Sulphur vulcanises very slowly and accelerators are not effective.

Properties

The rubber possesses very high tensile strength (3500–4000 psi) even without the use of carbon black. Crystallising is an inherent property of neoprene. It stands well to aging, resists aliphatic hydrocarbons and is flame-proof.

Neoprene is better than natural rubber, butyl or SBR as regards to oil resistance but inferior to nitrile rubber. The dynamic properties of this rubber are superior to those of most other synthetic rubbers though slightly less than those of natural rubber.

Applications

It is used in the manufacture of industrial hoses, coal mine belts, wire and cable coatings, and solid tyres. Because of a combination of polarity and crystallinity, neoprene occupies an important place among rubbers for use in adhesives. It finds application in aircraft, automobiles, furniture, shoes and industrial components. Neoprene latex is used in the manufacture of gloves and coated fabrics.

NITRILE RUBBERS

Manufacture

This rubber is obtained by copolymerisation of butadiene and acrylonitrile by emulsion polymerisation using free radical catalyst.

$$\underset{\substack{\text{Butadiene} \\ (60\text{–}85\%)}}{CH_2 = CH - CH = CH_2} + \underset{\substack{\text{Acrylonitrile} \\ (40\text{–}15\%)}}{CH_2 = CHCN} \longrightarrow$$

$$(- CH_2 - CH = CH - CH_2 - CH_2\ \underset{\substack{| \\ CN}}{CH} -)_n$$

Acrylonitrile can be prepared by passing acetylene and hydrogen cyanide over heated carbon.

Compounding

Basically nitrile rubbers are compounded in much the same way as natural rubber or SBR. Reinforcing fillers are necessary to obtain optimum properties. Vulcanisation can be achieved with sulphur and accelerator or peroxides when transparent compounds are obtained.

Properties

Oil resistance is the most important property of nitrile rubber. It is available in several grades of oil resistance based on the acrylonitrile content of the polymer. As the acrylonitrile content increases oil resistance, tensile strength, hardness, abrasion resistance, heat resistance and gas impermeability increases while low temperature flexibility, resilience and compression set decreases. It resists both aliphatic and aromatic hydrocarbons and is second only to sulphide rubbers in this respect.

Applications

Because of oil and fuel resistance characteristics, nitrile rubber finds uses in applications where these properties are required such as gasoline hoses for automotive, marine, aircraft, industrial air hose, carburettor parts, oil-drilling industry, adhesives, cements and brake linings.

ACRYLIC RUBBERS

Manufacture

Acrylic rubbers are obtained by emulsion polymerisation of methyl or ethyl acrylate and chloro ethyl acrylate or chloro ethyl vinyl ether.

$$\underset{\text{Ethyl acrylate (95\%)}}{-CH_2-\underset{\underset{OC_2H_5}{|}}{\underset{CO}{|}}{CH}} + \underset{\text{Chloro ethyl vinyl ether (5\%)}}{CH_2-\underset{\underset{CH_2CH_2Cl}{|}}{\underset{O}{|}}{CH}} \xrightarrow{\text{Peroxide catalyst}}$$

Chlorine is introduced to ease vulcanisation which can be carried out by:

(1) polyamine + sulphur

(2) peroxides

(3) bases (sodium metasilicate).

Properties

The elastomer has excellent combination of heat and oil resistance. It resists heat up to 200°C and is second to silicones only. It also possesses resistance to oxidation and ozone. Its resistance to sulphur bearing oils is also good. Its shortcomings are that it is attacked by acid, alkali, water and steam. Its low temperature performance is also not good.

Compounding

Acrylic rubbers need reinforcing agents for useful properties as they do not have high gum strength. They have a saturated backbone to need incorporation of copolymerised reactive curve sites.

Applications

Finds use in hose, tubing, belting and tank linings.

POLYSULPHIDE RUBBERS (THIOKOLS)

Manufacture

Polysulphide rubbers are obtained by condensation polymerisation of aliphatic dihalides and sodium polysulphides.

$$n\text{Cl(CH}_2)_2\text{Cl} + n\text{Na}_2\text{S}_x \rightarrow [(\text{CH}_2)_2\text{S}_x]_n + 2_n\text{NaCl}$$

Variation in properties is obtained by using different dihalides or their mixture and Na_2S_x of different ranks where x is known as rank.

Properties

The outstanding properties of these type of rubbers are their excellent resistance to both aliphatic and aromatic solvents (depending on sulphur content), very good low temperature properties and because of lack of unsaturation, high ozone resistance and low weathering. Some of the drawbacks are difficulty in processing, bad odour, poor compression set and moderate physical properties.

As sulphur content increases in the rubber, oil resistance increases but odour gets bad.

Compounding

Curing can be brought about with a metal oxide such as zinc oxide or lead oxide or with peroxides. Reinforcement is best obtained with various grades of carbon black.

Applications

The main use of these rubbers is in lining of fuel storage tanks, petrol and paint spray hoses, as a diaphragm of gas meters and in rollers for printing and lacquer coating industries.

Types of Thiokol (Polysulphide Rubbers)

There are various types of polysulphide rubbers marked as listed in Table 5.4.

Table 5.4. Types of polysulphide rubbers.

Thiokol A	*Thiokol FA*	*Thiokol ST*
Preparation		
$Cl(CH_2)_2Cl + Na_2S_4 \rightarrow$	$Cl(CH_2)_2Cl$ + Dichloroethane + $Na_2S_{1.8} \rightarrow$	Dichloroethane formal + small amounts of trifunctioned halide $\left[\begin{array}{l} CH_2Cl \\ CHCl \\ CH_2Cl \end{array}\right] + Na_2S_{2.25} \rightarrow$
High molecular weight linear compound, terminal OH	High molecular weight linear compound, terminal OH	Branched chain high molecular weight polysulphide
Curing		
ZnO or diisocyanate	ZnO or diisocyanate	Lead peroxide + paraquinone dioxide
Vulcanisation by terminal linkage only.	Vulcanisation by terminal linkage only	Vulcanisation by cross-linking
No cross-linking. Only entanglement	No cross-linking. Only entanglement	
Compounding		
All require filler, poor gum strength, black filler (36–60 phv) used. Non-black filler not satisfactory. No antioxidant needed. Extremely resistant to ozone.		
1 Solvent resistance best	Second for solvent resistance	Third for solvent resistance
2 Least permeable to vapour and gases	Easier processing	Good compression set
3 Difficult to handle	Smells less	Easier processing
4 Bad odour	Low temperature performance better than Thiokol A	Odour all right
		Best low temperature properties (–65 to 350°F)

Thiokol LP

It is a low molecular weight liquid polymer and is cured by lead dioxide or organic peroxide. It is used to impregnate leather, fabric and wood to make them water-proof and is also used as sealer for fuel tanks and for rocket binders.

POLYURETHANE RUBBERS (VULKOLLAN)

Polyurethane rubbers contain repeating urethane groups. Other groups such as urea, ester also may be included. Such a diversity of structure makes a very wide range of polymer properties possible.

Manufacture

There are three stages in their preparation:

(1) Production of a polyester by condensing adipic acid with a glycol until a molecular weight approximately 5000 is reached. Several other glycols can be used.

$HOOC(CH_2)_4COOH$ + Ethylene glycol 70%
Propylene glycol 30%
Mixture used to control m.p.
Excess glycol used so that terminal groups of the product are OH

$\longrightarrow$ $HOR\,[OCO(CH_2)_4COOR] - OH$
or
HO ~~ OH
Hard waxy product
m.p. 60–70°C
Mol. wt 2000–5000

(2) To prepare a prepolymer by reaction with diisocyanate. Excess isocyanate over that calculated is used to ensure that end groups are isocyanate.

$R(NCO)_2 + HO \sim OH \longrightarrow OCN \sim\sim$

2, 4 toluene diisocyanate (CH_3, NCO, NCO) or 1,5 naphthalene diisocyanate (NCO, NCO)

~ ~ OOCNHRNHCOO ~ ~ NCO urethane linkage. Product rubbery mol. wt. 70,000–80,000. Can be worked in rubber processing machinery.

(3) Further linking together the molecules by the action of chain extender a glycol, diamine or water on the terminal iscocynate groups (Table 5.5).

```
RNCO  ——→  RNH
+ HOH       |
RNCO       C=O + CO2
            |
           RNH
```

Table 5.5. Chain extenders used in urethane elastomers.

Abbreviation	*Chemical name*	*Structure*
TDI	2, 4-tolylene diisocyanate (80%) 2,6-tolylene diisocyanate (20%)	(CH_3, NCO, NCO substituted benzene); (CH_3, OCN, NCO substituted benzene)
HDI	Hexamethylene diisocyanate	OCN $(CH_2)_6$ – NCO
MDI	Diphenylmethane-p-p' diisocyanate	ONC–C_6H_4–CH_2–C_6H_4–NCO
NDI	1,5, Napthelene diisocyanate	(naphthalene with NCO, NCO)

Vulcanisation or conversion of long chains into 3 dimensional structure is brought about by the isocyanate group in one chain reacting with imino group of a urea group in another chain to give a methane bridge.

```
RNH + OCNR NH CONH  ——→  RNCONHR
 |                          |
C = O                      CO
 |                          |
RNH                     NHRNHCONH
```

Properties

Outstanding properties of this class of rubber are: (i) high abrasion resistance and tear strength even without carbon black; (ii) resistance to aliphatic solvents; and (iii) non-inflammable.

Applications

They are widely used in foams, for making thread for elastic garments and sports goods, top lift for ladies high heel soles and also engineering components of resilient type.

SILICONE RUBBERS (SILASTOMERS)

Properties

These are special purpose costly rubbers. The outstanding properties of these rubbers are:

(1) Thermal and oxidation stability upto as high as 260°C (Table 5.6).

(2) Retention of flexibility even up to as low as-100°C.

(3) Inertness to weathering, chemicals, oils and solvents.

(4) Desirable electrical properties.

(5) Non-toxicity.

(6) Absence of taste and odour.

Table 5.6. Estimated service life of silicone rubber.

90°C (194°F)	40 years
121°C (250°F)	10–20 years
150°C (300°F)	5–10 years
200°C (392°F)	2–5 years
250°C (482°F)	3 months
315°C (600°F)	2 weeks

Silicones are inert because of lack of unsaturation, absence of plasticisers and their elemented silicon constituent. These rubbers are extensively used in air craft applications and about 250 kg of these rubbers find use in a single bomber. They had spectacular growth during the past 50 years since the produce was introduced commercially. They are also used in wire and cable insulation and are valuable in surgical and prosthetic devices. These polymers may be straight-chain, or cross-linked with benzoyl peroxide or other free radical initiator, with or without catalyst. Stable over temperature range from –50 to +250°C. Very low surface tension; extreme water repellency; high lubricity; excellent dielectric properties; resistant to oxidation, weathering, and high temperature; permeable to gases. Soluble in most organic solvents; unhalogenated types are combustible. Non-toxic.

Manufacture

Their manufacture involves following steps :

(1) Preparation of chlorosilane

$$CH_3Cl + Si \xrightarrow[230°C]{Cu\ powder} CH_3SiCl_3 + (CH_3)_2SiCl_2 + (CH_3)_3SiCl$$

(2) Hydrolysis

$$R_2SiCl_2 + 2H_2O \longrightarrow OH - \underset{\substack{|\\R}}{\overset{\substack{R\\|}}{Si}} - OH + 2HCl$$

Silicol

(3) Polycondensation with loss of water

$$- \underset{\substack{|\\R}}{\overset{\substack{R\\|}}{Si}} - O - \underset{\substack{|\\R}}{\overset{\substack{R\\|}}{Si}} - O - \underset{\substack{|\\R}}{\overset{\substack{R\\|}}{Si}} - O + nH_2O$$

Molecular Weight

The polymer varies from 3,00,000–800,000 in molecular weight and has 4,000–11,000 silicone atoms per molecule.

Vulcanisation

Vulcanisation is carried out with the help of peroxides. R can be CH_3 or a mixture of CH_3 and C_6H_5 and depending on that different set of properties are obtained.

If R = CH_3 rubber obtained is best as regards heat resistance and can resist temperature upto 260°C.

If R = mixture of CH_3 + C_6H_5 – 5%, rubber obtained can resist extremely low temperature i.e., upto –100°C.

If R = – $CH_2CH_3CF_3$ (trifluoropropyl), oil resistance improved.

FLUORO CARBON RUBBERS (FLUOREL / VITON)

These are exciting new group of elastomers and serve needs for fuel and chemical resistant rubbers for service under extremely low or high temperatures.

Properties and Applications

Properties with introduction of fluorine imparts are: (i) heat and chemical resistance; (ii) flame retardance and; (iii) electrical characteristics. Some of these rubber retain their flexibility even up to 500°C and resist even fuming nitric acid. They are the costliest among all the rubbers and have best resistance of all the rubbers to heat, chemicals and solvents.

In fluorocarbons, stability of the molecules is due to primary forces between carbon and Fluorine. Fluorine forms a much stronger bond to Carbon than either hydrogen, chlorine, silicon or carbon to carbon and hence imparts the highest order of stability. Also small fluorine atoms fit nicely and uniformly around carbon chain and form a protective shield which effectively prevents chemical attack on weaker C–C linkage.

Vulcanisation

Curing or vulcanisation is carried out with the help of peroxides. Examples of a few fluoro rubbers are:

(1) Polyfluoroprene from 2 fluoro butadiene

$$(CH_2 - \underset{|}{\overset{F}{C}} = CH - CH_2)_n$$

(2) Fluorinated olefins: $(CH_2 - CF_2)_n$

(3) Fluorinated silicones which are known for best temperature flexibility i.e., from –50°C to 500°C.

These elastomers are prepared in emulsion using peroxy compounds as initiators.

Monomers must be pure. They range in molecular weight from 5,000 to 200,000. There are many types of fluorocarbon elastomers available with special properties to fit particular requirement as listed in Table 5.7.

POLYMERS RELATED TO RUBBERS

Polyethylene

Ethylene can be manufactured by the thermal and catalytic cracking of a variety of hydrocarbons such as ethane derived from natural gas to fuel oil. About half of it is used to produce ethylene polymers and copolymers. The consumption of polyethylene exceeds any other plastic material in the world. Polyethylene can be obtained using two methods :

(1) By high pressure which yields low density material.

(2) By low pressure which gives high density material.

Table 5.7. Comparison of the properties of fluorocarbon polymers with other polymers.

Property	*Polytetra fluoro ethylene*	*Silicone*	*Poly-ethylene*	*Polystyrene*	*Nylon*
Density (g/cm^3)	2.2	1.4–2.0	0.95	1.05	1.14
Tensile strength (lbs/inch^2)	2,00	200–1,100	1,850	7,000	> 10,000
Elongation (%)	50–400	75–200	550	3	80–100
Heat resistance (°C)	230–300	150–260	100	95	150
Dielectric strength (kV/cm)	200	150–225	400	400	160
Dielectric constant	2.0	3.0	2.3	2.5	4.0
Acid resistance	good	good	good	poor	poor
Alkali resistance	good	good	good	good	good
Solvent resistance	good	fair	good	poor	good

High pressure polyethylene

The high pressure technique uses pressures, between 1,000–3,000 atmosphere and temperatures up to 250°C. Free radical initiators such as benzoyl or diethyl peroxides and hydroperoxides or traces of oxygen (0.01%) can start the reaction. Ethylene must be pure and impurities viz. hydrogen and acetylene should be absent otherwise they act as chain transfer agents and it is difficult to get high-molecular weight products.

Properties

Low density (or high pressure) polyethylene is partially crystalline (50–60%) having melting point 105–115°C and density 0.91–0.94. It does not dissolve in any solvent at room temperature but is soluble in many solvents at temperatures above 100°C. The polymer has both long and short branches. It is inert chemically and has good resistance to acids and alkalies. It is used in containers for acids including hydrofluoric acid. The electrical properties of polyethylene are exceedingly good. It ~ges on exposure to light or oxygen with loss of strength and other properties.

It has weight average molecular-weight of 50,000–2,50,000 and number average molecular-weight 5,000–20,000.

Applications

The main use of low density polyethylene is in film (less than 0.01-inch thickness) and sheeting. This is because of its unique combination of properties such as flexibility (even without a plasticiser), high tear strength, moisture and chemical resistance. Nearly three-fourth of polyethylene manufactured is used in packaging, while a good portion goes for application in agriculture in canal or tank liners.

The other major use of polyethylene is in electric insulation because of its chemical inertness and light weight. About 10% of the material is used for toys, pipe, containers etc.

High-density (linear) polyethylene

There are two processes for manufacture of this material: (i) Phillips process; and (ii) Ziegler process.

(1) The Phillips process uses a chromium oxide catalyst on a silica support for polymerisation of ethylene at low pressure in paraffinic solvent. The pressure used is about 500 p.si and temperature 60–200°C. The product has density of 0.925–0.965 and molecular-weight 25,000–1,00,000.

(2) The Ziegler process uses an organometal catalyst comprising of Group IV to Group VI metal halides such as titanium tetrachloride and an aluminium alkyl such as triethylaluminium for polymerisation of ethylene in a solvent such as heptane. A temperature of 50–70°C and slight pressure are used for polymerisation.

Linear polyethylenes are over 90% crystalline having melting point above 125°C and density in the range of 0.95–0.97. There is less than one side chain per 200 carbon atoms in the main chain. The weight-average molecular weights are in the range of 100,000–200,000.

Properties

A comparison of the properties of high and low density polyethylenes is given in Table 5.8.

Crystallinity and molecular-weight determine the properties of polyethylene. High-density polyethylene of 95% crystallinity and 0.96 density exhibits higher melting point, tensile strength and hardness than a low-density polyethylene of 65% crystallinity and 0.92 density. Chemical properties such as resistance to oils, solvents and permeation by water vapour and gases are also superior in case of high density polyethylenes.

Applications

Forty per cent of linear polyethylene is used in the manufacture of bottles and other containers by blow moulding. About 25% of linear polyethylene finds use in the injection moulding of crates, tubs and housewares.

Other uses of linear polyethylene are in film and sheet, wire and cable insulation and pipe and conduct.

Table 5.8. Comparison of the properties of various polyethylenes.

Type	*Crystal-linity %*	*Melting point °c*	*Density*	*Tensile strength psi*	*Elongation at break*	*Hardness shore D*
High density	95	134–137	0.96	4,400	25	68
Low density	65	114–116	0.92	2,000	500	52

High molecular weight polyethylenes

High-molecular-weight polyethylene has weight-average molecular weight of 300,000–500,000 and has high impact and tensile strength and crack resistance. Ultra high-molecular-weight polyethylene has weight-average molecular weight of 3,000,000–6,000,000 and has exceptional abrasion and impact resistance. It finds use in bearings, gaskets, conveyor belt parts etc.

Fabrication equipment

Extruders, Banbury Mixers and Injection Moulding Machines are used widely while Roll mills and Calenders are used in some cases. Extruders are used for pelletising and for wire and cable coating, production of pipe, sheet and film and for feeding blow-moulding units in the production of bottles and other containers. Polyethylene must be extruded at a high temperature and because of high temperature requirement for a crystalline material, polyethylene extruders must be capable of high temperature operation and high-heat input.

The polyethylene extruders are mostly electrically heated. In general, their length to diameter ratios are 20/1 to 30/1 compared to 8/1 to 10/1 for machines designed for rubber and non-crystalline plastics. Because of long barrel design, there is adequate melting, mastication and proper metering of the polymer, which is particularly important for high speed extrusion of high-density polyethylene. Stock temperatures range between 150–250°C. Screw diameters of 1–12 inches are common and output varies between a few kg to several tons per hour.

Injection moulding machines for polyethylene are similar to those used throughout the plastic industry. In this operation, the polymer is melted on the barrel of the machine and forced into a mould under pressure. Most machines have hydraulically operated rams and mould clamps. Ram pressures are in the range of 10,000–30,000 psi. Injection moulding machines are used for making housewares such as waste paper baskets, garbage cans, dishes and automobile parts.

Roll mills, Banbury mixers and Calenders are similar to those used in the rubber industry. Banbury mixers are used for homogenising and for incorporation of pigments etc. Roll mills have provision of high temperature operation. Calenders are used for lamination and sheeting. In some cases special type of film or tape are derived from calendering. However, most laminating film or sheet production involves extruder-feed units.

Film and sheeting

Material less than 0.01 inch thickness is called film, while above this dimension it is known as sheeting. The polyethylene film can be produced by, tubular, chill-roll or water-quench process. In the tubular process, an extruder feeds molten material to a heated die designed to give a tubular shape. Air pressure is maintained to provide a bubble of the desired size. The melt is cooled in air. The tubing is pulled from the die and wound up on rolls. It may be used directly or may be slit to provide flat films up to 20 feet or more in width.

In the chill roll process, the extruder feeds molten polymer to a slit-tyre of die from which the melt extrudes as a thin sheet or web. The die is placed near the polished chill roll and the polymer is pulled under and over a serves of rolls onto a wind-up roll. Speeds up to 500–1000 feet per minute are common and film clarity is excellent.

Fresh polyethylene film surfaces are non-receptive to printing ink and must be treated by flame or electric discharge to create a mild oxidation on the surface. For production of sheet, the most common practice is to extrude molten web on a polished chill roll. Polyethylene sheet is used for the production of trays, boats, chairs etc.

Injection moulding

Equipment of this type are expensive and the technique is economic only for continuous runs in which thousands of identical parts are manufactured.

High density resins are slowly gaining acceptance in this area as they provide greater stiffness and so thinner walls can be produced. However, low density polyethylene is also used in great quantity for houseware and for general purpose containers. If pigments are added, coloured items are obtained. Concentrates of pigments and polymer are included along with pellets fed to the machine.

Blow moulding

Blow moulding is an important application for high density polyethylene. Bottles, cans and containers can be produced using this technique. Large tanks, box containers, large toy cars and animals can also be manufactured with its help.

A molten tubing (parison) is extruded into an open mould. By means of compressed air or steam, the plastic is then blown into the configuration of the mould. The solidified part is ejected from the mould and trimmed if needed. Blow moulding is one of the fastest growing field of the polyethylene industry.

Wire and cable

Polyethylene wire and cable coatings are applied by means of cross-head dies which are fed by extruders placed at right angles to the path of the wire. The wire is cleaned, preheated and straightened before entering the die. It is coated by molten polymer at speeds up to 2000 ft. per minute. As the wire comes out of the die; it is cooled and electrically tested for the integrity of the coating. In case of cable coating, the principle is applied but the diameter and thickness are greater and linear rates are lower.

Pipe

Recently the trend has been towards the use of medium and high density polyethylene for this product, although previously low-density polyethylene was the chief material for its production. Low density polymer has lower cost while high density polymer gives better performance and permits use of thinner walls. The best grades of pipe which can withstand high pressure are produced from high density polymer having high molecular weight.

The main outlet for polyethylene pipe is in farm and outdoor applications such as sprinkler system. Usually pipe having 1–2 inch diameter is manufactured in coils of 100 feet in length.

The polyethylene pipe can be produced by many techniques such as external-sizing tube, internal sizing, positive-pressure dies and biaxial orientation. The molten polymer is fed through a suitable die with the help of an extruder in all these techniques.

Polyvinyl Chloride

Polyethylene, polyvinyl chloride (PVC) and polystyrene are the three major plastic materials produced in the world today. PVC is unique among them because of its excellent range of properties and low cost.

Monomer

PVC is produced by the polymerisation of vinyl chloride monomer which is a gas boiling at –14°C. Vinyl chloride is manufactured by the chlorination of Ethylene to 1, 2-dichloroethane followed by its pyrolysis. The pyrolysis creates HCl as a by-product. Oxychlorination uses this HCl along with oxygen to react with ethylene to produce more dichloroethane. Following reactions are involved:

$$CH_2 = CH_2 + Cl_2 \rightarrow ClCH_2CH_2Cl$$

$$[ClCH_2CH_2Cl \rightarrow CH_2 = CHCl + HCl] \times 2$$

$$\underline{CH_2 = CH_2 + 2HCl + \tfrac{1}{2}O_2 \rightarrow ClCH_2CH_2Cl + H_2O}$$

$$2CH_2 = CH_2 + Cl_2 + \tfrac{1}{2}O_2 \rightarrow 2CH_2 = CHCl + H_2O$$

Manufacture

PVC can be manufactured by three polymerisation techniques—suspension, bulk emulsion and solution. Suspension polymerisation is used for the production of 80% of PVC. Small amounts are made by other methods.

The suspension method consists of polymerising monomer droplets in water which also acts as the heat transfer medium to remove the large heat of polymerisation.

$$nCH_2 = \underset{\displaystyle Cl}{\underset{|}{CH}} \xrightarrow[40\,-\,70°C]{\text{Free radical initiation}} \left[CH_2-\underset{\displaystyle Cl}{\underset{|}{CH}}- \right]_n$$

By vigorous agitation, vinyl chloride is dispersed into small droplets. Hydroxypropyl methyl cellulose, hydrolysed polyvinylacetate or gelatin are used as suspension stabilisers. Buffers are added to react with HCl liberated from vinyl chloride peroxide formed during the polymerisation reaction. Initiators such as peroxydicarbonate or peroxyester are used. The polymerisation takes 4–16 hours. Polymerisation temperature controls molecular-weight which can be between 20,000–120,000.

Structure

Crystallinity in PVC is quite low because it is a partially syndiotactic material with sufficient irregularity of structure.

Stability

PVC is unstable to heat and light and stabilisers are used to improve these properties. Organo-metallic compounds and metallic salts of lead, barium, tin or cadmium are some examples. Epoxy resin and fatty-acid salts are very effective stabilisers. All these materials take up HCl formed during degradation of PVC, or react with unstable sites along the polymer chain.

Plasticiation

PVC is a rigid plastic and needs a plasticiser to improve its properties. It lowers the glass-transition temperature. Tricresyl phosphate was the first important plasticiser to be used with it but has been replaced by other esters. Dibutyl phthalate, dibutyl sebacate and tributyl phosphate were also used but have been replaced. Now various low-molecular-weight polymers such as polypropylene glycol esters, dioctyl phthalate, dioctyl sebacate and adipate and trioctyl phosphate are used as plasticisers. A large proportion (nearly 30% by weight) of the plasticiser is added on hot rolls or in a Banburry.

Applications

PVC has excellent range of properties such as: (i) chemical, moisture and oxidative resistance; (ii) transparency and ease of colouring; (iii) flame retardance; (iv) strength and toughness; and (v) ease of fabrication. It has replaced many natural materials viz., wood, rubber, leather etc.

Some PVC is used as rigid resin while the rest is plasticised. In the rigid field, nearly 40% of the production is used as pipe for water supply, agricultural irrigation, drain and conduits for electrical and telephone cables.

PVC is also used in building construction for purposes such as window frames, gutters and siding, in wire and cable insulation, flooring, wall coverings, footwear, food packaging, bottles, phonograph records, toys etc.

Copolymers

Copolymerisation of small amounts of vinyl acetate with vinyl chloride results in improvement in many properties such as stability, lower softening point, higher solubility, colour clarity and flexibility. Compositions containing 5–40% of vinyl acetate are available in the market. Sound records (filter less), rain coat and watch strap are made out of copolymers of vinyl chloride and vinyl acetate.

Fabrication

Moulding

Organosols and plastisols (pastes) of PVC are applied by dip, slush, and rotational moulding. In dip moulding, an article such as a glove, wire rack, or tool handle is coated by dipping a heated form into a plastisol, gelling the required amount of compound, and then transferring this preformed article to a finishing oven set at a higher temperature to complete fusion.

In slush moulding, organosols or plastisols are transferred to open moulds, and then heated to complete the plasticisation of the PVC. The moulds are then cooled, and the articles removed. Products such as toys, shoes, vinyl boots, and traffic cones are formed by filling the mould with the proper compound, heating the mould to gel the outer layer, emptying the mould of excess compound, and fusion of the finished object.

Rotocasting

Rotocasting is a method of forming large, hollow parts or toys by rotation of the mould during the loading and fusing of the PVC plastisol. Machines with a number of arms containing moulds spin the plastisol to form the desired shape while the mould is programmed through its heating and cooling cycle. Boat bumpers, dolls, and vinyl balls are all produced by this method.

THERMOPLASTIC ELASTOMERS

The term thermoplastic elastomers is used for materials which have elastomeric properties, at room temperature but process like thermoplastics thus removing the need of vulcanisation. They can thus be processed on plastic equipments such as injection moulder, blow moulder, sheet extruder etc.

The advantages offered are significant and have resulted in a variety of applications. The commercially important thermoplastic elastomers are:

(1) Polystyrene/elastomer block copolymers such as with butadiene or isoprene.
(2) Polyurethane block copolymers.
(3) Polyester block copolymers.
(4) Polypropylene/ethylene propylene copolymers.
(5) Polyamide block copolymers.

There are two phases present in the thermoplastic elastomers. One phase is a hard polymer at room temperature but becomes fluid when heated. The second phase is a soft rubbery polymer. In the block copolymers, the two phases are formed from segments of the same chain. If X represents hard palstic and Y rubbery polymer, the structure X–Y–X is the simplest arrgangement of a three block system. Both ends of the rubbery Y segment must end in the hard X segment.

The polyurethane, polyester and polyamide block copolymers have the structure X–Y–X–Y–X. The hard segments are polyamide while soft rubbery segments are polyethers and polyesters.

The polypropylene/ethylene propylene copolymers are multi-phase systems and are cross-linked.

Manufacture

An alkyl lithium initiator is made use of for initiating the polymerisation of a monomer X. The resulting product can be used to initiate polymerisation of a second monomer Y thus giving a block copolymer X_n–Y_n. The polystyrene/elastomer block copolymers are made by this technique. Anionic polymerisation is employed to prepare block copolymers of styrene, butadiene and isoprene.

The polyester thermoplastic elastomers are prepared by condensation polymerisation of three monomers. A long-chain glycol such as polytetramethylene ether glycol is made to react with dimethyl terephthalate to give the elastomeric segments. Butanediol is made to react with dimethyl terephthalate to give hard segments. The polyurethane and polyamide block copolymers are manufactured following the same technique.

Polypropylene/ethylene propylene (EPDM) are mixed by intensive shear-mixing, resulting in grafting to produce polypropylene/ethylene propylene copolymer.

Relationship between molecular structure and properties

Molecular weight

Low molecular weight polymers are easily processed but do not give so good properties.

Ratio of hard to soft phase

The resulting product becomes stiffer with the increase in hard phase e.g., styrene content. When styrene content is about three fourth, the product is glass-like hard and possesses high impact resistance.

Application

The products based on polystyrene/elastomer block copolymers can be processed by conventional thermoplastic machines such as injection and blow moulders which work fast. The major use of polystyrene/elastomer block copolymer is in footwear. When mixed with resins and oils, tack increases and so the products find use in adhesives, coatings and sealants.

Polyester, polyurethane and polyamide block copolymers can replace conventional vulcanised rubber. They can be extruded or moulded just like thermoplastics at temperatures 175–225°C. Scrap can be used again. They are flexible, tough and solvent resistant. Some of them can also be used as hot-melt adhesives. Their hardness varies between 60 Shore A to 70 Shore D.

Polypropylene/ethylene-propylene copolymer can also replace conventional vulcanised rubber. They find use in wire insulation. Their hardness varies between 60 Shore A to 50 Shore D.

COMPOUNDING

The term compounding in rubber technology is used for selection and incorporation of various additives into rubbers so as to give a homogeneous mix for processing. Every recipe contains a number of components, each having a particular function to perform either during processing, reinforcement or vulcanisation and ultimately in the end use of the product.

Ingredients

The materials used in compounding can be categorised under the following heads.

Elastomers

These form the backbone and are the basic components. They can be natural or any synthetic rubber or thermoplastic elastomer and are selected to give specific physical properties to the final product e.g., oil, heat or aging resistance.

Fillers

These are used for reinforcing or reducing the cost.

Processing acids

As the name suggests, these materials help during processing which may include mixing, extrusion, calendering or moulding.

Vulcanising agents

These substance bring about cross-linking resulting in a marked improvement in the physical properties of the rubber products.

Accelerators

These ingredients reduce the time of vulcanisation by accelerating the rate of vulcanisation.

Activators

Their addition is essential for deriving the maximum benefit from vulcanisation. They form complexes with accelerates.

Anti-oxidants

They slow down the process of deterioration in service of the rubber compounds by checking oxidation,

Anti-ozonants

These ingredients slow down the deterioration due to attack by ozone.

Miscellaneous additives

These substances are not normally incorporated in the rubber compounds but may be used for specific purposes. Examples are blowing agents, colours, odorants, retarders etc.

Classification of Elastomers or Rubbers

Rubbers are classified into following three major classes:

General purpose

They are for general purpose where specific resistance to the action of petroleum or heat is not required. Examples are natural, SBR, polyisoprene polybutadiene, EPDM, butyl rubbers.

Solvent resistance type

They are used where specific resistance to the action of petroleum solvents is required. Examples are polysulphide, nitrile, neoprene, polyurethene rubbers.

Heat resistance type

They are used where effect of exposure to high temperatures can be tolerated. Examples are silicone and fluoro rubbers.

Table 5.9 lists some of the physical and chemical properties of important elastomers. The properties depend to a large extent on compounding reinforcement and vulcanisation and the values assigned are for general guidance only.

Table 5.9. Properties of elastomers.

Property	*Natural rubber*	*SBR*	*Butyl*	*Chloro-sulphonated polyethylene*	*EPDM*	*Fluorinated rubber*
Tensile strength (psi)	4000	3500	3000	2800	3000	2400
Elongation (%)	700	700	700	500	300	400
Modulus (psi at 300–400% elongation)	2500	2500	1000	–	–	250
Tear resistance	Excel.	Good	Excel.	Good	Good	Fair
Abrasion resistance	Fair	Good	Good	–	–	–
Adhesion	Excel.	Excel.	Good	Excel.	Excel.	Good
Electrical properties	Excel.	Fair	Excel.	Fair	Excel.	Excel.
Gas permeability	High	High	Low	–	–	–
Upper use temp. (°C)	80	110	100	120	150	230
Lower use temp. (°C)	–50	–50	–50	–50	–40	–40
Weather resistance	Fair	Fair	Excel.	Excel.	Excel.	Excel.
Ozone resistance	Poor	Fair	Excel.	Excel.	Excel.	Excel.
Oil resistance	Poor	Poor	Poor	Good	Poor	Excel.
Water swelling	Excel.	Good	Excel.	Excel.	Excel.	Excel.

Vulcanising Agents

Sulphur is the most common vulcanising agent. General purpose rubbers which form the bulk of rubbers used contain unsaturation and vulcanisation with sulphur is possible. Sulphur brings about cross-linking and converts the rubber to form three-dimensional structure. Thus a weak material is transformed into a strong elastic product having many desirable properties. Usually 0.5–5.0 phr. of sulphur is used.

The saturated rubbers cannot be cross-linked with sulphur and an organic peroxide is required for vulcanisation. Free radicals are generated when a peroxide decomposes and these make polymer chains to combine. Usually 2.5 phr of a peroxide is used. Dicumnyl peroxide is most prominent.

Accelerators

The vulcanisation with sulphur is slow and can be speeded up many times by the addition of small amounts of organic or inorganic compounds known as accelerators (–0.5–2 phr). Most common accelerators are listed in Table 5.10.

Table 5.10. Classification of accelerators.

Types	*Examples*	*Properties*
Amines	Hexamethylene tetramine	Delayed action
Aldehyde-amine	Butra aldehyde-aniline	In self curing adhesives
Guanidines	Diphenyl guanidine	Secondary accelerator
Thiazoles	2-mercapto benzo-thiazole-MBT	Medium speed, scorchy, most commonly used
	Benzothiazole disulphide-MBTS	safe processing, general purpose, less scorchy, expensive
Thiourams	Butramethylthioram disulphide-TMTD	Safe, fast curing give good aging & heat resistance
Sulphonamides	N-cyclohexy-2 benzothiazole sulphonamide-CBS	Safe processing, delayed action, used for heavily loaded compounds e.g., tyres
Dithiocarbonates	Zinc dimethyl dithiocarbonate-ZDC	Fast, used in latex
Xanthates	Dibutyl xanthogen disulphide	General purpose, low temp. use 70°C, used in latex

Activators

These compounds activate the accelerators to work more effectively and increase the cure rate. Most common example is zinc oxide which is used (3–5 phr) along with a fatty acid such as stearic acid (1–3 phr).

Antioxidants

Rubbers are affected by aging process when in service resulting in lowering in physical properties. Anti-oxidants are added to slow down and improve the ageing behaviour and to enhance the service life. There are three types of compounds which can be used for protection.

1. Secondary amines $\begin{matrix} R \\ R \end{matrix}\!>\!NH$ – Phenyl β naphthamine

2. Phenolics R$(OH)_2$ Phenol aldehyde condensate, styrenated phenol
3. Phosphites $(RO)_3P$

Amines tend to discolour (staining) but are powerful antioxidants. Phenols are less powerful and less staining and are used in light coloured products. Phosphites are less effective in natural rubber and more in SBR. Used 2–3 phr.

Antiozonates

These compounds are effective to check the action of ozone. They are mostly derivatives of paraphenyldiamine. The structural formula of anti-ozonates is $RNH.NHR_2$, where R can cyclohexyl compound and R_2 can be phenyl.

RNH NHR$_2$
|
R can be cyclohexyl
|
R_2 can be phenyl

Waxes are also used for this purpose.

Processing Aids

These compounds are used usually 2–10 phr and help proper mixing. The processing aid must be completely compatible with the rubber and other compounding ingredients. Incompatibility will result in improper processing or bleeding in the final product.

The aids can be petroleum or coal-tar products such as mineral oils, coal-tar pitch, cumar resins, vegetable oils, pine products such as resin, pine tar or resins such as coumarone-indene, phenol-formaldehyde, shellac or waxes, fatty acids, esters etc.

Miscellaneous Ingredients

These substances are used for some specific effect in a vulcanisate.

Blowing agents

For manufacture of micro-process rubber or blown sponge, these chemicals are added to generate gases during vulcanisation. Examples are carbonates and azo compounds.

Flame retardants

To reduce flammability, chlorinated hydrocarbons, phosphates and antimony compounds may be added.

Colourants

For increasing customer appeal, these ingredients are added. They should be stable and fast in colour.

Odorants

For masking odour of rubber compounds, these compounds (essential oils) are added generally in wearing apparel and drug sundries.

Retarders

These chemicals reduce the effect of accelerators and prevent scorching during processing and storage. These are generally organic acids e.g., salicyclic acid.

Mixing Methods

For incorporation of compounding ingredients into rubber, either a mill or a Banbury (internal) mixer is used. These are available in many sizes e.g., mills having size from 6" dia. × 13" length provided with motor of 7.5 H.P. capable of taking 1 kg material to 28" dia. × 84" length provided with motor of 250 H.P. capable of taking 150 kg and Banbury capable of taking batch of 500 g (H.P. of motor 7.5) to 500 kgs. (H.P. of motor 1500).

The rolls of a mill rotate in opposite directions and can be heated or cooled. They are made of chilled cast iron which is very hard. The front roll rotates at a slower speed compared to the back roll, while it is kept at a cooler temperature. For safety, a safety fuse called breaker plate or breaker nut is provided which is a fragile member between bearing block and screw. The two roll mills have the advantages that there is ease of processing, temperature does not rise to cause scorch and premature cross-linking while the shortcomings are that it is slow in operation. Degree of dispersion depends on manual efforts and high temperature operation is difficult.

The internal mixer has an enclosed chamber in which two rotors turn in opposite directions. There is greater efficiency in carrying out shearing action, mixing is automatic, there is reduction in mixing time, dispersion is better and it can produce highly compounded stock. Cooling water or steam is circulated through the rotors. The shortcomings are that cleaning is difficult and sliding door has tendency to jam.

REINFORCEMENT

Reinforcement implies incorporation into rubber of substances which give to the vulcanisates high abrasion resistance, tensile strength, tear resistance and enhancement in stiffness and hardness. These properties can be upgraded by the addition of certain fillers into the rubbers before vulcanisation.

Classes of Fillers

Fillers can be divided into two classes viz., inert fillers such as clay and whiting which have little effect on the physical properties and reinforceing fillers which improve these properties substantially.

Black filler

Carbon black

Carbon black is the most outstanding reinforcing filler for both natural and synthetic rubbers and is inexpensive. Table 5.11 shows the effect of incorporation of carbon black on the properties of natural rubber.

The first carbon blacks used in rubber were made using gas as the source of carbon. The channel black process produced small particle size blacks which were acidic and worked all right with natural rubber. However, the channel and early gas furnace processes were not efficient and polluted the atmosphere.

Table 5.11. Properties of raw, vulcanised and reinforced + vulcanised natural rubber.

Property	*Raw rubber*	*Vulcanised rubber*	*Reinforced + vulcanised rubber*
Tensile strength (psi)	300	3000	4500
Elongation at break (%)	1200	800	600
Modulus (psi at 400% elongation)	–	400	2500
Permanent set	Large	Small	Small
Water absorption	Large	Small	Small
Solvent resistance (Hydrocarbon)	Soluble	Swells only	Swells only

The oil furnace process produced blacks which were alkaline and higher in structure. This process replaced the channel and gas furnace processes and most of blacks used today are from this technique. They range from

200 to 900 Å in particle diameter and use low grade petroleum feed stock for source of carbon.

For obtaining maximum benefit of carbon black reinforcement, it should be thoroughly dispersed into the polymer. Small particle size (high surface area) carbon blacks are difficult to incorporate. High surface area carbon blacks give more reinforcement in a rubber compound. The particle size affects the abrasion resistance, resilience, tensile strength and tear resistance of the rubber compounds. Carbon black structure has more effect on modulus, hardness and extrusion die swell.

Electron microscopy shows that carbon blacks consist of irregular aggregates of spherical sub-units fused together in chain-like structures similar to that of graphite. As such unsatisfied bonds are present which are responsible for reinforcement. Some workers believe milling causes rupture of rubber molecules into free radicals, which react with carbon black to form a cross-linked network. The surfaces of the particles contain adsorbed material including hydrogenated and oxygenated structures such as quinone, hydroquinone, lactone and aromatic carboxyl. The pH of a water slurry of the black may be between 2.5–11.0.

The three parameters viz. particle size, pH and extent of agglomeration or chain structure play important role in determining the reinforcing behaviour of the blacks. Tensile strength, tear resistance, abrasion, hardness and toughness increase with decreasing particle size whereas resilience and ease of processing become inferior.

SBR gum vulcanisates have very poor physical properties. Tensile strength of SBR can be increased nearly ten fold by carbon black reinfrocement making it nearly equivalent to that of natural rubber. Similar is the case with tear resistance. Natural rubber has very good physical properties of its gum vulcanisates, but even then tensile strength can be increased nearly 50% by carbon black reinforcement.

As regards abrasion resistance, which is very important for heavy duty rubber such as tyre tread, it can also be upgraded 5 times by proper reinforcement. However, resilience of rubber decreases with increasing loading of carbon black resulting in heat build up and hysteresis loss. Thus a proper balance has to be struck. Increasing amount of reinforcing filler causes continuing improvement in properties upto certain limit till the optimum is reached beyond which the properties of the rubber compounds deteriorate.

Carbon blacks are generally electrically conductive. If structure of surface area increases, electrical conductivity also increases. A comparison of three types of carbon blacks is given in Table 5.12.

Carbon black increases tensile strength of natural rubber, SBR nitrile rubber and even the polysulphide rubber but has no effect in enhancing the tensile strength of neoprene or butyl rubber. This may be due to two types of opposite effects it causes: (i) it introduces many weak fix points resulting in strengthening and stiffening the structure; (ii) it may interfere with the ability of the polymer to crystallise resulting in lowering in tensile strength in case of neoprene and butyl rubber effect it appears to predominate.

Table 5.12. Comparison of various carbon blacks.

Channel		*Furnace*		*Thermal*	
pH	Acidic 4–4.5	Alkaline	8–10	Neutral	7–8
Effection curve	Retard cure	Scorchy		No tendency to scorch	
Particle size	Conductive 200Å	SAF	200Å	Max. physical FT	1200–2000Å
	HPC Hard processing 225Å	ISAF	250Å	Properties MT	3000–5000Å
	MPC medium processing 285Å	HAF	300Å	Most widely used	
	EPC easy processing 300Å	FEF	500Å		
		SRF	900Å		

(Å Angstrom unit = 10^{-10} m)

Non-black fillers

The non-black fillers used in rubber industry are clay, calcium carbonate and silica. Clays are used to the largest extent followed by calcium carbonates and silicas.

Non-black fillers include non-enforcing to highly reinforcing. The particle size plays the most important role in deciding the nature of the filler. While silicas (highly reinforcing) may have particle size of 0.01 micron, calcium carbonates (inert) have particle size of 10 microns i.e. 1000 fold.

Clays

Clays account for largest volume of non-black filler used in rubber industry. They are second to carbon black in this respect. The reasons for their wide use are processing benefits in extrusion and calendering, medium reinforcement and low cost. There are many types of clays but kaolins are used mostly in rubbers. The clays after mining are dried, pulverised and refined by air or water separation process to remove impurities and to produce clay of desired particle size. They are termed soft or hard clays. Particles of less than 2 micron constitute hard clay. Clays can be treated to bring about improvements in properties. Calcining removes bound water by high temperature treatment. The product so obtained has very good dielectric properties.

Clay reduces the cure rate which requires the addition of an activator such as triethanolamine or diethylene glycol.

Calcium carbonates

Calcium carbonates are derived from limestone which occur widely. However, only the white and bright variety is used as fillers. There are various grades available. Ultrafine grades have average particle size 0.03–0.1 micron while wet ground have average particle size 3 micron and dry ground 7 micron. Surface-treated grades containing 1–2% calcium stearate are used for easier incorporation and better dispersion.

Calcium carbonate is mostly a non-reinforcing type of filler. The one having finer particle size behave better in this respect. It gives high tear strength but low modulus.

Silicas

Silicas are noted for imparting a combination of tear strength, abrasion, resistance, adhesion and age resistance. Precipitated silicas offer the highest reinforcement out of all non-black fillers, smaller particle size varieties being the best in this respect. They may vary in pH (measured from aqueous slurry) from slightly acidic to slightly alkaline.

By the addition of silicas to rubber, viscosity rises more rapidly than most fillers which is true especially for silicas with high surface area. This results in reduction in optimum cure time and state of cure. Diethylene glycol or triethanolamine at 2 phr can reduce Mooney viscosity by about 30%.

Compared to carbon black of similar particle size, silicas and silicates impart low modulus to rubbers although tensile strengths obtained are similar. With decrease in particle size, tensile strength and abrasion resistance increase. Silicas impart high tear strength to the rubbers.

Reinforcement can be increased by the use of coupling agents such as mercapto silane (2%) which provide adhesion between silica and rubber resulting in increase in reinforcement.

Silicas can be used with advantage for reinforcing silicone rubbers. Fumed silicas give highest strength and electrical properties. Precipitated silicas provide moderately good tensile strength, tear strength and electrical properties.

Organic reinforcement

If fillers above 80 phr are used, processing is difficult, there is tendency for scorch and heat build up. Rubbers can take up much higher loadings of organic than of inorganic fillers. There are various types of organic fillers which can be used :

(1) *High styrene resin:* It is used in shoe compounds. Styrene content may vary from 50–85%. The resulting compound has high toughness.

(2) *Phenolic resins:* They can be incorporated up to equal parts of rubber and the mixture can be further reinforced with carbon black. Their incorporation results in ease of processing of unvulcanised stock and the vulcanised stocks give increase in tensile strength and hardness. Being phenolic, they have antioxidant properties. Due to presence of polar groups, adhesion to metals and textiles improves. Also solvent resistance goes up. Find use in high quality battery, boxes, V belting, roller compounds and hoses.

(3) *Cyclised rubber:* It behaves like high styrene resin.

(4) *Lignin:* It behaves like phenolic resins.

(5) *Graft polymers:* Monomers such as methylmethacrylate and styrene can be grafted to natural rubber. The resulting tensile strength is same as with black loading but resilience is higher.

VULCANISATION

The chemistry of vulcanisation is very complex and has not been well understood as yet. The process transforms an elastomer with weak thermoplastic properties into a strong, elastic and tough substance with useful properties and introduces a network of cross-links into the elastomer.

If rubber is heated with small amount of sulphur, then, subsequently, it has been found neither sulphur nor heat is essential for vulcanisation. There is one group of rubbers the thermoplastic elastomers (TPE), which do not require vulcanisation. Sulphur is used 0.5–5 phr for bringing about vulcanisation. If higher amounts (30–40 phr) of sulphur are used ebonite is obtained which is very hard. It is thus clear cross-linking should not be too much.

Tensile strength is maximum at optimum cure. Beyond this point, the stock is said to be overcured. It gets stiffer and harder but weaker and less extensible.

Sulphur Vulcanisation

Vulcanisation takes place in the presence of sulphur by heating but the process is quite slow. When small amounts of substances called accelerators are used, the process can be made faster many times. Also the presence of activators is essential to get full benefit. Thus the sulphur vulcanisation system should consist of :

(1) sulphur (0.5–5 phr) rhombic form normally used.

(2) an organic accelerator (0.5–2 phr).

(3) a metallic oxide such as zinc oxide (3–5 phr).

(4) a fatty acid such as stearic acid (1–3 phr).

It is believed during vulcanisation rubber-soluble metallic soap is formed by the reaction of metallic oxide and fatty acid. Sulphur vulcanisation is used with rubbers which are general-purpose type and contain unsaturation. With sulphur cross-links and cyclic structures are created as shown in Fig. 5.2.

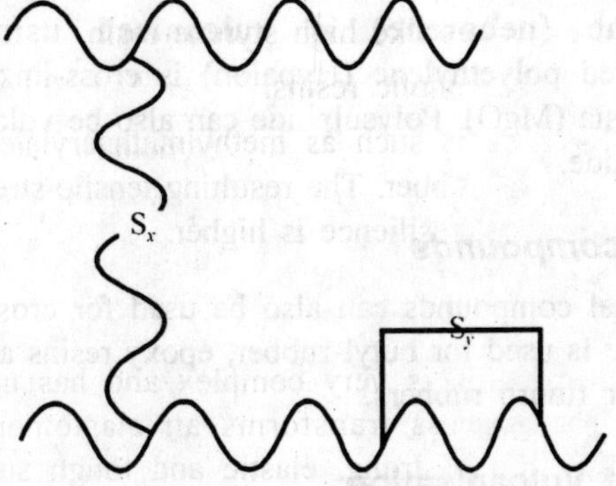

Fig. 5.2. Sulphur cross-links cyclic structure.

In efficient accelerated curing system, x is 1–2 with no cyclic groups formed while in inefficient systems, x can go up to 8 and many cyclic structures are formed. For most rubbers, one cross-link for 200 monomer units in the chain is sufficient to produce a suitable vulcanised product (molecular weight between cross-links equals ca 8000 to 12,000). Selenium or tellurium can be used in place of sulphur where excellent heat resistance is needed.

Non-sulphur Vulcanisation Peroxides

The saturated rubbers such as silicone or fluoro cannot be cross-linked by sulphur and accelerators. Organic peroxides are used for their vulcanisation. The peroxides decompose forming free radicals on the polymer chains. These chains may then combine to form cross-links (Fig. 5.3).

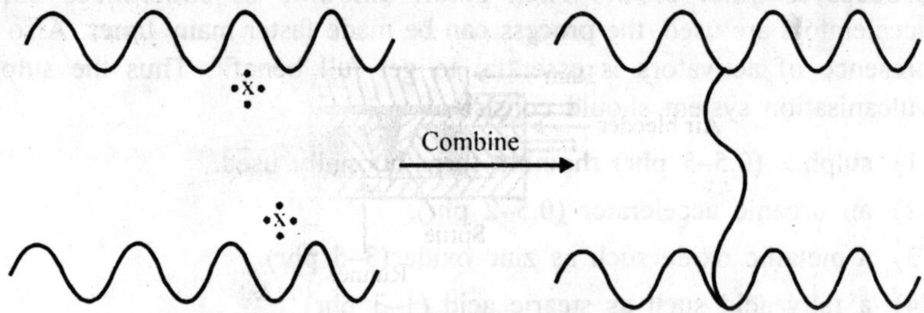

Fig. 5.3. Peroxides decompose forming free radicals on the polymer chains.

Carbon to carbon bonds are thus created which are very stable. Peroxides are used generally 2–5 phr. Dicumyeperoxide is most common.

Metal compounds

Polychloroprene (neoprene) is vulcanised using zinc oxide. Chlorosulphonated polyethylene (Hypalon) is cross-linked using litharge (PbO) or magnesia (MgO). Polysulphide can also be vulcanised using zinc oxide or lead oxide.

Difunctional compounds

Some difunctional compounds can also be used for cross-linking rubbers. Quinone dioxime is used for butyl rubber, epoxy resins are used for nitrile and diamines for fluoro rubbers.

Techniques of Vulcanisation

Vulcanisation can be carried out using several techniques (Fig. 5.4).

(1) *Compression moulding:* This is the most common method used in the industry. Here the blank is placed in a two piece mould, one part of

which is movable and the other stationary. The mould is closed and heat and pressure are applied so that the material flows to fill the mould. A little excess of the material is used to ensure that the mould is completely filled.

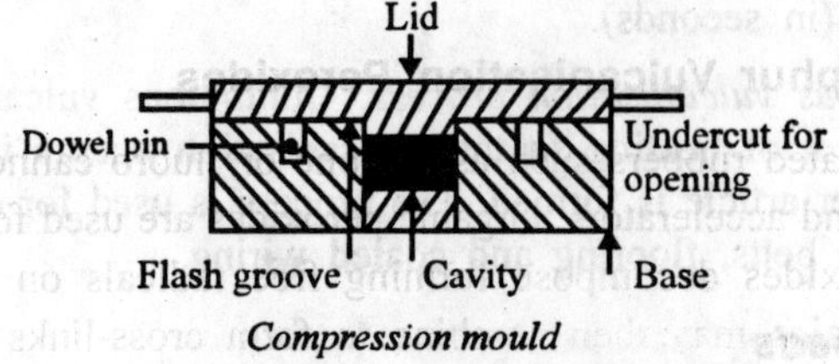

Compression mould

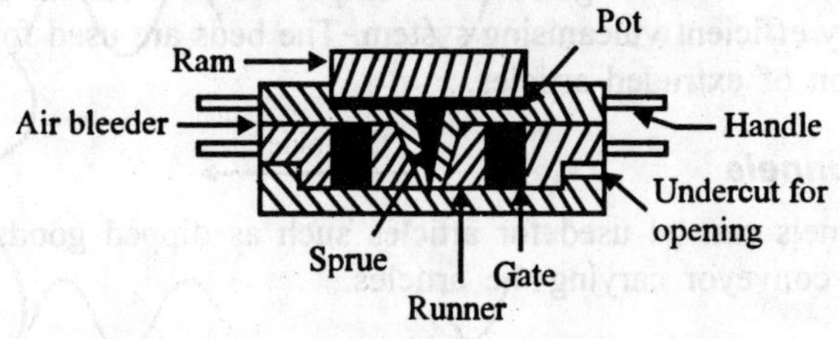

Transfer mould

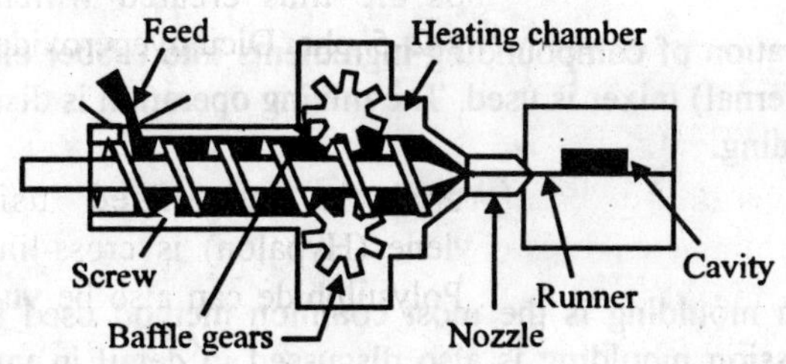

Injection moulding

Fig. 5.4. Compression, transfer and injection moulding techniques.

(2) *Transfer moulding:* This technique is a variation of compression moulding. The blank is placed in a separate chamber called the pot, preheated to a temperature below cure temperature and transferred into the heated mould where cure takes place. Although in this technique, moulds are more expensive, the process takes shorter cure time due to use of higher temperature and there is better heat transfer.

(3) *Injection moulding:* Injection moulding is used mostly for plastics. However, rubber articles can also be moulded using this technique. The polymer is preheated in a cylindrical chamber to a temperature at which

it will flow and is then forced into a comparatively cold, closed mould cavity by means of high pressure.

A great advantage of the process is the speed with which the articles can be produced (in seconds).

(4) *Continuous vulcanisation process:* Continuous vulcanisation involves some form of heating so that vulcanisation occurs immediately after the rubber article is formed. The process is used for extruded articles, conveyor belts, flooring and coated wiring.

Fluidised beds

Fluidised beds consist of glass beads suspended in a stream of heated air and are very efficient vulcanising system. The beds are used for continuous vulcanisation of extruded articles.

Hot-air tunnels

Hot-air tunnels can be used for articles such as dipped goods by passing an endless conveyor carying the articles.

PROCESSING METHODS

Mixing

For incorporation of compounding ingredients into rubber either a mill or a Banbury (internal) mixer is used. The mixing operation is discussed in detail in compounding.

Moulding

Compression moulding is the most common method used in the industry. The compression moulding is also discussed in detail in vulcanisation.

Extrusion

In the extrusion process, polymer or rubber is propelled continuously along a screw through regions of high temperature and pressure where it is melted and compacted and finally forced through a die shaped to give the final product. Several shapes can be given by extrusion such as rods, tubing, hose, sheeting and film.

The screw of an extruder can be divided into three zones. The feed zone picks up the polymer from a hopper and propels it into the main part of the extruder. In the compression zone, the loose feed is compressed when air escapes and material melts homogeneously. Some external heat is applied but much heat is generated by friction. The metering zone gives uniform flow rate. It serves as metering pump from where molten mass goes to the die at constant volume and pressure.

Modern trend in extruder design include twin-screw or mutipple screw in which screws turn side by side in opposite directions (Fig. 5.5).

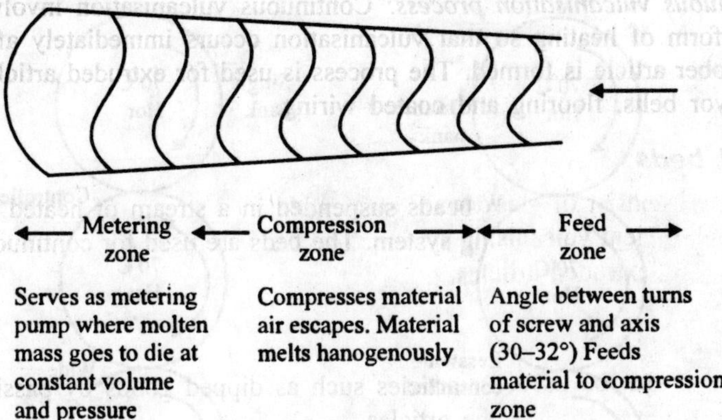

Fig. 5.5. Modern trend in extruder design include twin-screw or multiple screw.

Spider is the device that supports pin in extrusion. Breaker plate is a perforated steel plate. Its function is to deliver polymer at even pressure and correct temperature to the die. Supports fine screen (30–40 mesh) on the side where screw ends. The screen filters out unmelted solid particles. Bullet shape given to upstream of spider so that volume of polymer diminishes gradually and progressively. For good extrusion:

(1) Particles having flat plate like structure give good surface and easier flow e.g., china clay and mica.

(2) Rubber content should be between 10–70%. Above 70% rough surface obtained.

(3) S.P. (Superior rubber) rubber or high filler loading or high styrene resin gives better performance and speeds extrusion.

(4) Factice increases rate of extrusion.

(5) Mixing in internal mixer is recommended. Tough stock is good.

(6) There should be rest period between mixing and extrusion.

(7) Water should be absent in the stock or porosity develops.

(8) Die should not be too close to spider otherwise spider marks will come.

(9) There should be no area where rubber can stay in head. It will vulcanise and cause trouble.

(10) The material should be supplied at constant temperature and rate.

Calendering

Calendering is a process used for: (i) the manufacture of smooth sheet of uniform controlled thickness; (ii) to apply rubber to textiles or; (iii) embossing. Rubber in the process is passed between pairs of highly polished heated rolls under pressure (Fig. 5.6(a–d)).

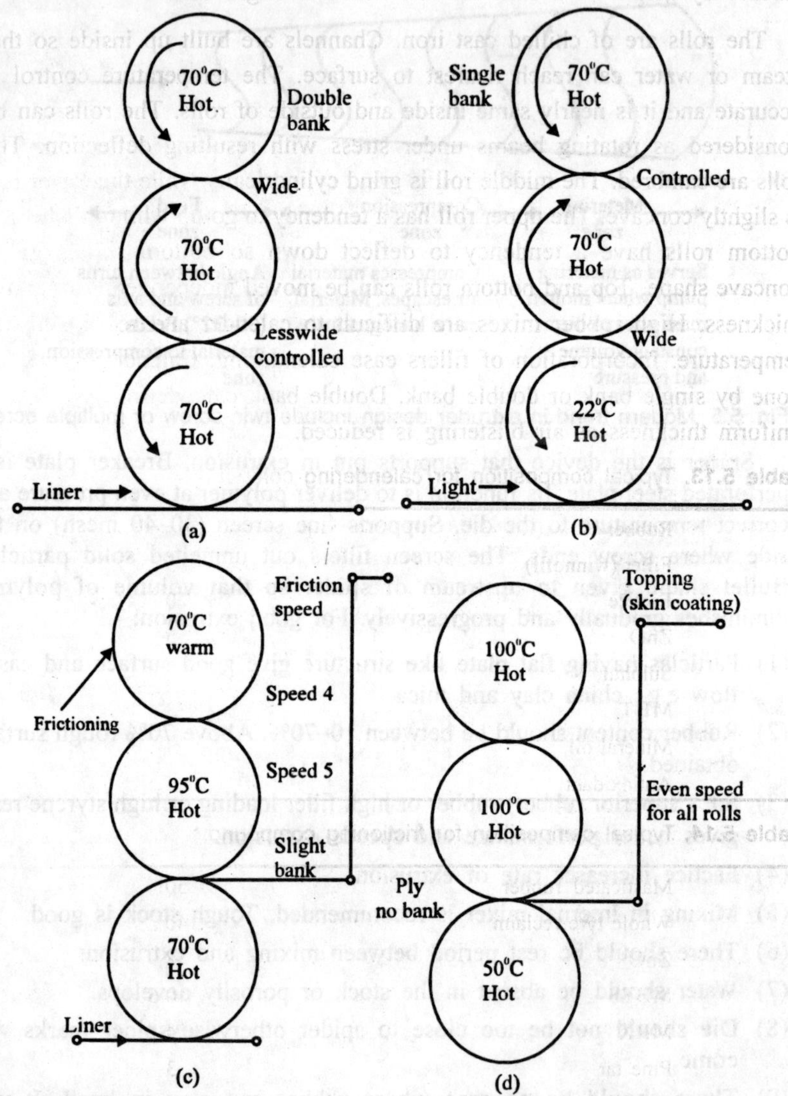

Figs. 5.6(a–d). Calendering process used for manufacture of smooth sheet of uniform controlled thickness.

Proper calendering requires precise control of roll temperature, pressure and speed of rotation. By maintaining a slight speed differentiation between a roll pair, it is possible to impart an exceedingly high gloss to the film or sheet surface. It is also possible to produce an embossed design on the surface by means of a calender roll which is engraved.

The rolls are of chilled cast iron. Channels are built up inside so that steam or water can reach nearest to surface. The temperature control is accurate and it is nearly same inside and outside of rolls. The rolls can be considered as rotating beams under stress with resulting deflection. The rolls are cambred. The middle roll is grind cylindrically while the lower one is slightly concave. The upper roll has a tendency to go up while middle and bottom rolls have a tendency to deflect down so bottom roll is given concave shape. Top and bottom rolls can be moved independently to adjust thickness. High rubber mixes are difficult to calender and require higher temperature. Incorporation of fillers ease calendering. Calendering can be done by single bank or double bank. Double bank calendering gives more uniform thickness as air blistering is reduced.

Table 5.13. Typical composition for calendering compound.

Rubber	100
Filler (winnofil)	85
Factice	20
ZnO	5
Sulphur	2
MBT	0.8
Mineral oil	3
Antioxidant	1.5

Table 5.14. Typical composition for frictioning compound.

Masticated rubber	50
Whole tyre reclaim	40
ZnO	8
MBTS	0.6
MRX	4
Pine tar	3
Sulphur	2.5

Table 5.15. Typical composition for topping compound.

Masticated rubber	75
Whole tyre reclaim	50
ZnO	3
CBS	0.8
GPF black	45
Mineral oil	5
Sulphur	2.5
PBox	1.5

TESTING

Rubbers are complex materials and the primary consideration before designing any article or component based on them, is to test the properties of vulcanised rubber.

To properly consider which properties should be measured and how the test should be carried out, it is necessary to first clearly identify why we are testing, because the requirements for each of the reasons will be different. The need of the testing is to prove performance—that is material has to pass through various specifications in a process, where it is mutually agreed for the particular properties of material. For example, oil seal, it has to be oil resistant, dimension stability, of all these are effected to the product.

Need/Purpose of Testing

Thus, while designing an article or a component for any engineering use, the first point to be considered is 'the material to be used' for its fabrication. For example, to design a metal part to be used in a machine, one would first think of the metal to be used, such as mild steel, steel alloy, brass, copper, stainless steel, etc. Such metallic materials are fairly standard materials and have definite and well-known physical properties. Therefore, to make the right choice of metal suitable for the specific end use is not all difficult for the design engineer.

However, if the article or the component is based either on rubber or plastic, the specific requirements of the component are very difficult to specify because one can design a rubber or a plastic product having desired properties with continuous variations over a very wide range in the composition of the product, durability and the cost of the product.

Plasticity Test

Various operations such as mixing, extrusion, calendering and moulding require different levels of shear. Tests to determine the processing characteristics of a compound should expose the test sample to shear level similar to one encountered in the actual operation. However, it is not possible to do so.

The shearing disc viscometer developed by Melvin Mooney has become very popular in the rubber industry. In this instrument, a flat disc rotates in a mass of rubber contained in a grooved cavity under pressure as shown in the Fig. 5.7. The torque required to rotate the disc at 2 rpm at a temperature of 120°C is termed Mooney Viscosity.

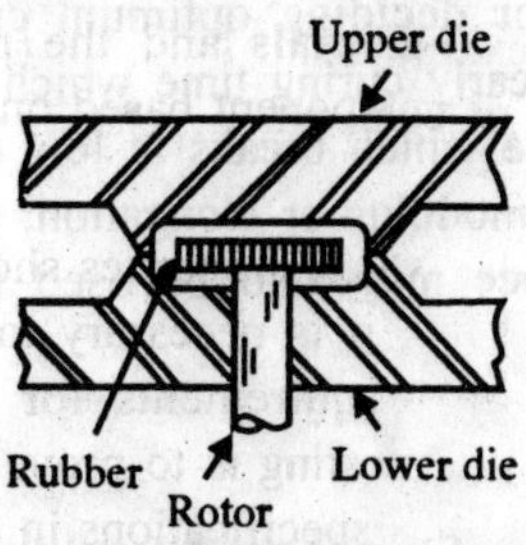

Fig. 5.7. Mooney chamber and rotor.

Two discs of diameter 1.75" are cut from rubber sheet not less than 0.4" thick. A hole 0.5" in diameter is made in centre of one of the discs which is kept below the rotor. The platens are electrically heated and the temperature is controlled at the desired level. The top platen is pneumatically lowered. The sample is allowed to warm up for one minute after the platens are closed. The motor is then started and the torque required to turn the rotor at 2 rpm is recorded after the prescribed time.

There are two rotors for the Mooney viscosity test a large one (1–5" dia) for general use and a smaller one (1.2" dia) for tough rubbers. Reading taken after 4 minutes is called Mooney Plasticity. The result is expressed as follows:

$$40 - \text{ML } 1 + 4 \ (120°\text{C})$$

where 40-M is the Mooney Plasticity number, L indicates that large rotor has been used, 1 is the time in minute for which the specimen has been preheated, 4 is the time in minutes after starting the motor at which the reading has been taken and 120°C is the temperature of the test. Scorch time is taken as the time when plasticity is 5 units higher than minimum value.

Evaluation of state of cure

As already mentioned, optimum cure is the one which optimises the properties desired in the product. Conventionally, tensile strength, modulus and ultimate elongation data have been used to estimate the best state of cure. Often the time at which maximum tensile strength is obtained is taken as the optimum period for cure.

If the optimum cure time is to be based on modulus and the modulus shows either a peak or a plateau at its maximum value the cure time required to reach 90% of that value is taken as optimum.

Ultimate elongation is also taken into consideration along with tensile strength and modulus for deciding optimum cure time. Many times it reaches maximum at an early curing time which is taken as the optimum cure time as a vulcanisate which breaks at low elongation is undesirable. Besides tensile strength, modulus or elongation, any other property which is of greatest consequence may also be used for determining time for optimum cure.

Cure metres

Curing for separate ranges of temperatures and carrying out separate tensile tests is a labourious business and so cure metres have been developed. Modulus change is monitored in the instruments. Several useful properties such as maximum and minimum stiffness, scorch time, cure time are obtainable from the data. The most popular cure metre used in the industry is oscillating-disk rheometre. Many versions of the instrument have been evolved.

Physical testing of vulcanisates

No tests should be performed until at least 16 hours after vulcanising as significant post vulcanisation changes in structure of the meterial may occur during this period.

Tensile tests

These tests are the most widely used test in the rubber industry. They are used to determine rate of cure and optimum cure, quality check on the compound, effect of immersion in liquids etc. Two shapes of test specimen are most popular-dumb bells and rings.

The dumb bell so called because of its shape has tabbed ends for gripping in the test machine tapers to a central constricted section of uniform width (Fig. 5.8). Bench mark 1" apart is stamped on the constricted section to facilitate observation of the elongation during test. The dumb bell specimen is clicked out with the help of a die.

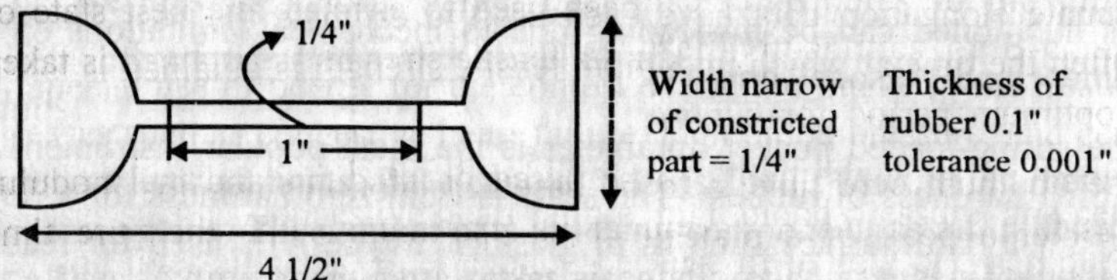

Fig. 5.8. Dumb bell specimen.

Advantages of using dumb bell specimen are ease of preparation and breakage in the area meant for following the elongation:

The ring specimen may have the following size: (i) inner dia - 44 mm; (ii) outside dia-52.6 mm; (iii) radial width-4 mm; and (iv) thickness of specimen-5 mm.

The ring is held by lubricated rollers in the machine so that the stress is equalised. Advantages of using ring specimen is possibility of using clamp separation as good measure of elongation and distribution of stress over a greater length. The speed maintained for testing is machine jaw or spindle separation rate of 20 ± 2 inch per minute. Dumb bell specimen gives result higher than ring specimen.

Tensile strength

Tensile strength is the maximum tensile stress reached in stretching a test piece to its breaking point. The force required is expressed as force per unit area of the original cross-section of the test length.

Elongation

Elongation is the extension between bench marks produced by a tensile force applied to the test piece and is expressed as a percentage of the original distance between the marks. Elongation at break or ultimate elongation is the elongation at the moment of rupture.

Modulus

Modulus is the stress required to produce a certain elongation. The standard temperature for conditioning and testing the samples is 23 ± 2°C. The tests should be made along the grain in the rubber i.e., parallel to direction of passage through the mill.

Hardness

Hardness, as applied to rubber, may be defined as the resistance to indentation under conditions that the rubber is not punctured. Hardness is one of the most useful and often quoted properties of rubber. However, the values mentioned can be misleading due to mechanical limitations of the instrument and personal error.

The spring-loaded pocket durameter is the most popular instrument for measuring hardness of rubbers. The scale is from zero hardness for a liquid to 100 for a hard surface such as of glass. The type A is used for reading up to 90. Type D is used for hardness above 90. It has a different indenter and stiffness spring. The instrument is portable and can be used for even irregular surfaces. However, differences of 5 units in reading are common.

Dead-load hardness testers are less convenient than durameters but the measurements are more reproducible. Here a dead load is applied to the indentor for a specified time and the hardness is obtained from the depth of the indentation.

Rebound tests

Resilience is the ability of a rubber vulcanisate to return the energy used to deform it during its recovery from deformation. It is expressed in per cent. The simplest test for resilience is the falling ball rebound 3. If the drop height of the steel ball is divided into 100 equal parts, the rebound height is equal to the resilience.

In a pendulum rebound test, the specimen is held at the rest position of the pendulum (zero degree) and is impacted by the centre of percussion of the arm. The angle of rebound is noted on a scale and resilience calculated by the formula:

$$R = \frac{1-\cos \text{angle of rebound}}{1-\cos \text{angle of fall}} \times 100$$

Lupke pendulum is popular in Europe. The rubber is subjected to only one half cycle of deformation. In a rebound test, the specimen should be preconditioned with six impacts before a reading is taken as the rebound height is then stabilised. The method is applicable to elastomers with hardness ranging from 30–85 IRHD only.

Tear test

Tear test results depend on type of specimen used, the rate of tearing and the temperature. The method is therefore useful for only laboratory comparison and cannot be applied for service evaluation.

Different types of dies can be used to prepare specimen as shown in Fig. 5.9.

No nick has to be cut in ASTM Die C specimen, while nicks of prescribed lengths are cut into the region of desired stress concentration.

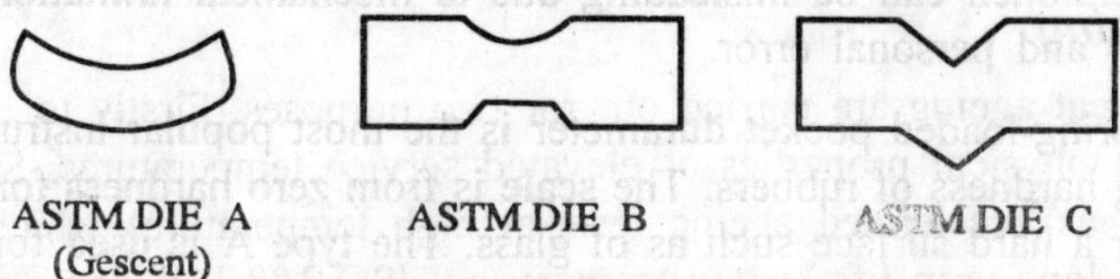

Fig. 5.9. Different types of dies.

The rate of stretching is kept at 20 inches per minute. The results are expressed as the force required to tear a specimen of given thickness while in ASTM type, it is dependent on width as well as thickness.

Abrasion tests

Abrasion is the progressive wearing away of a rubber article in service. In many applications, resistance to wear is one of the most important properties of a rubber article although it is one of the most difficult to measure. The type of wear undergone by an automobile tyre on a road is very different from the wear on a hose used for sand-blasting.

Abrasion resistance is a complicated phenomenon and is dependent on many properties such as resilience, stiffness, resistance to cutting and tearing etc. and different applications require these properties in varying proportions. Laboratory tests consist of rubbing the test piece against abrasive surface. These conditions are difficult to correlate with actual service conditions. Many well known names such as these of Akron, Dunlop, Du Pont and Pico have been associated with abrasion testers. The Akron tester is described below.

Akron

A moulded circular test piece of known weight is mounted on a motor driven spindle. The periphery of the test piece is brought into contact with the periphery of an abrasion wheel, which is also mounted on a spindle. Rotation of the test piece makes the abrasive wheel to rotate and the two are held together under a known force. The axes of the test piece and of the abrasive wheel are not in the same plane which results in a rubbing action. The abrasion resistance is calculated from the loss in weight after a specified number of revolutions of the abrasive wheel. Results are compared with those from a standard rubber and expressed as abrasion resistance index. The apparatus is suitable for specimen having hardness 57–80 BS degree, e.g. tyre-tread.

Ageing Tests

Heat ageing

Tests for heat ageing are carried out for two purposes. Firstly to establish changes in physical properties at elevated service temperatures. Secondly they serve as accelerated ageing tests at high temperatures which try to predict the long-term life at low temperature. ISO 188 lays down methods using an air oven or an oxygen pressure chamber.

The test consists of ageing test pieces for a given period at a specified temperature and then measuring the physical properties such as tensile strength, modulus, elongation at break and hardness. Comparison is thereafter made with non-aged test pieces.

In the oven method, the test pieces are exposed to air at atmospheric pressure. The total volume of the test pieces should not exceed 10% of the free air space in the oven. The air flow must be steady and between 3–10 complete changes of air per hour. No copper or its alloys should be used in the construction of the oven because they may have an accelerating effect in ageing. The temperature may be as required. However, 70°C or 100°C are most commonly used. The duration of the test is recommended as 1, 3, 7, 10 or a multiple of 7 days.

In the oxygen chamber method, the temperature is maintained at 70°C and a pressure of 2.1 MPa (300 psi). The ageing period should not be too long to prevent determination of ultimate physical property. After vulcanisation of specimen, a minimum period of 24 hours and maximum of 14 days prior to commencement of ageing should be allowed.

Resistance to Liquids

Elastomer when in contact with a liquid may absorb the liquid or the liquid may extract some soluble constituents from the elastomer or there may be chemical reaction between the two. Absorption is normally greater than extraction resulting in increase in volume which is called swelling. The International Standard for resistance to liquids in ISO 1817 which specifies choice of liquid, conditioning, temperature and duration of test.

The size of the test pieces when measuring change in volume should be 1–[illegible] cm with a uniform thickness of 2 mm and should not be longer or broader than 50 mm.

The test consists of completely immersing the test piece in the liquid in a stoppered glass bottle at the required temperature. The volume of the liquid must be at least 15 times the volume of the test piece.

Compression Set

Rubbers deform under load and the difference between the original and final dimensions is known as permanent set and can be measured in compression.

Small cylindrical disks of size 13 mm diameter by 6.3 mm thickness or 29 mm by 12.5 mm (Fig. 5.9) are compressed to a fixed height in a jig which consists of two or more flat parallel rigid metal plates of proper size to hold the test pieces. The plates are clamped together with nuts and bolts. Steel spacers of appropriate thickness in the form of rings around each belt are placed between the plates to control the thickness of the test pieces while compressed.

Three test pieces should be used for each determination and the average value should be taken. The bolts are tightened so that the plates are drawn together uniformly until they are in contact with the appropriate spacers sized to give a compression of 25%.

The duration of the test is 72 hours at 23°C or 24 hours at elevated temperature of test which can be one of the nine temperatures between 70°–250°C and for which the compressed test pieces have to be kept in the central part of an oven maintained at the test temperature. After the specific time, the test pieces are removed and allowed to recover at 23°C for 30 minutes before the thickness is measured.

The compression set is the difference between the original thickness of the test piece and the one after the recovery expressed as a percentage of the initially applied compression as shown in Fig. 5.10.

The compression set at constant strain $= \dfrac{t_a t_b}{t_a t_c} \times 100$

where t_a = original thickness of the test piece

t_b = thickness of the test piece after recovery

t_c = thickness of the spacer

Compression set value provides an indication of the degree of vulcanisation and unsuitable compounds can be identified if the value comes in the high range. Typical values for a few elastomers are given in Table 5.16.

Table 5.16. Compression set values of some elastomers.

	Temperature °C				
	20	*70*	*100*	*150*	*200*
Natural rubber	5	20	45	–	–
Neoprene	5	15	25	–	–
Nitrile	10	20	40	–	–
Silicone	10	10	10	15	40
Fluorosilicone	10	10	10	20	70
Fluorocarbon	10	10	15	20	25

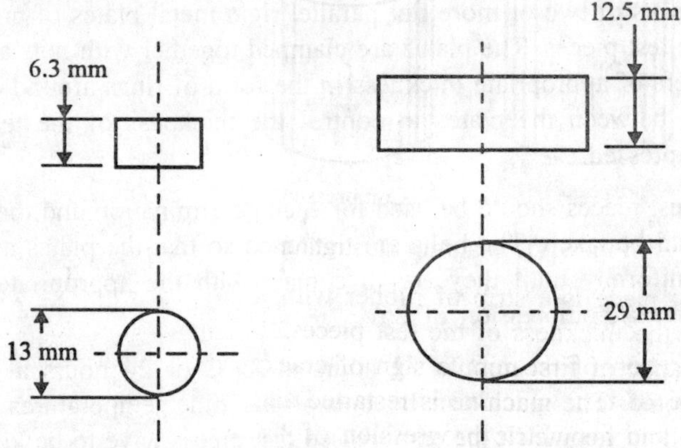

Fig. 5.10. Compression set test pieces actual size.

Fatigue-flex Cracking

After prolonged action of stress or strain on rubber, there is a change in its properties which is termed as 'Fatigue'. Flex cracking test is a fatigue test which strains the test piece in flexure and this type of deformation is experienced in service by footwear, tyre and belting.

Flex cracking is influenced by many factors such as type of rubber, additives such as fillers and protective agents, degree of vulcanisation or cross-linking and flaws which are present initially. Shape and size of the article, nature and magnitude of deformation and environmental factors such as temperature, humidity and also atmospheric oxygen and ozone also play important role. Ozone in particular can cause severe cracking.

The De Mattia machine (Figs. 5.11 and 5.12) is the most widely used apparatus for this test. In it, the upper grip is fised and holds one end of tl e test piece stationary. A second grip is activated by an eccentric drive and imparts a reciprocating motion to the other end of the test piece. The

ISO specification lays down a distance of 75 mm between the grips, a travel of 57 mm and a reciprocating movement of 5 Hz. Six to 12 test pieces should be used in the machine at one time.

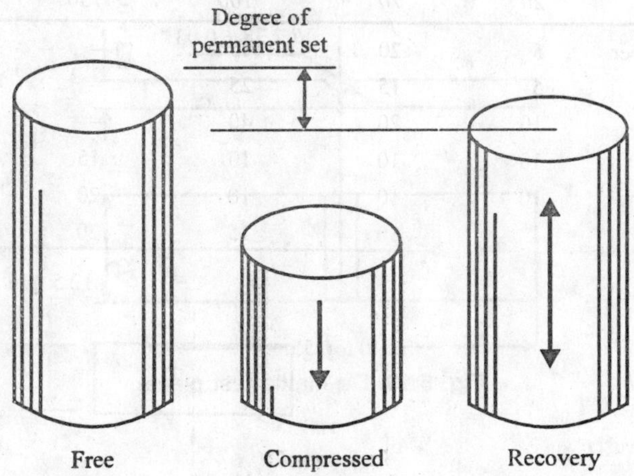

Fig. 5.11. Diagram illustrating compression set.

The test piece is a strip of rubber with a moulded groove. The results depend on the thickness of the test pieces. The test pieces are flexed until the appearance of first minute sign of cracking and the number of flexing cycles is noted. The machine is restarted and the number of flexing cycles is increased in geometric progression of 1.5 each time (Fig. 5.13).

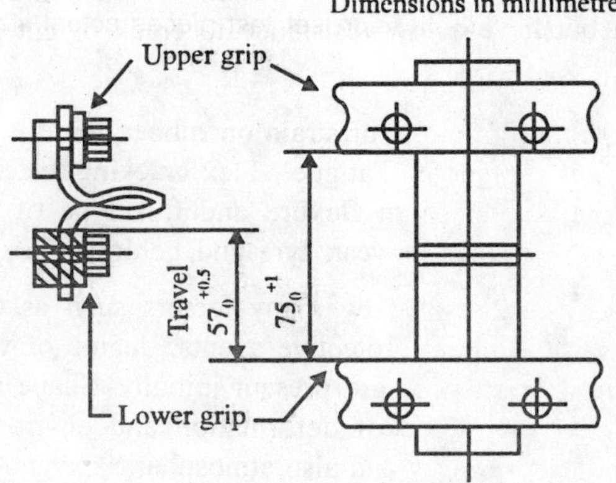

Fig. 5.12. Principle of the De Mattia machine.

Inspection is carried out thereafter. The cracking is graded at each inspection into six categories ranging from less than ten pinpricks which can be seen by the naked eye to a length of the largest crack greater than 3 mm.

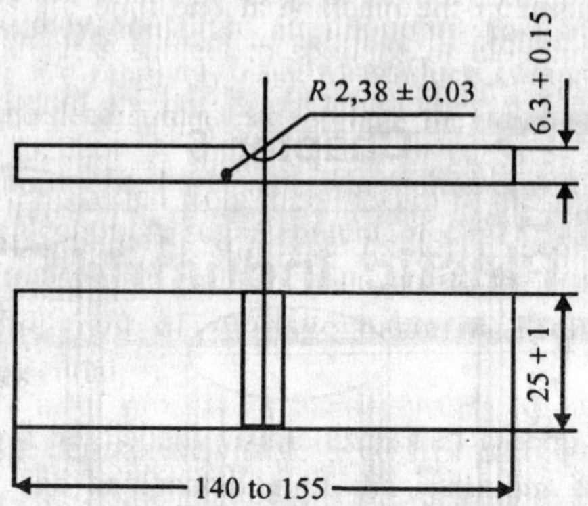

Fig. 5.13. De Mattia test piece.

Cut Growth

ISO 133 covers a variation of the flex cracking method covered by ISO 132 and is known as 'Crack or Cut growth' method. In this case, a 2 mm cut is made through the rubber and the length of the cut is measured with the aid of a magnifying glass at intervals. The number of cycles for the cut to extend by 2 mm and further extensions of 6 mm and 10 mm are recorded.

It is better to carry out both flex cracking and cut growth tests because certain vulcanisates viz., of styrene-butadiene rubbers show resistance to crack initiation but have a low resistance to crack or cut propagation.

Chapter 6

Plastic Industries

INTRODUCTION

Before knowing what are plastics, it is worth while to understand what is plasticity. Thus, plasticity is the property by virtue of which a material undergoes permanent deformation under stress. A common example is putty, which on pressing between fingers stays in the shape it is pressed. Plasticity is quite different from elasticity which is the property by virtue of which a material undergoes deformation under stress but reverts to its original shape on the removal of stress. A rheological property of solid or semi-solid materials expressed as the degree to which they flow or deform under applied stress, and retain the shape so induced, either permanently or for a definite time interval. It may be considered reverse of elasticity. Application of heat and/or special additives is usually required for optimum results.

Plastics are an arbitrary group of artificial materials, generally of synthetic organic origin which at some stage in manufacture are in a plastic condition during which they are shaped, often with the aid of heat and pressure and often in a mould.

This definition needs a little explanation. When it is said that plastics are an arbitrary group it means that there is no code or specification with which a material must comply in order to be called a plastic. For example, though synthetic rubber fulfils all the conditions laid down in the above-mentioned definition yet it is not included in the list of plastics.

By "artificial materials" is meant those materials which have been made by compounding and are quite different from naturally occurring materials such as wood, sandstone or silver.

"Generally of synthetic organic origin" indicates that most of the materials are made from simple organic chemical raw materials. By the term "organic" is meant only those chemical compounds in which element carbon is the principal structural unit. The word "generally" denotes that certain naturally occurring organic substances such as bitumin, casein and cellulose are also employed for the manufacture of plastics.

According to the definition, the plastic at some stage in their manufacture are truly plastic and hence capable of being shaped. It follows that plastic or plastic articles at ordinary temperature are not plastic but quite solid.

The plastics, according to the definition, are "shaped" by means of heat and pressure in a mould and the process is called moulding. In the definition it is mentioned, "shaped, often with the aid of heat and pressure and often in a mould". The world "often" is used because some plastics are shaped by casting in open moulds and baking without pressure, some by machining, and some by hot extrusion through a shaped die or nozzle.

SYNTHETIC RESINS

A synthetic resin is a resinous material made or synthesised by chemical means from simple organic raw materials. Many synthetic resins are used for plastics, for example phenol formaldehyde and polystyrene resins are used for making phenolic plastics and polystyrene plastics. However, there are plastics which are not made from synthetic resins, for example nitrocellulose plastics and casein plastics. Synthetic resins are of two types: (i) thermosetting and (ii) thermoplastic. Their definitions will be discussed in the classification of plastics.

Difference between Synthetic Resins and Plastics

All plastics are based on synthetic resins. A synthetic resin may be used as such for making plastics or it may be mixed with other ingredients such as fillers, plasticisers, dyes and pigments, catalysts and lubricants to form plastics. For example, polyvinyl chloride and polyethylene may be used as such for shaping articles out of them. Phenol formaldehyde resin is mixed with other ingredients to form plastics. So a synthetic resin is an essential ingredient of a plastic and it may or may not be mixed with other ingredients to form plastics.

Polymers are classified into three groups, viz. thermoplastics, thermosets and elastomers. In addition, thermoplastics are divided into amorphous, crystalline and liquid crystalline polymers.

Thermoplastics are linear or branched polymers which can be melted upon the application of heat. They can be moulded and remoulded using conventional techniques.

Thermosetting plastics or thermosets are heavily cross-linked polymers which are normally rigid and intractable. They consist of a dense three-dimensional molecular network and like rubbers, degrade rather than melt on the application of heat.

Elastomers or rubbers are polymeric materials which display elastomeric properties, i.e., they can be stretched easily to high extensions and will spring back rapidly when the stress is released.

Crystalline polymers are thermoplastic materials having regularly arranged molecular structure. In these polymers the chemical structure allows the polymer chains to fold on themselves and pack together in an organised manner. The resulting organised regions show the behaviour characteristics of crystals.

Amorphous polymers are thermoplastic materials having irregularly arranged molecular structure. When the molecules solidify they acquire random coily structure as in the molten state. Molecular packing will be very poor in these materials. All the crystalline polymers have amorphous regions between and connecting the crystalline regions. For this reason, crystalline polymers are often called semicrystalline polymers.

Liquid crystalline polymers are best thought of as being a separate and unique class of plastics. The molecules are stiff, rod-like structures that are organised in large parallel arrays or domains in both the melt and solid states. These large, ordered domains provide liquid crystalline polymers with unique characteristics compared to those of crystalline and amorphous polymers.

Organic and Inorganic Polymers

Further, polymers are divided into organic polymers and inorganic polymers. Organic polymers are formed by reacting organic monomers. The main chain will predominently contain carbon atoms. Inorganic polymers are formed by reacting inorganic monomers. The main chain will contain atoms like Si and Ge.

Commodity, Transition and Engineering Plastics

Based on applications, plastics are further classified as commodity plastics, transition plastics and engineering plastics. Commodity plastics are low strength materials, predominently used in non-critical consumer applications. Engineering plastics are hgh strength materials mainly used in engineering applications. Transition plastics are having strength in between that of commodity and engineering plastics. Some plastics have exceptional properties like high temperature resistance, chemical resistance and very high strength. Those plastics are known as speciality thermoplastics. There are some thermoplastic materials which are having elastomeric properties and they are classified as thermoplastic elastomers.

Polymer and Monomer

A polymer is a substance which consists of molecules which are, at least approximately, multiples of low molecular weight units, (In Greek *polys* means many and *meros* means part or unit). The low molecular weight unit is monomer.

Polymerisation

The linking of small molecules (monomers) to form large molecules (polymers or macromolecules) is called polymerisation.

Homochain Polymers and Heterochain Polymers

Also polymers are divided into homochain polymers and heterochain polymers. In homochain polymers, only one type of atoms combine together to form the main chain. They are formed by the addition polymerisation technique. In heterochain polymers, more than one type of atoms combine together to form the main chain. They are formed by the condensation polymerisation technique.

Homopolymer

If only one monomer is polymerised the product is called a homopolymer.

Copolymer

The polymerisation of a mixture of two monomers of about equal reactivity leads to the formation of a copolymer and the two types of monomer units enter the chain in a random fashion, as shown below :

A—B—A—A—B—B—B.

where A and B are two types of monomer units.

Polyblend

A mixture of macromolecules containing only one type of monomeric units with another macromolecules containing essentially only another type of monomeric unit is called a polyblend or a mixture of homopolymers.

Block Copolymer

It is a copolymer which contains longer stretches (or chain) of two or more monomeric units linked together by chemical valencies in one single chain.

A—A—A—A—B—B—B—B—A—A—A—A

where A and B are two different monomes.

Graft Copolymer

A graft polymer contains branches of varying length made up of different monomeric units on a common backbone or trunk chain.

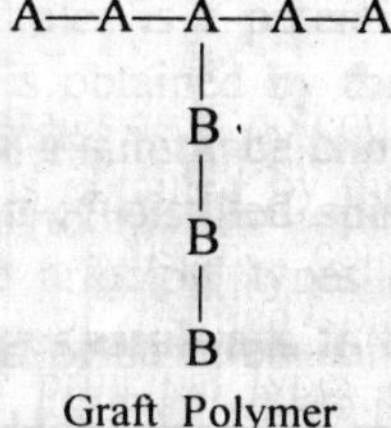

Graft Polymer

The detailed structure of block and graft copolymers is determined by the relative reactivity ratios of the various building units.

Isotactic Polymer

This is a formation of a polymer in which the monomers have entered the chain in a regular fashion.

Atactic (Random) Polymer

This is a formation of a polymer in which the monomers have entered the chain in a random fashion.

Syndiotactic Polymer

In this formation of a polymer the monomers have entered the chain in alternating fashion. It may be noted here that isotatic, atactic or syndiotactic polymers are formed depending upon the presence of complex catalysts.

CLASSIFICATION OF PLASTICS

The plastics are divided into two groups: (i) thermosetting materials and (ii) thermoplastic materials.

Thermosetting Materials

Thermosetting materials are those plastics which require heat and pressure to mould them into shape. When heat is applied, they first become soft and plastic and on further heating they undergo chemical change and set hard. The process is called thermosetting or thermohardening. When a material is thermoset, it is permanently set and and does not soften to any appreciable extent when again heated. However, intense heating will bring about the breakdown of the material by burning. The following are some of the thermosetting materials :

(1) Alkyds and Polyesters
(2) Aminos (urea and formaldehyde resins and plastics)
(3) Casein
(4) Epoxides
(5) Phenolics
(6) Silicones.

Thermoplastic Materials

Thermoplastic materials are those plastics which soften on the application of heat, with or without pressure but they require cooling to set them to shape. As the hardening in thermoplastic materials is not due to any chemical action, so the shaped articles from thermoplastic materials will resoften on heating. The following are some of thermoplastic materials :

(1) Cellulose acetate
(2) Cellulose acetate butyrate
(3) Cellulose nitrate
(4) Ethyl cellulose
(5) Nylon 6 : 6
(6) Nylon 6 : 10
(7) Nylon 6
(8) Nylon 10
(9) Polyurethanes
(10) Polyethylene
(11) Polypropylene

(12) P.T.F.E. or Polytetra-fluoroethylene

(13) Polystyrenes

(14) Styrene acrylonitrile

(15) Polymethyl methacrylate

(16) Polyvinyl chloride.

Thermoplastics

The general-purpose thermoplastic resins such as acrylonitrile-butadiene-styrene (ABS), vinyl acetate and acrylonitrile-styrene and the engineering thermoplastics such as polycarbonates, acetals, and nylons thermoplastics. Some of the important ones are described below.

Polyethylene

Industrially polyethylene is manufactured by three distinct processes, namely high, medium and low pressure. The products of these processes exhibit different properties as shown in Table 6.1. It may be appended that polymerisation can be conducted in gas phase or liquid phase (solution/emulsion/slurry).

Table 6.1. Properties of polyethylene produced by different methods.

	High pressure	*Low pressure*	*Medium pressure*
Crystallinity %	54–68	80–90	95
Density Kg/m^3	915–925	940–965	950–965
Melting point °C	108–110	120–140	140–143
Brittle point °C	**–70 & lower**	**–70 (lower)**	
Tensile strength kg/cm^2	**120–16**	**220–450**	**400–450**
Elongation % (cross-wise)	**400–700**	**100–1000**	**20**
Dielectic strength Kv/mm	**45–60**	**45–60**	
Organic solvents not affecting	**Paraffins Aromatics dissolve serviceable upto (50°C)**	**Paraffins Aromatics serviceable upto 80°C**	
Branching and unsaturation per 1,000 C-atoms (as methyl groups)	**20–30**	**7–10**	**2**
Melt index	**0.2 to 0.3**	**0.1 to 4.0**	
Resistant to heat	**Upto 98°C**	**Upto 110°C**	

High pressure polyethylene (low density polyethylene, LDPE)

The polymerisation is carried out under a very high pressure (pressures ranging upto 350 MPa) in a temperature range of 200°C to 240°C. The reaction is credited to free radical mechanism. This being a chain reaction, the exothermic heat of 100 kJ/mole has to be successfully dissipated. Otherwise the reaction mixture goes uncontrollable at a rate of temperature rise of 10° to 12°C per 1% ethylene converted. Chain reaction to be perpetual, ethylene of onecrous purity is essential. Such purity of immense nature obviously craved for the solvent/emulsion polymerisation, which can go with less purity stocks.

Process description

Fig. 6.1 represents the process flow sheet. The main steps involved in the process are :

(1) Purification and compression of ethylene in two stages (50 bars and 300 bars).

(2) Introduction of free radical initiators.

(3) Raising the reaction mixture to a temperature of 200°C and maintaining around 250°–270°C.

(4) Release of pressure in a number of stages.

(5) Separation of polymer from unreacted ethylene which is recycled.

Polymer grade ethylene (99.9%) mixed with 600–800 ppm oxygen is compressed mainly in two stages c_2 and c_1 upto 35 MPa. The compressed gas is passed into a tubular reactor : (i) (the reactor is tube-in-tube type), where heat dissipation is conductively carried out by circulating dowtherm or hot water. Another method of heat removal is to heat the incoming gas by out going mixture. The temperature of the reactants in the entering zone is around 150°C, but at the end of the coil, the temperature reaches 240°C. The temperature in the reactor outlet should never drop below 200°C, as the crystallisation of polyethylene begins, giving obstruction to flow. The reaction products are directed to pass into high pressure separator; (ii) the pressure of the gases, obtained from this stage is still above 35 MPa, hence, this after purification can go directly to second stage compression (c_1). From the high pressure separator, the melt goes into a second stage separator (low pressure); (iii) where the remaining ethylene is expelled and routed to 1st stage (c_2) compressor, through purification system. In fact, after first high pressure separator, there may be a series of separators to catch polymer dust. The gases are rich with ethylene and oxygen but contain some carbon dioxide and formaldehyde. Hence purification is essential before the gas enters second stage compression. A safety system is desirable while handling such high pressure gas, to actuate exit automatically, if disturbance in heat balance or flow obstruction appears in the reactor. It is always desirable to blow the whole system with nitrogen before the polymer gas enters the reactor. As high as 80% ethylene is recycled due to inklingly low conversions per pass. LDPE formed is collected from low pressure and high pressure separators. The gases leaving out of the system pass through a separator where some liquid polymers gets condensed.

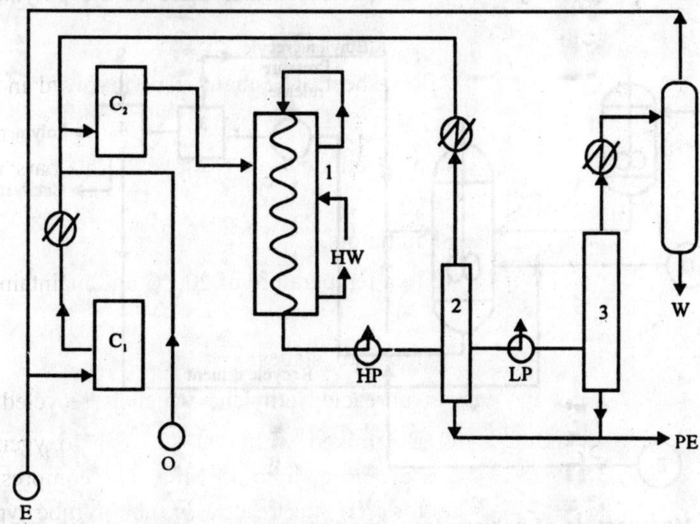

Fig. 6.1. Manufacture of polyethylene (low density). (E) Ethylene; (O) Oxygen; (W) Waxes and liquid polymers; (PE) Polyethylene; (HW) Hot water circulation; (HP) High pressure reliever; (LP) Low pressure reliever; (C_1 & C_2) Compressors; (1) Reactor (tubular); (2) Separator & cyclone (high pressure); (3) Separator; (low pressure).

LDPE can also be produced by solution polymerisation, where the pressure and temperature are aminably low order, 100 MPa, and temperature 190°C. The solvent constitutes a equivolume mixture of benzene and water. Oxygen as in the former case acts as initiator which is introduced into ethylene stream. After reaction, the products are separated, and unconverted ethylene goes back. This process gives a high conversion rate upto 20% per pass.

Medium pressure polyethylene

Flow sheet of medium pressure process based on chromium oxide catalyst is shown in Fig. 6.2. The polymerisation is conducted at a pressure of 5 MPa in solvent medium; either in a slurry fashion or in a solution. Phillips solution catalysts is prepared by muffling chromium-impregnated silica-alumina supports at selected conditions. Usually hexavalent chromium oxide on any inert base (silica-alumina) is a poorer solvent for polyethylene; will serve as polymer catalyst. Some active ingredients like calcium, barium, tungsten oxides are also used. Some times halogenated chromium oxide is prevalent in the industry to increase the molecular mass of the polymer.

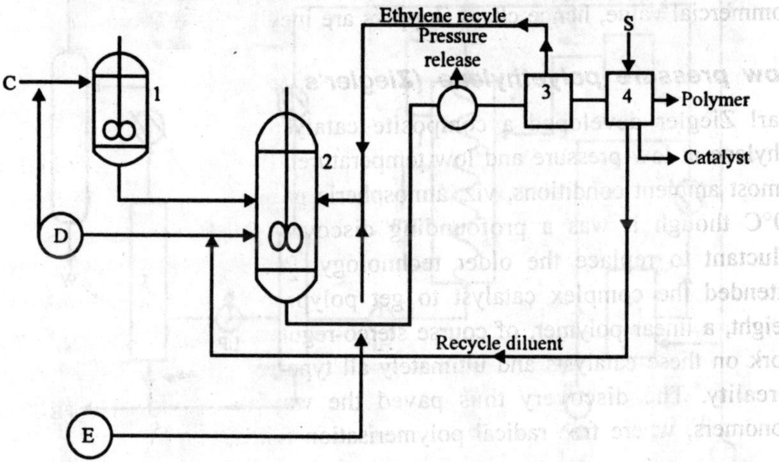

Fig. 6.2. Manufacture of polyethylene (medium pressure process). (1) Catalyst tank; (2) Reactor; (3) Separator; (4) Washer-evaporator; (C) Catalyst; (D) Diluent; (E) Ethylene; (S) Solvent.

The polyethylene obtained from this system is linear and highly crystalline. There are other catalysts like nickel, cobalt, molybdenum oxides. With different catalysts, the conditions of the reactor are likely to vary.

Process description

Medium pressure process employs a pressure of 35 MPa and a temperature of 130–170°C. 0.1% Cr_2O_3 treated with HF gives a molecular mass of 45,000 otherwise the polymer mass does not exceed 20,000. In this process cyclohexane is used as solvent and the concentration of ethylene in solution is kept around 5%. The solvent serves the dual role as a solvent to ethylene and polymer, and permit the chain length to grow irrespective of the presence of chain breakers. Viscosity, heat transfer problems and control of ethylene consumption are also attended successfully by solvents.

In recent versions, solution process gives high solids (40%) content with no de-ashing problem. In the slurry process higher yield of product per given reactor is an advantage. Further, because of poor solvency nature, the polymer or copolymer separates from the diluent in fine particles. Thus the viscosity of solution does not increase as seen in solution process. Of course slurry process takes a longer time (2–4 hours). Further all slurry processes are credited with yielding very high mole weights, perhaps of no commercial value, hence chain stoppers are inevitably sent into the system.

Low pressure polyethylene (Ziegler's process)

Karl Ziegler developed a composite catalyst system for polymerising ethylene at low pressure and low temperature. The conditions required are almost ambient conditions, viz. atmospheric pressure and a temperature of 60°C though it was a profounding discovery the manufacturers were reluctant to replace the older technology. Providentially, Giulio Natta extended the complex catalyst to get polypropylene of high molecular weight, a linear polymer, of course stereo-regular. This prompted others to work on these catalysts and ultimately all types of polymers have become a reality. The discovery thus paved the way to polymerise different monomers, where free radical polymerisation failed.

Production of HDPE (high density polyethylene)

Fig. 6.3 shows the outlines of the process. In the catalyst vessel a fresh and dilute solution of composite catalyst (max. 0.1 gm/litre) in pentane or hexane is made and this acts as a reservoir too. The solution is fed continuously into reactor where solvent and ethylene streams enter. A stirrer in the reactor is helpful for homogenising the mixture. The polymer mixture after reaction, is allowed to enter a flasher, where unreacted monomer and inerts are vented out. The solid material is given a thorough wash with water to decompose the catalyst. The strippers separate the organic solvent from water-hydrocarbon azeotrope and solvent is recycled. The polymer material is washed many times, either with water or with methanol, after which it is filtered and dried. Catalyst is obtained in a slurry and is recovered for usage. Additives for preserving the quality of polymer are added at the stage of granulation. It is found that by alcoholic wash, the ash content of the polymer decreases very much. Before start of the operation all units are submitted to nitrogen breathing, and is essentially important to see nowhere water or air seeps in.

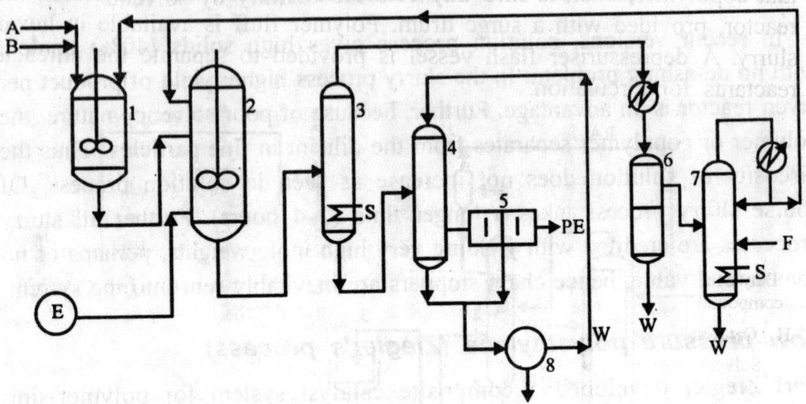

Fig. 6.3. Manufacture of polyethylene (Ziegler process). (1) Catalyst tank; (2) Reactor; (3) Solvent stripper; (4) Washing column; (5) Filter; (6) Solvent-water separator; (7) Solvent fractionater; (8) Spent catalyst recovery unit; (A) Alkyl aluminium; (B) Titanium halide; (E) Ethylene; (S) Steam; (W) Water; (P) Solvent recycle; (PE) Polythylene (HD); (F) Fresh solvent.

DU Pont process

This is a solution process. In Du Pont process soluble Ziegler catalysts are used, although it differs from Phillips process in a true solution way, i.e. all things remain in solution. These catalysts are comprised of $TiCl_4$, VCl_5, $VOCl_3$ and the residence time for polymerisation is 3 to 4 minutes. Molecular weight can be easily controlled by hydrogen and temperature. All solution processes are generally run at moderately high pressures and high temperatures. This gives rise to increased efficiency of catalyst (500 kg/gm). A polymer concentration of 35–40% is common in such solution process. Du Pont process is also suitable for co-polymer production. These polymers from this process are found to have less ash content because the catalyst can be dislodged very easily and effectively.

Solvay process

The catalysts are formed by reacting transition metal compound ($TiCl_4$) with one of three different supports at a time, followed by activation with organo-aluminium compounds. Polymerisation is carried out in a loop type reactor using n-hexane as diluent. Reactor conditions are moderate; 25–35 bars at 50–90°C. The resultant slurry is centrifuged. Wide range of molecular weights are possible with this process. Fig. 6.4. shows flow sheet of the process. The reactor is tubular provided with a jacket for circulating cold water. Hydrogen is incorporated in the reacting mixture, so that copolymerisation is smoothly controlled. Further, this is a closed loop reactor, provided with a surge drum. Polymer fluff is available in hexane slurry. A depressuriser flash vessel is provided to separate the unreacted reactants for circulation.

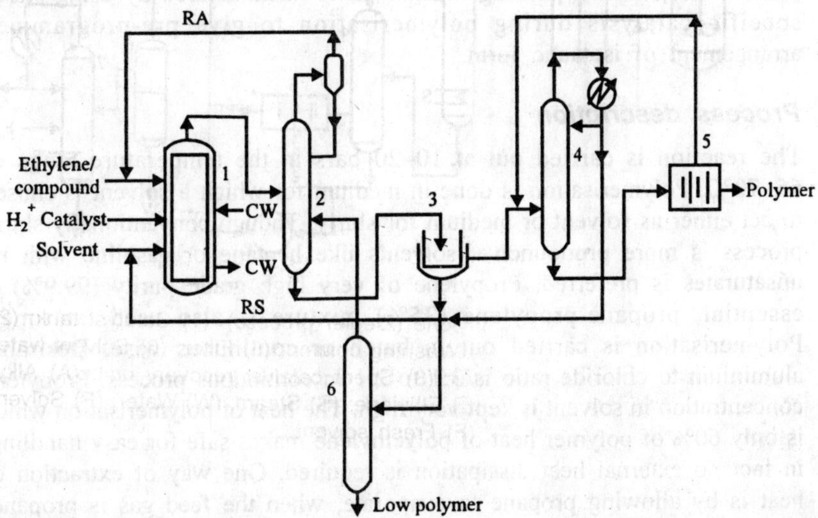

Fig. 6.4. Solvay process for manufacture of HDPE. (CW) Cold water; (RS) Recycle solvent; (RA) Recycle olifins; (1) Reactor water cooled; (2) Flash separator; (3) Centrifuge; (4) Solvent stripper; (5) Filter; (6) Solvent purifier.

Polypropylene

Polypropylene is commercially produced by only one method that is anionic polymerisation. The process is accomplished in hydrocarbon solvents with the same Ziegler type composite catalyst, but with a difference, titanium trichloride is opted over tetrachloride. Polypropylene produced by tetrachloride contains less isotactic polymer (35–40%), hence trichloride has been recommended which yields about 80–85% stereo regular polymer. $TiCl_3$ exists as porous crystals exhibiting active sites uniformly. $TiCl_3$ diethyl aluminium chloride yields minor atactic form only; 2:1 mole ratio of DEAC to $TiCl_3$ is used industrially to give good amount of stereo regular polymer. Higher amount of $TiCl_3$ spoils the active sites of catalyst. Production based on traditional catalysts involved after polymerisation, removal of catalyst dust and separation of isotactic polymer from atactic polymer. During deashing the unreacted monomer and solvent get poisoned, hence extensive purification is appended. The new catalyst allow polymer structure, yield, selectivity and rheology to be controlled resulting in simplification of process.

Biaxially oriented polypropylene film (BOPP) is produced by orienting long chain molecules in two directions. BOPP has a configuration ABA or BAB where A is P and B is PP copolymer. This exhibits a density of 0.905 gm/cc. Because of its excellent clarity and high barrier for moisture, it has become an excellent packing medium. It is manufactured by using stereo specific catalysts during polymerisation to give pre-programmed arrangement of isotactic form.

Process description

The reaction is carried out at 10–20 bars in the temperature range of 65–70°C. Polymerisation is done in medium for which a solvent is chosen to act either as solvent or medium for slurry. Though conventionally slurry process is more pronouncive, solvents like heptane or gasoline with no unsaturates is preferred. Propylene of very high grade purity (99.9%) is essential, propane-propylene (35%) mixture is also used, at times. Polymerisation is carried out in batch or continuous wise. Desirable aluminium to chloride ratio is 3:1 to 2:1 in continuous process. Propylene concentration in solvent is kept very high. The heat of polymerisation which is only 60% of polymer heat of polyethylene makes safe for easy handling, in fact no external heat dissipation is required. One way of extraction of heat is by allowing propane to evaporate, when the feed gas is propane-propylene.

Fig. 6.5 illustrates the process. A 10% solution composite catalyst is made in a solvent usually heptane and kept in a tank: (i) The solution is proportionally charged into reactor; (ii) where olefin and some more solvent are admitted. The reactor is kept under required pressure and temperature for about 3 to 4 hours to complete the reaction. Feed gas and catalyst thoroughly mix due to agitation. After the reaction, products are transferred into a stabiliser; (iii) where pressure is decreased, to let off unreacted gas. The catalyst is decomposed by means of alcohol. All the polymer product is centrifuged; (iv) and the solvent-alcohol mixer is separated by steam stripping in a stripper; (v) recovered solvent is distilled for purification; to begin circulation. A separate stripper-purifier section is essential for this. The heavy polymer with adhering solvent is stripped in another stripper; (vi) by means of steam, and the recovered solvent mixture is processed for alcohol and solvent in section; (vii) polypropylene collected from the bottom of the stripper is further centrifuged, mixed with stabilisers and sent to storage. Polypropylene requires about 1.1 tons of monomer per each ton of polymer.

Polypropylene being less denser than polyethylene (0.91 to 0.92), it has become an attractive thing for making water-floating objects. Ropes made out of polypropylene exhibit the same strength as steel wires, still floating and non-corroded, hence appealed for water floating ropes, tugs and vessels. Fibres of polypropylene can be knitted easily, all these advantages have made it as an impregnable rival to polyethylene.

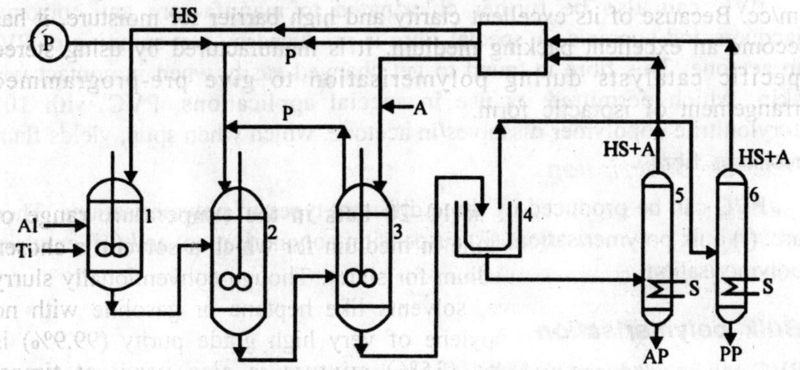

Fig. 6.5. Manufacture of polypropylene. (Al) Aluminium (Triethyl); (Ti) Titanium halides; (P) Propylene; (A) Alcohol; (HS) Hydrocarbon solvent; (S) Steam; (AP) Atactic polymer; (PP) Polypropylene.

Polyvinyl Chloride (PVC)

This is one of the most widely used big tonnage thermoplastic credited with diverse applications. Two types of homopolymers are produced one being flexible and the other a rigid type. Flexible polymers have a good elasticity possessing a density of 1150 to 1600 Kg/m^3. Rigid type polymers exhibit a density of 1150 to 1800 kg/m^3 with a good tensile strength 42–60 MPa. Rigid polymers are suitable for making articles of hard type sheets, papers, gears etc. Both the types show low crystallinity due to random orientation of chlorine atom. Polyvinyl chloride is obtained by radical polymerisation leading to head to tail configuration as shown here.

$$—CH_2—\underset{\displaystyle Cl}{\underset{|}{CH}}—CH_2—\underset{\displaystyle Cl}{\underset{|}{CH}}—CH_2—$$

PVC is moulded into phonograph records and is good for coatings and it is also used in the form of sheets. It can be used with or without plasticisers. It is white amorphous solid, soluble in chlorinated solvents, further it is very easily welded at 200°C, with rods made of PVC. However, it suffers from low ultimate temperature (65°C) at which articles made of it can be used. To overcome this difficulty vinyl chloride is copolymerised with vinylidene chloride or acrylonitrile. Vinylidene chloride is a comonomer upto 20%, in copolymer with vinyl chloride, and the product is known as Saran. These copolymers soften around 185–200°C and they are not easily soluble in organic solvents. Processing of vinylchloride copolymers is much easy, and are suitable for making fibres such as saran, or floorings etc.

PVC can also be further chlorinated to manufacture anti-corrosive lacquers and enamels. A special fibre is produced by wet spinning of PVC in acetone. This fibre is found to get changed easily when in contact with skin, which permitted its use in special applications. PVC with 10% acrylonitrile copolymer dissolves in acetone, which when spun, yields flame resisting fibres.

PVC can be produced by three different types of polymerisations. These are: (i) bulk polymerisation; (ii) suspension polymerisation; and (iii) emulsion polymerisation.

Bulk polymerisation

PVC can be produced by bulk polymerisation initiated by sunlight. However, due to the handling difficulties of the physical form of polymer, and high heat disposal rates, the conversions never exceeds 60% and the resultant polymers are of low molecular mass. In this method, first pre-polymerisation is carried out in essentially liquid medium (autoclave) to 7% conversion. The next operation consists of polymerising this prepolymer by dilution with monomer and 'beads' grow to high molecular product. By this way polymers of any desired type are produced. Presently this technology is not used anywhere.

Suspension polymerisation

Superficially it resembles emulsion polymerisation. Monomer is dispersed as fine droplets (50 to 250 μm) in the aqueous phase. Agitation is required to keep the monomer in a finely dispersed state, and to protect colloidal formations. Catalyst is soluble in aqueous phase. Peroxides of benzyl, laural, p-chlorobenzoyl or isopropyl carbonate or azo compounds initiate the polymerisation. Suspension agents are water soluble agents, like polyvinyl alcohol, starch, gums, geleatins etc. and are found useful. Polymer is not soluble in monomer, while monomer is soluble in polymer. The reaction mechanism and kinetics in all suspension polymerisations is similar to bulk polymerisation.

Emulsion polymerisation

Monomer is dispersed into water, which is emulsified with surface active agents like ammonium salts of higher fatty acids, sulphonates, palmitates and stearates. A solution of the following type is used. Water soluble initiator, like persulphate is also added into the aqueous phase along with a buffer to maintain pH around 4.

Water	100 parts
Emulsifying agent	0.5–1
Buffer	0.05–0.1
Ammonium persulphate	0.25
Sodium hydrosulphite	0.2
VCM	55–90

Aqueous phase consisting of emulsifying agent with soluble initiator and buffers is charged into an autoclave. VCM is discharged into the emulsion in droplets (1 μm) at 200–1000 kN/m^2 pressure and the temperature kept as low as possible. The reaction goes on for 6 hours. Unpolymerised material and latex containing (35–40% polymer) are later discharged into coagulator, where electrolytes coagulate the polymer from latex. The solvent is stripped off the polymer, washing, filtering, drying of the polymer are finishing operations to keep the monomer in liquid state. Fig. 6.6 shows the production of PVC by emulsion polymerisation. The polymerisation is done in autoclaves of capacity 40 to 100 cubic metres. Autoclaves are cooled to remove the polymer heat by water. After the reaction is over the contents are degassed, unreacted monomer is purified and recycled. Polymer is separated by centrifugation. Lower temperatures give rise to high molecular weight polymer.

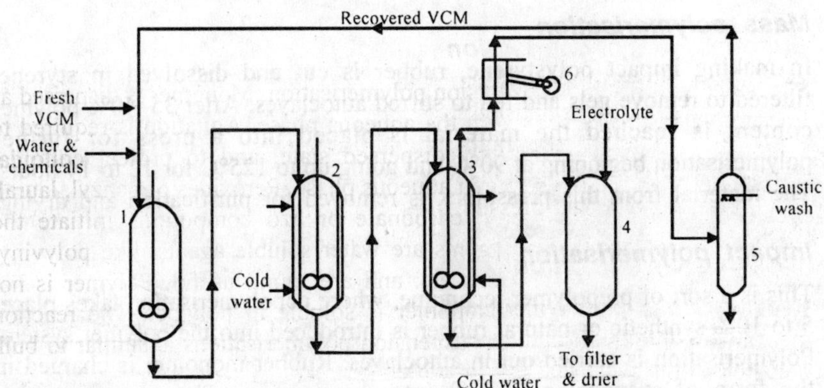

Fig. 6.6. Manufacture of polyvinyl chloride (PVC). (1) Tank for emulsion preparation; (2) Polymerisation reactor; (3) Degassing operation; (4) Coagulation tank; (5) Impuritiés eliminator; (6) Compression system.

Polystyrene

Next to polyethylene and polypropylene, polystyrene is the widely used plastic.

Polystyrene can be made in a number of ways to suit the end use. As the molecular weight increases, it slowly turns from liquid to solid state. High strength polystyrene is usually having a molecular weight more than 100,000; these are difficult to mould, the intermediate range ploystyrene 70,000–100,000 is easy to mould. As the molecular weight drops below 70,000; the tensile strength, impact strength and elongation fall of rapidly: however, these are classified as general purpose polystyrenes. Polystyrene has got many favourable properties, like ease of fabrication, high thermal stability, low specific gravity (900 kg/cu.m) etc. Homopolymers of polystyrene are extremely brittle and rigid hence copolymers are opted.

Methods of polymerisation

Polystyrene production is carried out by free radical initiation or by co-ordinated catalyst. Bulk, suspension and emulsion polymerisations are in use. The polymer produced by initiators is atactic, while Ziegler's polymers are isotactic. Polystyrene is a linear polymer, though the phenyl groups often prevent ordered arrangement of macromolecules and crystalline structure. At 80°C and above amorphous polystyrene passes into rubbery state till it reaches 150°C, whereby it becomes plastic. Further heating above 300°C results in decomposition. It dissolves in most of the aromatic solvents. Isotactic polymer melts at 220°C. Its brittleness decreases with the addition of plasticisers, thus lowering its glass transition temperature.

Mass polymerisation

In making impact polystyrene, rubber is cut and dissolved in styrene, filtered to remove gels and fed to stirred autoclaves. After 35–40% polymer content is reached the material is placed into a press for further polymerisation beginning at 90°C and going up to 125°C for 12 to 14 hours. The material from this pressblock is removed for purification and drying.

Impact polymerisation

This is a sort of prepolymer technique, where copolymerisation takes place. 5 to 10% synthetic or natural rubber is introduced into the polymer system. Polymerisation is carried out in autoclaves. Rubber monomer is charged in the form of suspension (water) into the reactors. The temperature of polymerisation is kept around 90°C to 130°C in gradual steps for about 10 hours. Some times two catalysts are introduced, like benzyl peroxide and ditertiary butyl perbenzoate. A slight deviation in the impact process is known as crystal process, where pre-polymerisation is not called in. In suspension polymerisation the raw material is converted into polymer unlike continuous mass polymerisation. It gives a thorough control over the molecular weight.

Emulsion polymerisation

This is more popular version of polymerisation. Latex or suspension polymerisation can be affected to produce different granular polymers. In emulsion polymerisation initiators are persulphates and emulsifiers are soaps or sodium petroleum sulphonates. Fig. 6.7 shows the outlines of the process. Polymerisation is accomplished in 3 to 5 m^3 enamelled reactors fitted with a water jacket stirrer and reflux condenser. The monomer is suspended in aqueous phase using a stabiliser. Sodium sulphate is added to control pH. A thorough agitation keeps the monomer suspended in medium. The aqueous phase emulsified and mixed with monomer. The emulsion is sent into reactor which is heated and kept at 60°C. There may be number of reactors in series, polymerisation is carried out in nitrogen atmosphere. Catalysts and chain transfer agents are added into the reactor. The reaction takes place for 3 to 6 hours, after which it is sent into coagulator. The polymer mass is coagulated by heating to 120°C later allowed to settle. The polymer from the latex is separated by centrifugation. The polymer is washed and sent to dryer.

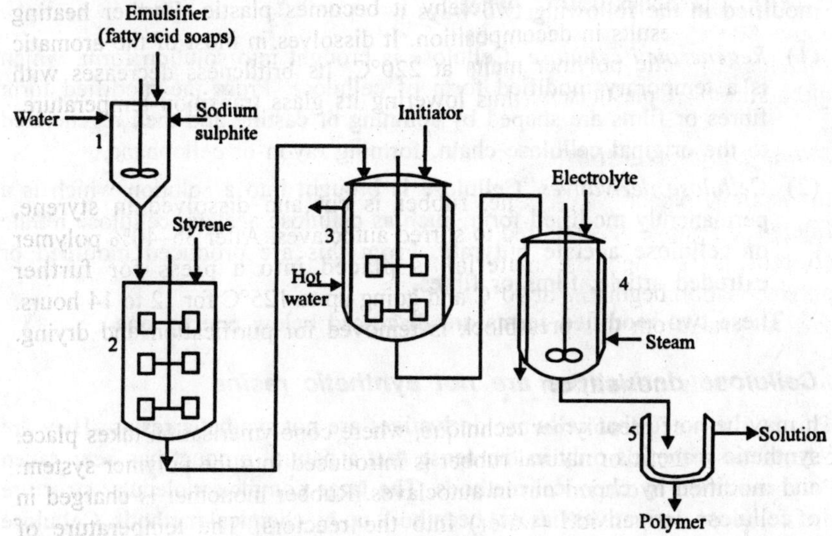

Fig. 6.7. Polymerisation of styrene.

Modified styrenes

Modified styrene also be polymerised in a similar way. Allyl styrenes like α-methyl styrene can be polymerised with difficulty, although copolymerisation is relatively easier. O, m, p methyl styrenes are used in polymer processes. Ethyl toluene derivatives with styrene give on polymerisation ion-exchange resins. Divinyl benzene (0.5 to 15%) is used for production cationic exchange resins. Similarly halogenated styrenes, like dichlorostyrenes are polymerised to give improved heat resistance and fire resistance polymer. Other low molecular weight resins are produced from low grade styrene and phenol or polyesters. Copolymers with styrene are plenty, like methyl methacrylate-styrene, polyphenylene-styrene, acrylonirile-styrene, acrylic acid-butadiene styrene graft polymer etc.

Cellulose

Cellulose is the name given to a long chain of atoms consisting of carbon, hydrogen and oxygen arranged in plant life, especially in cotton and wood.

Cellulose and Cellulose Derivatives

Cellulose, as it occurs in plant life, cotton or wood cannot be shaped, moulded, cast or extruded. To carry out these processes cellulose is modified in the following two ways :

(1) *Regenerated cellulose.* Cellulose is brought into solution form, which is a temporary modified form of cellulose. From the modified form fibres or films are shaped by spinning or casting and then regenerated to the original cellulose chain, forming rayon or cellophane.

(2) *Cellulose derivatives.* Cellulose is brought into a solution which is a permanently modified form, such as cellulose acetate, cellulose nitrate or cellulose acetate butyrate. From this are produced moulded or extruded article, films or fibres.

These two modified forms are discussed below one by one :

Cellulose derivatives are not synthetic resins

It may be noted that cellulose derivatives are not synthetic resins. They are synthetic materials only in the sense that a natural product has been taken and modified by chemical methods. The large complex molecular structure of cellulose derivatives has not been built up by chemical methods. Cellulose from which its derivatives are formed, has already complex large molecular structure formed by nature and this complex structure of the cellulose contributes to the outstanding physical properties of the cellulose products.

Manufacture of cellulose in its pure form

Cellulose is wide distributed in nature. It is the essential constituent of trees, corn stalks and grasses. In its purest from it is found as cotton fibres. The commercial sources of cellulose are wood and cotton fibres.

(1) *From wood.* Trees are debarked, chipped and treated chemically to remove the unwanted lignin and other structural constituents of wood which are mixed with cellulose in its natural form. About 60% of cellulose is recovered from the wood in the form of chemical wood pulp. It is supplied to the manufacturers of cellulosics in the form of large rolls or sheets of blotting paper or as compressed bales.

(2) *From cotton linters*. The small fibres left on the cotton seed after the long fibres are removed are known as cotton linters. Long fibres are used for the manufacture of thread for textile purposes but the short fibres are useless for textiles. These small fibres are removed from the cotton seed in special delintering machines. Then unwanted dirt, wax, seed coat cells etc. are removed from these fibres by mechanical and chemical treatment. About 82% of purified cellulose is recovered from raw linters. Cellulose from cotton linters is generally supplied in compressed bales.

Regenerated cellulose

If cellulose is brought into a solution by a chemical reaction and it is reprecipitated out of the solution to give essentially the same chemical structure but in different form (film or fibre), it is said to be regenerated cellulose. In this transformation of shape, chemical and physical properties do not change much. The following are the two forms of the regenerated cellulose.

Cellophane

Cellulose obtained from wood pulp or cotton linters is taken and is treated with caustic forming soda cellulose. The chemical reaction is represented by the following equation were R stands for cellulose and only one OH is shown for simplification.

$$\underset{\text{Cellulose}}{[C_6H_7O_2(OH)_3]_n} + NaOH + H_2O \rightarrow \underset{\text{Sodium cellulose}}{[C_6H_7O_2(OH)_2ONa]} + H_2O$$

Soda cellulose is shredded into pieces and is aged under controlled conditions of temperature and humidity. During ageing, the original high polymer molecule breaks down a little. Breakdown is indicated by better solubility in solvents.

Aged and shredded pieces of soap cellulose are dissolved by treating with carbon disulphide followed by an aqueous solution of caustic soda. A viscous orange solution which is known as viscose is formed.

Viscose solution is filtered to remove foreign bodies, and then vacuum debubblising is done to remove bubbles etc. Finally ageing is done to adjust the final viscosity.

Viscose solution is then extruded through a narrow slot into an acid solution (sulphuric acid). Acid bath acts as a coagulating bath and cellulose is regenerated from the viscose solution.

Due to the extrusion through a narrow slot, the regenerated cellulose forms a film called cellophane. This film is thoroughly washed, desulphurised, bleached and re-washed to remove the traces of regenerating and treating chemicals. The washed film is treated with a softening agent which is generally glycerine and finally dried.

Rayon

Rayon is the generic name given to fibres (filaments cut into small lengths) and continuous filament yarns made of regenerated cellulose. The method of manufacture is the same as in the manufacture of cellophane. The only difference is that the viscose solution is extruded through spinnerets (small holes) into the acid bath instead of extruding it through a narrow slot into the acid bath. Fibres are formed. The fibres of regenerated cellulose are similarly washed, desulphurised, bleached and re-washed to remove traces of all the chemicals and finally dried.

If fibres are dried and sold in the form of continuous filament yarn, it is called viscose rayon yarn. If fibres are cut into short lengths before drying, they are called staple fibre. These staple fibres are very useful for blending with other fibres and spinning into threads.

Cellulose derivatives

Cellulose derivatives are of two types as already described: (i) cellulose esters, and (ii) cellulose ethers.

Cellulose esters

If cellulose is dissolved by chemical reaction, for example, by forming an ester and this ester is precipitated, the resulting compound, that is, the ester, gives entirely different physical and chemical properties. For use this material is re-dissolved in some solvent or compounded with other materials for moulding or casting into desired shapes.

Cellulose esters of commercial importance are nitrocellulose (or cellulose nitrate), cellulose acetate, cellulose acetate butyrate, cellulose propionate and cellulose acetate phthalate.

(1) *Nitrocellulose (or cellulose nitrate) and celluloid.* Nitrocellulose is prepared from cellulose by treating it with concentrated nitric acid in the presence of sulphuric acid as catalyst. Sulphuric acid acts as dehydrating agent and it also hydrolyses the cellulose.

Application of cellulose nitrate depends on its nitrogen content. Theoretically cellulose mononitrate contains 6.77% nitrogen, cellulose dinitrate has 11.16% nitrogen and cellulose trinitrate has 14.16% nitrogen. The following Table shows the main applications of different forms of cellulose nitrate with different nitrogen content :

Nitrogen content	*Main applications*
12.2 to 13.8%	Sporting Powders, Explosives
11.5 to 12.2%	Lacquers
11.2 to 11.7%	Films, Rayon
10.5 to 11.5%	Celluloid (plastics)

Cellulose nitrate suitable for plastics is made by using the following acid mixtures :

Sulphuric acid	75.5%
Nitric acid	15.75%
Water	9.25%

The ratio of acid to cellulose is 50 : 1. Nitrating temperature is 20°C. The cellulose nitrate obtained has nitrogen content of 10.9%. About 135 lb. of acids are used for every 100 lb. of cellulose nitrate manufactured. After nitration the whole mixture is centrifuged and excess acid is removed and brought up to strength and is re-used.

Cellulose nitrate obtained is dropped into a large tank of cold water where remaining acid is removed. In order to completely remove sticking acids, it is washed a number of times with water and is finally treated with boiling water.

Cellulose nitrate so obtained is bleached and again washed. To remove water it is washed with alcohol in a centrifuge. The resulting product is commercial cellulose nitrate and contains about 33% of alcohol.

This commercial form of cellulose nitrate is mixed with camphor which acts as a plasticiser, dyes or pigments in a mechanical heavy dough mixture. The dough obtained is filtered to remove any coarse particles etc. The material obtained is celluloid.

Celluloid is converted into sheets on hot rollers. These sheets contain large quantities of solvent which is removed by drying which takes a number of days. In drying the sheets are distorted. To remove this defect they are finally polished by placing them between polished nickel plates and subjecting them to heat and pressure.

Cellulose nitrate sheets (celluloid sheets) are used for moulding. For extrusion the sheets obtained (before polishing) are cut into pellets which are fed into the extruder. Tubes and rods are manufactured by extrusion.

(2) *Cellulose acetate butyrate.* These are white pelletes or granules. A thermoplastic resin. Sp. gr. 1.2; soluble in ketone; organic acetates; lactates; methylene; ethylene; propylene chlorides and high boiling solvents. Excellent weathering properties; low water absorption; low heat conductivity and high dielectric strength; high resistance to oil and grease. Combustible. Non-toxic. These can be prepared by the reaction of purified cellulose with acetic and butyric and hydrides with sulphuric acid as catalyst and glacial acetic acid as solvent. The ratio of acetic and butyric components may be varied over a wide range. These are used in moulding composition; film and sheet; photographic film, lacquers, protective coating solutions; water desalination membrane; piping and tubing; packaging; etc.

(4) *Cellulose acetate propionate:* This is similar to cellulose acetate butyrate but made with propionic anhydride instead of butyric anhydride. It is usually stable; requires less plasticiser; and is compatible with more plasticisers than the butyrate.

(3) *Cellulose acetate phthalate:* This is a reaction product of phthalic anhydride and cellulose acetate. It is used for coating of tablets and capsules.

Cellulose ether

The most commonly used cellulose ethers are ethyl cellulose and methyl cellulose.

(1) *Ethyl cellulose.* These are white; granular; thermoplastic solid; lowest density of the commercial cellulose plastics. Properties varies with extent to which hydroxyl radicals of cellulose have been replaced by ethoxy group. Standard commercial product has 47–48 per cent ethoxy

content; sp. gr. 1.07–1.18; refractive index 1.47; high dielectric strength; softening point 100–130°C. Soluble in most organic liquids and compatible with resins, waxes, oils, and plasticisers; inert to alkalies and dilute acids. Insoluble in water and glycerol. Non-toxic. Combustible.

These can be prepared from alkali cellulose and ethyl chloride or sulphate; from cellulose and ethyl alcohol in the presence of dehydrating agent.

These are used as hot melt adhesives and coatings for cables, paper, textile etc.; exclusion wire insolation protective coatings; pigment grinding basis; toughening agent for plastics, printing inks, moulding powders, proximity fuses; vitamin preparations; casting for rocket propellants; food and feed additives.

(2) *Methyl cellulose:* These are in the form of white powder. Swells in water producing clear to opalescent, viscous, colloidal solution. Insoluble in anhydrous alcohol, in ether and in chloroform. Combustible. Non-toxic.

These are used as food products (except confectionary). As thickening agent, stabiliser, emulsifier; thickener in paint stripping preparations.

Cellulose Acetate

Cellulose acetate is manufactured in the following operations :

(1) *Pretreatment of cellulose.* Pure cellulose is dried at a temperature of about 50°C to a moisture content of less than 4%. After this it is pretreated with small amounts of mineral acids in order to increase the reactivity of cellulose to break it down to make it more soluble.

(2) *Esterification with acetic anhydrides.* To 100 parts (dry) pretreated cellulose is added to a mixture of 200 parts of glacial acetic acid, 100 parts acetic anhydride and 100 parts concentrated sulphuric acid. The mixture is kept below 10°C. After two hours the mixture first becomes fluid and then starts thickening up. The process is finished when no cotton fibres remain. It takes generally about 5 hours.

(3) *Hydrolysis or ripening.* The material formed after acetylation is cellulose triacetate which is soluble only in chloroform, trichloro ethylene etc. It is also known as primary acetate. Water is carefully added so that there is no undue rise in temperature while residual acetic anhydride is converted into dilute acetic acid. Ripening is carried out by allowing the mixture to stand in large, open copper pans at about 25°C for 4 to 5 days in presence of dilute acetic acid. During ripening

acetyl groups are split off and the acetyl value of the cellulose acetate decreases. Ripening is allowed until the acetyl value has reached the corresponding value of cellulose diacetate. As the ripening continues the material becomes golden syrup. Degree to which primary acetate is ripened depends upon the purpose for which it is required. Cellulose triacetate has acetyl value of 62.5%, cellulose diacetate has 48.8%, and cellulose monoacetate 29.4%.

(4) *Precipitation.* Water is added to the syrup very carefully. Cellulose acetate is precipitated as a finely divided white material. It is centrifuged and washed.

(5) *Stabilisation and drying.* It is stabilised by boiling with very dilute mineral acid for about 2 hours. Sulphuric esters are removed by so doing. Cellulose acetate is then dried at low temperature until moisture content is below 3%. Then it is ground to a coarse powder.

Properties of cellulose acetate

Cellulose acetate is a white, odourless, tasteless, non-toxic solid. It is manufactured having substitutions ranging from 2.3 to 3.0 or from 53 to 62.5% combined acetic acid. The triacetate having 60.0 to 62.5% acetic acid is not used much on account of its poor solubility in ordinary solvents. Cellulose acetate which finds commercial application has 53 to 59% combined acetic acid. Cellulose acetate used for plastics has acetic acid value between 52 and 54% and for lacquers it has acetic acid value from 54 to 55%.

Polyvinyl Acetate

It is a thermoplastic resin and is a vinyl acetate polymer. Polyvinyl acetate is prepared from vinyl acetate which is a liquid boiling at 72°C and is obtained by the interaction of acetylene and acetic acid in the presence of a complex mercury catalyst.

Polyamide Resins

Polyamide resins are synthetic polymers which have recurring amide groups. Ther term nylon refers specifically to those synthetic polyamides which are capable of forming fibres.

Methods of preparation

The following are the brief general methods of their preparation.

(1) *By the reaction of dicarboxylic acids with diamines*. For example, nylon 6, 6 and nylon 6, 10 are products of the condensation reaction of hexa-methylene diamine with adipic acid and with sebacic acid.

(2) *By the condensation of amino acids*. For example, nylon 6 and nylon 11 are obtained by the self-condensation of ε-amino caproic acid and ω-amino undecanoic acid.

(3) *By the reaction of so-called polymerised vegetable oil acids with polyamines*. The unsaturated fatty acids in vegetable oils such as linoleic acid may dimerise or polymerise to low polymers through their unsaturated groups. The di or polycarboxylic acids obtained yield tough polyamides by condensation with di or polyamines.

Properties and uses

Polyamide resins are horny, whitish, translucent, high melting polymers. Polyamide resin can be essentially transparent and amorphous when their melts are quenched. On annealing or cold drawing, they become highly crystalline and translucent. These polymers are used for fibres, bristles, bearings, gears, moulded objects, coatings and adhesives. The term nylon refers specially to those synthetic polyamides which can be spun into fibres The following are the properties of nylon:

(1) Nylon is a thermoplastic material. The natural colour of nylon is milky white but coloured materials are also available.

(2) Nylon is used for injection moulding and extrusion has high melting points, 160°C to 264°C. Plasticised nylon flows easily.

(3) Nylon is insoluble in common organic solvents such as methylated spirits, benzene and acetone. It is soluble in phenol and formic acid.

(4) Nylon has low specific gravity, 1.1. It has high water absorption.

(5) Nylon mouldings and extrusions have good physical strength, especially high impact strength.

(6) Nylon 6 : 6 and nylon 6 : 10 are primarily used for fibres. However, they are also employed in moulding compositions.

(7) Nylon 6 and nylon 11 are mainly used as moulding compositions for the production of machine parts such as gears and bearings and for electrical mountings. They are also used as filaments, bristles and films. Nylon bearings and gears perform quietly and need little or no lubrication.

Unsaturated fatty acids polyamides are employed as coatings and adhesives and are also used in epoxy and phenolic resin formulations.

THERMOSETTING

Thermosettings are long chain molecules that give the polymer cross-links in three dimensional infusible structures. These are produced by condensation reaction. Setting is caused by unidirectional polymerisation under heat and pressure resulting in hard rigid masses; the way they differ from thermoplastics. Economically they are not regarded as important as there is no scrap value, hence they are not produced in such large tonnage as thermoplastics.

Cross-linking occurs, during curing or polymerisation or with cross-linking agents. The demands for strong and stiffest polymers for special uses cannot be over looked, hence they have been produced on a limited scale in the category of engineering plastics. Reinforcement to improve the quality of thermosets is a common practice. The most widely used material for reinforcement is glass fibre; others include synthetic fibres, cotton, paper.

Phenol-formaldehyde, urea-formaldehyde, polycarbonates are among the older engineering plastics. These plastics are exceptionally strong, dimensionally stable, and resist abrasion, wear and impact; and work well over a wide temperature range.

Sheet moulding compounds (SMC) are made by sandwiching resin, reinforcement fillers etc. with the aid of thickners between two sheets of plastics. Thermosetting resins (plastics) are used in several ways, like moulding powders casting resins, impregnating resins or manufactured goods, sheets, rods and tubes (extruded shapes). Polysulphones, polyacetals because of the transparent nature, toughness, high heat distortion temperature have penetrated into medical appliances and devices. New combinations of resins like styrene-maleic anhydride (SMA), impact modified nylons-6; ABS/nylons, ABS/PCs are also classified as new engineering polymers because of the high performance and rare properties they exhibit. And new developments to make them fire-proof to replace metals are in progress round the world.

Plastics are divided into the following categories depending upon the prominence of use.

Resins for Moulding Purpose

(i) Phenol-formaldehyde; (ii) phenol-furfural; (iii) urea-formaldehyde; and (iv) melamine formations.

Resins for Adhesives

(i) Urea-formaldehyde; (ii) Melamine-formaldehyde; and (iii) Phenol-formaldehyde.

Resins for Coatings, Varnishes etc.

(i) Phthalic anhydride-glycerine; (ii) maleic anhydride-glycerine; (iii) phenol-formaldehyde; (iv) urea-formaldehyde; and (v) melamine-formaldehyde.

Phenol-Formaldehyde Resins

These are widely used in electrical and automative industry. Phenol-formaldehyde polymers can be produced in a wide variety of forms because of bi and tri functional groups. The (condensation) reaction between phenol-formaldehyde results in two types of resins known as Novolacs and Resols. The condensation reaction proceeds in alkaline as well as in acid medium producing two distinct types of resins. When phenol and formaldehyde react in acidic medium with a slight excess of phenol or equimolal quantities of phenol and formaldehyde o- and p-hydroxy benzyl alcohols results.

These resins are soluble in many organic solvents and have a good demand in paint and varnish industry. To produce thermoset from this resin, it is heated and cured by mixing with a small amount of hexamethylene tetramine to effect cross-linking. Instead of acidic medium if alkaline medium with excess of formaldehyde is used, the condensation reaction leads to di, or trimethylol phenols in stages. In the first stage phenoxide is formed.

In the manufacture of phenolic resins, it is possible to produce a resin fit for many different applications, like resins for moulding, heat resistance duties, or coatings, enamels, or adhesives, specially for plywood industry.

Manufacture of phenol-formaldehyde resins

Technically speaking resins of three stages A, B and C are produced in industry. 'A' stage resenoid is primary fusible end product ($H_2C - C_6H_4(OH)\text{-}O\text{-}C_6H_5$). The resenoid can be heat hardened by using a catalyst. Resenoid of A, if produced from basic catalyst, can be very easily made into hardened 'C' stage. On the other hand, if 'A' resenoid is produced by acidic catalyst, 'C' stage resin will not be hard, hence suitable only for paints and varnishes. Commercial 'B' stage resins are those, which are sold to the moulder and contain A resin plus pigments, plasticisers, fillers, colourants, lubricants etc. and can directly go into moulds. After curing they become hardened. Fig. 6.8 shows the different routes by which these resenoids are differed for various designations. Phenol and formaldehyde along with the catalyst are charged into the reaction kettle equipped with heating coils and reflux condenser. Equal moles of phenol and formaldehyde with Na(OH) of 3% of wt of phenol, make a charge for base catalysed resenoids. The reaction is allowed to proceed in a kettle (1) kept at 100°C; during the reaction water separates out and floats as a layer. If the base is much then neutralisation is required (usually lactic acid is used). The bottom layer from the kettle contains the resin in its plastic state. It can be mixed with alcohols or other solvents and used as paints. The resins of this type exhibit hydrophobic properties. The fusible 'A' resin with fillers and dyes is gently heated to harden for rigid plastics. The mixture after hardening is ground in a grinder to powder which is sold as moulding powder. In one step resenoid production always hydrophilic type resin is favoured. In this case, a high amount of formaldehyde to phenol is taken with Na(OH) catalyst in excess quantity. Some water remains in resin in colloidal form hence they are called hydrophilic resins.

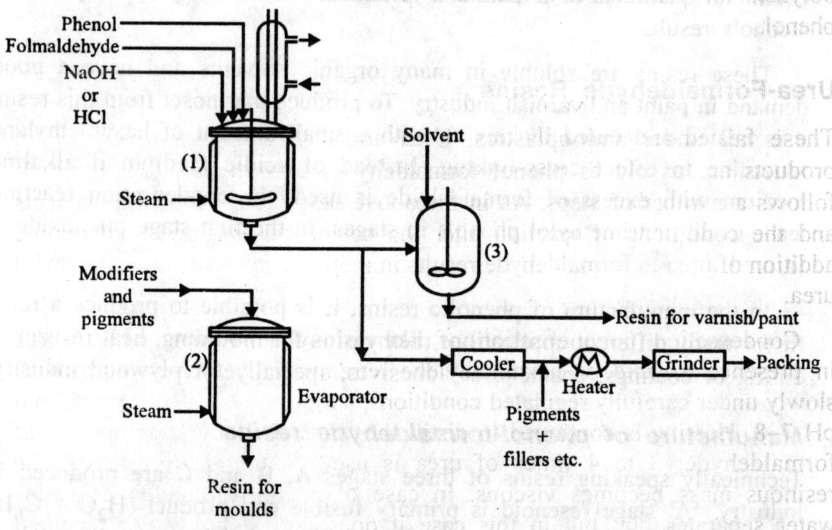

Fig. 6.8. Flow diagram for phenol formaldehyde.

Neutralisation of excess alkali has to be done by using an acid. The resin produced is soluble in alcohols, and the resin can also be plasticised with other reagents. Further, mixture is fed into castings and kept at 80–90°C for several days so that heat curing may take place, then they become influsible. Alternatively resin from the kettle is run down into pans to evaporate (2) the water and the material is allowed flow on to the floor to harden into thick mass. The resin is now in fusible state. Resin produced by acid catalysation also adopts the same technology except change in catalyst.

Acid catalysed reaction follows electrophilic substitution leading to novolac resins, while base catalyst gives a solid resin known as resol softening at 65–75°C. Resol when heated polycondensation sets-in resulting a cross-linked polymer, which now becomes a thermoset and resists heat up to 280°C. At higher temperatures degradation occurs. Novolacs can be transformed into resols by adding formaldehyde and resols can be converted into novolacs by adding phenol. Novolacs are mixed with hexamethylene diamine (initiator) before moulding powders are produced. Laminated plastics for fabricating very strong articles are obtained by impregnating layers of various fillers with resin solution. The solvent is removed and the sheets are compressed in press at 80–100 kg/cm^2 until they harden. Moulds are cured at a temperature of 140–160°C for general purpose as well as shock resistant materials. A good heat fusion and soluble polymer for varnishes and paints is produced by substituting cresol for phenol.

Urea-Formaldehyde Resins

These fall under aminoplastics. Urea and foamaldehyde condensation products are as old as phenol formaldehydes and condensation reaction follows a number of steps. As in the above case the ratio of the reactants and the conditions of reaction alter the course of reaction. Nucleophilic addition of urea to formaldehyde results in monomethylol urea or dimethylol urea.

Condensation (polymerisation) of methylol urea takes place when heated in presence of acid or hexamethylene tetramine. The reaction proceeds slowly under carefully regulated conditions, temperature is kept at 40°C and pH 7–8. Heating is continued to distill the water. Usually large amount of formaldehyde; 3 to 4 times of urea is used. After water removal the resinous mass becomes viscous. In case of phenol-formaldehyde resins water separates out, but in this case it does not. Hence water repellent materials like thiourea or cellulose are added to aid the separation of water from the product. Highly viscous resin is allowed to set gradually after fillers are added. Acid catalysts like bromine, tartaric acid are added before the resin is kneaded. Under vacuum the acid catalysts are driven off, subsequently heat promotes the polymerisation. Thermosetting resin is mixed with ammonium chloride, which works as a hardener, before moulding.

While it is only urea that can give a polymer with aldehyde, other substituted ureas cannot give, as methylene bridges with nitrogen are absent ($-N=CH_2$) i.e. NH_2 group acts as a primary amine and an amide.

Phenol-Furfural Resins

Phenol-furfural resins do not have features of phenol-formaldehyde resins which exist in A, B and C stages. They can be kept in liquid state for a long period and brought into presses or moulds as heavy liquids. The resin in cold state is mixed with fillers and used as powder, as we see in phenol formaldehyde resin. The heat hardening temperature is approximately 180°C that is why they are resistant to heat. The fillers, lubricants accepted by phenol-formaldehyde resins are also accepted here.

Urea-Furfural Resins

These resins have been known only for chemical interest and confined to laboratory only.

Melamine-Formaldehyde Resin

Melamine is produced from urea under high pressure in an atmosphere of ammonia. Melamine reacts with formaldehyde through a nucleophilic addition to give methylolamine. A variety of methylols are possible with methylamine because of the availability of six replaceable hydrogens.

Amino compounds, primarily the melamines, can be modified chemically or by use of fillers. Melamines are moulded at 150–175°C under a pressure of 150–350 bars. Thermosetting moulding compounds are produced with α-cellulose as the filler.

Articles made from amine resins appear water-clear, hard, strong but can be broken. They are used like urea formaldehyde resins in laminations. These are good adhesives and can be safely used for moulds too. Textile treatment is recommended because of the less water absorption capacity. The single greatest use of melamine moulding compounds is in the production of dinnerware. Like other amino resins these can be used for binding purposes, such as abrasive paper, wood boards, flooring and furniture assembly.

Epoxy Resins

Epoxy resins are termed as general purpose polymers and contain at least two epoxide groups made by reacting diphenol with epichlorohydrin. These can be produced in forms ranging from low viscosity liquids to high melting solids just by varying the repeating units in polymer molecule. Epoxy resins are cross-linked by reaction with hardners, leaving excellent wetting adhesive properties and can give composites with fillers like glass fibres, asbestos, paper fibres, cotton etc. Infact most important attribute of epoxies which they share with polyester resins is their ability to convert from liquids to thermosets without pressure and without any volatile by-products. Resins having melting point range of 64–76°C or 97–103°C are used for coatings. Commercially these resins are made by reacting bisphenol-A and epichlorohydrin. Condensation takes place in basic medium at 60–70°C. It is a ring opening reaction; a neucleophilic substitution. The acid liberated is neutralised by base present in the medium. 4% caustic acts as catalyst. Toluene is added to separate the salt from the crude product. Toluene can be later removed by distillation.

Most of these resins are soluble in organic solvents like acetone, toluene, MEK. They are having good mechanical properties, exhibit good electrical insulation properties even in presence of water. High adhesive strength, good chemical and heat resistance have made them useful in composites with different fillers. Further delicate electrical circuits and instruments are encapsulated in epoxy castings. Obviously properties of these can be modified by chlorine blends, curing agents, modifiers etc. Most of the glycidyl type epoxy resins are made of mono, bi- or poly-nuclear phenols. Epoxy resins are extensively used to bond concrete blocks and in other structural works; matrix materials in high performance composites, such materials are used in aggresive environments. Moisture affects badly the strength of resin. These resins can be made with aliphatic polyols or aromatic polyols.

Aliphatic polyols of ethylene, propylene and pentaerithrytol have been used in aliphatic resins. Epichlorohydrin reacts with polyols in presence of Lewis acids to form chlorohydrin ether which on hydrodechlorination in presence of sodium aluminate gives epoxy resins. Epoxy-phenolics, epoxy modified silicones are also in service for better performance.

Epoxy esters

Epoxy resins when modified with fatty acids produce vehicles for surface coatings. Hydroxyl groups are esterified with two or three fatty acid groups by heating epoxy resin with acid at 150–200°C under an inert gas. Water produced in the reaction is removed by azeotropic distillation. Chain lengthening techniques are used to prepare light molecular weight epoxy ester directly from a low mole weight resin. Carboxylic anhydrides like pyromellitic dianhydride (PMDA) curing agent for glycidyl ethers of bisphenol results in high heat resistant plastics.

Epoxy acrylates

By using monobasic unsaturated acid and epoxy resin vinyl ester resins are formed. Carboxylic acids are used as curing agents. Primary or secondary amines Lewis acids (BF_3), and halogenated amines are also used as curing agents. These are cured by oxidative polymerisation catalysts, like cobalt naphthenate. Araldites available in market are such adhesives. Liquid resins from bisphenol, epichlorohydrins are polymerised by oxidation.

ABS Plastics

These are two phase systems consisting of acrylonitrile; butadiene-styrene copolymers and truly thermoplastic type. There are two basic types of ABS polymers. One type is just a mechanical blend of styrene-acrylonitrile copolymer and butadiene-acrylonitrile rubber. In the other type styrene acrylonitrile copolymer grafted on butadiene-acrylonitrile copolymer of varying amounts. Acylonitrile gives the blend, the desired properties of chemical resistance and tensile and flexural strength styrene gives good thermoplastic characteristics. The heart of the process is grafting. The properties of ABS resins are given in Table 6.2.

Table 6.2. ABS resins properties.

Melt flow (G/Min) 2–45	
Tensile strength (kg/cm^2)	350–540
Flexural strength (kg/cm^2)	520–800
Rockwel hardness	94–110
Flexural modulus (kg/cm^2)	18,000–27,000
Impact strength (kg-cm/cm)	10–40
Specific gravity	1.03–1.05

Styrene and acrylonitrile monomers are added to a latex of polybutadiene and the mixture is agitated at 50°C. This allows the monomer to be absorbed by the polymer latex. Potassium persulphate is used as a catalyst in this system. The emulsion is coagulated after copolymerisation is over. The mix is washed and dried and compounded.

Production of copolymers can be done by usual methods like mass, suspension, emulsion and solution polymerisation. Free radical producing catalysts are common initiators. Emulsion polymerisation is used, where potassium persulphate acts as catalyst and sodium stearate acts as emulsifying agent. Modifiers are usually mercaptans. The reaction is carried out in absence of oxygen at around 0°C. ABS resins may be fabricated by most methods used for thermoplastics, injection moulding gives housings for radios, telephones etc. extrusion gives pipes and fittings and also sheets.

Decorative pieces are added attractions; can be metallised, plated. Although these are flammable; are serviceable in a temperature range of –40°C to 107°C.

Fig. 6.9. shows the production flow sheet of ABS plastics. It contains two sections. First section gives butadiene-acrylonitrile polymer, while the second gives acrylonitrile styrene polymer. These two polymers are then blended to yield ABS polymer.

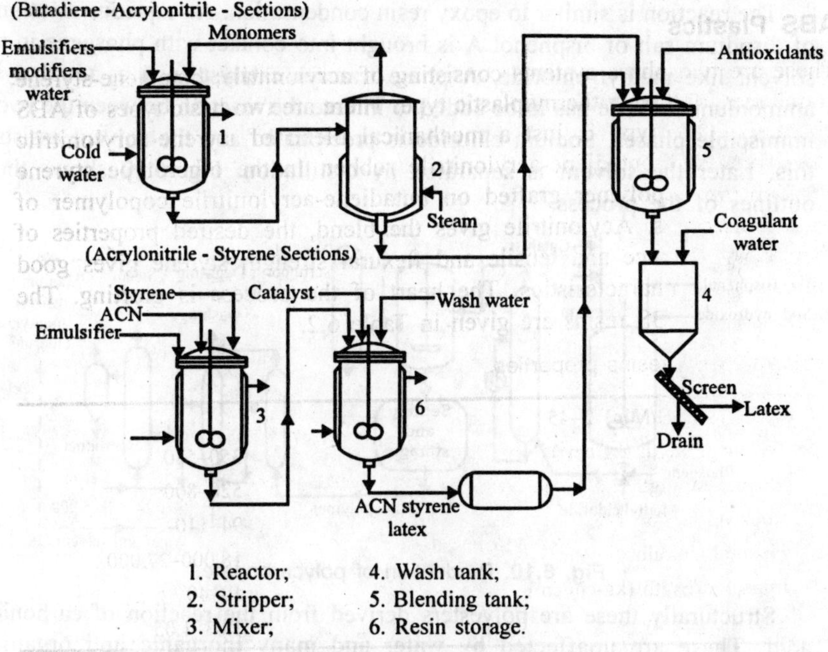

Fig. 6.9. Styrene-butadiene-acrylonitrile plastics.

Reactor (1) contains the water-emulsion, modifiers, short stop and monomers. The polymerisation is carried out and the polymer is stripped off (2) to separate monomers for recirculation. From mixer (3) the acrylonitrile and styrene monomers in emulsion-water is carried to polymerisation reactor.

The polymer material is separated in the form of latex. This latex and latex from the other section are sent proportionally into another graft-polymeriser (5) reactor where antioxidants and modifiers are added. The crumb is washed with water in a tank (4) to remove catalyst and chemicals and sent to drier and mill section.

Polycarbonates (PC)

These are relatively transparent and expensive thermoplastics. These are esters of carbonic acid HOCOOH, and are produced by condensation polymerisation of sodium salt of bisphenol A and phosgene. They have excellent impact resistance with high heat distortion temperature. The credit of transparency and chemical resistance, are widely used in specialised applications.

The reaction is similar to epoxy resin condensation. An aqueous solution of disodium salt of bisphenol A is brought into contact with phosgene in a solvent like methyl chloride. A phase transfer catalyst such as tetrabutyl ammonium chloride has to be added to initiate the reaction between the two immiscible phases. Sodium chloride is precipitated and the solvent assists this. Later the solvent is separated by distillation. Fig. 6.10 shows the outlines of the process.

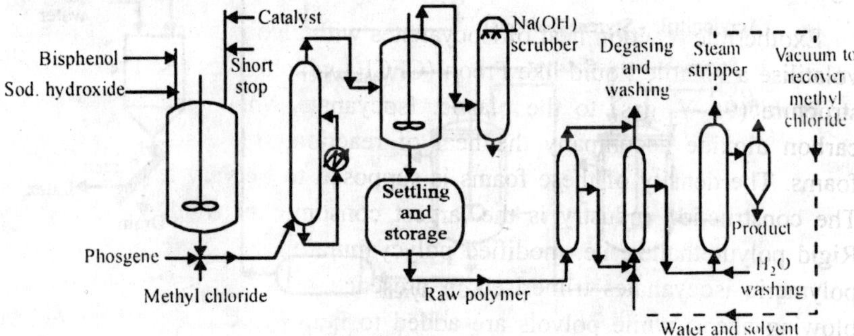

Fig. 6.10. Production of polycarbonate.

Structurally these are polyesters derived from the reaction of carbonic acid. These are unaffected by water and many inorganic and organic solvents. Aromatic dihydroxy compounds when react with phosgene in an inert solvent give rise to interfacial polymerisation resulting in resins. Transesterification is also common. Catalysts like alkaline metal oxides, or alkalies are used during condensation reaction that takes place around 260–300°C. The reaction product shows a molecular weight of 100,000. These are partially crystalline and transparent with high impact strength. Because of transparency, window screens, pans, lenses etc. are made with these and can suitably moulded at a high temperature 500–600°C. They are self-extinguishing type also.

Polyurethanes

Polyurethanes are generally used for load bearing, abrasion resistance works. Their low conductivity and lightness with strength has been widely accepted in refrigeration and packaging industry respectively. Diisocyanates react with any compound having an active hydrogen, such as hydroxyl terminated epoxies, polyesters, furans and cellulose, rubber and groups like SH, NH, NHR, $CONH_2$, substituted carbamide CONHR, SO_2, NH_2, substituted sulphonamide SO_2NHR, thionides, substituted thioamides etc. Only not very active hydrogen containing groups require the assistance of alkaline catalysts rather than metallic or acidic catalysts. Isocyanates of all types are extremely reactive, with water they produce carbon dioxide, when a large amount of heat given. The produced urea again reacts with isocyanates to form substituted ureas to form prepolymers. High molecular weight polyethenes, polypropylene glycols, polyesters are used in the soft foams. Rigid foams do not require high purification of reactants. Polymethylene poly phenyl isocyanate is the most widely used isocyanate in rigid foams.

Exothermic reaction heat of isocyanates with alcohols usually is made to volatilise a volatile liquid like Freon ($CFCl_3$) which gives rise to the foamy structure (95 Y. gas) to the plastic. Isocyanate when reacts with water, carbon dioxide accompany the heat of reaction, this also provides rigid foams. The density of these foams is supposed to be very less (32 kg/m^3). The construction industry is the largest consumer of rigid polyurethanes. Rigid polyurethanes, i.e. modified polycyanurate (PIR) foams comprise of polymeric isocyanates trimerised in presence of a surfactant catalyst and blowing agent; while polyols are added to help processing, i.e. to reduce internal friability and improve adhesion substrates.

Urethane paper based insulations with aluminium foils, produced by continuous laminations are used in buildings for insulation purpose. Glass fibre reinforced laminates yield fire-proof insulation material. The compressive strength of these goes up to 350 KPa with TS touching to 200 KPa. Majority of the laminates contain either TDI or MDI (Diphenyl methane diisocyanate) and NDI (Naphthalene 1, 5 di-isocyanate).

Cast polyurethanes

Castings are done by two shot methods, i.e. first producing prepolymer and then polymerising prepolymer. Excess of diisocyanate at 130°C with polyol gives liquid prepolymer. This liquid prepolymer, is heated to remove the gases. This polymer mixed with chain extenders, and curing agents is then cast into moulds. Curing takes place at 200 to 250°C for 15 to 20 hours.

Polyimides

Polyimides are heat resisting plastics and stable up to 500°C. Polyimides are produced by two stage condensation process. The initial condensation of pyromellitic anhydride with diaminodiphenyl ether in anhydrous dimethyl acetamide yields a soluble acid. Later this is heat cured to an insoluble infusible poly aromatic imide. These can be used for filled polymers, and other industrial applications like piston rings, valve seats, bushings and glass sheets.

Poly Acetals

Polyvinyl alcohol films and fibres are made water resistant by immersion in aqueous formaldehyde. The acetal reaction takes place between hydroxyl groups and aldehyde group resulting vinyl formal form. This form is soluble in many organic solvents.

These are suitable for coatings, and laminating purpose. Some polyacetals derived from butyraldehyde can be plasticised and can be utilised in coating textiles.

EQUIPMENTS USED FOR POLYMER PROCESSING

Polymer processing, the technology of forming plastics, was derived from the long established metal-processing industry, which explains some of the basic similarities of the two processes. Since the early days of polymer processing the understanding of the behaviour of polymers has grown, and this has resulted in the process becoming more refined and hence more efficient. The equipment used for polymer processing tends to be more complex, mainly because molten plastic is a very viscous liquid whereas molten metal is far more free flowing. In simple compression-moulding processes this is of little consequence as the raw plastic is in the form of granules which are introduced directly into the mould cavity. In injection or transfer moulding, however, the granules are plasticised (changed to a molten state) before moulding takes place and viscosity then becomes an important consideration.

All polymer processing techniques require the input of heat to soften the polymer and enable the forming process to produce the required shape. The temperature and period of heating are critical, and are the main source of processing defects. If the polymer is exposed to excessive or prolonged heating, thermal degradation will take place. This will show itself in a reduction of viscosity, oxidation or hydrolysis and, in extreme cases, depolymerisation.

Polymers can be broadly categorised as thermoplastic, thermosetting and elastomers, the names being synonymous with their main characteristic. Thermoplastic polymer softens under the application of heat and so it is heated to make it mouldable, then cooled to hold the shape. The process is reversible and there is virtually no limit to the number of times that this softening/hardening cycle can be carried out. Consequently clean scrap can easily be recycled.

Thermosetting polymer hardens under the application of heat and so it is heated to hold its shape. The polymer undergoes an irreversible chemical change during the heating cycle and cannot therefore be reused.

Elastomers exhibit a highly elastic, rubber like behaviour. They are really a sub-group of either thermoplastic or thermosetting polymers, but behave differently in that they take up their original shape upon removal of a deforming force.

Polymerisation is the reaction in which the raw materials (the monomers) are converted into the products (the polymers). As the reaction takes place, the material is constrained by a mould or die so that the desired shape or form is achieved. Monomers are the relatively pure chemical products of the petrochemical industry and after conversion, the name is usually changed by simply adding the prefix poly :

Monomer	Polymer
Vinyl chloride	Polyvinylchloride
Ethylene	Polyethylene
Propylene	Polypropylene

There are exceptions to this, however. Cyclohexane is the monomer from which nylon is made and the mixture of the monomers acrylonitrile, butadiene and styrene results in the copolymer acrylonitrile butadiene styrene (ABS).

The material used in all the moulding processes is supplied in the form of liquids, sheets, granules or powder. This makes the handling of them very convenient and problems do not arise until the material is being manipulated in the mould. The liquids are monomers which are mixed and allowed to polymerise *in situ* in the mould. The sheets, granules and powders are polymers, the plastics manufacturer having already mixed the monomers to form a polymer resin and then ground this up into a powder or granule. Polymers display an unusual characteristic in that at times they behave like elastic solids and at others like viscous liquids. This visco-elastic nature of polymers is the root cause of a number of problems encountered in the formation of plastic components.

Both viscosity and elasticity can be reduced by increasing the working temperature, but there is an upper limit at which the polymer will begin to degenerate in some way. In addition different types of polymers have different working temperature ranges. PVC for example has a very narrow range in comparison to polythene, whose viscosity can be varied by a factor of ten. Viscosity can also be reduced by the addition of plasticisers. These are liquids which enter the structure of the polymer and enable the molecular chains to slide more easily over one another. The visco-elastic characteristic can also be modified by compounding two or more different polymers.

Apart from altering the visco-elastic nature of the polymer, a number of additives can be introduced either at the mixing stage or during processing, to achieve other desired effects. The type of additive and its effect are summarised in Table 6.3.

Table 6.3. Types if additives and their effects.

Type of additive	*Purpose*
Gases	Expanding and foaming.
Liqids	Plasticisers, lubricants, stabilisers.
Pigments and dyes	Protection of material, protection of packaged contents, decoration.
Fillers (wood, paper pulp, asbestos, chalk, talc, graphite)	Modification of properties, e.g. flammability. Reduce brittleness, increase stiffness. Reduce material cost by acting as extenders.
Metallic	Modification of electrical and magnetic properties and heat transfer characteristics. Protection, e.g. lead-filled PVC radiation screen increase stiffness.
Flexible cotton and synthetic fibres	Improvement in mechanical strength, stiffness and impact resistance.
Rigid asbestos and glass fibre	Considerable increase in Young's modulus and tensile strength.
Metallic fibres	Extends use in high temperature applications.

Compression Moulding

Compression moulding is normally used for thermosetting plastics, although certain thermoplastic articles are made by this process. A measured amount of polymer is charged to the mould, which is then closed and heated. The polymer can be in the form of a loose material charge (i.e. powder or granules) or be a preformed shape, as shown in Fig 6.11. The preform is a low density moulding that has been lightly compacted in a cold mould.

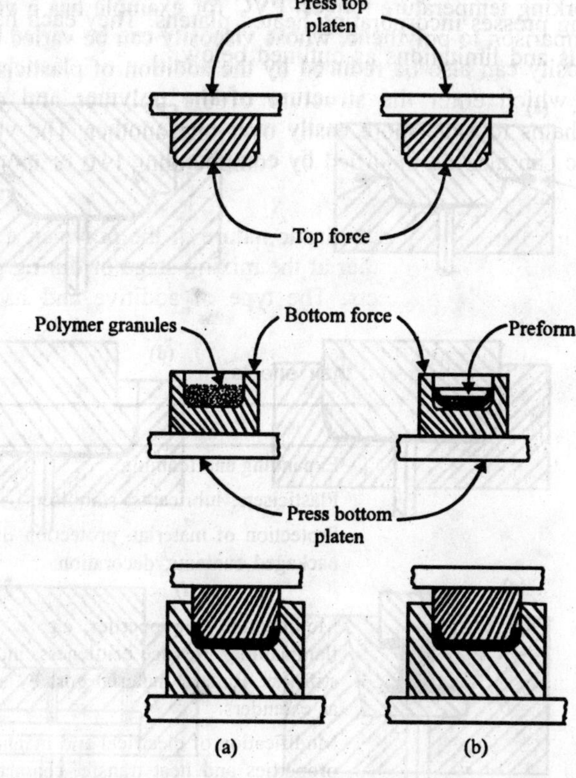

Fig. 6.11. Compression moulding using (a) a loose material charge and (b) a low density preform.

The pressure exerted by the plunger is usually pre-determined to give the required density of the finished article. The heat and pressure cause the polymer to plasticise and flow throughout the cavity. The pressure required may be governed initially by certain additives in the polymer. Heavy, coarse fillings will cause poor flow characteristics which may be impossible to counteract by temperature alone. The shape of the component may also affect the polymer flow. This is particularly true of intricate shapes and deep sections. For this reason, compression moulding is better suited to relatively shallow components having no intricate shapes or inserts. Preforming helps to overcome the flow problem as the uniform, low density preform will not need to flow as much as loose material charged directly into the heated mould. The design of compression moulds varies quite markedly as shown in Fig 6.12. All these would be used in upstroking or downstroking presses incorporating heated platens. They each have various advantageous and limitations as outlined below.

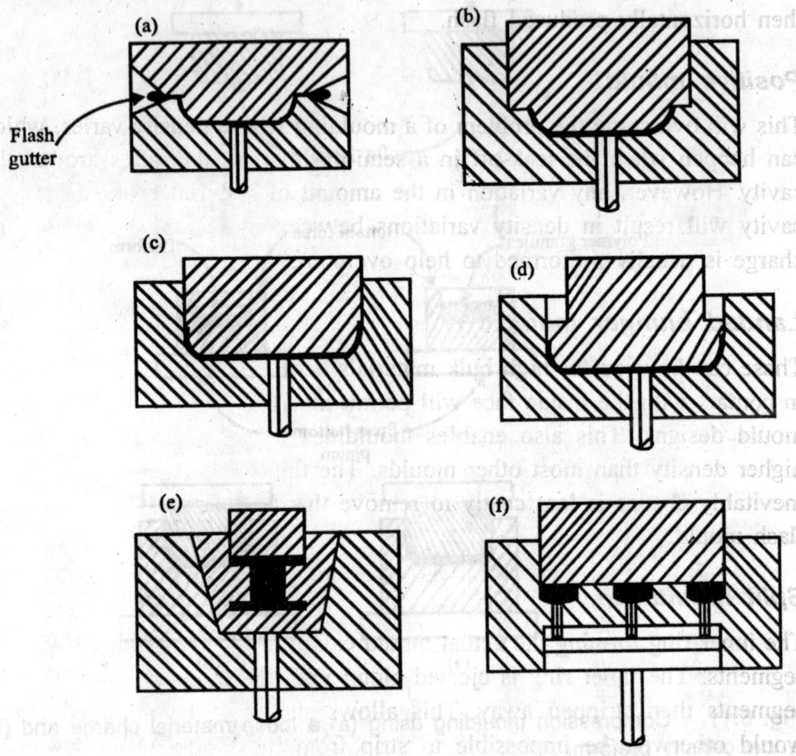

Fig. 6.12. Types of compression mould all with a bottom ejector rod; (a) flash mould; (b) semi-positive mould; (c) positive mould; (d) landed plunger mould; (e) split mould; and (f) gang mould.

Flash mould

The cavity is charged with slightly more material than is needed to produce the finished component, which results in the excess being forced into the flash gutter. A pressure will thus be set up in the cavity balancing the pressure from the press ram, so ensuring the mouldings are all of uniform density. This is the simplest type of mould and therefore the cheapest to manufacture.

Semi-positive mould

Also known as positive flash moulds, because excess material will escape vertically until the plunger engages with the cavity walls. Consequently the mould can be overcharged to ensure the mouldings are all of uniform density. This type of mould is particularly suitable for high bulk factor materials or deep mouldings, and the vertical flash is usually easier to trim then horizontally produced flash.

Positive mould

This will overcome the problem of a moulding whose density varies, which can happen when the leak-off in a semi-positive mould varies around the cavity. However, any variation in the amount of material charged into the cavity will result in density variations between individual moulding. The charge is usually preformed to help overcome the problem.

Landed plunger mould

These can be used for high bulk materials because the reduced land width in contact with the inside face will permit much deeper cavities than other mould designs. This also enables mouldings to be produced of a much higher density than most other moulds. The thin line of vertical flash that inevitably ensues is less costly to remove than a horizontal line from the flash mould.

Split mould

The inner ring forming the actual mould cavity is usually made up of three segments. The inner ring is ejected along with the moulding and the ring segments then stripped away. This allows shapes to be moulded which would otherwise be impossible to strip from the mould. This design of mould is a much cheaper way to produce such components than alternative methods (e.g. machining).

Gang mould

This is an economical method of producing a number of small components in a single shot. The cavities can be machined in a single block as shown, or as separate mating blocks enabling cheaper replacements to be made if damage occurs. The mouldings are connected by a thin line of flash if the measured charge is correct, but if this cannot be controlled accurately, the land between adjacent cavities is kept to an absolute minimum.

Transfer Moulding

This overcomes the main disadvantage of compression moulding by having a separate heated chamber outside the mould cavity. The polymer is first charged to the heated transfer pot and when sufficiently molten, forced by the plunger into the mould below. After cooling the moulding has to be broken from the sprue, the feed pipe from the transfer gate (Fig 6.13). Unfortunately this creates an increase in the downtime of the process, as the material left in the transfer gate (the cull) has to be removed after each moulding cycle and discarded. This adds to the curing time of between one and five minutes for which the polymer has to remain in the mould.

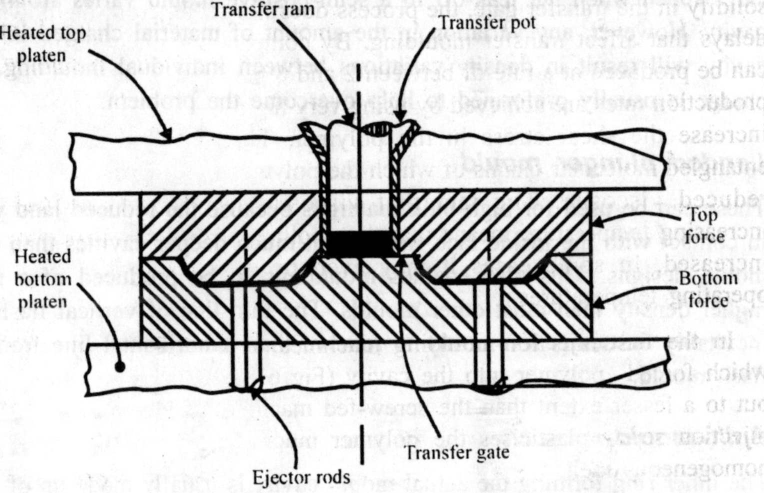

Fig. 6.13. A transfer mould.

The main criterion in the design of the mould is the positioning of the transfer gate. For articles of consistent shape the gate should be positioned such that the polymer is forced to travel the shortest possible distance in filling the mould cavity. However, due to the polymer being plasticised before it enters the mould cavity, thick sections do not suffer from density variations. As this leads to transfer moulds being used for articles having both thick and thin sections, it is better to place the transfer gate above a thick rather than thin section. Just like the multiple or gang compression mould, a multi-cavity transfer mould can be used. This has the cavities arranged radially around the central transfer pot.

The closure of the transfer mould can be done by operating the transfer pot with a separate press cylinder to the ram, or allowing the ram pressure to hold the mould cavity closed. In the latter case the area of the transfer chamber should be at least 25 per cent greater than the area of the contact face of the mould, otherwise there is a danger of the internal pressure forcing open the cavity.

Injection moulding

This is basically a form of transfer moulding which is used for thermoplastics. The polymer granules are forced by a ram or screw, through a heated chamber into a split mould. As the polymer does not solidify in the transfer gate, the process does not suffer from the downtime delays that affect transfer moulding. By comparison, injection mouldings can be produced at a rate of between 2 and 6 per minute. The much higher production rates are achieved by using very high injection pressures, which increase the shear stress in the polymer. This tends ·to drew out the entangled molecular chains of which the polymer is composed, resulting in reduced viscosity. The fact that viscosity can be reduced without increasing temperature means that the cooling cycle for the mould is not increased. In some cases it may even be decreased by using lower operating temperatures.

In the first injection-moulding machines; a simple plunger was used which forced polymer into the cavity (Fig 6.14). This is still used today but to a lesser extent than the screw-fed machine shown in Fig 6.15. The injection screw plasticises the polymer much quicker and gives a more homogeneous melt.

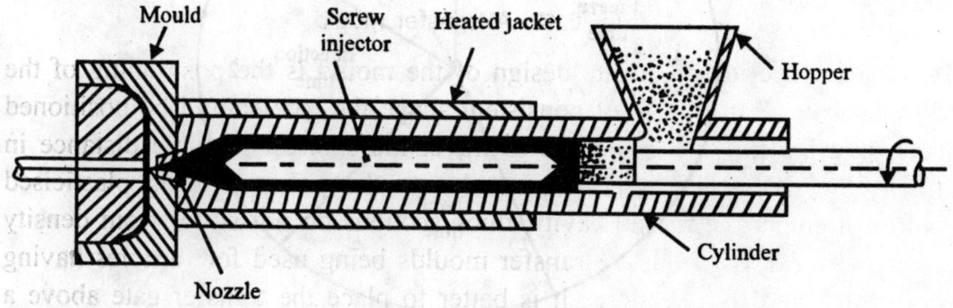

Fig. 6.14. A plunger type injection moulding machine.

The process involves five basic steps in its cycle which is summarised in Fig 6.16.

(1) The split mould closes and is clamped against the injection nozzle.

(2) The plunger or screw draws material from the hopper and pushes it

into the heated barrel. Further along the barrel, plasticised material is injected into the mould.

(3) The pressure is maintained while the polymer is cooling and starts to solidify.

(4) The mould remains closed until the polymer is solid enough for safe removal (during this time the plunger moves back ready to draw in more material—unnecessary with the screw feed).

(5) The mould opens and the moulding is ejected.

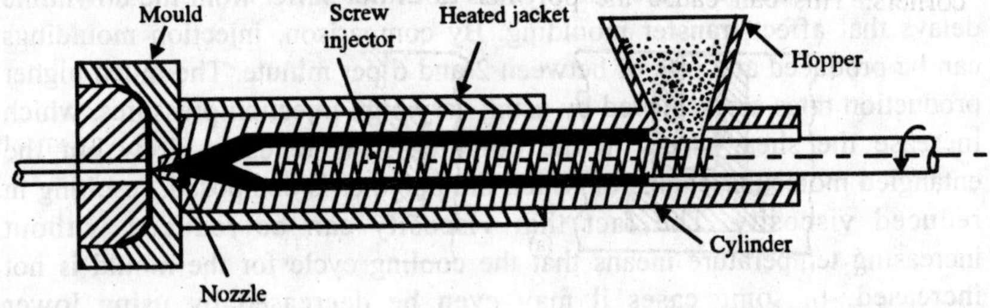

Fig. 6.15. A screw type injection moulding machine.

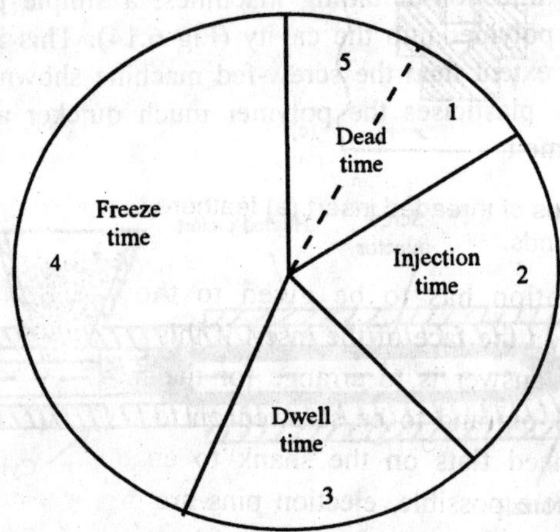

Fig. 6.16. Diagram of the injection moulding cycle.

Inserts in plastics

A fairly common feature of polymer components is some form of insert which has been incorporated during the forming process. Most inserts are metallic and are usually of the form of an external pin or internal bush. The most common type of insert is the threaded insert for although threads can be easily formed in the polymer, the material is often too weak to accommodate the forces exerted on the thread form.

Three different designs for a threaded insert are shown in Fig 6.17. Each illustrates a different method of preventing the insert from being pulled out or turning. Feathered edges alone are to be avoided as the thickness of polymer preventing the insert from being pulled out can easily be sheared. Similarly square or hexagonal inserts can cause the stress, which is developed as a result of the tendency to rotation, to be concentrated at the corners. This can cause the polymer to crack or split.

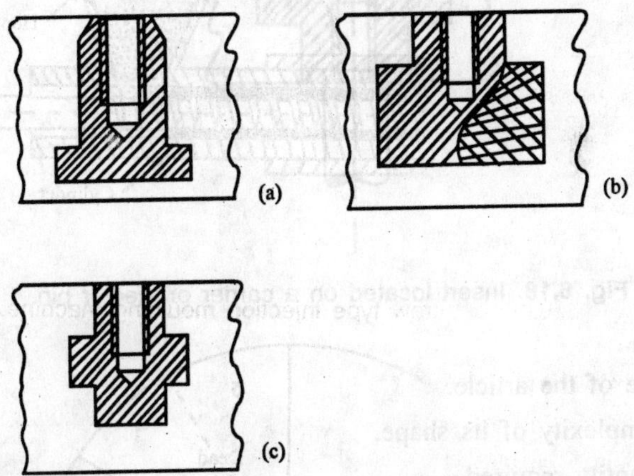

Fig. 6.17. Types of threaded insert (a) feathered edge; (b) knurled periphery; and (c) squared ends.

Consideration has to be given to the possibility of the polymer encroaching on the face of the insert if this is required to be flush with the surface. One answer is to arrange for the insert to be the same diameter as the carrier pin and to be screwed on to it. In this situation, the carrier pin would need flats on the shank to enable it to be unscrewed after ejection. Where possible, ejection pins are arranged to push on the insert carrier, or are designed as a combined insert carrier/ejector pin (Fig 6.18).

Selection of production processes

The correct choice of production process for a particular component often requires a deep understanding of polymer processing and a wide range of experience. A number of factors and variables have to be considered, such as:

(1) The type of polymer – thermosetting or thermoplastic.

(2) The form – liquid, granule, powder or sheet.

(3) The type of additives–solid, liquid, gaseous or metallic.

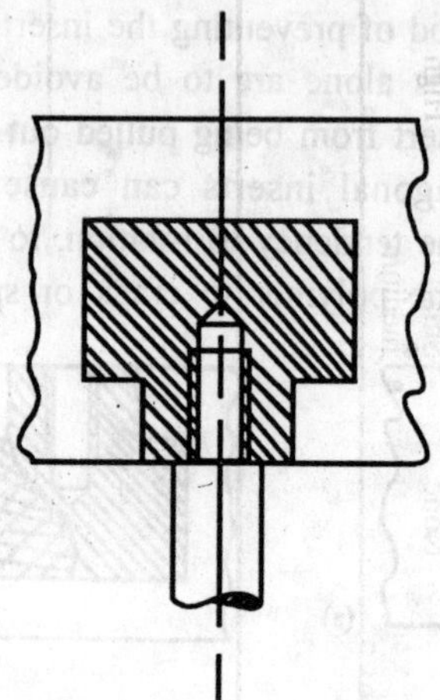

Fig. 6.18. Insert located on a carrier or ejector pin.

(4) The size of the article.

(5) The complexity of its shape.

(6) The quantity required.

(7) Whether inserts are required.

(8) The design of the tooling.

(9) The tool cost.

Table 6.4 gives an approximate guide to the relative merits of the three basic processes. It should be appreciated, however, that no process is strictly limited, as certain factors can sometimes be modified to give greater freedom and adaptability. Injection moulding, for example, appears more suited to very high output of small articles. Despite this the process has been successfully applied to the production of car bodies and boat hulls.

After the production process has been chosen and put into operation, there are numerous faults that can occur in the component. The most common faults and their probable causes (★) are shown in Table 6.5.

Table 6.4. A guide to the three basic mouding processes.

Process	Factor Polymer type	Output rate	Complex shapes	suitable for inserts	Waste material	Component size	Machine capital	Mould cost
Compression moulding	Mainly thermosetting	Low or medium	No	No	Little or none	Medium to large	Medium	Low
Transfer moulding	Mainly thermosetting	Low	Yes	Yes	Considerable	Medium to large	Medium	Medium
Injection moulding	Thermo Plastic	High	Yes	Yes	None	Small to medium	High	High

Table 6.5. Common injection-moulding faults.

Fault	Operating temperature	Injection pressure	Probable cause Injection time	Cooling time	Mould locking force	Gate design
Brittleness	★	★				
Bubbles		★	★			★
Crazing	★	★				
Distortion	★	★	★	★		
Flashing	★	★			★	
Gloss	★					
Hollows	★	★		★		★
Sticking	★	★				

Other Processing Techniques

Compression, transfer and injection moulding are the three most widely used methods of polymer processing. There are a number of other basic techniques used by industry for producing the vast array of consumer products, not the least of which is packaging, which accounts for so much of the polymer that is manufactured.

Calendering

Three or four heated rolls in a single vertical stand are used for producing continous thermoplastic sheet. The sheet can be used as the raw material for further processing or as a finished product.

Extrusion

This uses a machine similar to the screw injection machine illustrated in Fig. 6.15 with an extrusion die in place of the split mould. The solid sections, tubes of various diameters can also be extruded, including tubular film for making bags and sacks.

Blow moulding

Tube is extruded intermittently from an extrusion machine mounted above a split mould as shown in Fig. 6.19. The closing of the mould traps the bottom of the tube which can then be blown to the shape of the mould be means of an air nozzle incorporated with the extruding die. The technique is widely used for bottles and containers, and a threaded neck can be formed by the mould if required.

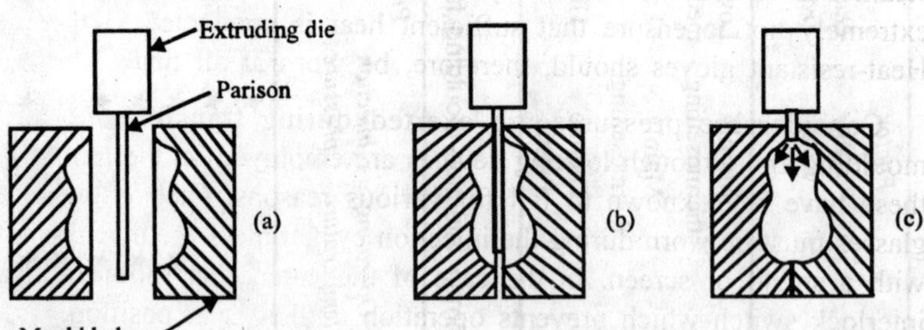

Fig. 6.19. Blow moulding a bottle (a) a tube of plastic (a parison) is extruded between the halves of a split mould; (b) the mould halves close, pinching and sealing the end of the parison; and (c) air is blown into the parison, stretching and forming it to the shape of the mould cavity.

Thermoforming

This describes a number of techniques in which a heated thermoplastic sheet is clamped around its circumference above a male or female die. Air pressure is invariably used to force the sheet to the die shape, either by increasing the air pressure above the sheet or by creating a vacuum inside the die cavity. The process is used for items such as buckets, bowls and wide-necked containers.

Slush casting

A female mould is heated and charged with the slush, a fine polymer powder suspended in a liquid. After allowing time for the slush to coat the inside of the mould, the excess is poured out. Upon solidification, the polymer can be stripped from the inside of the mould. The outside surface of the article reproduces the mould detail to a high degree and the technique is used extensively for manufacturing toys such as dolls and balls.

Dip coating

This is very similar to the slush casting process but is the inverse, in that a male mould is dripped into the slush. Items such as gloves are made in this way, and metal components can be heated and plastic coated in a similar manner.

Safety in plastics moulding

The most dangerous aspect associated with polymer moulding is in the use of power presses with compression and transfer moulds. In addition, when presses are used for polymer moulding the platens are heated, usually by steam but sometimes electrically. In both cases the platen's surfaces are extremely hot to ensure that sufficient heat is conducted to the moulds. Heat-resistant gloves should, therefore, be worn at all times.

Considerable pressures are exerted during transfer and injection moulding and although locking devices are employed on the mould halves, these have been known to fail for various reasons. Face visors or safety glasses must be worn during the injection cycle unless the machine is fitted with a shield or screen. In the case of the latter, this should activate an interlock switch which prevents operation until it is in position.

Suitable ventilation or extraction equipment is necessary with all the moulding processes. When the polymer is curing in the moulds, fumes are given off which at the least are very unpleasant, but in certain polymer compounds can be highly toxic. If mould venting is carried out the fumes are released in great quantities.

Consolidation

Monomers are the raw material from which the polymers are made. They are supplied in the form of liquids, sheets, granules and powders for processing by a variety of techniques of which compression, transfer and injection moulding are the most widely used for engineering components.

Compression moulding is used mainly for thermosetting plastics and employs a wide range of mould designs to counteract some of the many problems inherent in the process. The main problem is the poor flow of the polymer in the cavity and although this is lessened to some extent by preforming, it is still the main disadvantage.

Transfer moulding overcomes this disadvantage by plasticising the polymer outside the mould cavity and then forcing it through the transfer gate in a much more fluid condition. This in itself creates a problem, however, as the thermosetting plastic cures in the gate along with moulding. This cull has to be removed and discarded, lengthening the production time and adding to the costs.

Injection moulding is a very similar process which is used for thermoplastics. By keeping the injection nozzle short and recycling the sprue, however, it does not suffer form the same problems. The screw extruder is more widely used than the ram injector and is more suited to continuous production techniques.

Moulded inserts overcome many of the weaknesses of plastic components, by allowing parts such as bushes and threads to be incorporated in more suitable materials.

It should be borne in mind that there are many other polymer-processing methods which, although having limited use for engineering components, are widely used for consumer products and packing. The most commonly used processes are these: *calendering*—the use of heated rolls for producing continous plastic sheet; *extrusion*—using a screw extruding machine, similar to the screw fed injection moulder, to produce continuous lengths of material; *blow moulding*—extruding a tube into a closed, split mould and then inflating the tube to the shape of the cavity; *thermoforming*—forcing a softened sheet of plastic on to a male form tool by air pressure, or into a female mould by creating a vacuum; *slush casting*—coating the inside of a heated mould with fine polymer powder suspended in a liquid (a slush); *dip coating*—dipping a heated former into a slush (the inverse of the slush casting process).

Chapter 7

Oils, Fats and Waxes

INTRODUCTION

Oils, fats and waxes belong to the class of compounds called lipids, which are insoluble in water, but soluble in ether and other organic solvents. The lipids have long been considered important as one of the three major classes of food materials. They differ from the other two, proteins and carbohydrates, in that they provide more than twice the number of calories per unit weight when converted to carbon dioxide and water. Hence, they were thought to be primarily products for storage of energy in plants and animals. More recently, lipids have been shown to be extremely important in the organisation of the living cell.

Different metabolic functions are carried out in specific parts of the cell, and cell membranes which isolate the parts from one another are made up of lipids and proteins. Although not all of the functions of lipids in cell membranes are well understood, the selective permeability of lipid-membrane constituents certainly plays a role in the isolation of cell functions.

Oils and fats are usually distinguished from one another by their melting points and, to some extent, by their sources. Oils are generally liquid at ordinary temperatures while fats are solid or semi-solid. The fats are usually of animal origin while oils are extracted from plant tissue, fish, or marine animals. Both fats and oils are composed largely of glycerol esters of fatty acids. The esters are formed by combination of carboxylic acids and alcohols with loss of water. Since glycerine (glycerol) is a trihydric alcohol, one molecule can combine with three molecules of fatty acid to form a triglyceride.

$$
\begin{array}{lcl}
CH_2OH & & \\
| & & \\
CHOH + 3C_{17}H_{35}COOH & \longrightarrow & CH_2OOCC_{17}H_{35} \\
| \qquad \text{Stearic acid} & & | \\
CH_2OH & & CHOOCC_{17}H_{35} + H_2O \\
\text{Glycerol} & & | \\
 & & CH_2OOCC_{17}H_{35} \\
 & & \text{Glycerol tristearate} \\
 & & \text{(tristearin)}
\end{array}
$$

Fats and oils obtained from animal or vegetable tissue are predominantly triglycerides with traces of mono- and diglycerides, free fatty acids, phospholipids, sterols, and other minor constituents. Procedures for refining and processing crude oils remove most of the trace constituents.

Waxes differ from fats and oils in that a mono- or dihydric alcohol has replaced glycerol in combination with the acid. Wax alcohols may be either aliphatic or alicyclic. In mammals, cholesterol esters of fatty acids are found, particularly in the adrenals, the liver, and blood plasma. These cholesterol esters are of interest in nutrition because of the correlation of the incidence of atherosclerosis with their levels in blood plasma. Both the acids and alcohols in waxes show considerable variation in type.

In addition to cholesterol (illustrated), other sterols are found in minor amounts in fats either as esters or in an uncombined form. Some of the common sterols found in vegetable oils are stigmasterol, ergosterol, β-sitosterol, and lanosterol. Vegetable oils may also contain xanthophylls and carotenes which impart colour to the oil, and which in many oils is considered desirable, from the standpoint of appearance.

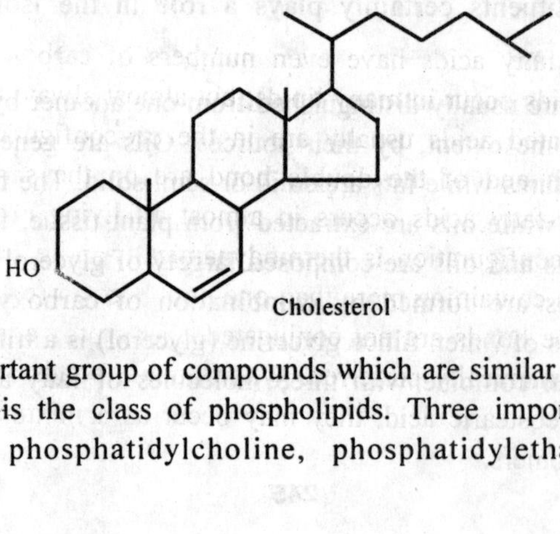

HO

Cholesterol

One important group of compounds which are similar in structure to the triglycerides is the class of phospholipids. Three important phosphatidic esters are phosphatidylcholine, phosphatidylethanolamine, and phosphatidylserine. Phosphatidylinositol is also often present with sugar moieties attached. Phosphatides are separated by the addition of small amounts of water to crude oils. This degumming process produces "lecithin" from soyabean oil, a product which has some commercial value because of its surface-active properties.

$$\begin{array}{ll}
RCOOCH_2 & Y = -CH_2CH_2\overset{\oplus}{N}Me_3\overset{-}{O}H \quad \text{3-phosphatidylcholine} \\
\quad | & \\
RCOOCH & Y = -CH_2CH_2NH_2 \quad \text{3-phosphatidylethanolamine} \\
\quad | & \\
CH_2O\overset{O}{\overset{\|}{P}}\underset{OH}{\underset{|}{}}-OY & Y = -CH_2\overset{CO_2H}{\overset{|}{C}}H-NH_2 \quad \text{3-phosphatidylserine}
\end{array}$$

FATTY ACIDS

The physical and chemical properties of fats and oils are essentially determined by the fatty acid composition of their triglycerides. Most of the common fats and oils are principally mixed triglycerides of five common fatty acids; palmitic, stearic, oleic, linoleic, and linolenic acids. These five acids are unbranched carboxylic acids and except for palmitic acid, they all have eighteen carbon atoms in the chain. Their structures and properties, along with those of some of the many fatty acids appearing less commonly are illustrated in Table 7.1. The fatty acids of this table are only representative of the many hundreds of such acids which have been isolated and identified, but they illustrate some rules, which generally apply in their chemistry.

The common fatty acids have even numbers of carbon atoms. Odd-numbered fatty acids occur in many lipids, but almost always in very minor amounts. Unsaturated acids usually are in the *cis*-configuration, i.e., the hydrogens at each end of the double bond are on the same side. This preference for *cis*-fatty acids occurs in almost all living organisms, even though the *trans*-configuration is thermodynamically more stable. Also, in the common acids containing more than one double bond (e.g., linoleic and linolenic acids) the bonds are not conjugated, though the conjugated acids are also thermodynamically preferred. When double bonds are conjugated, however, as in eleostearic acid, they may occur as a mixture of *cis-trans* or as all *trans* isomers.

Table 7.1. Some important fatty acids.

No. of carbon atoms	*Systematic name*	*Common name*	*Melting point, °C*	*Boiling point, °C/mm*	*Methyl ester (BP°C/mm)*	*Common Sources*
			Unbranched Saturated Fatty Acids			
4	Butanoic	Butyric	–5.3	164	103	Milk fats
6	Hexanoic	Caproic	–3.2	206	151	Milk fats and some seed oils
8	Octanoic	Caprylic	16.5	240	195	Milk fats and *Palmae* seed oils
10	Decanoic	Capric	31.6	271	228	Sheep and goats' milk, palm seed oils, sperm head oil
12	Dodecanoic	Lauric	44.8	130/1	262	Coconut oil
14	Tetradecanoic	Myristic	54.4	149/1	114/1	Palm and coconut oils
16	Hexadecanoic	Palmitic	62.9	167/1	136/1	Palm oil
18	Octadecanoic	Stearic	70.1	184/1	156/1	Animal fats
20	Eicosanoic	Arachidic	76.1	204/1	188/2	Some animal fats
22	Docosanoic	Behenic	80.0	–	206/2	Various seed oils incl. peanut oil
24	Tetracosanoic	Lignoceric	84.2	–	222/2	Minor amounts in some seed oils
26	Hexacosanoic	Cerotic	87.8	–	237/2	Plant waxes
28	Octacosanoic	Montanic	90.9	–	–	Beeswax and other waxes
30	Triacontanoic	Mellissic	93.6	–	–	Beeswax and other waxes
			Branched Saturated Fatty Acids			
5	3-Methylbutanoic	Isovaleric		–51.0	177	– Dolphin and tortoise fats
19	10-Methylstearic	Tuberculostearic		11	–	– Lipid of Tubercle bacillus
			Monoenoic Fatty Acids			
10	9-Decenoic	Caproleic	–	142–8/15	115–6/12	Milk fats
14	9-Tetradecenoic	Myristoleic	–	–	108-109	Some seed fats–milk fats
16	9-Hexadecenoic	Palmitoleic	–0.5 to 0.5	–	134–5/1	Many fats and marine oils

No. of carbon atoms	Systematic name	Common name	Melting point, °C	Boiling point, °C/mm	Methyl ester (BP°C/mm)	Common Sources
18	6-Octadecenoic	Petroselinic	30–33	208–10/10	–	Parsley seed oil
18	9-Octadecenoic	Oleic	16.3	153/0.1	152.5/1	Almost all fats and oils
22	13-Docosenoic	Erucic	33.5	241–3/5	169–170/1	Rapeseed oil
10	2, 4-Decadienoic	Stillingic	–	–	–	Stillingia oil
18	9, 12-Octadecadienoic	Linoleic	–5	202/1.4	149.5/1	Many vegetable oils
22	13, 16-Docosadienoic	–	–	–	–	Rapeseed oil
			Trienoic fatty acids			
16	6, 10, 14-Hexadecatrienoic	Hiragonic	–	180–190/15	–	Sardine oil
18	6, 9, 12-Octadecatrienoic	–	–	–	–	Seeds of the evening primrose family
18	9, 12, 15-Octadecatrienoic	Linolenic	–11 –	157–158/ 0.001/0.002	184/4 –	Linseed oil
20	5, 8, 11 Eicosatrienoic	–	–	–	–	Brain phosphatides
20	8, 11, 14-Eicosatrienoic	–	–	–	–	Shark liver oil
18	9, 11, 13-Octadecatrienoic	Eleostearic	–	–	–	Tung oil
			Fatty acids of more unusual structure			
18	12-Hydroxy-9-octadecenoic	Ricinoleic	–	–	–	Castor oil
20	14- Hydroxy-11 -eicosenoic	Lesquerolic	–	–	–	*Lesquerella* seed oil
18	12, 13-Epoxy-9-octadecenoic	Vernolic	–	–	–	Some *Compositae* seeds
18	8, 9-Methylene-8-heptadecenoic	Malvalic	–	–	–	*Malvaceae seeds*
19	9, 10-Methylene-9-octadecenoic	Sterculic	–	–	–	*Sterculiaceae seeds*
18	13-(2-Cyclopentenyl) tridecanoic	Chaulmoogric	–	–	–	Chaulmoogra oil

Isolation and identification of fatty acids has been facilitated greatly by the development of gas-liquid chromatography. Triglyceride esters are not distillable, but they can be readily converted to methyl or ethyl esters by alcohol interchange, and it is as the methyl esters that acids are usually separated and identified. Even triglycerides may be partially separated by high-temperature gas chromatographic techniques according to the number of carbon atoms and the number of double bonds present. Other techniques such as selective enzyme-catalysed hydrolysis allow determination of the position of fatty acids on the glycerol moiety.

The physical properties of fatty acids are functions of their chemistry. Generally, melting points and boiling points rise with the number of carbon atoms. However, double bonds, particularly those of *cis*-configuration, lower the melting point so that saturated C_{18} stearic acid melts at 70.1°C, while monoenoic C_{18} oleic acid melts at 16.3°C and the diunsaturated C_{18} linoleic acid melts at –11°C. This profound effect on melting point comes about because double bonds break up the regular alternating structure of the hydrocarbon chain which allows the molecule to readily fit into a regular crystal lattice. The boiling points of the acids and their esters are not as greatly affected by double bonds, and separation by distillation is not ordinarily practicable. Branching of the carbon chain usually lowers the boiling point and melting point for acids with the same number of carbons, while polar groups, such as hydroxyl functions, attached to the chain increase the boiling point and the viscosity, though they may decrease the melting point.

TRIGLYCERIDES

Each molecule of triglyceride has three fatty acids and, in general, high-melting fatty acids produce high-melting glycerides. However, in natural fats properties are influenced to some extent by the distribution of isomers in the mixture. Consider a mixture of triglycerides with only two fatty acids, A and B. The following sequences are possible since the middle 2-position of glycerine differs from the 1 and 3 positions:

AAA, AAB, ABA, ABB, BAB, BBB

These may all occur; distribution may be random, but biosynthesis is usually directed so that a given triglyceride may appear in higher concentration than would be expected from strictly statistical considerations. Most oils have many more fatty acids than two, so that the distributional possibilities are numerous. A result of this multiplicity of molecular species is that fats do not possess sharp melting points; instead they freeze or melt over a wide temperature range. The first molecules to form solid crystals during cooling are the ones with the highest saturated fatty acid content.

These may then be removed by filtration to leave an oil with higher unsaturate content. Such a process, called winterisation, is used in making salad oils which can be kept under refrigeration. There are a few exceptions to the rule that fats behave as mixtures of many types of molecules. Castor oil contains 90 per cent ricinoleic acid and it can be considered under many circumstances to behave as triricinolein (glycerol triricinoleate). Cocoa butter which is used in confectionary coatings melts in the rather narrow range of 89–95°F. Its major ingredient is 2-oleopalmitostearin.

$$\begin{array}{l} CH_2OOCC_{17}H_{35} \\ | \\ CHOOC(CH_2)_7CH{=}CH(CH_2)_7CH_3 \\ | \\ CH_2OOCC_{15}H_{31} \end{array}$$

2-Oleopalmitostearin

Although a wide variety of molecular species exist in natural fats, the oils from many plant-seed sources show little variation in over-all fatty acid composition within a given species, especially if the plants are grown under similar climatic conditions. Interesting genetic and climatic variations do occur, however. Ordinary safflower oil usually contains about 75 per cent linoleic acid and 12–15 per cent oleic acid. In a new variety these ratios are essentially reversed. Sunflower oil exhibits considerable variation of fatty acid content with climate. Sunflower seeds grown in the Southern United States have a much lower linoleic to oleic acid ratio than seeds from the same variety grown in India. The composition of depot fats of non-ruminant animals is to a great extent a function of the fats ingested. If a high percentage of unsaturated oils is eaten, depot fats will contain more unsaturated fatty acids than if saturated fats are predominant in the diet. Ruminant animals show more consistency in fatty acid composition, since bacterial flora in the rumen modify diet fats.

EDIBLE VEGETABLE OILS

Soyabean Oil

A representative composition of soyabean oil is: saturated acids, 14 per cent; oleic acid, 28 per cent; linoleic acid, 50 per cent; and linolenic acid, 8 per cent. Iodine value ranges from 120–141. Soyabean oil which has not been hydrogenated tends to oxidise to give reverted flavours, often described as "painty" or "grassy." Consequently it is not commonly used for cooking. It can be partially hydrogenated to an oil of greater stability, and, in this form it is a major component of salad oils (after winterising), shortenings, and margarines.

Cottonseed Oil

This is still a major oil in India, although it is a by-product with production dependent upon the amount of cotton grown. Most cottonseed oil is winterised and used for salad oil. Another important use of unhydrogenated cottonseed oil is in the frying of potato chips and other snack foods. A typical composition of cottonseed oil is about 21 per cent palmitic acid, 3 per cent other saturated acids, 27 per cent oleic acid, and linoleic acid comprising most of the other 49 per cent. Since it contains no linolenic acid it is not readily subject to flavour reversion. Hydrogenated cottonseed oil is very important in the manufacture of shortening.

Peanut Oil

Peanut oil is also a by-product oil, generally coming from low-grade nuts. It contains 13 per cent saturated fatty acids, 60 per cent oleic acid, 22 per cent linoleic acid, and minor amounts of fatty acids with more than eighteen carbon atoms. It is an excellent cooking oil, used by manufacturers of snack foods because of its good flavour.

Sunflower Oil

Sunflowers grow well in cold climates; consequently sunflower seed is a major source of oil in northern countries such as the USSR and Canada. As noted previously, the composition of sunflower oil changes when the seed is grown in warmer climates. Northern sunflower oils contains about 75 per cent linoleic and 14 per cent oleic acid, while southern sunflower oils may have over 50 per cent oleic acid and as little as 35 per cent linoleic. Respective iodine values are 130 for northern sunflower oils and 106 for oil from plants grown in some southern states.

Olive Oil

Olive oil is desired for its flavour; hence the best oil, called virgin oil, is pressed from the fruit pulp of the olive and not refined. Oil extracted with solvent, refined, and deodourised is called pure olive oil and is quite bland. Olive oil is expensive, and is therefore a speciality item in India used primarily for its flavour. It has been blended with other oils for use in salad dressings. A typical olive oil may contain 8 per cent palmitic acid, 2 per cent stearic acid, 82 per cent oleic acid, and 8 per cent linoleic acid. It is relatively low in polyunsaturation, with iodine values ranging from 80 to 90.

Safflower Oil

Safflower is an ancient crop, grown in the Middle East since prehistoric times. It has the highest linoleic acid content of any commercial oil, but it contains no linolenic acid. It is therefore not subject to the severe flavour reversion problems of soyabean oil. With the increasing demand for polyunsaturated oils in the diet, safflower has found use in soft margarines, salad oils, dietetic imitation ice creams, etc. Because of its high polyunsaturate content, it is not a good frying oil, since it tends to produce off-flavours and polymeric products when heated in air.

In composition it is quite similar to olive oil, with a low polyunsaturate content and high oxidative stability at frying temperatures. It can be used as a salad or cooking oil. Regular safflower oil contains 75–80 per cent linoleic acid and 13 per cent oleic acid, while in the new variety these values for the two fatty acids are reversed. Safflower grows well in hot, dry climates where soyabeans and most other commercial oilseeds are not well adapted.

Coconut Oil

This oil differs from those previously described because it contains large percentages of saturated fatty acids with less than eighteen carbon atoms. Its composition gives it a high thermal and oxidative stability so that it is used in filled milks and imitation dairy products to replace butter fat and for other uses in which off-flavours develop quickly from oxidation. Coconut oil, however, is prone to hydrolytic rancidity, since it contains significant amounts of caprylic(C_8) and capric(C_{10}) fatty acids which, when converted from their glyceride esters to free acids have strong, unpleasant odours and flavours. The fatty acid composition of coconut oil is: caprylic acid, 8 per cent; capric acid, 7 per cent; lauric acid, 48 per cent; myristic acid, 18 per cent; palmitic acid, 8 per cent; palmitoleic acid, 1 per cent; stearic acid, 2 per cent; oleic acid, 6 per cent; linoleic acid, 2 per cent.

Rapeseed Oil

Rapeseed oil is unusual in that it contains a large percentage of the 22-carbon unsaturated acid, erucic acid. Rape plants are adapted to cold climates. Mustard seed and a new oilseed, Crambe, produce oils of similar composition. Before refining, these oils contain sulphur compounds which give a strong cabbage-like odour. Although rapeseed oil has been used in margarine for some time, current trends are toward the use of oils containing fatty acids of lower carbon number, in edible products, especially. A zero-erucic rape has been developed in which the erucic acid has essentially been replaced by oleic. A typical analysis for ordinary rapeseed oil is as follows: palmitic acid, 1 per cent; stearic acid, 1 per cent; oleic acid, 17 per cent; linoleic acid, 14 per cent; linolenic acid, 8 per cent; erucic acid, 51 per cent; other C_{20}-C_{22} unsaturated fatty acids, 4 per cent; and C_{20}-C_{24} saturated acids, 4 per cent.

INDUSTRIAL OILS

The oils described above are used primarily for edible purposes although all of them find some application in industry as sources of fatty acids, soaps, detergents, etc. Soyabean and safflower oils, being semi-drying oils, are used in paints and coatings. Some oils, however, are produced almost exclusively as industrial raw materials to produce a variety of products.

Linseed Oil

Linseed oil comes from varieties of the flax plant which are grown for oil production. It contains over 50 per cent linolenic acid, more than any other common oil. As already noted, the unique structure of linolenic acid makes it susceptible to oxidation, a reaction which is responsible for the drying action of polyunsaturated oils. Initially, drying oils such as linseed oil react with the oxygen in the air to form hydroperoxides, but these, being thermally unstable, break down quite rapidly to undergo a series of complex reactions, including a free-radical polymerisation process. Such polymerisation takes place in oil-based paints and varnishes; it is often accelerated by adding to coating formulations catalysts such as cobalt naphthenate or other transition-metal compounds which increase the rate of peroxide decomposition. A representative composition for linseed oil is: saturated fatty acids, 10 per cent; oleic acid, 20 per cent; linoleic acid, 17 per cent; and linolenic acid, 53 per cent.

Tung Oil

This oil, produced from tung nuts, contains the conjugated trienoic acid eleostearic acid, which constitutes more than 80% of its component fatty acids. The susceptibility of eleostearic acid to oxidative polymerisation makes it an excellent drying oil.

Castor Oil

Castor oil is an interesting raw material in that it has two chemical "handles," the double bond and the hydroxyl group of ricinoleic acid, which comprises over 90 per cent of the fatty acid in the oil. Castor oil is characterised by a high viscosity resulting from hydrogen bonding of its hydroxyl groups. It is a raw material for paints, coatings, lubricants, cosmetics, and a variety of other products. In France, a commercial nylon is produced from castor oil. The castor plant (*Ricinus communis*) produces an extremely toxic protein, ricin, several potent allergenic materials, and an alkaloid, ricinine, which all appear in castor pomace. Thus, the residual meal cannot be used as an animal feed without further treatment, a fact which has been a deterrent to economic production in this country.

FATS OF ANIMAL ORIGIN

Lard

Lard is the fat rendered from the fatty tissues of the pig. For many years it was the major shortening product, but it has been supplanted to a great extent by hydrogenated vegetable oils. A great deal of lard is used in commercial frying operations. Although it has a relatively low content of unsaturated acids (iodine value 55–60) it is often hardened further by hydrogenation and blended into shortenings. Approximate composition of lard is: myristic acid, 1 per cent; palmitic acid, 27 per cent; palmitoleic acid, 3 per cent; stearic acid, 14 per cent; oleic acid, 43 per cent; linoleic acid, 10 per cent; and minor constituents, 2 per cent.

Tallow

Tallow is rendered from the fatty tissues of cattle in a manner similar to that use in the production of lard. After rendering, it is usually deodorised and used in shortenings and bakery margarines. It is also an industrial raw material, being used for the manufacture of soap and fatty acids, particularly industrial oleic acid. Tallow is somewhat higher in saturated fatty acids than lard. A typical composition is: myristic acid, 3 per cent; palmitic acid, 30 per cent; palmitoleic acid, 3 per cent; stearic acid, 19 per cent; oleic acid, 44 per cent; and minor constituents, 1 per cent.

Marine Oils

Marine oils are processed from herring, menhaden, pilchard, and sardines. They are highly unsaturated, with fatty acids often containing five or six double bonds. Such polyunsaturation makes them highly susceptible to oxidation and formation of fishy flavours and odours.

PROCESSING AND REFINING FATS AND OILS

The purpose of oil extraction is to obtain the maximum amount of good quality oil while getting maximum value from the residual meal. Oil is extracted from oilseeds by pressing, solvent extraction, or a combination of both. When brought in from the field, seeds are cleaned to remove stones, pieces of metal, and trash. They may then be stored for processing. Moisture content during storage is kept low to prevent microbial spoilage and spontaneous combustion. The seeds are cracked and, with some oilseeds, complete or partial decortication (removal of hull) is accomplished by a combination of air classification and screening or, in a few cases, flotation. The seeds are reduced by grinding or rolling, then steam cooked and flaked. The heat treatment coagulates the proteins and raises the moisture content so that extraction is accomplished most easily.

Cooked seeds may then be conveyed to screw presses for continuous pressing. The press cake emerging from a screw press will contain 8 to 14 per cent residual oil if solvent extraction is to follow, or 2 to 4 per cent if screw pressing is the only extractive process. Solvent extraction is done in a countercurrent procedure in which intimate contact is achieved between seeds and solvent. The apparatus may be the basket type arranged vertically or the rotary cell type in which containers rotate horizontally. The common solvent for edible oils is hexane, or actually a hexane-type naphtha boiling in the range of 146–156°F. Extraction equipment must be enclosed and thoroughly protected because of solvent flammability. There has been interest for many years in extraction with non-flammable solvents, especially trichloroethylene, but the advantages of using a non-flammable solvent have been more than offset by the toxicity problems generated by chlorinated solvents.

After extraction, the solvent is recovered by distillation from the miscella (mixed solvent and oil). Solvent is also removed from the oilseed pomace by steam stripping. Recovered solvent is re-used for extraction while the extract oil is usually combined with the press oil for refining, except in cases such as olive oil in which the virgin press oil has special flavour qualities. While most oilseed meals are used for animal feed, especially soyabeans, there is increasing use of the high-protein meal for human consumption. To produce protein isolates there must be as little denaturation as possible. Improvements are still needed in processing meal to keep temperatures low and protect the protein.

Marine oils are often solvent-extracted from fish meal; but animal fats such as lard or tallow, as well as many marine oils, are rendered by either dry or moist heat. Dry rendering in steam-jacketed tanks followed by pressing is satisfactory for inedible fats, but most edible fats are produced by wet rendering, using steam at 40 to 60 psi. In the latter process 99.5 per cent or more of the fat is recovered from the raw material.

Edible oils have a number of trace impurities, most of which are undesirable. Refining with caustic soda is used to eliminate free fatty acids, phosphatides, and other impurities which may cause off-flavours and undesirable colour. Usually the amount of free fatty acids is determined by analysis of the crude oil. The oil is then contacted with enough alkali solution to neutralise the free fatty acids plus an amount determined empirically to produce the desired effects with minimum loss. Reaction of caustic soda with the free fatty acids produces soapstock which can be continuously removed by centrifugation. Phosphatides are soluble only in the anhydrous oil, so these are taken out with the soapstock. If prior removal of lecithin is desired, it is accomplished before alkali treatment by degumming with water or steam. After separation of the soapstock, the oil may be water-washed and again centrifuged to bring soap concentration to a satisfactory low level. The refining of vegetable oils is usually a continuous process, though it may be done batchwise. Good alkali-refining reduces free fatty acid content to 0.01–0.03 per cent. In addition to free fatty acids, other undesirable components, such as gossypol (a somewhat toxic dark coloured material of cottonseed oil) are removed by contact with the alkali solution.

Other refining methods, such as acid refining, are used for specialised purposes. Some solvent-extracted oils may also be miscella refined with alkali before the solvent is removed. Better contact with alkali solution and better separation of soapstock are among the claimed advantages.

Bleaching

Some pigmented materials such as chlorophyll, which are not eliminated by alkali refining, lend themselves to removal by absorption processes. Activated carbon has been used for bleaching, but acid-activated clay is less expensive and quite effective. The amounts of activated clay used depend upon the oil; for vegetable oils, 0.25–2.0 per cent by weight quantities are common. Atmospheric bleaching at 212–250°F is common, but changes can occur in bleached oils due to oxidation. If these changes are undesirable, vacuum bleaching at temperatures of 170–180°F may be employed. A contact time of clay with oil for 10–15 minutes is usually required. Continuous vacuum bleaching equipment is commercially available. Many refineries discard the clay without recovering the retained oil, although this results in a loss of oil almost equal to the weight of clay used. A number of processes are used to remove oil from spent bleaching clay. While the recovered oil is usually low in quality, in some circumstances it may still be worth enough to justify its recovery.

Hydrogenation

Edible oils are often hydrogenated to produce hard or plastic fats or to enhance the oxidative stability of the oils by lowering their content of polyunsaturated acids. In the hydrogenation process, one or more carbon-carbon double bonds in a triglyceride molecule are converted to single bond by addition of hydrogen as illustrated below:

$$CH_3(CH_2)_7CH{=}CH(CH_2)_7COOR \xrightarrow[\text{Catalyst}]{H_2} CH_3(CH_2)_{16}COOR$$

The hydrogenation reaction requires hydrogen at moderate pressures and a catalyst, which is usually finely divided nickel. Such catalysts may be the type prepared by reducing nickel salts which have been precipitated onto an inert support of high surface area, or they may be Raney nickel-type prepared by dissolving aluminium, from a nickel-aluminium alloy using strong alkali. Variables affecting hydrogenation include the catalyst, the temperature, hydrogen pressure, and the amount of agitation. If only partial hydrogenation is desired, the selectivity of the hydrogenation system is important. Consider, for example the molecule of oleolinoleolinolenin illustrated below:

$$\begin{array}{l} CH_2OOC(CH_2)_7CH\overset{a}{=}CHCH_2CH\overset{b}{=}CHCH_2CH\overset{c}{=}CHCH_2CH_3 \text{ (linolenic)} \\ | \\ CHOOC(CH_2)_7CH\underset{d}{=}CHCH_2CH\underset{e}{=}CH(CH_2)_4CH_3 \text{ (linoleic)} \\ | \\ CH_2OOC(CH_2)_7CH\overset{f}{=}CH(CH_2)_7CH_3 \text{ (oleic)} \end{array}$$

If a light, partial hydrogenation is desired to increase oxidative stability, it would be preferable to first reduce all the bonds labelled *c* in the linolenic acid moiety. Because no catalyst is entirely selective, any of the other double bonds may be reduced. Reduction of one of the bonds labelled *a* or *b* would produce an "isolinoleic" acid from the linolenic acid portion of the molecule. Reduction of bonds *d* or *e* would produce isooleic or oleic acid from the linoleic acid and hydrogenation of bond *f* will form a stearic acid moiety from oleic acid. Not only is no catalyst completely selective, but hydrogenation catalysts may also cause isomerisation of an unsaturated bond from a *cis* to a *trans* configuration, or the migration of double bonds up and down the fatty acid chain. Some copper catalysts are reported to be considerably more selective for the hydrogenation of linolenic acid in the presence of other fatty acids than the commonly used nickel catalysts.

Hydrogenation is a very versatile reaction allowing considerable variability in the properties of plastic fats formed an important factor in the preparation of the final products, such as margarines and shortenings. When very light hydrogenation is used to produce a liquid oil of relatively high stability, some of the fat molecules may become quite saturated. These can be removed by winterising (chilling) the oil and removing the solidified portion, called stearine.

Deodorisation

Bleached and hydrogenated oils often have undesirable off-odours which result from the presence of free fatty acids, aldehydes, or other slightly volatile materials. Such oils are deodorised by steam stripping under high vacuum at elevated temperatures (400–470°F). The oil must be carefully cooled before it is contacted with air, because undesirable oxidation can be quite rapid at the high temperatures in the deodoriser. Equipment for deodorisation as well as most of the other processing steps is increasingly being built of stainless steel, since iron is a pro-oxidant at elevated temperatures. Copper is an especially strong oxidation catalyst, so that copper pipes or brass valves should not be used in oil processing. At the end of the deodorisation step antioxidants and metal sequestrants suitable for food use are often added to oils.

ANALYTICAL PROCEDURES

There are a large number of procedures, though many of these have value only for testing properties which have special applications. It is useful, though, to describe a few of these tests because of their general applicability and wide use.

Iodine Value

Iodine value is defined as the number of centigrams of iodine absorbed by one gram of fat. Thus, it is a measure of unsaturation in fats and oils. In practice, it is determined by using iodine chloride or iodine-bromide as a reagent, rather than free iodine. Even with these more reactive compounds absorption by double bonds is not usually complete, so that iodine value must be regarded as an empirical test which is of value to the processor. Since gas chromatography is now used widely, it is the practice in many research laboratories to first determine the composition of the fatty acid and then calculate the iodine value from that composition. Some iodine values of free fatty acids are presented in Table 7.2.

Table 7.2. Iodine values of unsaturated fatty acids.

Fatty acid	*No. of C atoms*	*No. of double bonds*	*Iodine value*
Decenoic	10	1	149.1
Dodecenoic	12	1	128.0
Tetradecenoic	14	1	112.1
Hexadecenoic	16	1	99.78
Oleic	18	1	89.87
Linoleic	18	2	181.04
Linolenic	18	3	273.52
Eicosenoic	20	1	81.75
Erucic	22	1	74.98

Saponification Value

The saponification value is the weight in milligrams of potassium hydroxide needed to completely saponify one gram of fat. It can give an indication of the average molecular weight of the fatty acids in a fat, but presents less useful information than the fatty acid composition determined by gas chromatography.

Melting Point

When applied to fats the term melting point has little significance, since melting usually occurs over a wide range of temperature. Some useful information may be gained, though, if melting points are determined by rigidly reproducible methods. A number of different methods are used, including the FAC melting point, the Wiley melting point, the congeal point, and the titer. In addition to these values, the solid fat index (SFI) is often used for specifying oil properties. The SFI is related approximately to the percentage of solids in a fat at a given temperature. When determined at a number of specified temperatures it can be especially useful to margarine manufacturers or other processors who need to control the characteristics of their manufactured products by blending.

Peroxide Value

Oxidation of fats is desirable in drying oils, but undesirable in most other products. The first step in oxidation is the formation of hydroperoxides.

These hydroperoxides have no odours or flavours, but they readily break down to produce aldehydes, hydrocarbons, ketones, and other volatile products which are characteristic of oxidative rancidity. Peroxides react with potassium iodide to liberate free iodine which may then be titrated. The peroxide value is expressed as milliequivalents of iodine formed per kilogram of fat. The peroxide value is used as an indicator of oxidative rancidity. Organoleptic evaluation by a trained odour panel can often detect rancidity at peroxide levels as low as 1 meq/kg.

Active Oxygen Method (AOM)

This method was developed to predict the shelf-stability of fats. It is useful with natural animal fats, but hydrogenated fats and oils which often are stabilised by the addition of antioxidants do not correlate ,well with AOM values. In this test, oils are heated at 97.8°C while air is blown through the sample. The AOM value is reported as the number of hours to reach a peroxide value of 100 meq/kg. A number other tests have been devised to predict oil stability, but none of them have yet shown accurate correlations for all types of oils.

Free-Fatty Acid (FFA)

Free fatty acids result from hydrolysis of fats. Oilseeds contain lipase enzymes which catalyse such hydrolysis when seeds are damaged or become moist and begin to germinate. Lipases are also found in animal tissues. Consequently, if seeds are improperly stored, or if animal tissues are allowed to stand for a while before rendering, the FFA values will be high. Deodorisation lowers the FFA to 0.05 per cent or less. To determine FFA, the sample is titrated with sodium hydroxide. It is calculated as free oleic acid and reported as a percentage.

Smoke Point

Smoke point is defined as the temperature at which a fat will produce continuous wisps of smoke. Impurities such as free fatty acids lower the smoke points drastically. A 1 per cent FFA content may lower the smoke point of a shortening from 425 to 320°F.

Colour

Colour with a few exceptions, light colour is desirable in oils. Colour is best measured by recording the visible absorption spectrum. However, there are empirical methods which are quite satisfactory for quality control. For this purpose oils may be compared in a tube of standard depth with the colour of glass standards, using the Lovibond Tintometer. Colour is reported as yellow and red values, the former usually being about 10 times the latter. Reflectance values are measured for solid fat products.

TRENDS IN OIL PROCESSING

As with most manufacturing processes, innovation is almost constantly occurring. For example, new hydrogenation units have been built which use commercial liquid hydrogen rather than stored gaseous hydrogen; or hydrogen is manufactured at the site. Efforts are continuing to make processing more automated and more continuous. One factor, which is only beginning to have an impact now but will be increasingly important is the need to eliminate pollution in plant effluents. Such effluents may be both organic and inorganic. A large amount of pollutants result when soapstock is acidified (acidulated). Treatment of soap with sulphuric acid liberates free fatty acids, most of which are recovered, and large amounts of sodium sulphate. The aqueous phase from acidulation will have a high BOD, due to the unrecovered acids, and a high concentration of inorganic salts, both of which place a burden on the local sewage system. Considerable effort is now being expanded to reduce both the quantity and the concentration of effluents from oil refineries. This effort will no doubt have a large influence on oil processing in the future.

Chapter 8

Soaps and Detergents

INTRODUCTION

A soap is actually a specific type of salt, the hydrogen of the fatty acid being replaced by a metal, which in common soaps is usually sodium. Soap lowers the surface tension of water and thus permits emulsification of fat-bearing soil particles. A typical commercial cleansing soap is made by reacting sodium hydroxide with a fatty acid. The lower the hydrogen content of the acid, the thinner the soap. The by-product of the reaction is glycerol. Many different carboxyl-containing substances are used, including rosin, tall oil, vegetable and animal oils and fats (stearic, palmitic, and oleic acids). Olive oil is used for Castile soap; transparent soaps are made from decolourised fats. The specific gravity of soaps is slightly more than 1.0; inclusion of air gives a floating product. Water solutions of sodium soaps in bar, chip, or powder form are universally used as mild emulsifying detergents for washing textiles, skin, paint, etc. Medically soap is used as an antidote for poisoning by ingestion of mineral acids or heavy metals. Liquid green soap is made with potassium hydroxide and a vegetable oil.

Heavy-metal soaps (loosely called metallic soaps) are those formed by metals heavier than sodium (aluminium, calcium, cobalt, lead, zinc). These soaps are not water-soluble; specific types are used in lubricating greases, gel thickeners, and in paints as driers and flatting agents. Napalm is an aluminium soap.

Detergent is a substance that reduces the surface tension of water; specifically, a surface-active agent which concentrates at oil-water interfaces, exerts emulsifying action, and thus aids in removing soils. The older and still widely used types are the common sodium soaps of fatty acids, which are relatively weak. The much stronger synthetic detergents are classed as anionic, cationic, or non-ionic, depending on their mode of chemical action. The latter functions by a hydrogen bonding mechanism. The most widely used group comprises linear alkyl sulphonates (LAS), often aided by "builders". LAS are preferable to alkyl benzene sulphonates (ABS), because they are readily decomposed by microorganisms (biodegradable). LAS are straight-chain compounds having ten or more carbon atoms in the chain; the branched-chains characteristic of ABS resist decomposition; these have been largely replaced by LAS because of water pollution.

The comparison is shown in Table 8.1 which indicates the significant differences in the processes used to manufacture detergents and soaps and also the differences in chemical composition, resulting in differences in action. Soaps give precipitates, hence are not efficient in hard or acid waters, whereas detergents are. Furthermore, although ordinary soaps do vary somewhat in chemical composition, they are essentially and simply the sodium or potassium salts of various fatty acids. On the other hand, detergents are very complex mixtures of many substances, all chosen to impart a particular action in cleaning. The modern concept of surface-active agents, or surfactants, includes soaps, detergents, emulsifiers, wetting agents, and penetrants. They all owe their activity to a modification of the properties of the surface layer between two phases in contact with another. Most surface-active agents have water-attracting, or hydrophilic groups, on one end of the molecule, water-repelling, or hydrophobic groups, on the other.

Table 8.1. General differences between soaps and detergents in composition and continuous manufacture.

To make synthetic detergents (continuous process):	
Tallow + mathanolysis ⟶	Detergents
Tallow methyl ester + high-pressure hydrogenation ⟶	
Tallow fatty alcohol + sulphonation ⟶	
Sulphated fallow fatty alcohol + NaOH ⟶	
Sodium salt of sulphated tallow fatty + builders, etc. ⟶	
To make soap (continuous process):	
Tallow + hydrolysis (splitting fats) ⟶	Soaps
Tallow fatty acid + NaOH ⟶	
Salt of fatty acid + builders, etc. ⟶	

Detergents excel in possessing these special properties to effect soil removal. There are light and heavy-duty detergents, and Table 8.2 differentiates between these in composition.

Table 8.2. Basic composition of three types of dry phosphate-based detergents (granules).

Ingredient	Function	*Ingredient on dry-solids basis, wt%*		
		Light-duty high sudsers	*Heavy-duty controlled sudsers*	*Heavy-duty high sudsers*
Surfactants				
Organic active, with suds regulators	Removal of oily soils cleaning	25–40	8–20	20–35
Builders				
Sodium tripolyphosphate and/or tetrasodium pyrophosphate	Removal of inorganic soil, detergent-building	2–30	30–50	30–50
Sodium sulphate	Filler with building action in soft water	30–70	0–30	10–20
Soda ash	Filler with some building action	0	0–20	0–5
Additives				
Sodium silicate having $2.0 \leq SiO_2/Na_2O \leq 3.2$	Corrosion inhibitor with slight building action	0–4	6–9	4–8
Carboxymethyl cellulose	Antiredeposition soil	0–0.5	0.5–1.3	0.5–1.3
Fluorescent dye	Optical brightening	0–0.05	0.05–0.1	*ca.* 0.1
Tarnish inhibitors	Prevention of silverware tarnish	0	0–0.2	0–0.02
Perfume and sometimes dye or pigment	Aesthetic, improved product characteristics	0.1	0.1	0.1
Water	Filler and binder	1–5	2–10	3–10

Recently, the composition of detergents underwent rapid changes because of environmental considerations. Evidence indicated that phosphates from detergents may contribute to the eutrophication of lakes, so the use of phosphates in detergents was banned in some areas of the country. Many different substitutes were formulated into detergents, but some of these were found to be unsafe and were then banned. The position taken by the detergent industry has been that phosphates in waste-water can be removed by special treatment in sewage plants and, in view of the proved lack of toxicity of phosphates, their replacement may not be the most desirable solution. The soap and detergent industry and its suppliers face an enormous task in testing new materials for all possible effects on the environment, and extensive research will be needed before this complex problem can be solved.

DETERGENT RAW MATERIALS

A large volume of active organic compounds, or surfactants, for both detergents and soap are manufactured in final form by soap and detergent companies. Examples are linear alkylbenzene sulphonate (LAS) and fatty alcohol sulphate, which these companies manufacture in hundreds of millions of pounds. The same is true for fatty acids, the basic materials for soaps. Most of the inorganic materials are purchased, such as oleum, caustic soda, various sodium phosphates, and a large number of additives, the last-mentioned amounting to 3% or less of the total product weight.

Surfactants

These embrace "any compound that affects (usually reduces) surface tension when dissolved in water or water solutions, or which similarly affects interfacial tension between two liquids. Soap is such a material, but the term is most frequently applied to organic derivatives such as sodium salts of high molecular weight alkyl sulphates or sulphonates." The surfactants of both soap and synthetic detergents perform the primary cleaning and sudsing of the washing action in the same way through the reduction of surface tension. The cleaning process consists of (i) thoroughly wetting the dirt and the surface of the article being washed with the soap or detergent solution, (ii) removing the dirt from the surface, and (iii) maintaining the dirt in a stable solution or suspension (detergents). In wash water, soaps or detergents increase the wetting ability of the water so that it can more easily penetrate the fabrics and reach the soil. Then soil removal begins. Each molecule of the cleaning solution may be considered a long chain. One end of the chain is *hydrophilic* (water-loving); the other is *hydrophobic* (water-hating, or soil-loving). The soil-loving ends of some of these molecules are attracted to a soil particle and surround it. At the same time the water-loving ends pull the molecules and the soil particles away from the fabric and into the wash water. This is the action which, when combined with the mechanical agitation of the washing machine, enables a soap or detergent to remove soil, suspend it, and keep it from redepositing on clothes.

Classification of Surfactants

In most cases the *hydrophobic* portion is a hydrocarbon containing 8 to 18 carbon atoms in a straight or slightly branched chain. In certain cases, a benzene ring may replace some of the carbon atoms in the chain, for example, $C_{12}H_{25}$—, $C_9H_{19}.C_6H_4$—. The *hydrophilic* functional group may vary widely and may be anionic, e.g., —$OSO_3^{\ominus}$ or —$SO_3^{\ominus}$; cationic, e.g., —$N(CH_3)_3^{\oplus}$ or $C_5H_5N^{\oplus}$ —; zwitterionic, e.g., —$N^{\oplus}(CH_3)_2(CH_2)_2COO^{\ominus}$; semi-polar. e.g., —$N(CH_3)_2O$; or non-ionic, e.g., —$(OCH_2CH_2)_nOH$. In the anionic class one finds the most used compounds, namely linear alkylbenzene sulphonates from petroleum, and alkyl sulphates from animal and vegetable fats (Fig. 8.1). Other examples are alkylbenzene-ether sulphonate, fatty alcohol-ethylene oxide sulphate, alkylglycerol-ether sulphonate, alkyl esters of isothionate, and methylalkyl laurates. The last mentioned are generally employed in newer applications, such as liquids and bars, where special properties are needed and /here a premium price will be paid. Soap is also anionic in character. Quaternary trimethylalkylammonium halides, of which cetyltrimethylammonium bromide is an example, are the most common *cationic* surfactants. Dialkyldimethylammonium chloride is a cationic fabric softener. Being generally weak in detergent power, although they have good lubricating, antistatic, and germicidal properties, they are not used as household detergents. Anionics and cationics are not compatible, since the; combine to form an insoluble precipitate. For the same reason cationics are not compatible with soap. Alkyl betaines are representatives of the *zwitterionics*; dimethylalkylamine oxides are *semi-polar*; and ethylene oxide condensates of fatty alcohol illustrate the molecular structure of truly *non-ionic* surfactants, of which there are many excellent soil-removing types and which are chiefly emulsifying agents but low sudsers, hence useful in drum-type automatic clothes and dish washers. In general, the hydrophilic nature of these functional groups decreases from ionic to non-ionic. Much of the organic detergent chemical research carried out in recent years has dealt with the synthesis of new surfactant molecules as an extension of the following reactions.

Anionic, giving, in solution, surface-active ions bearing a negative charge:

$$C_{12}H_{25}OH + SO_3 \longrightarrow C_{12}H_{25}OSO_3H \xrightarrow{NaOH} C_{12}H_{25}OSO_3^{\ominus}\,Na^{\oplus} \text{ (detergent)}$$

$$\text{Fat} + H_2O \longrightarrow C_{17}H_{35}COOH \xrightarrow{NaOH} C_{17}H_{35}COO^{\ominus}Na^{\oplus} \quad \text{(soap)}$$

Cationic, yielding, in solution, surface-active ions positively charged:

$$C_{12}H_{25}Cl + N(CH_3)_3 \longrightarrow C_{12}H_{25}N(CH_3)_3^{\oplus}Cl^{\ominus}$$

Semi-polar:

$$C_{12}H_{25}N(CH_3)_2 + H_2O_2 \longrightarrow C_{12}H_{25}-\underset{CH_3}{\overset{CH_3}{\underset{|}{\overset{|}{N}}}} \longrightarrow O + H_2O$$

Zwitterionic:

$$C_{12}H_{25}N(CH_3)_2 + ClCH_2\overset{O}{\overset{\|}{C}}ONa \longrightarrow C_{12}H_{25}N^{\oplus}(CH_3)_2CH_2\overset{O}{\overset{\|}{C}}O^{\ominus} + NaCl$$

$$C_{12}H_{25}N(CH_3)_2 + \underset{O-\!\!-C=O}{CH_2-\!\!-CH_2} \longrightarrow C_{12}H_{25}N^{\oplus}(CH_3)_2CH_2CH_2\overset{O}{\overset{\|}{C}}O^{\ominus}$$

Non-ionic:

$$C_{12}H_{25}OH + n\overset{O}{CH_2\;\;CH_2} \longrightarrow C_{12}H_{25}(OCH_2CH_2)_nOH$$

Suds Regulators

A suds regulator—either a stabiliser or suppressor—is often used with a surfactant. These materials have no common chemical relationship and are often specific for certain surfactants. Examples of stabiliser surfactant systems are lauric ethanolamide-alkylbenzene sulphonate and lauryl alcohol–alkyl sulphate. Suds suppressors are generally hydrophobic materials, some examples of which are long-chain fatty acids, silicones, and hydrophobic non-ionic surfactants.

Builders

Builders boost detergent power, and complex phosphates, such as sodium tripolyphosphate, have been used most extensively. These are more than water softeners which sequester water-hardening calcium and magnesium ions. They prevent redeposition on fabrics of soil from the wash water. Proper formulation with *complex phosphates* has been the key to good cleaning and a full partner with surfactants in making possible the tremendous development of detergents. Polyphosphates (e.g., sodium tripolyphosphate and tetrasodium pyrophosphate) have a synergistic action with the surfactant and hence reduce the overall cost, together with an enhanced effectiveness. The rapid rise in the acceptance of detergents stemmed from the building action of the polyphosphates. Surfactants, suds regulators, and detergency builders make up the basic surfactant detergent formulation; however, small amounts of additives are necessary (3% or less).

Additives

Corrosion inhibitors, such as sodium silicate, protect metal and washer parts, utensils, and dishes from the action of detergents and water. Carboxymethyl cellulose has been used as an *anti-redeposition agent. Tarnish inhibitors* carry on the work of the corrosion inhibitor and extend protection to metals such as German silver. Benzotriazole has been used for this purpose. *Fabric brighteners* are fluorescent dyes which make fabrics look brighter because of their ability to convert ultraviolet light to visible light. Two materials used for such dyes are sodium 4(*2H*-naphtho[1,*2-d*] triazol-2-yl)stilbene-2-sulphonate and disodium 4,4'-bis(4-anilino-6-morpholino-S-triazin-2-ylamino)-2,2', stilbene disulphonate.

Bluings improve the whiteness of fabrics by counteracting the natural yellowing tendency. The ingredients used for this purpose can vary from the long-used ultramarine blue to new dye materials. *Anti-microbial* agents are ingredients not now in general use in household products; there are several, including carbanilides, salicylanilides, and cationics. Peroxygen-type *bleaches* can also be employed in laundry products, but their use in this country has been limited. Peroxygen bleach containing detergents are common in Europe, where hypochlorite bleaches are uncommon, because of the greater effectiveness of the former under high-temperature European washing conditions. *Perfume* has become an accepted ingredient, and compositions and fragrances vary widely. However, the soap and detergent industries are the largest consumers of perfumes among the industries. Colouring is a means of emphasising certain properties of a product and of obtaining distinctiveness.

MANUFACTURE OF DETERGENTS

The most widely used detergent, a heavy-duty granule, is presented in Fig. 8.1., with the quantities of materials required. The reactions are:

Linear alkylbenzene sulphonation:

1. Main reaction

$$R\text{-}C_6H_5 + H_2SO_4 \cdot SO_3 \longrightarrow R\text{-}C_6H_4\text{-}SO_3H + H_2SO_4$$

Alkylbenzene Oleum Alkylbenzene sulphonate Sulphuric acid

$\Delta H = -180$ Btu/lb of alkylbenzene

2. Secondary reactions

$$R\text{-}C_6H_4\text{-}SO_3H + H_2SO_4 \cdot SO_3 \longrightarrow (SO_3H)C_6H_3\text{-}SO_3H + H_2SO_4$$

Alkylbenzene sulphonate Oleum Disuphonate Sulphuric acid

$$R\text{-}C_6H_4\text{-}SO_3H + R^1\text{-}C_6H_5 \longrightarrow R\text{-}C_6H_4\text{—}SO_2\text{—}C_6H_4\text{-}R^1 + H_2O$$

Alkylbenzene sulfonate Alkylbenzene Sulphone 1% Water

Fatty alcohol sulphation:

1. Main reaction

$R\text{–}CH_2OH + SO_3 \cdot H_2O \rightleftharpoons R'OSO_3H + H_2O$ $\Delta H = -140$ to -150 Btu/lb fatty alcohol

2. Secondary reaction

$R\text{–}CH_2OH + R'\text{–}CH_2\text{–}OSO_3H \longrightarrow R\text{—}CH_2\text{—}O\text{–}CH_2\text{–}R' + H_2SO_4$

$R'\text{–}CH_2\text{–}CH_2OH + SO_3 \longrightarrow R'\text{—}CH{=}CH_2 + H_2SO_4$

$R\text{—}CH_2OH + SO_3 \longrightarrow RCHO + H_2O + SO_2$

$R\text{—}CH_2OH + 2SO_3 \longrightarrow RCOOH + H_2O + 2SO_2$

This presentation is supplemented by Table 8.2, which gives the basic constituents in more detail for the three types of detergent granules. The continuous flowchart in Fig. 8.1 can be broken down into the following coordinated sequences:

Sulphonation-sulphation. The alkylbenzene (AB) is introduced continuously into the sulphonator with the requisite amount of oleum, using the dominant bath principle shown in Fig. 8.3 to control the heat of *sulphonation* conversion and maintain the temperature at about 130°F. Into the sulphonated mixture is fed the fatty tallow alcohol and more of the oleum. All are pumped through the *sulphator*, also operating on the dominant bath principle, to maintain the temperature at 120 to 130°F, thus manufacturing a mixture of surfactants.

Neutralisation. The sulphonated-sulphated product is neutralised with NaOH solution under controlled temperature to maintain fluidity of the surfactant slurry (Ch and Op). The surfactant slurry is conducted to storage (Op).

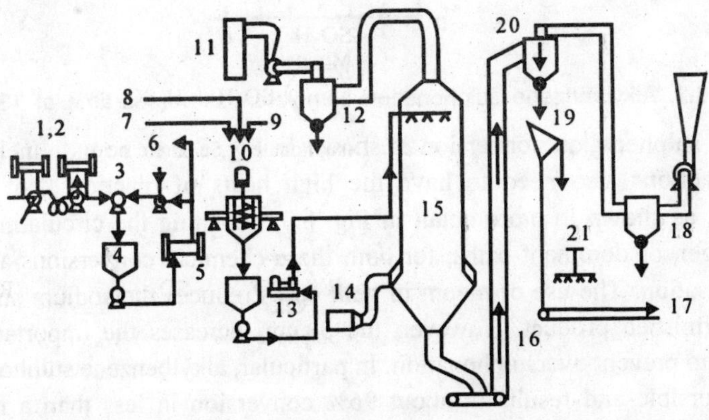

Fig. 8.1. Simplified continuous flowchart for the production of heavy-duty detergent granules. 1. Sulphonator; 2. Sulphator; 3. Mixer; 4. Surge tank; 5. Cooler; 6. Neutraliser; 7. Phosphate; 8. Silicate; 9. Dry scrap; 10. Crutcher; 11. Stack; 12. Cyclones; 13. H.P. pump; 14. Furnace; 15. Spray tower; 16. Air lift; 17. To packing; 18. Filter; 19. Screen; 20. Cyclone; 21. Perfume.

The surfactant slurry, the sodium tripolyphosphate, and most of the miscellaneous additives are introduced into the *crutcher*. A considerable amount of the water is removed, and the paste is thickened by the tripolyphosphate hydration reaction:

$$\underset{\text{Sodium tripolyphosphate}}{Na_5P_3O_{10}} + 6H_2O \longrightarrow \underset{\text{Sodium tripolyphosphate hexahydrate}}{Na_5P_3O_{10}\cdot 6H_2O}$$

This mixture is pumped to an upper story, where it is *sprayed* under high pressure into the 80-ft-high spray tower, counter to hot air from the furnace. Here are formed dried granules of acceptable shape and size and suitable density. The dried granules are transferred to an upper story again by an air lift which cools from 240°F and stabilises the granules. The granules are separated in a cyclone, screened, perfumed, and packed.

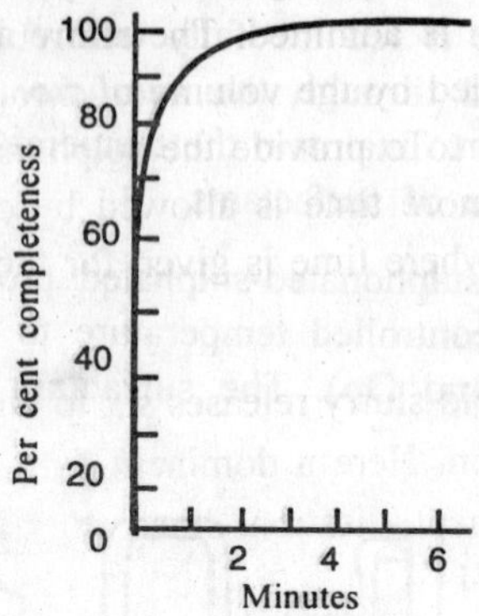

Fig. 8.2. Alkylbenzene sulphonation completeness versus time at 130°F.

The sulphonation conversion is shown in Fig. 8.2 to be extremely fast. The reactions also need to have the high heats of reaction kept under control, as shown in more detail in Fig. 8.3, depicting the circulating heat exchanger, or dominant baths, for both these chemical conversions and for neutralisation. The use of oleum in both cases reduces the sodium sulphate in the finished product. However, the oleum increases the importance of control to prevent oversulphonation. In particular, alkylbenzene sulphonation is irreversible and results in about 96% conversion in less than a minute when run at 130°F with 1 to 4% excess SO_3 in the oleum (Fig. 8.2).

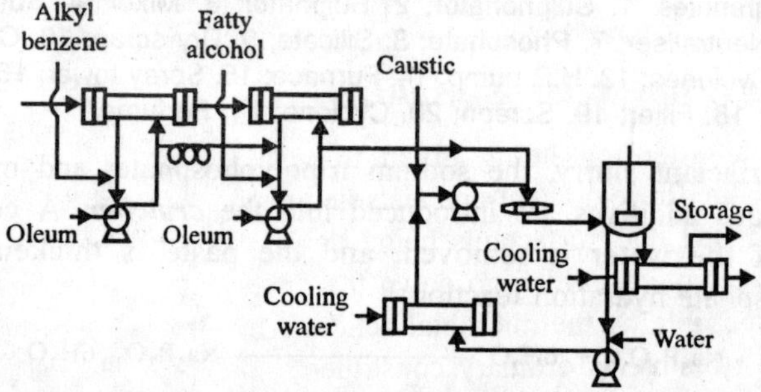

Fig. 8.3. Continuous series sulphonation-sulphation, ending with neutralisation, in the circulating heat-exchanging dominant bath to control heat.

A certain minimum concentration of SO_3 in the oleum is necessary before the sulphonation reaction will start, which in this case is about 78.5% SO_3 (equivalent to 96% sulphuric acid). As both these reactions are highly exothermic and rapid, efficient heat removal is required to prevent oversulphonation and darkening. Agitation is provided by a centrifugal pump, to which the oleum is admitted. The recirculation ratio (volume of recirculating material divided by the volume of throughput) is at least 20:1 to give a favourable system. To provide the sulphonation time to reach the desired high conversion, more time is allowed by conducting the mixture through a coil (Fig. 8.3), where time is given for the sulphonation reaction to go to completion.

Neutralisation of the acid slurry releases six to eight times as much heat as the sulphonation reaction. Here a dominant bath (Fig. 8.3) is employed which *quickly* effects the neutralisation, since a partly neutralised acid mix is very viscous.

Other surfactants are being used in increasing amounts, especially for detergents in the form of liquids and bars. These are more rapidly biodegradable than branched-chain benzene sulphonates. These are: Benzenoid, non-benzenoid, amphoteric, *anionic*, carboxylic acids (and salts thereof), Phosphoric and polyphosphoric acid esters (and salts thereof), sulphonic acids (and salts thereof), sulphuric acid ester (and salts thereof), *cationic*, amine oxide and oxygen-containing amines (except those having amine linkages), amines and amine oxides having amine linkages, amines not containing oxygen (and salts thereof), quaternary ammonium salts not containing oxygen. *Non-ionic total.* Carboxylic acid amides, carboxylic acid esters, ethers.

BIODEGRADABILITY OF SURFACTANTS

In view of the attention being given to water pollution control and abatement, product-development chemists and chemical engineers in recent years have realised that surfactants being developed for use in household and industrial detergents which go down the drain to the sewer must be as readily decomposable by the microbial action: (i) of sewage treatment, and (ii) of surface streams, as ordinary constituents of household wastes. This new parameter has been added to the performance, efficiency, and cost factors the detergent industry must consider in developing new products. Some surfactants, like tetrapropylene-derived alkylbenzene sulphonate, degrade *slowly*, leaving a persistent residue. Others are more readily decomposable by micro-organisms and leave practically no persistent residues. The ease with which a surfactant is decomposed by microbial action has been defined as its *biodegradability*. Tests are being developed and standards are being established for biodegradability. To have broad application, such standards must recognise the breadth of variation in environmental conditions. Materials which may be only partly degraded in inefficient treatment processes can be completely decomposed by more sophisticated biological treatment systems. Tests that measure dieaway in river water, or simulate the biological processes employed in sewage treatment plants, represent only part of the extensive testing used in measuring biodegradability.

Based on research which had been under way for several years, the detergent industry set target date for more readily degradable detergents. Complete conversion was the overall objective, and the most important step was replacing tetrapropylene benzene sulphonate (TPBS). This detergent material had been the workhorse of the detergent industry.

TPBS is produced by alkylating benzene with a propylene tetramer, followed by sulphonation of the benzene ring. The propylene tetramer consists of a multitude of branched isomers and has few, if any, straight-chain alkyl groups. Research on a more readily degradable surfactant developed a straight-chain hydrocarbon to make the alkylbenzene. The straight-chain material gives more readily degradable detergents and has fit easily into the detergent formulations where it has replaced TPBS.

STRAIGHT-CHAIN ALKYLBENZENE

With the endeavour to improve the biodegradability of detergents, chemical and engineering efforts made possible supplanting of various branched-chain alkylbenzenes by straight-chain alkyl products along the following lines, as illustrated in Fig. 8.4. These compounds are sulphonated after being made into normal, or straight-chain, alkylbenzenes. The straight-chain α-olefins (preferably C_{12}) are made:

(1) By polymerising ethylene to α-olefins, using Ziegler-type and other polymerisations, e.g., the Alfol process (Fig. 8.5.), furnishing high-purity, but not cheap, products.

(2) By molecular sieve and other means to extract straight-chain paraffins from kerosene (low-priced).

(3) By cracking paraffins such as wax (cheap α-olefins).

(4) By obtaining fatty alcohols by improved hydrogenation of natural fats, namely, from methyl esters of fatty acids.

α-Olefins, or paraffin halides, can be used to alkylate benzene through the Friedel-Crafts reaction, employing hydrofluoric acid or aluminium fluoride as a catalyst. Fatty alcohols can be sulphated for use in detergents as outlined in Fig. 8.1. An example of a biodegradable linear (C_{12}) alkylbenzene sulphonate is:

$$CH_3—CH_2—CH_2—CH_2—CH_2—CH_2—CH_2—CH_2—CH_2—CH_2—CH—CH_3$$

(with the CH carrying a benzene ring bearing SO_3Na)

FATTY ACIDS AND FATTY ALCOHOLS, FOR MANUFACTURE OF DETERGENTS AND SOAPS

Fatty alcohols and fatty acids are mainly consumed in the manufacture of detergents and soaps, as presented in this chapter. However, fatty acids, both saturated (e.g., stearic acid) and unsaturated (e.g., oleic) have long been employed in many industries as both free acids and, more frequently, as salts. Examples are:

(1) Zinc or magnesium stearates in face powders.

(2) Calcium or aluminium soaps (insoluble) employed as water repellants in water-proofing textiles and walls.

(3) Triethanolamine oleate in dry cleaning and cosmetics.

(4) Rosin soap consumed as a sizing for paper.

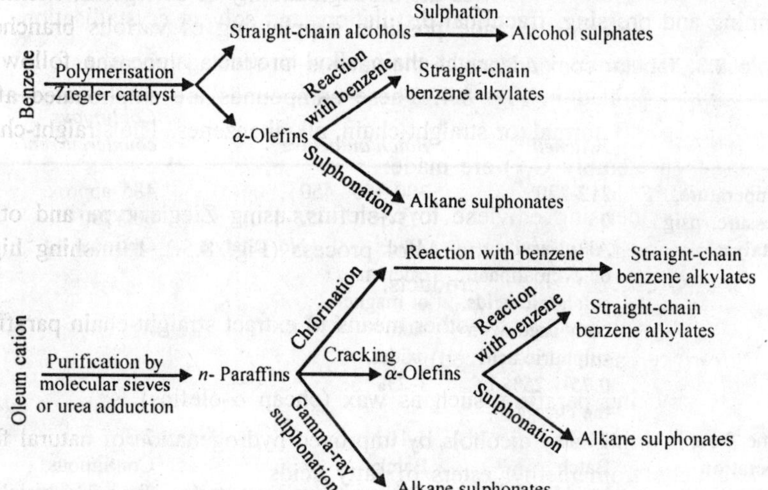

Fig. 8.4. Some possible paths to soft-detergent components.

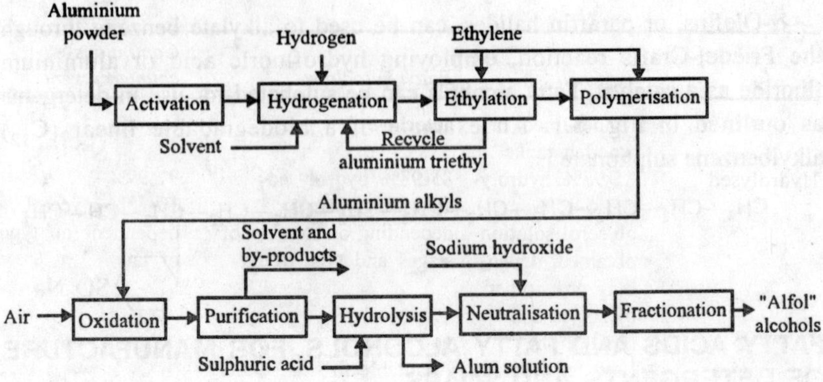

Fig. 8.5. Fatty alcohols made by means of the organometallic route have carbonchain lengths ranging from 6 to 20.

Manufacture of Fatty Acids

Long-consumed basic raw materials such as oils and fats been very extensively supplemented by improved chemical processing and by synthetic petrochemicals. A selection from these processes is given here. Table 8.3. compares three processes long used for splitting fats. The high-pressure hydrolysis mostly used in the soap industry is catalysed by zinc oxide. Fatty acids are drawn off from the distillate receiver for sale or further conversion to fatty acid salts (calcium, magnesium, zinc, etc.). Several older and less used separation methods for purifying fatty acids are panning and pressing, fractional distillation, and solvent crystallisation.

Table 8.3. Tabular comparisons of the various fat-splitting processes.

	Twitchell	*Batch autoclave*		*Continuous countercurrent*
Temperature, °F	212-220	300-350	450	485 approx.
Pressure, psig	0	75-150	425-450	600-700
Catalyst	Alkyl-aryl or cycloaliphatic sulphonic acids, both used with sulphuric acid 0.75-1.25% of the charge	Zinc, calcium, or magnesium oxides, 1-2%	No catalyst	Optional
Time, hour	12-48	5-10		2-3
Operation	Batch	Batch		Continuous
Equipment	Lead-lined,	Copper or stainless-steel		Type 316 stainless

(Contd...)

	Twitchell	*Batch autoclave*	*Continuous countercurrent*
	Monel-lined, or wooden tanks	autoclave	tower
Hydrolysed	85-98% hydrolysed 5-15% glycerol solution obtained, depending on number of stages and type of fat	85-98% hydrolysed 10-15% glycerol, depending on number of stages and type of fat	97-99% 10-25% glycerol, dependent on type of fat
Advantages	Low temperature and pressure; adaptable to small scale; low first cost because of relatively simple and inexpensive equipment	Adaptable to small scale; small scale than continuous process; faster than Twitchell	Small floor space; uniform product quality; high yield of acids; high glycerine concentration; low labour cost; more accurate and automatic control; lower annual costs
Disadvantages	Catalyst handling; long reaction time; fat stocks of poor quality must often be acid-refined to avoid catalyst poisoning; high steam consumption; tendency to form dark-coloured acids; need more than one stage for good yield and high glycerine concentration; not adaptable to automatic control; high labour cost	High first cost; catalyst handling; longer reaction time than continuous processes; not so adaptable to automatic control as continuous; high labor cost; need more than one stage for good yield and high glycerine concentration	High first cost; high temperature and pressure; greater operating skill

Manufacture of Fatty Alcohols

The Ziegler catalytic procedure is used for converting α-olefins to fatty alcohols and the methyl ester hydrogenation. The text presented under soap for the modern continuous hydrolysis of fats to furnish fatty acids which may be hydrogenated to fatty alcohols. The Ziegler procedure is an important one for manufacturing C_{12} to C_{18} α-olefins and fatty even-numbered straight-chain alcohols for detergents. To summarise, the Ziegler reactions are

Aluminium triethyl preparation:

$$Al + \frac{3}{2} H_2 + 2Al(C_2H_5)_3 \longrightarrow 3Al(C_2H_5)_2H$$

$$3Al(C_2H_5)_2H + 3C_2H_4 \longrightarrow 3Al(C_2H_5)_3$$

Build-up reaction (random or Poisson distribution of chain lengths):

$$\underset{\text{Aluminium triethyl}}{Al(C_2H_5)_3} + \underset{\text{Ethylene}}{(x + y + z)C_2H_4} \xrightarrow[250°F]{2{,}000\ psig} \underset{\text{Aluminium "trialkyl"}}{Al \begin{cases} (C_2H_4)_x + 1H \\ (C_2H_4)_y + 1H \\ (C_2H_4)_z + 1H \end{cases}}$$

ΔH is exothermic, liberating approximately 1000 Btu/lb of ethylene reacted. The "1H" is the terminal alkyl (C_2H_5) of the chain.

Oxidation reaction:

$$\underset{\text{Aluminium "trialkyl"}}{Al \begin{cases} R \\ R^1 \\ R^2 \end{cases}} + \frac{3}{2}\ \underset{\text{Dried air}}{O_2} \xrightarrow[90°F]{100\ psig} \underset{\text{Aluminium trialkoxide}}{Al \begin{cases} OR \\ OR^1 \\ OR^2 \end{cases}}$$

ΔH is exothermic, liberating about 1100 Btu/lb of alkyl oxidised. Its conversion is 98% to 90°F in about 2 h.

Acidolysis (hydrolysis):

$$\underset{\text{Aluminium trialkoxide}}{Al \begin{cases} OR \\ OR^1 \\ OR^2 \end{cases}} + \frac{3}{2}\ \underset{\text{23-26\% sulphuric acid}}{H_2SO_4} \xrightarrow{190°F} \underset{\text{Aluminium sulphate}}{\frac{1}{2}Al_2(SO_4)_3} + \underset{\text{Alkyl alcohols}}{ROH + R^1OH + R^2OH}$$

Neutralisation of slight excess sulphuric acid:

$$H_2SO_4 + 2NaOH \longrightarrow Na_2SO_4 + 2H_2O$$

These reactions furnish a substitute for the C_{12} to C_{14} chain length and for the fatty acids in imported coconut oil or copra. Such reactions are run non-catalytically at 2,000 psig and 250°F in an inert hydrocarbon solvent such as heptane or benzene. In these solvents aluminium "trialkyl" is no longer pyrophoric at less than 40% concentration. Under the above conditions, it takes approximately 140 min to build up to a C_{12} average chain length when reacting 5 mol of ethylene for each 3/2 mol of aluminium triethyl. The main side reaction is a thermal displacement reaction between aluminium alkyl and ethylene which produces α-olefins. α-olefins can react with either aluminium triethyl or aluminium alkyls in a build-up-type reaction to form straight-chain alkyls or, to lesser extent, somewhat unstable branched alkyls.

The aluminium alkyl is *oxidised* to aluminium alkoxide by dried air at moderate temperatures of 60 to 120°F and moderate pressures in the presence of a solvent. This oxidation apparently proceeds by the addition of a molecule of oxygen to form a peroxide, which then rearranges to the alkoxide either internally or with another molecule of alkyl. The solvent and any olefins are distilled off from the aluminium alkoxides. *Acidolysis* completes the conversion of the aluminium alkoxide to fatty alcohols. Water gives a bad gel, but dilute sulphuric acid forms the soluble aluminium sulphate, which is readily marketable if kept iron-free. *Neutralisation* with dilute caustic prepares the product for the final steps of dehydration, recovery of a small amount of C_2H_5OH and C_4H_8OH, and fractionation first under atmospheric conditions, and then under vacuum, to separate and obtain the fatty alcohols (Fig.8.5). It is based on the Ziegler reactions.

Fatty Alcohols from Methyl Esters

Fats have long been basic raw materials for soaps and detergents. Such fats as are available are glyceryl esters of fatty acids (C_{12} and C_{16}) and have been hydrolysed to the acids for soap and reduced to the alcohols by catalytic hydrogenation for detergents. The methyl esters of fatty acids are also hydrogenated to fatty alcohols. These esters are prepared by reacting methanol with coconut or tallow triglyceride, catalysed by a small amount of sodium. Previously, the refined oils dried by flashing at 300°F under a vacuum of 25 in., or otherwise it will consume relatively expensive sodium and also form soap. The methyl exchange esterification takes place in about an hour, when the reaction mix is settled and separated into an upper layer rich in ester and methanol and the lower layer rich in glycerine and methanol. The ester layer is washed countercurrently and in a continuous manner to remove excess methanol, to recover glycerine, and to remove the catalyst, which would poison the hydrogenation. Yields of fatty alcohols are 90 to 95%.

Hydrogenation of methyl esters is catalysed by an Atkins-type catalyst (made from copper nitrate, chromic oxide, and ammonia, with final roasting) and is carried out at approximately 3,000 psig and 500 to 600°F. The continuous equipment consists of three vertical reactors 40 ft high using 30 mol of heated hydrogen per mol of ester; the hydrogen serves not only for reducing but also for heating and agitation. The crude alcohols are fractionated to the specified chain length.

SOAP

Soap comprises the sodium or potassium salts of various fatty acids, but chiefly of oleic, stearic, palmitic, lauric, and myristic acids. For generations its use has increased until its manufacture has become an industry essential to the comfort and health of civilised man. We may well gage the advance of modern civilisation by the per capita consumption of soap and detergents.

Soap Manufacture

Soap manufacture is carried out in kettle process, however, it is mainly used by smaller factories or for special and limited production. As soap technology changed, *continuous* alkaline saponification was introduced Even now computer control allows an automated plant for continuous saponification by NaOH of oils and fats to produce in 2 hours the same amount of soap (more than 300 tons/day) made in 2 to 5 days by traditional batch methods.

The present procedure involves *continuous splitting,* or *hydrolysis,* as outlined in Table 8.1. After separation of the glycerine, the fatty acids are neutralised to soap.

The basic chemical reaction in the making of soap may be expressed as *saponification;*

$$\underset{\text{Caustic soda}}{3NaOH} + \underset{\text{Glyceryl stearate}}{(C_{17}H_{35}COO)_3C_3H_5} \longrightarrow \underset{\text{Sodium stearate}}{3C_{17}H_{35}COONa} + \underset{\text{Glycerine}}{C_3H_5(OH)_3}$$

The procedure is to split, or *hydrolyse*, the fat, and then, after separation from the valuable glycerine, to *neutralise* the fatty acids with a caustic soda solution:

$$\underset{\text{Glyceryl stearate}}{(C_{17}H_{35}COO)_3C_3H_5} + 3H_2O \longrightarrow 3\ \underset{\text{Stearic acid}}{{}_{.7}H_{35}COOH} + \underset{\text{Glycerine}}{C_3H_5(OH)_3}$$

$$\underset{\text{Stearic acid}}{C_{17}H_{35}COOH} + \underset{\text{Caustic soda}}{NaOH} \longrightarrow \underset{\text{Sodium stearate}}{C_{17}H_{35}COONa} + H_2O$$

It should be pointed out that the usual fats and oils of commerce are not composed of the glyceride of any one fatty acid, but of a mixture. However, some individual fatty acids of 90% purity or better are available from special processing. Since the solubility and hardness of the sodium salts (Table 8.4) of the various fatty acids differ considerably, the soapmaker chooses the raw material according to the properties desired, with due consideration of the market price.

Table 8.4. Solubilities of various pure soaps.

	Stearate	*Oleate*	*Palmitate*	*Laurate*
	(In grams per 100 g of water at 25°C)			
Sodium	0.1[a]	18.1	0.8[a]	2.75
Potassium	....	25.0	....	70.0[a]
Calcium	0.004[b]	0.04	0.003	0.004[b]
Magnesium	0.004	0.024	0.008	0.007
Aluminium	*i*	*i*	*d*	

[a] Approximate
[b] Solubility given at 15°C only.
i Indicates that the compound is insoluble
d Indicates decomposition

Raw Materials

Tallow is the principal fatty material in soapmaking; the quantities used represent about three-fourths of the total oils and fats consumed by the soap industry. It contains the mixed glycerides obtained from the solid fat of cattle by steam rendering. This solid fat is digested with steam, the tallow forming a layer above the water, so that can easily be removed. Tallow is usually mixed with coconut oil in the soap kettle or hydrolyser in order to increase the solubility of the soap. *Greases* (about 20%) are the second most important raw material in soapmaking. They are obtained from hogs and smaller domestic animals and are an important source of glycerides of fatty acids. They are refined by steam rendering or by solvent extraction and are seldom used without being blended with other fats. In some cases, they are treated so as to free their fatty acids, which are used in soap instead of the grease itself. *Coconut oil* has long been important. The soap from coconut oil is firm and lathers well. It contains large proportions of the very desirable glycerides of lauric and myristic acids. Free *fatty acids* are utilised in soap, detergent, cosmetic, paint, textile, and many other industries. The acidification of "foots," or stock resulting from alkaline refining of oils, also produces fatty acids. The important general methods of splitting are outlined in Table 8.4. The Twitchell process is the oldest. Continuous countercurrent processes are now most commonly used.

The soapmaker is also a large consumer of chemicals, especially caustic soda, salt, soda ash, and caustic potash, as well as sodium silicate, sodium bicarbonate, and trisodium phosphate. Inorganic chemicals added to the soap are the so-called *builders*. Important work by Harris of Monsanto and his co-workers demonstrated conclusively that, in particular, tetrasodium pyrophosphate and sodium tripolyphosphate were unusually effective synergistic soap builders. Of considerable economic importance was the demonstration that combinations of inexpensive builders, such as soda ash, with the more effective (and expensive) tetrasodium pyrophosphate or sodium tripolyphosphate, were sometimes superior to the phosphate used alone. It was further shown that less soap could be used in these mixtures to attain the same or more effective soil removal.

In *continuous, countercurrent* splitting the fatty oil is deaerated under a vacuum to prevent darkening by oxidation during processing. It is charged at a controlled rate to the bottom of the hydrolysing tower through a sparge ring, which breaks the fat into droplets. These towers, about 65 ft high and 2 ft in diameter, are built of Type 316 stainless steel. The oil in the bottom contacting section rises because of its lower density and extracts the small amount of fatty material dissolved in the aqueous glycerine phase. At the same time deaerated, demineralised water is fed to the top contacting section, where it extracts the glycerine dissolved in the fatty phase. After leaving the contacting sections, the two streams enter the reaction zone. Here they are brought to reaction temperature by the direct injection of high-pressure steam, and the final phases of splitting occur. The fatty acids are discharged from the top of the splitter or hydrolyser to a decanter, where the entrained water is separated or flashed off. The glycerine-water solution is discharged from the bottom of an automatic interface controller to a settling tank. Fig. 8.6 for glycerine processing.

Purification of Fatty Acids

Although the crude mixtures of fatty acids resulting from any of the above methods may be used as such, usually a separation into more useful components is made. The composition of the fatty acids from the splitter depends upon the fat or oil from which they were derived. Those most commonly used for fatty acid production include beef tallow and coconut, palm, cottonseed, and soyabean oil. Probably the most used of the older processes is panning and pressing. This fractional crystallisation process is limited to those fatty acid mixtures which solidify readily, such as tallow fatty acid. The molten fatty acid is run into pans, chilled, wrapped in burlap bags, and pressed.

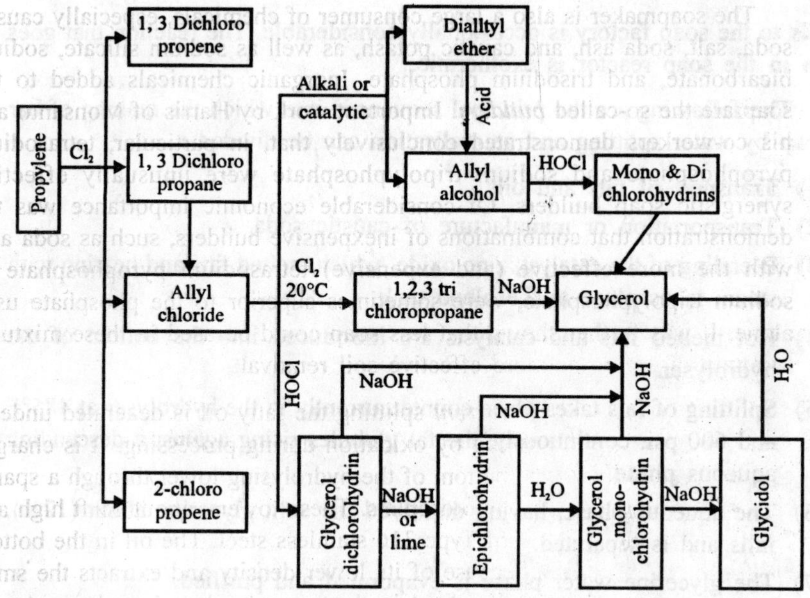

Fig. 8.6. Various possible processing steps for making glycerine from allyl chloride.

This expression extracts the liquid *red oil* (mainly oleic acid), leaving the solid stearic acid. The total number of pressings indicates the purity of the product. To separate fatty acids of different chain lengths, distillation is employed, vacuum distillation being the most widely used. Three fractionating towers of the conventional tray type, are operated under a vacuum. Preheated, crude fatty acid stock is charged to the top of a stripping tower. While it is flowing downward, the air, moisture, and low-boiling fatty acids are swept out of the top of the tank. The condensate, with part of it redrawn as a reflux, passes into the main fractionating tower, where a high vacuum is maintained at the top. A liquid side stream, also near the top, removes the main cut (low-boiling acids), while overheads and non-condensables are withdrawn. The liquid condensate (high-boiling acids) is pumped to a final flash tower, where the overhead distillate is condensed and represents the second fatty acid fraction. The bottoms are returned to the stripping tower, reworked, and removed as pitch. The fatty acids may be sold as such or converted into many new chemicals.

The *energy* requirements that enter into the cost of producing soap are relatively unimportant in comparison with the cost of raw materials, packaging, and distribution. The energy required to transport some fats and oils to the soap factory is occasionally considerable. The reaction that goes on in the soap reactor is exothermic.

The following are the principal sequences into which the making of bar soap by water splitting and neutralisation can be divided :

(1) Transport of fats and oils.

(2) Transportation or manufacture of caustic soda.

(3) Blending of the catalyst, zinc oxide, which melted fats and heating with steam takes place in the blend tank.

(4) Hot melted fats and catalysts are introduced into the bottom of the hydrolyser.

(5) Splitting of fats takes place countercurrently in the hydrolyser at 485°F and 600 psi, continuously, the fat globules rising against a descending aqueous phase.

(6) The aqueous phase, having dissolved the split glycerine about (12%), falls and is separated.

(7) The glycerine water phase is evaporated and purified.

(8) The fatty acids phase at the top of the hydrolyser is dried by fluashing off the water and further heated.

(9) In a high-vacuum still the fatty acids are distilled from the bottoms and rectified.

(10) The soap is formed by continuous neutralisation with 50% caustic soda in high-speed mixer-neutraliser.

The neat soap is discharged at 200°F into a slowly agitated blending tank to even out any inequalities of neutralisation. At this point the neat soap analyses: 0.002 to 0.10% NaOH, 0.3 to 0.6% NaCl, and approximately 30% H_2O. This neat soap may be extruded, milled, flaked, or spray-dried, depending upon the product desired.

These finishing operations may be detailed: the pressure on the neat soap is raised to 500 psi, and the soap is heated to about 400°F in a high-pressure steam exchanger. This heated soap is released to a flash tank at atmospheric pressure, where a partial drying (to about 20%) takes place because the soap is well above its boiling point at atmospheric pressure. This viscous, pasty soap is mixed with the desired amount of air in a mechanical scraped-wall heat exchanger, where the soap is also cooled by brine circulation in the outer shell from 220 to about 150°F. At this temperature the soap is continuously extruded in strip form and is cut into bar lengths. Further cooling, stamping, and wrapping complete the operation. This entire procedure requires only 6 h, as compared with over a week for the kettle process. The main advantages of soap manufactured by this process as compared with the kettle process are: (i) improved soap colour from a crude fat without extensive pretreatment, (ii) improved glycerine recovery, (iii) flexibility in control, and (iv) less space and labour. Intimate molecular control is the key to the success of this continuous process, as, for example, in the hydrolyser, where the desired mutual solubility of the different phases is attained by appropriate process conditions.

Typical Soaps

The main classes of soap are toilet soaps and industrial soaps. These different soaps can frequently be made by one or more of the procedures described. Detergents have replaced over 80% of all soaps. Except for the purest of toilet soaps, various chemicals are added as builders to improve economically the overall cleansing quality. Practically all soap merchandised contains from 10 to about 30% water. If soap were anhydrous, it would be too hard to dissolve easily. Almost all soaps contain perfume, even though it is not apparent, serving merely to disguise the original soapy odour. Toilet soaps are made from selected materials and usually contain only 10 to 15% moisture; they have very little added material, except for perfume and perhaps a fraction of a per cent of titanium dioxide as a whitening agent. *Shaving* soaps contain a considerable proportion of potassium soap and an excess of stearic acid, the combination giving a slower-drying lather. "Brushless" shaving creams contain stearic acid and fats with much less soap.

Another type of bar soap is *milled toilet soap*. The world milled refers to the fact that, during processing, the soap goes through several sets of heavy rolls, or mills, which mix and knead it. Because of the milling operation, the finished soap lathers better and has a generally improved performance, especially in cool water. The milling operation is also the way in which fragrant perfumes are incorporated into cold soap. If perfume were mixed with warm soap, many of the volatile scents would evaporate. After the milling operation, the soap is pressed into a smooth cylinder and is extruded continuously. It is then cut into bars, stamped, and wrapped.

Crystal phases in bar soap. The physical properties of bar soap are dependent upon the crystalline soap phases present and the condition of these phases. Any of three or more phases may exist in sodium soaps, depending upon the fat used, the moisture and electrolyte composition of the system, and the *processing* conditions. Milled toilet soaps are mechanically worked to transform the omega phase, at least partially, to the translucent beta phase producing a harder, more readily soluble bar. Extruded floating soaps contain both crystals formed in the freezer and crystals that grow from the melt after it leaves the freezer. Processing conditions are adjusted for an optimum proportion of crystallised matrix, which adds strength and rigidity to the bar. If necessary, the bar may be tempered by reheating to strengthen it.

GLYCERINE

Glycerine is supplied in several grades. The pure grade contains not less than 95% glycerol, and is suitable for resins and other industrial products. Yellow distilled is used for certain processes where higher-purity types are not essential, e.g., as a lubricant in tyre molds. Glycerine is employed in making, preserving, softening, and moistening a great many products such as alkyds; tobacco; Cellophane; explosives; drugs and toilet goods (including toothpaste); monoglycerides and goods; urethane foams; triacetin; and miscellaneous.

Glycerine Manufacture

Glycerine may be produced by a number of different methods, of which the following are important: (i) the saponification of glycerides (oils and fats) to produce soap, (ii) the recovery of glycerine from the hydrolysis, or splitting of fats and oils to produce fatty acids, and (iii) the chlorination and hydrolysis of propylene and other reactions from petrochemical hydrocarbons. The new processes (ii and iii) have resulted in lowering production costs for glycerine.

In recovering glycerine from soap plants, the *energy* requirements are mostly concerned with heat consumption involved in the unit operations of evaporation and distillation. The breakdown of natural and synthetic procedures for glycerine is:

Glycerine from sweet water from hydrolyser	*Glycerine from petroleum*
Evaporation (multiple effect) for concentration	Purification of propylene
Purification with settling	Chlorination to allyl chloride
Steam vacuum distillation	Purification and distillation
Partial condensation	Chlorination with HOCl
Decolouration (bleaching)	Hydrolysis to glycerine
Filtration or ion-exchange purification	Distillation

Glycerine recovery and refining

Practically all natural glycerine is now produced as a coproduct of the direct *hydrolysis* of triglycerides from natural fats and oils. Hydrolysis is carried out in large continuous reactors at elevated temperatures and pressures with a catalyst. Water flows counter-current to the fatty acid and extracts glycerol from the fatty phase. The sweet water from the hydrolyser column contains about 12% glycerol. *Evaporation* of the sweet water from the hydrolyser is a much easier operation compared with evaporation of spent soap lye glycerine in the kettle process. The high salt content of soap lye glycerine requires frequent shutdown of the evaporators to remove crystallised salt. Hydrolyser glycerine contains practically no salt and is readily concentrated. The sweet water is fed to a triple-effect evaporator, where the *concentration* is increased from 12% to 75 to 80% glycerol. Usually, no additional heat (other than that present in the sweet-water effluent from the hydrolyser) is required to accomplish the evaporation. After concentration of the sweet water to *hydrolyser crude*, the crude is settled for 48 h at elevated temperatures to reduce fatty impurities that could interfere with subsequent processing. Settled hydrolyser crude contains approximately 78% glycerol, 0.2% total fatty acids, and 22% water. The settled crude is *distilled* under a vacuum (60 mm Hg) at approximately 400°F. A small amount of caustic is usually added to the still feed to saponify fatty impurities and reduce the possibility of codistillation with the glycerol. The distilled glycerine is condensed in three stages at decreasing temperatures. The first stage yields the purest glycerine, usually 99% glycerol, meeting CP specifications. Lower-quality grades of glycerine are collected in the second and third condensers. Final *purification* of glycerine is accomplished by carbon bleaching, followed by filtration or ion exchange.

Synthetic glycerine from petroleum

The growing market for glycerine, and the fact that it was a coproduct of soap and dependent upon the latter's production, had been the incentives for research into methods for producing this trihydroxy alcohol. The propylene is chlorinated at 510°C at 1 atm to produce allyl chloride in seconds in amounts greater than 85% of theory (based on the propylene). Vinyl chloride, some disubstituted olefins, and some 1:2- and 1:3-dichloropropanes are also formed. (The reaction producing allyl chloride was new to organic synthesis, involving as it does the chlorination of an olefin by substitution instead of addition.) Treatment of the allyl chloride with hypochlorous acid at 38°C produces the glycerine dichlorohydrin ($CH_2Cl{\cdot}CHCl{\cdot}CH_2OH$), which can be hydrolysed by caustic soda in a 6% Na_2CO_3 solution at 96°C. The glycerine dichlorohydrin can be hydrolysed directly to glycerine (Fig. 8.6), but this takes two molecules of caustic soda; hence a more economical procedure may be to react with the cheaper calcium hydroxide, taking off the epichlorohydrin as an overhead in a stripping column. The epichlorohydrin is easily hydrated to monochlorohydrin and then hydrated to glycerine with caustic soda. The reactions are:

$$CH_3{\cdot}CH{:}CH_2 + Cl_2 \longrightarrow CH_2Cl{\cdot}CH{:}CH_2 + HCl \quad \text{(85\% yield)}$$

$$CH_2Cl{\cdot}CH{:}CH_2 + HOCl \longrightarrow CH_2Cl{\cdot}CHCl{\cdot}CH_2OH \quad \text{(95\% yield)}$$

$$CH_2Cl{\cdot}CHCl{\cdot}CH_2OH + 2NaOH \longrightarrow CH_2OH{\cdot}CHOH{\cdot}CH_2OH + 2NaCl$$

The overall yield of glycerine from allyl chloride is above 90%.

Another process for obtaining glycerine from propylene, involving the following reactions, where isopropyl alcohol and propylene furnish acetone and glycerine (through acrolein) and in good yields:

$$CH_3{\cdot}CHOH{\cdot}CH_3 + \text{air} \xrightarrow{\text{Catalyst}} CH_3{\cdot}CO{\cdot}CH_3 + H_2O_2$$

$$CH_3{\cdot}CH{:}CH_2 + \text{air} \xrightarrow{\text{Catalyst}} CHO{\cdot}CH{:}CH_2 + H_2O$$

$$CHO{\cdot}CH : CH_2 + H_2O_2 \longrightarrow CHO{\cdot}CHOH{\cdot}CH_2OH \xrightarrow{H_2} CH_2OH{\cdot}CHOH{\cdot}CH_2OH$$

Recent process for making glycerine

The most recent process for making glycerine is based on a charge of propylene and oxygen coproducing glycerine and large amount of acetone. Propylene is first hydrated to isopropanol using a standard process. The isopropanol is oxidised in the liquid phase by bubbling pure oxygen through a liquid mixture of isopropanol and hydrogen peroxide at a pressure of about 2.5 atmospheres and at temperatures between 90 and 140°C. The reaction mixture is diluted with water, stabilised, and fractionated to yield hydrogen peroxide solution, acetone, and unreacted isopropanol, which can be recycled.

A mixture of propylene with slightly more than an equal amount of steam is reacted with 25 per cent oxygen based on the weight of the propylene. The reaction is carried out over a fixed-bed of catalyst based on cuprous oxide, supported on silicon carbide or some other thermal conductivity support. Close control of reaction temperature is necessary. The reaction pressure may be from 1 to 10 atmospheres and the temperature between 300 and 400°C. The reaction mixture is cooled and fractionated to give acrolein, unreacted propylene, and tarry by-products. The yield of acrolein is near 86 per cent of the propylene consumed.

The purified acrolein is mixed with pure isopropanol from the propylene hydration step and the mixture vapourised. The mixed vapours containing 2–3 moles of alcohol per mole of acrolein are passed through a catalyst bed containing both uncalcined magnesium oxide and zinc oxide, with the magnesia predominating. The reaction takes place at about 400°C and yields about 77 per cent allyl alcohol (based on acrolein charged) and an additional quantity of acetone which is added to that formed in the isopropanol oxidation step.

Purified allyl alcohol is agitated with a water solution of hydrogen peroxide containing a small amount of tungsten trioxide in solution. The effective catalyst is a 0.2 per cent pertungestic acid solution in two molar aqueous hydrogen peroxide. The reaction temperature is 60–70°C; reaction time is about two hours, producing glycerine in water solution.

The reaction mixture is pumped to receivers and distilled to yield high-purity glycerine. The recovered catalyst solution is recycled. The yield of the final step is 80–90 per cent based on the allyl alcohol charged.

The control of both the propylene oxidation step and the hydroxylation step must be exceptionally good for acceptable yields. The conditions quoted are approximate, and the practical operating limits are doubtless much less broad.

The overall yield of useful products (glycerine and acetone) is not known. The overall glycerine yield on propylene charged to the oxidation is in the range of 50–60 per cent.

Chapter 9

Essential Oils

INTRODUCTION

The manufacture of perfume, cologne, and toilet water, collectively known as the fragrances. The creation of a good perfume for use in a consumer product, be it a perfume or a detergent, requires an intimate knowledge of all fragrance materials available. Fragrance compounds are complex mixtures prepared by blending together various fragrance materials such as essential oils (which may themselves be mixtures of several organic chemicals) and synthetic aroma chemicals. The fragrance compounds have no direct or absolute counterparts in nature and are incorporated as raw materials in soaps or colognes to get the desired moles. Perfume oil essence, perfume base and bouquet are some of the terms used to describe fragrance compounds. Not many people realise how complex the creation of an acceptable fragrance has become; requires professional knowledge, skill, and experience, coupled with specialisation in synthetic chemistry's technical problems, followed by consumer panel testing. This change has resulted from a number of factors, among which are: (i) increase in the number of available raw-material ingredients, both natural and synthetic; (ii) a variety of new types of products requiring fragrances; (iii) innovations in packaging, especially aerosol sprays, virtually non-existent before, and other new forms of despensing such as perfume powders, cream sachets, gels, lotions, and sticks; (iv) postwar prosperity, which has widened the spending habits of the public as well as increased the consumer potential; (v) broadened channels and methods of distribution, including door-to-door selling; and (vi) phenomenal growth in men's toiletries. These numerous recent developments are in contrast to the two main changes which took place before the prewar era, the introduction of synthetics and improved methods of obtaining true oils.

PERFUMES

Perfumery is the art of producing fragrances through the combination of odouriferous substances. The word perfume is derived from the Latin meaning "through smoke". Throughout history, perfumes have played an important role in human lives, and have been associated with notions of happiness, beauty, and satisfaction. Until this century, fragrance materials have been derived from natural sources, which has placed limitations on odour types and markets. The increased use of perfumes during the last few years would have been impossible without the development of the chemistry that allowed the invention of totally new odouriferous molecules as well as the synthesis of natural ones. Fragrances are no longer a luxury for the rich but today are incorporated routinely in a great number of products that are in daily use.

CONSTITUENTS OF PERFUMES

The constituents of perfumes are: vehicle solvent, the fixative and odouriferous elements.

Vehicle

The modern solvent for blending and holding perfume materials is highly refined ethyl alcohol mixed with more or less water according to the solubilities of the oils employed. This solvent, with its volatile nature, helps to project the scent it carries, is fairly inert to the solutes, and is not too irritating to the human skin. The slight natural odour of the alcohol is removed by deodorising, or "prefixation," of the alcohol. This is accomplished by adding a small amount of gum benzoin or other resinous fixatives to the alcohol and allowing it to mature for a week or two. The result is an almost odourless alcohol, the natural rawness having been neutralised by the resins.

Fixative

In an ordinary solution of perfume substances in alcohol, the more volatile materials evaporate first, and the odour of the perfume consists of a series of impressions rather than the desired ensemble. To obviate this difficulty, a fixative is added. Fixatives may be defined as substances of lower volatility than the perfume oils, which retard and even up the rate of evaporation of the various odourous constituents. The types of fixative considered are animal secretions, resinous products, essential oils, and synthetic chemicals. Any of these fixatives may or may not contribute to the odour of the finished product but, if they do, they must blend with and complement the main fragrance.

Animal fixatives

Of all animal products, *castor,* or *castoreum*, a brownish-orange exudate of the perineal glands of the beaver, is employed in the greatest quantity. Among the odouriferous components of the volatile oil of castor are benzyl alcohol, acetephenone, *l*-borneol, and castorin (a volatile resinous component of unknown structure).

Civet

Civet is the soft, fatty secretion of the perineal glands of civet cats, which are indigenous to many countries, and was developed in Ethiopia. The secretions are collected about every 4 days by spooning and are packed for export in hollow horns. The crude civet is disagreeable in odour because of the skatole present. On dilution and aging, however, the skatole odour disappears, and the sweet and somewhat floral odour of civetone, a cyclic ketone, appears.

Musk

Musk is the dried secretion of the preputial glands of the male musk deer, found in the Himalayas. The odour is due to a cyclic ketone called *muskone*, which is present to the extent of from 1/2 to 2%. Musk, the most useful of the animal fixatives, imparts body and smoothness to a perfume composition even when diluted so that its own odour is completely effaced. Musk is used for its own sake in heavy oriental perfumes.

Ambergris

Ambergris is the least used, but probably best known, of the animal fixatives. It is a calculus, or secretion, developed by certain whales. Ambergris is obtained by cutting open the captured whale, or it is obtained from whales stranded on a beach. It is waxy in consistency, softening at about 60°C, and may be white, yellow, brown, black, or variegated like marble. It is composed of 80 to 85% ambrein (triterpenic tricylic alcohol), resembling cholesterol and acting merely as a binder, and 12 to 15% ambergris oil, which is the active ingredient. It is employed as a tincture, which must be matured before it is used. The odour of the tincture is decidedly musty and has great fixative powers.

Musc zibata

Musc zibata is the newest animal fixative, derived from glands of the Louisiana muskrat. About 90% of the unsaponifiable material in muskrat glands consists of large, odourless cyclic alcohols, which are converted to ketones, increasing the characteristic musk odour nearly 50 times. It is a replacement for or an addition to Asiatic musk.

Resinous fixatives

Resinous fixatives are normal or pathological exudates from certain plants, which are more important historically than commercially. These are hard resins, e.g., benzoin and gums; softer resins, e.g., myrrh and labdanum; balsams, moderately soft, e.g., Peru balsam, tolu balsam, copiaba, and storax; oleoresins, oily materials, e.g., terpenes; extracts from resins, less viscous, e.g., ambrein. All these substances, when being prepared for perfume compounding, are dissolved and aged by methods passed down by word of mouth. If solution is brought about in the cold, the mixture is called a *tincture*. If heat is required to give solution, the mixture is an *infusion*. Alcohol is the solvent, sometimes aided by benzyl benzoate or diethyl phthalate. The most important of the soft gums is *labdanum*. The leaves of a plant growing in the Mediterranean area exude this sticky substance. An extract from this gum has an odour suggestive of ambergris and is marketed as *ambrein*, having extremely good fixative value. Of the harder plant resins used in perfumes, *benzoin* is the most important. The history of chemistry was influenced by this substance. The early source of benzoin was Java, where it was called *luban jawi*. Through various contractions and linguistic modifications, it became "banjawi," "benjui," "benzoi," "benzoin," and "benjamin." In early organic chemical history an acid isolated from this gum became known *benzoic acid*, from which compound the names of all benzo compounds of today are derived.

Essential-oil fixatives

A few essential oils are used for their fixative properties as well as their odour. The more important of these are clary sage, vetiver, patchouli, orris, and sandalwood. These oils have boiling points higher than normal (285 to 290°C).

Synthetic fixatives

Certain high-boiling, comparatively odourless esters are used as fixatives to replace some imported animal fixatives. Among them are glyceryl diacetate (259°C), ethyl phthalate (295°C), and benzyl benzoate (323°C). Other synthetics are used as fixatives, although they have a definite odour of their own that contributes to the ensemble in which they are used. A few of these are :

Amyl benzoate	Musk ketone	Heliotropin
Phenethyl phenylacetate	Musk ambrette	Hydroxycitronellal
Cinnamic alcohol esters	Benzophenone	Indole
Cinnamic acid esters	Vanillin	Skatole
Acetophenone	Coumarin	

Odourous Substances

Most odourous substances used in perfumery come under three headings: (i) essential oils, (ii) isolates, and (iii) synthetic or semisynthetic chemicals. The essential oils and isolates are discussed separately in this chapter. Synthetic and semi-synthetic chemicals are discussed below.

Synthetic and Semi-Synthetic Chemicals used in Perfumes and Flavours

More and more important constituents of perfumes and flavours are being made by the usual chemical synthetic procedures. Compositions containing predominantly inexpensive synthetics now account for more than 50% of the fragrances used in perfumes. Some constituents are chemically synthesised from an isolate or other natural starting materials and are classified as semi-synthetics. Examples are vanillin, prepared from eugenol from clove oil, ionone, from citral from lemon grass oils, and terpineols, from turpentine and pine oil.

Fragrance Raw Materials

Natural products

Essential oils are volatile materials produced from odourous plant material, generally by water or steam distillation. A concrete is an extraction almost exclusively from vegetable origin such as leaves, bark, flowers, and fruit. This is normally obtained by extraction with hydrocarbon solvents.

Absolutes are the alcohol-soluble portion of concretes, obtained by extracting the concretes with alcohol. Resinoids are perfume materials obtained by extraction of plant resinous substances with hydrocarbon solvents. Tinctures are alcoholic solutions. In perfumery, these are generally the solutions obtained by maceration of various odourous materials with alcohol.

Natural products used in perfume include ambergris, benzoin, castoreum, civet, clove leaf oil, galbanum, jasmine absolute, labdanum, mate, melilot, mimosa, musk tonquin, myrrh, oakmoss or mousse de chene, olibanum, opopanax, orris, patchouli, rosemary oil, sandalwood oil, vetivert oil, and violet leaves absolute.

Aroma Chemicals

During the last couple of years, there has been a rapid advance in the capabilities of instrumental techniques for the separation and identification of volatile organic substances. Of particular importance to the perfumery industry was the development of capillary gas chromatography of columns and the ability to use them directly in tandem with a mass spectrometer. Computer technology is used to interpret the vast amount of data generated by such a combination of instruments. These developments along with Fourier transform nmr spectroscopy have allowed discovery and identification of extremely minute odouriferous samples and have revolutionised not only the analysis of essential oils and extractives but also the direction of the synthesis of aroma chemicals.

Research in aroma chemicals can be divided into three general categories: (i) duplication of naturally occurring chemicals, for example, phenethyl alcohol, which occurs in rose oil; (ii) chemical modification of abundant, naturally occurring materials, e.g., acetylated vetivert oil ("vetivert acetate") from vetivert oil, and vanillin from lignin; and (iii) synthesis based on industrial organic feedstocks, e.g., nitro musks.

Odour Vocabulary

The descriptions and groups of fragrance raw materials are helpful in evaluating existing aroma chemicals or newly developed materials. To illustrate the use of the odour vocabulary, two well-known materials are: citronellol and lyral.

Classification of Perfumes

Nature offers the perfumer an unlimited number of models. The impressions, however eccentric and exclusive, always have a recognisable theme and category and are easily identified and described: straight florals/Muguet des Bois, floral bouquet, aldehydic floral, oriental, chypre/Ma Griffe, woody/Chamade, green/Bill Blass, citrus/Quartz, fougree/Maja, Canoe/Ambush, musk/Musk Oil, animal/Civet, and leather/Tabac Blond.

Men's perfumes can be classified in the following categories: green/Old Spice Herbal, citrus/Aqua Velva, fougere/Fougere Royal, Canoe/Canoe, spice/Old Spice, woody/Deep Woods, herbal/Polo, musk/Musk for Men, leather/Ted, oriental/Skin Bracer and chypre/Paco.

Manufacture

Compounding of fragrances

A compounding department is a scaled-up version of the perfumes' laboratory. All equipment in a modern compounding facility is made of stainless steel. The formula is usually rewritten into a working formula which means adding crystals first, more viscous materials second, and then part of the liquids until the mixture is homogeneous. The rest of the liquids are added, and then the valuable, delicate, and often more volatile materials are added. The time required for mixing usually is several hours, because crystals must dissolve without heating.

Quality control in perfumery

All fragrance raw materials exist in many different qualities. The name or description of any essential oil, natural product, isolate, or fragrance chemical has little meaning except to aid in identification. The odour quality of a given product becomes the most important criterion in determining its acceptability according to an established quality. Although there are other physical, chemical, and analytical tests, they have little significance unless the primary requirement of odour quality is satisfied.

The olfactory evaluation of fragrance materials is done on separate odourless slips of blotting paper that are dipped into the standard and the sample simultaneously and to the same depth so that the same amount of material is on both blotters.

Creation of fragrances

There are more than 6000 raw materials at the disposal of the perfumer. It is important to start with a common idea inspired by a new synthetic, a new natural material, or a newly discovered odour facet of an existing material. The perfumer creates an accord based on the inspiring note. The accord or theme can be used in an existing floral or woody floral fragrance, or the perfumer can create a floral composition around the new original theme. A fine fragrance must work on the skin because it has to blend with body odour. It also has to last, and be pleasant and diffusive. It must have the quality of genuine beauty and a signature that distinguishes it from any other fragrance.

A fragrance odour can be analysed as a blend with a top note continuing into a middle note to a final or end note. The materials with the greater vapour pressures evaporate faster. What the perfumer must do is to smooth the odour profile so that there are no discontinuities in odour impact as the different components volatilise.

Industrial Perfumery

Fragrances play an important role in detergents, soaps, liquid fabric softeners, tumble-dryer softeners, bleach products, shampoos, deodorants and antiperspirants, and talcs and powders.

Fragrance Safety

The fragrance industry has a long and clean record of safety, largely because of the nature, source, and methods of use of its products. However, with the wider-spread application and daily exposure to fragrances in all manner of products, companies have, in recent years, devoted much attention to a reappraisal of their safety criteria. As a minimum, the following tests are carried out; acute oral toxicity, skin irritation and sensitisation potential, and, where significant, eye irritation, phototoxicity and photosensitisation tests.

Odours and their Mixtures

Odour is that property of a substance that makes it perceptible to the sense of smell. Although it is certain that odour is caused by molecular structure, there remains little predictability about this correlation of odour with structures. It is known from experience that certain chemicals have certain types of odour, but even an expert cannot predict what an unfamiliar molecule or mixture will smell like.

Intensity perception

The measurable properties of odour perception are its intensity and its character. The measurement and expression of odour mixture intensity can be achieved in a number of ways, but they are almost all subjective. The simplest but the most subjective method is the use of a hedonic scale. The format possibilities are many, e.g., 0 = no odour, 1 = barely perceptible, 2 = distinct, 3 = strong, 4 = very strong, and 5 = overpowering.

One of the most objective measures of an odourant's intensity is its threshold which, however, reflects the intensity of only one specific odourant concentration, i.e., the weakest that can be detected.

All the senses, including smell, relate to the psycho-physical law (Steven's law) as expressed by the power equation :

$$\Psi = k\phi^{\beta}$$

where Ψ is the perceived intensity of the sensation, i.e., the odour; k is a constant; and ϕ is the physical intensity of the stimulus, i.e., the odourant.

Generally, the sensation varies as a power function of the stimulus. In olfaction, the power β is less than one.

Character perception

Odour character is purely subjective. Several chemicals have been experimentally characterised by scaling the degrees to which each chemical possesses subjective reference qualities, e.g., burnt, fruity, and spicy. But no method exists that can reliably characterise odours.

Mixture perception (modification)

Many odourous and non-odourous chemicals are used to control odours, but only those that work essentially by altering the way the nose perceives the character and intensity are true odour modifiers.

Products

Although the original intent of deodorisers was to reduce malodour, use for decorative purposes is becoming an important factor. The active ingredient of masks and counteractants are oils: essential oils, which are fragrance extracts from flowers, herbs, fruits, and trees; animal extractives; and aroma chemicals, which may be either natural or synthetic.

ESSENTIAL OILS

An essential oil is the predominantly volatile material isolated by some physical process from an odourous single-species botanical. Over 3000 oils have been identified from the vast number of plant species and several hundred have been commercialised. Of these, some are extremely rare and produced in only kilogram quantities, e.g., violet oil, concretes (flower extracts), and angelica root oil.

Essential oils are now generally manufactured close to the growing area. Product quality has suffered, however. Essential oils are easily adulterated and this technology has not been overlooked by isolated local producers.

Essential oils are isolated from various plant parts, such as leaves (patchouli), fruit (mandarin), bark (cinnamon), root (ginger), grass (citronella), wood (anyris), heartwood (cedar), gum (myrrh), balsam (tolu balsam), berries (pimenta), seed (caraway), flowers (rose), twigs (clove stems), and buds (cloves). These plants parts are processed to yield their quintessences or essential oils, which are mostly devoid of cellulose, glycerides, starches, sugars, tannins, salts, and minerals which also occur in these botanicals.

The common physical method for isolating essential oils from the botanical is steam distillation, wherein small amounts of non-volatiles may be carried over. Some oils are expressed. Many flower oils are extracted with a purified petroleum solvent.

Essential oils are generally liquid at room temperature; however, some are semi-solid, such as *Mentha arvensis* (Brazilian mint), and several are solid, e.g., oil of guaiac wood. The cultivation of essential oil-bearing plants has kept pace with modern agricultural methods. Hybrids are grown to yield oils of specific odour, flavour, or properties.

Essential oils are used as such for flavours and fragrances, but products derived from, or based on essential oils have large volume usage for specific applications. Essential oils are concentrated, rectified, extracted, or chemically treated to further isolate vital components, purify, adjust properties, or increase the concentration of significant flavour of fragrance components.

Composition

The volatile components of essential oils usually contain fifteen carbon atoms or less. However, seed oils contain long-chain fatty acids or esters and even glycerides that are carried over in distillation; the amounts of these components are very low.

Essential oils basically are made up of carbon, hydrogen, and oxygen, and occasionally nitrogen and sulphur. The largest class of components is the terpenes which have ten carbon atoms and are head-to-tail condensation products of two isoprene molecules. The terpenes may be aliphatic, alicyclic, or bi- and tricyclic of varying degrees of unsaturation up to three double bonds.

It is not uncommon for an essential oil to contain over two hundred components and often the trace substances (in ppm) are essential to the odour and flavour. The absence of even one component may change the aroma. The same species of botanical, grown in different parts of the world, usually has the same components; however, climatic and topographical conditions affect plants and can alter the essential oil quantitatively, but rarely qualitatively.

Production

In the botanical, the essential oil is present in oil sacs. It is isolated by comminution, and the action of heat, water, and solvents.

Steam distillation

Steam or hydrodistillation is the preferred method for producing essential oils, employing either water, wet steam, or steam.

Extraction

An essential oil that is sensitive to heat, eg., jasmine or tuberose, or contains an essential non-volatile constituent, e.g., piperine in black pepper, is extracted with a solvent.

Properties

Pungent taste, and odour, usually nearly colourless when fresh, but becoming darker and thick on exposure to the air, optically active; sp. gr. 0.850–1.100. Soluble in alcohol, carbon disulphide, carbon tetrachloride, chloroform, petroleum ether and fatty oils; insoluble in water, except for individual constituents of some oils which may be partially water-soluble, resulting in a loss of these constituents during steam distillation.

Synthetic Substitutes for Essential Oils

Fluctuations in the cost and availability of natural oils and the high cost of some oils have induced users to seek substitutes. Several former large-volume oils have been replaced by synthetics because of large volume demands. The expensive variations in cost are especially evident in essences such as rose, jasmine, violet, lilac, neroli, etc. Duplication of these items is economically worthwhile. Modern techniques and instrumentation offer the possibility for total analysis of these oils. This route has been undertaken by the primary fragrance and flavour companies throughout the world in the attempt to find economic and available substitutes. Nonetheless, there is a trend away from synthetic oils because complete duplications are in most cases not technically, aesthetically, or economically possible.

Health and Safety Factors

Most essential oils, since they are natural and have a history of use, are considered GRAS (generally recognised as safe). Safety and toxicity testing and evaluations and regulations are different for food additives than for fragrance oils and cosmetics. Some oils can be used for both purposes, including celery, rose, and black pepper. Some oils, such as cinnamon oil, can be used in flavours, but not in fragrances because of skin irritation.

Commercial Essential Oils

The commercial essential oils include allspice (pimenta berry), bitter almond, amyris, anise, star anise, sweet basil, bay (myrcia), bergamot, sweet birch, bois de rose (rosewood), camphor, cananga, caraway, cardamom, cassia, cedarwood, cinnamon, citronella, clove, coriander, eucalyptus, geranium, ginger, grapefruit, jasmine, juniper, labdanum, lavandin, lavender, lemon, distilled lime, Japanese mint, neroli, nutmeg, ocotea, bitter orange, sweet orange, origanum, orris root, palmarosa, patchouli, black pepper, peppermint, petitgrain bigarade, pine, pinus pumilio, rose, rosemary, dalmatian sage, sage clary (sage muscatel), East Indian sandalwood (santal), spearmint, spike lavender (spike), thuja (cedarleaf), thyme, turpentine, vetiver, wintergreen, and ylang ylang.

ESSENTIAL OILS USED IN PERFUMES

Essential oils in varying proportions, directly enter into some perfumery compounds and flavouring essences. Novel and wholesome natural blends in the form of essential oils of varying compositions, within the same species of a plant material, are now possible as a result of screening experiments conducted by scientists all over the world. If made available in sufficient quantities, these materials could be used directly for creating variety.

However, it is often desirable to 'rectify' the essential oils for specific uses and purposes. The process of rectification may consist of one of the following:

(1) Removal of colour, moisture, sediments and polymerised residues, for obvious reasons;

(2) Preparation of terpeneless or sesquiterpeneless oils with a purpose to remove undesirable odours, improve the shelf life and solubility of the material in alcohol;

(3) Isolation of main components in olfactorily desirable form; and

(4) Reconstitute the oils after regrouping of the components or replacing some of the components without loss in overall odour value of the product and so on.

NATURAL ISOLATES

Natural isolates reviewed here are a few among the many, cited as examples of (i) general methods of isolation; (ii) odour description and value; (iii) physico-chemical characteristics; (iv) suggested uses; and (v) their useful derivatives as an extended list of available range of odours.

For the purpose of a system, classification based on chemical functionals like alcohols, aldehydes, ethers, hydrocarbons, ketones and phenols, has been taken as the basis of groupings. Under each group, in addition to well-known commercially important and versatile isolates, some less known or less used isolates have also been included as examples to show how a large spectrum of odour resources are available in nature itself and how it can be further extended by appropriate derivative formations. Non-essential component of essential oils can be useful building blocks for the semi-synthesis of materials far more useful than the parent body. Further, the fact that non-oxygenated components of essential oils, terpenes and sesquiterpenes, have offered the building blocks for the semi-synthesis of almost all commercially important aroma chemicals, is too well-known.

Odour transformations, with the increase in the mole size of derivatives and change in active functional groupings, has been brought out for the guidance of a student in perfume chemistry. The functional uses of aroma chemicals in perfumes and flavours as fixative, extender, modifier, toner, fuser, mellower, rounder, lifter, substitute, supporter, perceptor, and so on have been indicated. Uses of aroma chemicals other than in perfumery have also been indicated. The natural isolates are classified as under :

Alcohols	Carverol, cedrol, citronellol, elemol, eudesmol, farnesol, fenchol, geraniol, guiol, linalool, menthol, myrtenol, nerolidol, perillyl alcohol, rhodinol, and vetiverol.
Aldehydes	Anisaldehyde, citral, citronellal, and cuminaldehyde.
Ethers	Anethole and safrole.
Hydrocarbons	Caryophyllene, cedrene, α-pinene, and β-pinene.
Ketones	Atlantone, carvone, menthone, piperitone, plugone, and thujone.
Phenols	Carvecrol, eugenol, isoeugenol, and thymol.

Carveol

Carveol, a cyclic terpene alcohol, $C_{10}H_{16}O$ occurs in nature, in small quantities, in caraway seed oil. It can be isolated from natural sources with the help of appropriate derivatives. Commercially it is obtained by the reduction of carvone with lithium-aluminium hydride. Carveol has an odour which is in between caraway and spearmint, but more towards caraway than spearmint. It is used in flavour compositions in support of minty and spicy notes.

Cedrol

Cedrol is a white crystalline, dextrorotatory tertiary sesquiterpene alcohol, and commercially known as cedar camphor or cypress camphor. Cedrol is found in the oils of cedarwood obtained from North American wood of *Juniperus virgiana, J. polycorpos, J. procera* Hocks and in oil of *Cupressus semipervirens*.

Cedrol can be isolated, from any of the aforesaid oils by: (i) fractional distillation under vacuum and bulking together of various fractions, so as to achieve cuts of desired purity; and (ii) a product of higher purity can be obtained by chilling, to lower temperatures, the cuts obtained by fractional distillation and subsequent separation of the solids by filtration. However, (iii) a product of still higher purity can be prepared by the crystallisation of crude solids through pet, ether or alcohol.

From the oil itself solids can be thrown out by agitating it with a small amount of alcohol followed by a quick filtration.

Cedrol, in pure form, exhibits a very mild odour, but lower grades smell cedar with balsamic sweetness. Being a high boiling product with mild odour, cedrol is used as a fixative in perfumes for soaps, detergents and household articles.

Being a tertiary alcohol, cedrol dehydrates into terpene, cedrene. Therefore, derivative formations are best carried out in hydrocarbon suspensions.

Citronellol

Citronellol, 2-6 dimethyl-1-octane-8-ol, $C_{10}H_{20}O$, is an alcohol of very wide application in perfumery. Citronellol is a colourless oily liquid, b.p. 97°C, commercially available under different grades and offered under the trade denotations, like pure, extra, prime and special, each differing from the other in total alcohol content and in odour value.

It can be isolated from Java citronella oil by: (i) repeated fractional distillation; (ii) treating the enriched fractions with formic acid at a higher temperature, so as to destroy any contaminated geraniol; (iii) isolated by the phthalate or chlorophosphoric acid derivative formation, followed by regeneration and distillation; (iv) large quantities of citronellol are, however, produced by the catalytic hydrogenation of citronellal obtained from Java citronella oil or *Eucalyptus citriodora* oil; (v) it can also be prepared by the hydrogenation of geraniol; and (vi) from pinene via myrcene, this being the present day method of manufacture.

Citronellol has a fresh rosy odour which varies with the source. Called rose alcohol, it is one of the most important aroma chemicals used in large quantities in perfumes of all types. Citronellol undergoes esterification and etherification.

Elemol

Elemol is a sesquiterpene alcohol, $C_{15}H_{26}O$, m.p. 51–52°C, b.p. 130°C, it occurs in 50% yield in the tail fractions (above 140°C/3 mm), of Java citronella oil and in Manila elemi oil (obtained from the resinous exude of *Canarium luzonicum* Gray, grown in Philippines). It can be isolated from these sources by: (i) fractional distillation; (ii) by derivative formations like benzoate or borate followed by regeneration; and (iii) better isolations are possible if the derivative formation is applied on saponified elemol rich fractions.

Importance of considering elemol here is to illustrate an example of complete utilisation of an essential oil, like the tail ends of Java citronella oil, which are often lost sight of or are left behind as pot residues if operating vacuum is not good.

Fractions rich in elemol are viscous oily liquid, b.p. 144°C. In cold, these fractions become a semi-solid mass. Elemol is rarely isolated in pure form. In fact these fractions rich in elemol find a large scale direct use in perfumes for soaps, detergents and technical products. Elemol rich fractions have a faint, sweet woody odour with a dry peppery character. It blends well with geraniol fractions, thus offering a technical product known in commerce as "Geraniol for soap". Being a high boiling product, it acts as a good fixative and can be usefully incorporated in a verbena compound.

Eudesmol

Eudesmol, a bicyclic sesquiterpene tertiary alcohol, $C_{15}H_{26}O$, normally consists of a mixture of isomeric forms, alpha and beta. It is generally a pale yellow crystalline mass. Eudesmol is found in (i) wood oil of Araucaria (*Callitropsis aracariodes*, grown in New Caledonia); (ii) Japanese wood oil of *Cryptomeria japonica* Don. It can be isolated from the tail fractions of eucalyptus oils, on chilling and further purification by crystallisation from alcohol.

Eudesmol has a fine, sweet woody odour and can be used as a fixative in such or other types of perfumes including the floral and spicy ones. In fact it goes well with all types of odours. Eudesmol can be converted into a very useful acetate derivative.

Farnesol

Farnesol, $C_{15}H_{26}O$, a cyclic primary sesquiterpene alcohol, is found in nature in the form of a mixture of sterioisomers. It occurs in the flower oils of jasmine, cassia, lily, cananga, neroli, rose and in the oils of ambrette seed, palmarosa, lemongrass and citronella and also in some balsam oils. In practice it is isolated from ambrette seed extract or prepared from natural nerolidol. Synthetically it is made from methyl heptanone. Enriched fractions, b.p. above 1.55°C, of the oil of ambrette seed on saponification followed by phthalisation offers a pure material.

Farnesol has a sweet floral green odour suggesting of linden bloom and reminiscent of muguet notes. It is used in high grade perfumes to fuse and enhance the sweet floral notes in cassia, rose, magnolia and neroli, etc.

Fenchol

Bicyclic terpene alcohol, fenchol, $C_{10}H_{18}O$, is found in root oil of *Pinus palustris* (yellow pine oil) and in oils of *Baeckea frutescens* and *Dalbergia parviflora*.

Fractions boiling around 200°C from yellow pine oil when converted into phthalate or borate derivative on regeneration yield fenchol in pure form. Fenchol is a pale yellow crystalline mass, m.p. 48–49°C. It can be prepared by the reduction of fenchone obtained from cedar leaf oil. Fenchol odour, though diffusive camphor like, is at the same time like sweet lime. Fenchol finds use in some flavours and is used in soap perfumes for the purpose of producing 'lift'. It is essentially used in low priced soap, bath salt, and pine type room spray perfumes.

Geraniol

Geraniol, $C_{10}H_{18}O$, unsaturated primary terpene alcohol, is found widely distributed in nature either free or in combined form as esters. Oil of palmarosa contains as much as 95% and *Cymbopogan khasianus* has established itself as a geraniol-rich *Cymbopogon* sp. with a geraniol content of 75–80%. Oil of citronella contains 30–40% geraniol and in small amounts it is found in lemongrass oil, coriander, lavender, linaloe oils, etc.

Geraniol can be isolated: (i) by fractional distillation of saponified oils of palmarosa, *Cymbopogon khasianus* or citronella oil; and (ii) in its chemically pure form it can be isolated from these sources by the formation of phthalate or calcium chloride derivatives.

Geraniol is one of the most widely used aroma chemicals. Though it possesses a fine rose aroma yet it finds a wide application in other perfume compositions. Geraniol is a colourless to pale yellow liquid, b.p. 110–112°C.

Geraniol, apart from its natural sources, can be prepared by the reduction of citral and also synthetically from pinene. It offers a large number of derivatives like esters, ethers, and other condensation products with a wide spectrum of odour patterns which are useful for a large number of themes and types of perfumes and flavours. By itself, geraniol has a bitter taste, therefore, it does not go into any flavour composition.

Guiol

Bicyclic sesquiterpene tertiary alcohol, Guiol, $C_{15}H_{26}O$, is found in Guiac wood oil. This oil is distilled from the wood of a lofty tree, *Bulnesia sarmienti* Lor, growing in the jungles of Paraguay and Argentina. Guiol, a solid, m.p. 91°C, can be isolated from higher boiling fraction, of guiac wood oil. It is, however, rarely isolated in pure form, instead, fractions rich in guiol, are directly used and as such converted into various derivatives. Guiol has a soft woody odour, though mild but very persistent. It can blend well with floral and woody odours giving carnation, chypre or modern aldehydic notes.

Linalool

An aliphatic tertiary terpene alcohol, linalool, $C_{10}H_{18}O$, is widely distributed in nature. It is found in Japanese Ho-oil (80-90%), linaloe wood and berry oil (60-80%), Bois-de-rose oil (80-90%). It occurs either free or in combined form as acetate. Linalool is one of the most important aromatic isolates used extensively in perfumes for soaps and cosmetics and in flavours.

By fractional distillation a sufficiently pure grade of linalool is manufactured. Some of the oils containing linalool as a principle component are directly used in blending. However, on careful boratisation, a purer grade of linalool can be isolated.

Menthol

Menthol, a cyclic saturated secondary alcohol, $C_{10}H_{20}O$, occurs in nature in various isomeric forms. It is found in the oils of *Mentha piperita* (50-65%) and *Mentha arvensis* (75-90%).

Large quantities of *M. arvensis* are processed for the isolation of menthol. Some quantity of menthol, in long crystal form, is deposited by itself, if the oils mentioned above are left undisturbed during winter months.

Commercially, oil free from moisture and suspended impurities or better rectified oil after the removal of lighter boiling terpenes and bottom residues, is gradually chilled and the crystals, thus deposited are centrifuged. The filtrate is rechilled to collect fresh crops of crystals and the process repeated till the maximum amount of menthol is recovered in the form of crystals.

Menthol has a cool refreshing sweet taste with a diffusive pungent odour. It is widely employed in pharmaceutical preparations as it acts as a local anaesthetic, relieves headache, acts as antiseptic in nasal and respiratory track. It also enters into products like toothpastes, mouth washes, chewing gums, pan masalas, cigarettes, and cosmetics for producing cooling effect. It is used in balsams and also as a general purpose flavouring agent.

Myrtenol

Bicyclic, unsaturated primary alcohol, $C_{10}H_{18}O$, myrtenol occurs in the form of its acetate in the oil distilled from *Myrtus communis* and in the tail fractions of Spanish *Eucalyptus globuls* oil. On the saponification of the higher boiling fraction of myrtle or eucalyptus oil, followed by isolation of myrtenol, by phthalisation. Synthetically it is prepared by the oxidation of pinenes or on reduction of myrtenal.

Myrtenol has a warm woody odour and gives good results in lavender, citrus, mens' colognes, oriental, chypre, spruce and pine type of perfumes.

Nerolidol

Sesquiterpene alcohol, nerolidol, $C_{15}H_{26}O$, occurs in oils of orange flower, neroli, ylang-ylang, verbena, etc. It is found as a main constituent (75-85%) in oil of Cabreuva (*Myrocarpus fastigiatus*).

Fraction of oils rich in nerolidol on saponification in alcoholic medium and further on rectification under vacuum give nerolidol. Apart from its isolation from natural sources, nerolidol can be synthesised from geranyl acetone and acetylene.

On the basis of versatile nature of nerolidol in its use, it is called "Linalool of Sesquiterpene Series". It has a long lasting, delicate, woody, floral green odour and goes well with oakmoss, sandalwood and clary sage.

It has a blending, fixing and rounding up capacity in florals like jasmine, neroli, lily, and lilac perfumes. In flavouring it goes well with certain range of flavours.

Perillyl alcohol

Perillyl alcohol, dihydrocuminyl alcohol, $C_{10}H_{16}O$, occurs in nature, as free or in the form of esters, in different optical isomeric forms. It is found in the oils of bergamot, lavender, perilla, spearmint and gingergrass. In gingergrass oil it exists along with geraniol and can be isolated by fractional distillation or by chemically destroying geraniol by formylation followed by distillation. It blends well with green floral notes in narcissus and lily imparting petal like effect to rose perfumes.

It can be suitably used in citrus, vanilla, minty and fruity flavour themes. Apart from its isolation from natural sources, this alcohol can be prepared by the reduction of perilla aldehyde obtained from Japanese Perilla leaf oil and also by the oxidation of β-phellanderene.

Vetiverol

Bicyclic and tricyclic, primary and tertiary sesquiterpene alcohol mixtures, $C_{15}H_{24}O$, is found in the oils of vetiver from Java, Reunion, Haiti and Bourbon have been given the name Vetiverols. However, a tricyclic primary alcohol, out of this mixture can be isolated by phthalisation and purified by distillation under vacuum. Vetiverol as a mixture of alcohols, can be isolated from the oil, by fractional distillation, preferably after saponification. Vetiverol is somewhat greenish coloured viscous material having a warm, sweet, mildly earthy, somewhat rooty, balsamic odour of good tenacity. A 50% pure product is used in soap perfumes.

p-Anisaldehyde

p-Anisaldehyde, $C_8H_8O_2$, is found along with anethole in oils of anise, fennel, star anise and in flower oils of cassia and vanilla beans. Anisaldehyde can be isolated from the above sources by bisulphite adduct formation. It is a colourless liquid having a strong odour of hawthorn. A 98% pure product, use in soap perfumes based on cassia, mimosa, lilac, heliotrope, hawthorn and new mown hay themes. Anisaldehyde finds a wide and versatile use in anise, apricot, butter cherry, chocolate, nuts and peach flavours.

Citral

Aliphatic terpene aldehyde, citral, $C_{10}H_{16}O$, is found in a large number of essential oils as a major or minor component. In lemongrass oil, it is found to the extent of 80% and in *Backhousia citriodora* oil about 90%. *Litsea cubeba* contains 75% and *Eucalyptus straigeriana* contains 35-45% citral. It has a powerful lemon odour and is commonly used in flavours like apple, grape, orange, lime and strawberry etc.

Citral on condensation, cyclisation, enolisation, oxidation, hydrogenation, acetal formation, etc. offers products of commercial importance. Many of the first step reaction products undergo many more reactions further to produce a still larger number of useful products.

Citronellal

Citronellal is not widely distributed in nature. It is found in three isomeric forms: d-, l- and dl-isomeric forms and as d-isomer it is a main constituent of (i) Citronella oils of Java, Formosa, Guatemala and India (35-45%), (ii) *Eucalyptus citriodora* oil (70-90%), and in small quantities in oils of lemon and rose. It can be isolated from oil of *Eucalyptus citriodora* by fractional distillation and enriched to 85-90% purity. Higher purity can be achieved by chemical purification by bisulphite adduct formation.

Citronellal is used in perfumes for soaps and all kinds of technical products. It is also used in creating artificial citrus flavours. Commercially it is obtained from pinene via myrcene and on reduction of citral. Chemically its aldehyde group and ethylenic linkage are both very active, the reason that a large number of aroma chemicals are obtained from citronellal.

Cuminaldehyde

The aromatic aldehyde, cuminaldehyde, $C_{10}H_{12}O$, is a main constituent of cumin seed oil and is also found in cassia, cinnamon, myrrh and some eucalyptus oils. It can be isolated from cumin seed oil by treatment with sodium bisulphite solution yielding as adduct from which the aldehyde can be regenerated by alkali.

Possessing a disagreeable pungent green odour, this product is a yellowish coloured liquid, and is commercially available in 95% purity. It is useful for soap perfumes and in hay, mimosa, herbal fantasy top notes. In small quantities it finds use in flavour compositions reminding of Indian curry, spice and condiment odours.

Anethole

Anethole, $C_{10}H_{12}O$, is found in nature as a main constituent of essential oils of anise, fennel, star anise and others.

Being a low melting solid, m.p. 21°C, anethole can be isolated from the above oils after enrichment by fractional distillation followed by separation on cooling. Anethole, is available in 99% purity, having flash point 92°C, with an anise type of odour with recommended use in flavours and mouth care products, etc. Various grades of anethole are available including the pharmaceutical grades. With its very sweet odour and taste, it finds use in all types of food products especially confectionery items. It goes into beverages, flavouring of pharmaceutical preparations, mouth washes, dentrifices, gargles, etc. It is prepared synthetically from: (i) estragole; (ii) anisole; and (iii) anisaldehyde.

Safrole

A phenol ether, safrole, $C_{10}H_{10}O_2$, is found to the extent of even 90% in the oil of *Illicium parviflorum* Mich. It is also found in American and Brazilian Sassafras oils, oil of star anise and as a minor component of a large number of essential oils. It can be isolated by fractional distillation or fractional distillation followed by chilling to lower temperatures of cuts. Safrole is a colourless liquid, congealing at 11°C. 98% pure product is commercially available, with a typical sassafras odour, having flash point 100°C, with a recommended use in cheap soap perfumes.

Safrol has a warm, woody, spicy floral odour and is widely used in beverages, pharmaceuticals, tooth and oral preparations and for scenting cheap soaps.

Cedrene

American cedarwood oil, *Juniperus virginiana* L., contains 60-70% of isomeric mixture of cedrenes, which is also found in J. *polycarpos* Kock, J. *sinensis* L. and in many other essential oils. Cedrene is a sesquiterpene hydrocarbon, $C_{15}H_{24}$, which can be isolated from the above mentioned sources by fractional distillation. When isolated from American cedarwood oil, it is a colourless syrupy oil.

Cedrene has a dry woody camphoraceous odour and is used in perfumes for soaps, household products, industrial masking agents and other technical products.

α-Pinene

α-Pinene, $C_{10}H_{16}$, a cyclic terpene hydrocarbon, of great commercial importance, is found in nature in various isomeric forms. It occurs in the form of d-isomer in oils of *Pinus halepensis* (Green turpentine), *Pinus sylvestris* (Russian turpentine), and in hinoki wood oil, being main constituent as 80, 70 and 60 per cent respectively. It mostly occurs as a mixture in many other essential oils. α-Pinene is isolated from the above sources by fractional distillation through efficient columns.

β-Pinene

β-Pinene, $C_{10}H_{16}$, occurs in nature accompanied by α-pinene in various turpentine oils. β-Pinene has a greater solubility in alcohol; therefore, α and β-pinenes can be separated by dilute alcohol.

It has a dry woody piney odour. Small amounts of β-pinene find use in creating artificial essential oils. Larger quantities are used as a starting material for the manufacture of a large number of industrially important aroma chemicals.

Atlantone

Sesquiterpene ketone, Atlantone, $C_{15}H_{22}O$, occurs in Himalayan cedarwood oil (*Cedrus deodara* Loud) and Atlas cedarwood oil. This ketone, in fact a mixture of α and γ forms, can be isolated by repeated fractionation of the oils of their origin and concentrated further by the boration of the fractions to remove alcoholic impurities and thus leaving behind enriched ketone. Atlantone is a pale yellow viscous liquid. Atlantone has a woody animal odour, with a sweet background. Himalayan cedarwood or Atlas cedarwood oils owe their respective odours essentially to atlantone.

Atlantone on hydrogenation yields Atlantol, which has a fine woody odour. In fact when deodar oil itself is hydrogenated and distilled under vacuum, the product of higher odour value is obtained.

Carvone

Cyclic terpene ketone, carvone, $C_{10}H_{14}O$, is available in nature in different isomeric forms. Caraway and dill seed oils contain about 50-60% d-carvone and spearmint oil consists of upto 70% carvone in 1-form. It is found in *Lindera sesicea* oil (Kuromoji oil) from Formosa and occurs in gingergrass oil also. Carvone forms an addition compound with sodium bisulphite from which it can be regenerated with alkali. It also forms an addition compound with hydrogen sulphide gas which can be applied to isolate carvone in pure form. At ordinary temperatures, carvone is a colourless liquid, b.p. 90°C, but at lower temperatures it congeals.

Carvone has a wide application in flavouring of pharmaceutical products, toothpastes, mouth washes, bread, pickles and spices, etc. With its warm herbaceous odour it is used as a fortifier of caraway and dill oils. It is used for flavouring beverages also.

Thujone

A cyclic terpene ketone, thujone, $C_{10}H_{16}O$, occurs widely distributed in nature. It occurs as a chief constituent of oils of Thuja. Thujone can be isolated from its naturally occurring sources on enrichment by fractional distillation or by treatment of the enriched fraction with sodium bisulphite. Thujone occurs in alpha and beta forms. Thujone is a colourless liquid, b.p. 63°C, having a peculiar penetrating odour, which is camphoraceous and herbal in character. As this material is considered toxic in nature, it does not have much application in perfumery. It is, however, useful in creating synthetic essential oils like sage and wormseed. It blends well with geranium, galbanum and oakmoss.

Carvecrol

Carvecrol, $C_{10}H_{14}O$, a commercially important phenol, occurs in oils of Organum of different origins like Syria, Spain and Morocco, and also in the oils of Savory and Majoram. It can be isolated from the natural sources with the help of water soluble sodium or potassium salts. Carvecrol is available as a 98% pure product, clearly soluble in 60% alcohol (1:4).

Carvecrol has a somewhat harsh penetrating dry medicinal odour. It is widely used as an antiseptic, germicide and disinfectant and is used as anaesthetic in toothaches. In perfumery it is used in preparation of artificial essential oils, scenting of soaps and perfumes for household products. In flavour industry it is essentially used in spice, meat and mouth wash flavours. Carvecrol is synthetically prepared from *p*-cymene.

Eugenol

Eugenol, $C_{10}H_{12}O_2$, is found as a main constituent in several essential oils like clove leaf, bud and stem oils, oils of cinnamon leaf, *Ocimum gratissimum*.

It can be isolated from the above sources by fractional distillation, which gives eugenol of 85% purity but a chemically pure product, 99-100%, can be prepared from the enriched fractions or directly from the crude oils by treatment with dilute sodium or potassium hydroxide solutions. The water soluble sodium phenate can be decomposed by acids to yield the phenol in pure form. Eugenol is widely used in pharmaceutical preparations, in perfumes for soaps and cosmetics and as a flavouring agent in all kinds of foods, meats, sausages and sauces, etc.

Isoeugenol

In nature isoeugenol, $C_{10}H_{12}O_2$, is found in oil of champaca, nutmeg and ylang-ylang, and commercially it is manufactured by the treatment of caustic soda or potash on eugenol at high temperatures. Isoeugenol can be isolated from its natural sources by the conventional method used to isolate phenols from the oils. Isoeugenol possesses mild, sweet, floral odour resembling carnation. It is extensively used in perfume compositions. Isoeugenol is a very important intermediate in the synthesis of vanillin.

Thymol

Thymol, $C_{10}H_{14}O$, is a commercially important aroma chemical found in nature in the essential oils of *Thymus vulgaris*, ajowain, *Ocimum gratissimum*, and *O. viride*, and in many other oils.

Thymol can be isolated from the natural sources by enrichment by fractional distillation or by direct isolation by conventional method of recovery of phenols from essential oils. Thymol, a colourless transparent hexagonal crystalline product. It can be economically synthesised from *p*-cresol. It has a powerful medicinal odour and finds more application in flavours than in perfumes. It has antibacterial, antiseptic and disinfectant properties for which it enters into dental preparations, oral hygiene products, in vermifuges, and in anti-gastrointestinal products.

FLAVOURS AND SPICES

Flavours

Flavour is that property of a substance (commonly a food or one used in food) that causes a simultaneous reaction or sensation of taste on the tongue and odour in the olfactory centre in the nose. Thus defined, taste and odour are descriptive of a sensation, whereas taste in its colloquial meaning is practically synonymous with flavour.

The preferred definition is "Flavour is the sum total of those characteristics of any material taken in the mouth, perceived principally by the sense of taste and smell and also the general senses of pain and tactile receptors in the mouth as perceived by the brain".

Food acceptance

There are four significant factors of food which when present in proper qualitative proportions, make a food generally acceptable. These are appearance, mouthfeel (a texture or kinesthetic factor evaluated by the skin or muscles in the mouth which includes smoothness, roughness, stickiness, slickness, and related surface characteristics, as well as brittleness and viscosity which are a food's physical characteristics), flavour, and nutritive value. When all four are interdependent, the sense of sight (appearance) takes precedence over the others and nutritive value is obviously not an important consideration.

Taste

Certain basic principles are involved in the physiology of flavour perception. It is generally accepted that there are four tastes (sweet, salty, bitter, and sour) although at various times three more have been proposed (alkaline, meaty and metallic).

There are several factors that affect the extent and character of the sensation. A hereditary or genetic factor may cause a variation between individual reactions. Thus phenylthiourea, which can cause a bitter taste sensation, may not be perceptible to certain individuals whose general ability to distinguish other tastes is not noticeably impaired. The variation of pH in saliva (which acts as a buffer) may influence the perception of acidity differently in individuals. Enzymes in saliva can cause rapid chemical changes in basic good ingredients, such as protein and carbohydrates, with variable effects on the individuals. Although the values cannot be considered absolute, approximate magnitudes or taste sensitivity has been measured (Table 9.1).

Table 9.1. Approximate taste thresholds.

Basic taste	*wt%*	*Test standard*
Sour	0.007	Hydrochloric acid
Salty	0.25	Sodium chloride
Sweet	0.50	Sucrose
Bitter	0.00005	Quinine

Several generalisations can be made regarding taste; there are four, or perhaps five, taste qualities (salty, sour, bitter and sweet, and possibly, metallic); the material must be in water solution (e.g., saliva) to have taste; only acids are sour with sourness not identical to chemical acidity or pH (which is a function of the hydrogen ion concentration), but which appears to be a function of the entire acid molecule (a combination of pH and acid concentration determines the actual degree of sour taste); only salts are salty, however not all salts are salty, some are sweet, bitter, or tasteless, with the salty taste exhibited by ionised salts and the anions contributing the greatest to their salty taste; and organic molecules are usually sweet or bitter, or a combination of these, or tasteless (probably owing to lack of water solubility).

Odour

Olfactory response is only observed when the substance contacts the olfactory membrane, also called the olfactory mucosa or olfactory epithelium, which occupies an area of ca 2.5 cm^2 in each nostril. Above the nasal passages, the two olfactory clefts are separated by the nasal septum. For a substance to have an odour, it must be capable of reaching the olfactory epithelium high up in the nose, i.e., it must possess sufficient volatility and low molecular weight so that, by inspiration, it may make contact in the nasal passages.

Flavour materials

Materials for flavouring may be divided into several groups, most commonly labelled artificial and natural. Under natural materials are spices and herbs; essential oils and their extracts, concentrates, and isolates; fruit and fruit juices; animal and vegetable materials and their extracts; and aromatic chemicals isolated by physical means from natural products, e.g., citral from oil of lemongrass, and linalool from bois de rose. Under artificial materials are found aliphatic, aromatic and terpene compounds that are made synthetically as opposed to those isolated from natural sources. Examples of each grouping are benzaldehyde made synthetically (artificial) or isolated from oil of bitter almond (natural); and 1-menthol made synthetically (artificial) or isolated from oil of *Mentha arvensis* (natural).

Compounding

In the compounding technique, constituents are selected or rejected because of their odour, and the results of flavour tests in water, syrup, milk, etc. After the compound is considered to be characteristic, it is tested as a flavour in the final product by an applications laboratory. Flavouring proportions used in high-fat foods must be greater than in low-fat food because flavours are absorbed by fat thereby lowering flavour impact.

The flavour chemist is responsible for the basic knowledge of sensory and application properties of each of the large number of raw materials. The tremendous number of possible combinations of these items to produce specifically flavoured finished compounds is readily apparent. It is not uncommon to develop a flavour that combines essential oils, plant extractives, fruit juices, and synthetics. As mentioned previously, the choice of materials depends upon the type of product, conditions of manufacture, and intended use.

Spices

A spice may be any of various aromatic plant materials used to flavour foods, or a variety of dried plant products (sometimes used in the form of extracts) that exhibit a pronounced aroma and flavour, but these descriptions are incomplete. Spices are derived from vegetable substances that have flavour and aroma from which no volatile or flavouring principles have been removed.

Spices can be divided into four categories: the so-called true or tropical spices, such as black and white pepper, cloves, cinammon, ginger, nutmeg, and mace; these may be buds, fruit, bark, roots or other parts of tropical plants; herbs, such as sage, rosemary, majoram and oregano; these are usually the leafy parts of plants that grow in the temperate zones; spice seeds, such as mustard, celery, caraway, dill, fennel, and anise which grow in both tropical and temperate zones; and dehydrated aromatic vegetables such as onion, garlic, parsley, leek and sweet pepper, many of which are new produced in India.

Condiments are sometimes classified as spices although they are usually employed in a different manner. Spices are used during the cooking or processing of food, condiments are added to food after it has been served at table (chutney, piccallili, catsup, and salad dressings might be mentioned in this category). Black pepper may be considered a spice or a condiment, depending on its use.

Use

The essential oils carry the aroma of the spices (or aromatic plant) in concentrated form. The essential oils are volatile and do not contain the very high boiling or non-volatile constituents that are responsible for the characteristic taste of certain spices. For example, piperine and zingerone, the biting principles of black pepper and ginger root, respectively, are completely lacking in the essential oils of these two spices. To isolate these compounds the spice has to be extracted with a volatile solvent (ethanol, acetone, hexane, etc). By removing the solvent by careful evaporation of the solution *in vacuo*, a viscous, semi-solid or sometimes solid residue is obtained (oleoresin). These products have lately achieved great importance in the food industry; they offer many advantages over straight spices, e.g., cleanliness and uniformity of flavour from lot to lot. This is especially important when foods are produced by an automated process; the use of oleoresins permits better distribution of flavour and a standardisation of the flavour level that is difficult to obtain with straight spices. Moreover, oleoresins, when properly prepared, are almost free of bacteria, moulds, spores, etc., which is a major advantage over crude spices and aromatic herbs, especially in prepared foods. Because of their high concentration, a few kilograms of oleoresins (or in many cases of essential oils, where the non-volatile portion is of minimal significance to the total flavour) can replace many kilograms of natural spices. Oleoresins are often added to the food product in a diluted form; the most common method is distribution on a dry edible carrier, such as table salt or dextrose. These products, commonly known as dry solubles, are usually equivalent to the corresponding natural spice.

Chapter 10

Surface Coating Industries

INTRODUCTION

Products of the surface-coating industries are essential for the preservation of all types of architectural structures, including factories, from ordinary attacks of weather. Uncoated wood and metal are particularly susceptible to deterioration, especially in cities where soot and sulphur dioxide accelerate such action. Besides from their purely protective action, paints, varnishes, and lacquers increase the attractiveness of manufactured goods, as well as the aesthetic appeal of a community of homes and their interiors.

PIGMENTS

Inorganic

Inorganic pigments are an integral part of decorative, protective, and functional coatings. They provide mass colouration for fibres, plastics, paper, rubber, elastomers, glass, cement, glazes, and porcelain enamels, and they are colourants in inks, cosmetics, and markers.

Pigments can retard corrosion and act as fungistats and antistatic agents. It can be coloured, colourless, black, white, or metallic. Most pigments are insoluble solids. When dispersed in a binder or medium, most pigments remain insoluble.

Properties

Colour results from the pigment's selective absorption of visible light. Large pigment particles also may scatter light and thereby influence the opacity of the binder. Because inorganic pigments involve both transition and non-transition elements, several different theories explain how colour is produced by these compounds.

The intense yellow colour of lead chromate, $PbCrO_4$, is attributed to a charge-transfer process between oxygen and chromium. The presence of two different oxidation states for iron in iron blue, Fe(III)NH_4–[Fe(II)(CN_6))], results in charge-transfer and the associated intense blue colour. The colour of cadmium sulphide is based on its semi-conductor properties and the charge transfer transitions between the valance band and the conductance band. The colour of ultra-marine blue results from electronic transitions within S_3.

Most inorganic pigments are crystalline. Polymorphism is displayed by pigments such as titanium dioxide (rutile, anatase, and brookite), lead chromate (light and primrose type) and calcium carbonate (calcite and aragonite).

The average ultimate particle size of most commercial pigments, excluding extenders, is 0.01–1.0 μm dia. (Table 10.1). The hiding power of a pigment depends primarily upon the ability of the dispersed particles to scatter light. Factors that influence hiding power are refractive index and particle size. The effective particle size for most pigments should be ca one-half the wavelength of visible light.

Microscopy, gas absorption, and sedimentation techniques are used to determine particle size. Particle shape can be noted by transmission electron microscopy.

Another property of pigments is surface area. Carbon blacks have surface areas of 6-1100 m^2/g, whereas most other pigments have areas of 1-100 m^2/g. To achieve complete dispersion, the surfaces of the pigment particles must be properly wetted. Surface-area determination can be made by the BET method, which is based on the absorption of nitrogen at extremely low temperatures and low pressures.

Table 10.1. Particle size of some inorganic pigments, μm.

Pigment	*Particle size, μm*
Iron blue	0.01-0.2
Titanium dioxide	0.2-0.3
Red iron oxide	0..3-4.0
Natural crystalline silica	1.5-9.0
Strontium chromate	0.3-20.0
Hydrated aluminium oxide	0.4-60.0
Micaceous iron oxide	5.0-100.0

Manufacture

Many pigments are manufactured by precipitation, i.e., lead chromate; or by precipitation and calcination, i.e. cadium sulphide; or by calcination, i.e. chromium oxide. Titanium dioxide is prepared by a vapour-phase reaction.

Care is taken in selection of raw materials and control of manufacturing parameters, i.e., temperature, concentration, and pH excesses, and agitation. Other process steps in manufacture include chemical end treatment, filtration, washing, drying, grinding, and packaging.

White Pigments

The principal white hiding pigment is titanium dioxide; others are zinc oxide, zinc sulphide, lithopone, lead pigments, and antimony oxide (Table 10.2).

Titanium dioxide. The stable crystal form of TiO_2 is rutile; anatase can be converted to rutile at ca 700–950°C (Table 10.3). Titanium dioxide, mp > 1800°C, is insoluble in water, organic solvents, alkalies, and most acids. It is attacked by sulphuric and hydrofluoric acids after long contact and at high temperature. Titanium dioxides are not ideal whites because they do not reflect all wavelengths of visible light equally. They absorb below 430 nm in the blue or violet end, of the spectrum.

Table 10.2. Characteristics of some white hiding pigments.

Pigment	*Chemical composition*	*Refractive index*[a]	*Average particle size, μm*
Titanium dioxide, anatase	TiO_2	2.55	0.2
Titanium dioxide, rutile	TiO_2	2.70	0.2-0.3
Zinc oxide	ZnO	2.01	0.2-0.35
Zinc sulphide	ZnS	2.37	0.2-0.3
Lithopone	28% ZnS, 72% $BaSO_4$	1.84	0.2-0.3
Lead carbonate, basic white lead	$2PbCO_3.Pb(OH)_2$	2.0	1.0
Lead sulphate, basic	$PbSO_4.PbO$	1.93-2.02	0.8
Antimony oxide	Sb_2O_3	2.1	1.0

[a] A common refractive index for a paint vehicle is 1.6.

Zinc oxide. Zinc oxide is manufactured from sphalerite. In the indirect process, zinc metal is vapourised, followed by oxidation and collection of the zinc oxide powder. In the direct process, zinc sinter and coal are burned and the resulting zinc vapour is oxidised. Zinc oxide, together with sulphur and organic accelerators, is used in the vulcanisation of elastomers. In addition, zinc oxide absorbs uv radiation and thereby protects organic binders from photodegradation and reduces chalking. Leaded zinc oxide is prepared by cofuming lead and zinc ores or by blending lead-free zinc oxide and basic lead sulphate or lead silicate. It commonly contains 12–55 wt% basic lead sulphate.

Table 10.3. Typical pigment properties of anatase and rutile TiO_2.

	Anatase	*Rutile*
Density, g/cm^3	3.8-4.1	3.9-4.2
Refractive index	2.55	2.76
Oil absorption, g oil/100 g pigment	18.30	16.48
Tinting strength, Reynolds method	1200-1300	1650-1900
Particle size (av), μm	0.3	0.2-0.3

Zinc sulphide. It is soluble in acids but insoluble in water. It is softer than TiO_2 and does not have the yellowish undertone associated with TiO_2. It is produced from $ZnSO_4$ and Na_2S. It has low toxicity.

Lithopone. Lithopone is a mixed pigment which originally consisted of 28–30 wt% ZnS and 70–72% $BaSO_4$; grades with higher zinc sulphide contents (ca 60 wt%) have been developed. It is prepared by coprecipitation of ZnS and $BaSO_4$. Lithopones are used in water-based paints and in rubber and plastics.

Lead pigments. White lead is the oldest white hiding pigment. Today's usage of lead pigments, e.g., basic lead sulphate, basic lead silicate, basic lead silicosulphate, and dibasic lead phosphite is limited because of restrictions on lead compounds.

Antimony oxide. Antimony oxide is an effective fire retardant controls chalking, and improves tint retention in exterior enamels. It is prepared by roasting stibnite, Sb_2S_3.

Zirconium oxide and zircon. These are used in porcelain or vitreous enamels. Vanadium-doped zircon is a blue pigment used in glazes and enameling.

Extender pigments

Extender pigments are inexpensive colourless or white pigments with refractive indexes < 1.7. Kaolin is a hydrous aluminosilicate mineral. It is used as a filler in the paper and paper-board industry and in paper coatings.

Calcined kaolins are white and hard; they are used in water-based and traffic paints.

Clays

Clays are graded according to particle size, i.e., fine, intermediate, and coarse. Mined clay is slurried at the open-pit mine site and the slurry is pumped to a degritting station and then stored in tanks. Slurries are blended, and particle-size distribution is controlled through processing. Attapulgus clay is a crystalline, hydrated magnesium aluminium silicate (Fuller's earth) used in paints as a flattening agent and in sealants.

Calcium carbonates

Natural calcium carbonate, a widely used pigment, is quarried or mined. Limestone is the raw material for precipitated grades. Synthetic calcium carbonates are available in several grades of calcite and aragonite. High calcium types are used in paints, plastics, rubber, paper, adhesives, and joint fillers.

Barium sulphate

Barium sulphate can be prepared synthetically or from the mineral barite. It is used mainly as a weighting material in drilling muds. It is an excellent extender pigment because of its inertness and high refractive index.

Silicates

Magnesium silicate (talc) is derived from natural sources such as soapstone, steatite, and asbestine. Talcs reduce chalking, blistering, and cracking of paint film.

Calcium silicate is derived from wollastonite. The synthetic material is prepared hydrothermally from diatomaceous silicas and lime. It is used in coatings for water-thinned emulsion paints where hiding is achieved by air voids. Natural silicas are obtained from tripolitic ores. They are used as abrasives, reinforcing fillers for silicone rubber, and in caulking compounds and porcelain insulators.

Diatomaceous silica pigments are hydrous silicas obtained by open or strip mining. Diatomaceous silica controls gloss and sheen of flattening paints and is used in water-thinned emulsion flat wall paints, flat varnishes, primers, concrete, and stucco finishes. Precipitated synthetic silicas have large surface areas (100–300 m^2/g) and high oil absorption. Pyrogenic silicas are extremely pure SiO_2 prepared by high temperature reaction. They are used in rubber and plastics. Aerogel and hydrogel silicas have surface areas of 100–400 m^2/g and high porosity; they are used as flattening agents in coatings.

Micas

Micas are complex aluminium potassium silicates; muscovite is the most important. They are used in wallpaper and coated paper, primer and corrosion coatings, and in joint cements and lubricants.

Other extender pigments

Hydrated aluminium oxides, manufactured by seed nucleation of a saturated alumina solution, are fine white pigments. Pumice is a glossy volcanic lava used to impart a rough non-slip surface. Asbestos, a fibrous magnesium silicate, functions as a viscosity-control agent. Calcium sulphate is used in composite pigments and as a filler in plastics and primer paints. Sodium silicoaluminate is an ultrafine synthetic pigment used as a filler in paper and as an extender of TiO_2. Sodium, potassium, aluminium silicate (nepheline syenite) is used in the glass and ceramic industries.

Coloured pigments

Iron oxides are characterised by low chroma and excellent lightfastness; they are non-toxic and inexpensive. They are processed from haematite, limonite, siderite, and magnetite, and provide a range of reds, yellows, purples, browns, and blacks. Raw sienna is a dark yellow that is converted to burnt sienna, a deep brown, by calcination. Ochres are yellow iron oxides, lighter in colour than siennas. Cyprus is the source of fine raw umber, commonly called Turkey umber. Burnt umber is produced by calcinating raw umber. The flake like crystalline structure of gray micaceous iron oxide resembles that of mica. It is used in metal-protective coatings.

Synthetic iron oxides are red, yellow, brown, and black. Advantages over the natural oxides include chemical purity, uniform particle size and size distribution, and the possibility of use in pre-dispersed vehicle systems by flushing techniques. Ferrous sulphate heptahydrate, $FeSO_4.7H_2O$ (copperas) is the main source of iron for these pigments. For example, calcination gives copperas red.

$$FeSO_4.7H_2O \xrightarrow{\Delta} FeSO_4.H_2O + 6H_2O$$

$$6FeSO_4.H_2O \xrightarrow{\Delta} 2Fe_2O_3 + Fe_2(SO_4)_3 + 6H_2O + 3SO_2$$

$$Fe_2(SO_4)_3 \xrightarrow{\Delta} Fe_2O_3 + 3SO_3$$

Direct precipitation of red iron oxides involves growth of iron oxide particles on specially prepared nucleating particles or seeds of Fe_2O_3.

Synthetic red iron oxides are prepared in various grades from light to dark reds and are sold as Indian red, Turkey red, and Venetian red. The Penniman-Zoph process for the manufacture of yellow iron oxides involves the preparation of a seed or nucleating particle by alkali precipitation of ferrous sulphate. Magnetic gamma iron oxide, used in recording tape, is prepared from an acicular yellow iron oxide precursor. Zinc and magnesium ferrites are tan pigments formed by the interaction of iron oxides with metallic oxides. The principal use of transparent iron oxides is in metallic automotive finishes. Chemically they are the same as their opaque counterparts but are much smaller in particle size.

Lead chromate and molybdate oranges

These pigments provide a gamut of hues from greenish yellow through orange to light red. Medium chrome yellows are essentially pure $PbCrO_4$ and provide the reddest yellow hue. Primrose chrome yellow and light chrome yellows are solid solutions to lead chromate and lead sulphate. Chrome orange are $PbCrO_4.PbO$ and the variation from light to dark orange is a consequence of change in particle size. Molybdate oranges are solid solutions of lead chromate, lead molybdate, and lead sulphate. They display hues from orange to red. They are used in industrial finishes; where the risk of health hazards is minimal. Medium-yellow shades are used in traffic-paint formulations. Blends with organic reds and violets produce inexpensive, durable automotive finishes. Heat-resistant types, are suitable as colourants in plastic materials.

Lead chromates are prepared by precipitation from soluble salts in aqueous media. Primrose and light chrome yellows require coprecipitation of lead chromate and lead sulphate. The light chrome yellows are precipitated hot and with excess lead. Chrome oranges are precipitated from alkaline solutions. Chrome yellows and molybdate oranges are soft textured. Their dispersibility results in proper pigment-grind development and high gloss suitable for topcoat finishes. Normal lead silica chromate is manufactured by coating a core of silica with a medium-yellow lead chromate. The pigment is used primarily for traffic paint.

Cadmiums

Cadmium pigments provide brilliant colours ranging from the greenish primrose yellow through orange, red, and maroon. They display clean, bright hues; heat stability up to and in some cases above 400°C; excellent bleed and alkali resistance; fair acid resistance; and good tint light fastness

when protected from moisture. They are used in artists' paints, heat-resistant coatings, printing inks, and latex paints. Mercadmium pigments are based on cadmium sulphide-mercury sulphide solid solutions; variation in chemical composition yields a hue range similar to those of the selenide oranges, reds, and maroons (Table 10.4).

Table 10.4. Typical chemical compositions and related hues for cadmium pigments.

	Compositions, wt %			
	CdS	*ZnS*	*CdSe*	*HgS*
Cadmium yellow, concentrated				
primrose	79.5	20.5		
lemon	90.9	9.1		
golden	93.4	6.6		
deep golden	98.1	1.9		
Cadmium sulphoselenide orange-reds, concentrated				
light orange	85.0		15.0	
light red	67.5		32.5	
medium light red	58.8		41.5	
medium red	51.5		48.5	
dark red	44.8		55.2	
maroon	35.0		65.0	
Mercadium orange-reds				
deep orange	89.0			10.9
light red	83.4			16.6
medium light red	81.0			19.0
medium red	78.5			21.5
dark red	71.6			23.9
maroon	73.5			26.5

Synthetic mixed metal oxides

These pigments are classified by crystal structure, e.g., mixed-phase rutile pigments or mixed-phase spinel pigments. They are produced by solid-state chemical reactions at high temperature. The dry or wet raw materials are mixed and blended and then calcined. These pigments show excellent heat stability as well as weather resistance and lightfastness. They are used in the glass and ceramics industries. Nickel titanates are colourants for PVC siding materials.

Iron blue

Iron blue, as first developed, was hard-textured and difficult to grind. The modern iron blues are ferric ammonium cyanides used for colouring paper, for bluing, and in printing inks. Their poor alkali resistance can be improved by treatment with nickel compounds. Preparation is based on the oxidation of Berlin white, $Fe(NH_4)_2[Fe(CN)_6]$, which is produced from sodium ferrocyanide, ferrous sulphate, and ammonium sulphate.

Ultramarine (Lapis lazuli blue) provides good brilliance, heat resistance, lightfastness, and alkaline stability. It is prepared from intimate mixtures of china clay, sodium carbonate, sulphur, silica, sodium sulphate, and charcoal or pitch. The intensity of the colour depends on reactant concentrations, residence time, and temperature. Manganese violet and cobalt violet are prepared by precipitation. The former is used in cosmetics and is also used to tone white pigments.

Chrome oxide green is a calcined pigment prepared by reduction of sodium bichromate with sulphur or carbonaceous materials. Because it reflects irradiation and is similar in reflective properties to chlorophyll, it is used extensively in camouflage coatings. Transparency permits formulations of polychromatic finishes. These pigments are manufactured by hydrolysing a complex chromium borate obtained by heating sodium bichromate with boric acid.

Chrome greens are blends of variable composition of primrose chrome yellow and iron blue. They are used in paints, printing inks, flooring materials, plastics, and paper goods because of their high chroma and excellent hiding power. Natural mercuric sulphide, HgS, can be derived from the mineral cinnabar. As the pigment vermilion, it occurs in a red crystalline or black amorphous form. Antimony vermilion, Sb_2S_3, varies in hues from orange to red. Van Dyke brown is a natural pigment, consisting of as much as 92 wt% organic matter, water, and traces of iron oxide, alumina, and alkali oxides. It is found in peat beds in the FRG.

Black pigments

Carbon black, one of the oldest pigments, is used in plastics, paints, and printing inks.

Speciality pigments

Speciality pigments include zinc yellow (corrosion-inhibitive primers), basic zinc chromate (a wash primer), strontium chromate (metal protection and corrosion resistance), red lead (protection of iron and steel surfaces), cuprous oxide (an antifoulant agent for marine paints and colourant for ceramic glazes and glasses), calcium plumbate (anticorrosion pigment for steel), basic lead silicochromate (electro-deposition of water-based automotive coatings), white molybdates (corrosion inhibitors), modified barium metaborate (mold and enzyme inhibition, corrosion and chalk resistance), zinc phosphate (corrosion-resistant steel undercoatings), and nacreous pigments (pearl essence for decorative effects).

Metallic pigments

Metallic pigments are prepared from metallic elements and their alloys, e.g., aluminium, copper, bronze, and zinc. Aluminium flake pigment is sold in a paste form, which typically contains 65 wt% flake metal and 35 wt% volatile hydrocarbon. Bronze pigments or powders are manufactured from granular alloys of copper and zinc; they are used for coating paper and cardboard. Gray metallic zinc powder or dust is widely used on iron and steel because of its cathodic effect. Luminescent pigments are based on zinc sulphide and zinc-cadmium sulphide that have been doped with an activating material; they are used for indoor decoration, safety signs, television and electronics, and military coatings.

Organic

Pigments are coloured, colourless, or fluorescent particulate solids which are usually insoluble in and essentially physically and chemically unaffected by the vehicle or medium in which they are incorporated. They alter appearance either by selective absorption or by scattering of light. Pigments may be organic or inorganic chemicals. For application, they are usually dispersed in vehicles, e.g., inks or paints. In some cases, the substrate may serve as a vehicle, e.g., in the mass colouration of polymeric materials.

Colour and structure

In organic compounds, colour is associated with double bonds. The unsaturated, conjugated double bonds in chromophores contribute to the selective absorption of light. Chromophores include nitroso, nitro, carbonyl, thiocarbonyl, azo, azoxy, azomethine, and ethenyl groups, whereas auxochrome groups include amino, alkylamino, dialkylamino, methoxy, and hydroxy groups.

Properties

The characteristics that control the performance of a pigment include its chemical composition; chemical and physical stability; solubility; particle size and shape; degree of dispersion or aggregation; crystal geometry; refractive index; specific gravity, absorption; extinction coefficients; surface area and character; and the presence of impurities, extenders, or modifying agents. The inherent strength of a pigment is controlled by its light-absorbing properties, which are related to molecular and crystalline structure. The intensity is a measure of brightness or cleanness, as opposed to dullness, or shade. Generally, a pigment having a molecular structure containing two or more chromophores is less intense than one containing a single chromophore.

Fastness defines the inherent ability of a pigment to withstand the chemical and physical factors to which it is exposed in its application. Durability defines the ability to withstand weather, light, water, gases, and industrial effluents. Dispersibility is measured by the effort required to develop the full tinctorial potential of a pigment in a vehicle; it affects the maximum surface area and homogeneity of the pigment system. Texture describes the ease of dispersion. The working properties include compatibility, oil absorption, contribution to rheology, ease of grinding, wettability, gloss, bronzing, hiding power, flocculation, etc.

Classification

Organic pigments are classified as azo and non-azo pigments. The former contain chromophore (—N=N—) groups and are subdivided into pigment dyes and precipitated azos. Pigment dyes are insoluble on formation; whereas precipitated azos include products containing salt-forming groups, principally sulphonic or carboxylic acids.

Azo pigments

The Hansa yellows are semi-opaque, intense, monoazo pigments which are used in emulsion paints and paper coatings. The intense colour of diarylide yellows provides greater tinctorial strength and resistance to bleeding and heat than is provided by the Hansa yellows. Nickel azo yellow is a greenish yellow, chelated nickel azo pigment with high transparency and excellent durability.

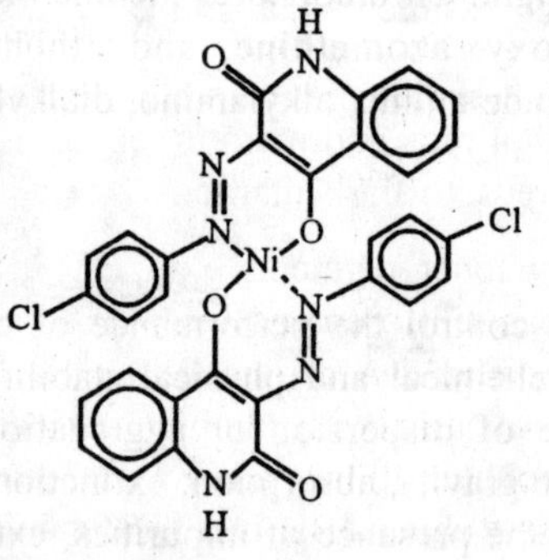

Nickel Azo Yellow

Hansa orange, Dinitraniline orange, and pyrazolone orange all exhibit intense, high strength colour with varying degrees of lightfastness and chemical resistance.

Azo reds and maroons

Toluidine red is one of the most popular red pigments for industrial enamels. It exhibits good bake and chemical resistance but poor lightfastness. Para red is similar but darker and bluer. Chlorinated para reds are intense, very light yellowish red. Parachlor Red is one of the more lightfast azo-pigment dyes. Lithol reds are intense reds of dark shades. They are formed by coupling diazotised Tobias acid to 2-naphthol and precipitated as the Ba, Ca, and Sr salts. Lithol reds are inexpensive and are used widely where extreme durability is not required.

The BON reds and maroons derive their name from β-oxynaphthoic acid as the second component in the coupling of various diazotised amines containing salt-forming groups. Lithol Rubine exhibits a wide range of intense shades suitable for industrial enamels. Permanent red 2B defines the calcium, barium, and manganese precipitations of the coupling from diazotised 2-chloro-4-aminotoluene-5-sulphonic acid with 3-hydroxy-2-naphthoic acid. They are used in printing inks, plastics, and automotive finishes. Yellow BON maroon is a lightfast, yellow maroon with superior solvent bleed resistance suitable for outdoor finishes. The manganese toner of Lithol Red 2G is intense in colour and is used mainly in blends with inorganic molybdate orange for durable reds; the calcium toner is used in printing inks.

The pyrazolone reds are disazo pigments which provide high colour intensity, excellent masstone lightfastness, high transparency and good bleed, bake, and chemical resistance. The search for pigments with improved properties has resulted in a group of higher molecular weight disazo products.

The monoazo naphthol reds and maroon pigments provide a wide range of colours from light reds to dark maroons.

Lakes are either dry toner pigments extended or reduced with a solid diluent or an organic pigment prepared by precipitation of a water-soluble dye on an adsorptive surface. Alizarine Red B or madder lake is a co-ordination complex of alizarin with alumina and calcium plus sulphated castor oil and a phosphate. It has an intense, deep bluish-red of fair lightfastness. Helio Fast Rubine 4BL is a precipitated quinizarinsulphonic acid derivative used to shade pigments for organic coatings.

A basic-dye pigment is the precipitated product of the reaction between a basic dye and either complex inorganic heteropoly acids or precipitants, e.g., tannic acid, tartar emetic, clay, or rosin soaps. Despite their high costs, these pigments are widely used for their high strength and intensity.

Phthalocyanines

Phthalocyanines are characterised by excellent lightfastness, intensity, bleed and chemical resistance, and heat stability. Copper Phthalocyanine Blue exists in a red-shade blue alpha form and the more stable green-shade beta form. Chlorination gives Copper Polychlorophthalocyanine Green which has excellent pigmentary properties. Copper polybromochlorophthalo cyanines provide yellower shades of green than the polychloro compound.

Quinacridone pigments offer outstanding fastness properties in the orange, maroon, scarlet, red, magenta, and violet ranges. An example of a violet with excellent fastness properties is Pigment Violet 23, a dioxazine pigment derived from 3-amino-*N*-ethylcarbazole.

Vat pigments

A number of vat dyes based on anthraquinone and substituted and/or condensed anthraquinones have found use as pigments because of fastness and durability characteristics. In general, vat dyes must be modified chemically and physically to develop pigmentary strength and colour intensity.

The thioindigos provide a broad range of hues from a yellow shade of red to violet. The perinone pigments are diimides of naphthalene-1,4,5,8-tetracarboxylic acid formed by condensation of o-diamines with the acid or acid anhydride. The perylene pigments are diimides of perylene-3,4,9,10-tetracarboxylic acid and generally are stronger and more resistant to chemicals, heat, and solvent bleeding than the thioindigos.

Among the anthraquinone vat dyes, indanthrone is redder but less intense than the phthalocyanine blue pigments, but as durable as Copper Phthalocyanine Blue. It is used in automotive and other high quality finishes. Partially chlorinated indanthrone blues are greener than the unchlorinated but not as bake resistant. Other anthraquinone pigments include Isodibenzanthrone Violet, Dibromoanthranthrone Orange, Flavanthrone Yellow, and Anthrapyrimidine Yellow.

Pigments can be standardised only in terms of performance, colour, durability, and working properties, and often only for the specific application or vehicle system for which they are intended. In general, organic pigments have no toxic effects.

Organic pigments are used for decorative and functional effects; they provide colour, hiding power, and high visibility and contribute to durability. They contribute to countless applications involving colouration.

Dispersed

A dispersed pigment concentrate is an extremely fine distribution of colour pigment in any medium; it is suitable for supplying colour to surface printing and coating or a complete mass colouring.

Dispersion

Reduction of the pigment to primary pigment particle size is necessary in order to develop the optimum visual performance and economic properties. The extent of the reduction is determined by the nature of the pigment, the dispersion system and processing equipment, product-use requirements, and economics.

The maximum aggregate size that adversely affects physical properties is a small fraction of the thickness of the film, coating, or thin mass-coloured material. The optimum size to which a pigment should be reduced is the primary pigment particle. In dispersed pigments, the term primary pigment particle refers to individual crystals and tightly held aggregates. Agglomerates are larger associations of the primary pigment particles, i.e., crystals and aggregates. Ideally, a dispersion consists mainly of primary pigment particles and a minimum of loosely held aggregates. The system of wetted primary particles must be stabilised in order to avoid reversal of the dispersion process.

Flushing is the direct transfer of pigments in an aqueous phase to an oil or non-aqueous phase without intermediate drying. Because of the difficulties involved in flushing inorganic pigments, it is often more practical to disperse them in dry form. Flushing is therefore confined mostly to organic pigments, e.g., Diarylide Yellow. It is usually carried out in a heavy-duty dough mixer and produces dispersions of better gloss, transparency, and strength than methods using dry pigments.

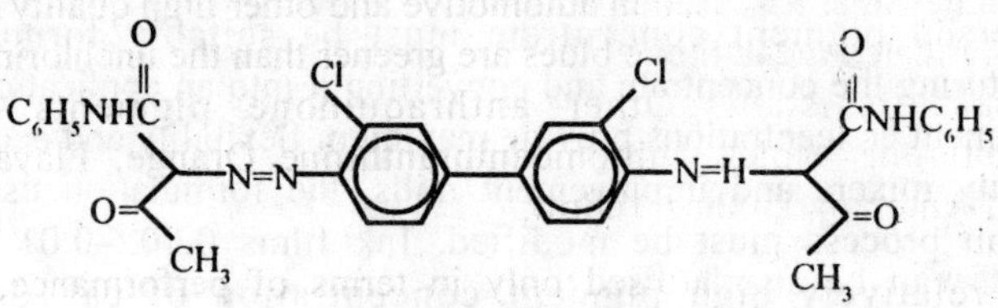

Diarylide Yellow

Equipment

Kneader dispersers process systems with extremely high plastic viscosity, i.e., up to 10^7 Pa.s (10^8 P). The Banbury mixer is the most efficient mixer for preparing large amounts of concentrated pigment dispersions in rubber and thermoplastic carriers. The flusher or heavy-duty mixer is used primarily in the ink and paint industries. The two-roll mill can be considered a kneader or internal mixer; clearance between the rolls is large (up to 15 mm) and the mass to be processed is subjected to kneading just before it enters the nip itself. These mills are used to prepare dispersions in elastomers and thermoplastic resins.

Close-tolerance mills can be classified as cylindrical roll mills, rotating-disk mills, and cone mills. Cylindrical rolls mills of several types are used for processing pigment dispersions in paste form for inks, paints, and other uses. Rotating-disk mills, cone mills, stone mills, and colloid mills essentially are rotors in the shape of disks or truncated cones which rotate at high speed (3000–5000 rpm) and at a small distance (10–100 μm) from a stationary surface.

Impeller and impingement high speed, fluid-energy mills are used for the preparation of fairly low viscosity mill bases for inks and paints.

Dispersion-process equipment employing a large number of dispersing surfaces, whose movement is not precisely controlled, are categorised into ball or pebble mills (media size > 3.0 mm) and bead, shot, and sand mills (media size 0.2–3.0 mm). The most common design of a ball or pebble mill is a horizontal cylinder, partially filled with balls or pebbles, which rotates on the horizontal axis at a rate that cascades the balls or pebbles rather than centrifuges them or allows them to slide. These mills operate more efficiently with small particle size media. They have been largely replaced by bead, shot, and sand mills, which are vertical or horizontal cylindrical containers filled with grinding medium under intense agitation.

Uses

A dispersed pigment concentrate must be suitably formulated for manufacturing the concentrate and converting it into an application product. High pigment concentrations provide maximum flexibility and economy. For heavy-duty mixers and impingement mills, the formulation used for the dispersion process must be modified. Ink films 0.002–0.01 mm thick require relatively high pigment concentration of ca 10–20 wt%. Manufacturers offer a wide variety of pigment concentrates in liquid, paste, and solid form.

In various countries, most letterpress and lithographic oil inks are manufactured from dispersed pigment concentrates, mainly in flushed form. Most gravure and flexographic liquid inks are made from dry pigment or presscake (Table 10.5). Gravure inks are characterised by very fluid viscosities (ca 0.1 Pa.s 1P) when press ready; these inks dry by solvent evaporation, leaving a film ca 10 μm thick. Flexographic inks, like gravure inks, are pigment-resin-solvent systems of low viscosity (ca 0.3 Pa.s = 3P) and film thickness of ca 8 μm; they dry by solvent evaporation.

Environmental and safety considerations will affect future development in ink manufacture. Concern for organic-solvent vapour emission favours aqueous or water-based inks. Today, flexographic ink is the only one based on water as a solvent.

Coatings are generally heavier films, a fraction of to a few millimetres thick; they require a lower pigment content than inks. Inorganic pigments are hydrophilic and are easily dispersed in water-based systems. Organic pigments, on the other hand, are hydrophobic and require wetting agents for dispersion.

Table 10.5. Distribution of usage of paste ink colourants by pigment form, %.

Type of Ink	*Flush*	*Dry*	*Other*
Letterpress news	95	5	
Heat-set letterpress	95	5	
Glycol letterpress	0	70	30
Other letterpress	90	10	
Lithographic news	95	5	
Heat-set web offset	95	5	
Quick-set sheet offset	90	5	ca 5
UV offset	0	100	
Conventional metal decorating	0	100	
Dry offset metal and plastic	50	50	

Dispersed-pigment or colour concentrates for use in plastics vary widely in pigment concentration and physical form. Some are manufactured as liquids and high viscosity gels which are usually metered directly into the processing equipment. Pellet concentrates are the most popular form. The effect on the plastic is a consideration in pigment selection. Acetals, ABS, fluoropolymers, and nylon present special problems.

Mass colouration, i.e., spin-colouration or dope-dyeing of synthetic fibres, particularly polyolefins, can be accomplished with pigment dispersions. In this process the colourant is incorporated before the fibre is formed.

Pulp colourants are water pastes derived from pigments, extended pigments, or lakes. The principal applications are in paper and textile printing, leather and paint.

PAINTS

Paint is a substance composed of solid colouring matter suspended in a liquid medium and applied as a coating to various types of surfaces.

The purpose of the coating may be decorative, protective, or functional. Decorative effects may be produced by colour, gloss, or texture. The protective coating may be the paint on a wooden boat which serves as a barrier against moisture and prevents rotting; the interior lining of metal cans or drums which prevents corrosion from foods or chemicals; the coating on electrical parts to exclude moisture; the fire-retardant paint which protects combustible surfaces; the coating on plaster or concrete which makes for ease of cleaning, etc. An example of the functional use of paint would be as a traffic paint which marks the centre or edge of a road for safer driving.

The paint industry serves two distinct types of markets, trade sales (shelf goods) and industrial sales (chemical coatings).

Trade sales is the large consumer-oriented portion of the business. Trade-sale paints are house paints and other products marketed through wholesale and retail channels to the general public and professional painters for use on new construction or for the maintenance of old buildings. Also included in this category are the paints sold to garages and repair shops for automobile refinishing, marine finishes for boats, paints for graphic arts such as signs, paints for refinishing machinery and equipment, and paints sold to government agencies, especially the traffic paints used for road-marking.

Industrial sales comprise the coatings sold directly to the manufacturer for factory application. The products on which paints are applied include durable goods such as automobiles, appliances, and house sidings, and non-durable goods such as cans for food and beverages.

All coatings contain a resinous or resin-forming constituent called the *binder*. This can be a liquid such as a drying oil, or a resin syrup that can be converted to a solid gel by chemical reaction. In some instances, where the binder is either a solid or is too viscous to be applied as a fluid film, a volatile solvent or *thinner* is also added. This evaporates after a film is deposited, causing solidification of the film. The binder plus solvent is known as the *vehicle*. Most paints also contain *pigments* which are described in the first part of this chapter.

In addition to pigment, binders, and thinners, a paint may contain many additives, such as defoamers, thickeners, flow agents, and driers, to improve specific properties.

Paints can be classified according to the binder used, e.g., alkyds, vinyls, and epoxies. Paints are also classified according to their properties or end use. Alkyd enamels, for instance, are gloss paints with good abrasion resistance and good cleanability, while alkyd flat wall paints are characterised by very low sheen and good film build.

Paints applied directly to the surface are called undercoaters or primers. Primers are used to aid the adhesion of the top coat to a surface and to prevent absorption of the top coat into a porous surface. Primers can also be used to prevent corrosion of a metal surface. Fillers or surfacers are types of primers that fill scratches and surface imperfections to give a smooth surface. Paints used as the final coat are referred to as finish coats or top coats. Some top coats are self-priming.

Binders

The protective properties of a coating are determined primarily by the binder. In the early days of paint technology, binders were limited to materials of natural origin, such as drying oils, congo resins, and asphalts. These still find some usage in the protective-coating industry.

Coatings in liquid form can be divided into two types, *solutions* and *dispersions*. In solution systems the resin binder is dissolved in a solvent. In dispersion systems the resin is in the form of tiny spheres (usually 10 microns or less in size) suspended in a volatile liquid carrier. If the liquid-in a dispersion system is water, the system is called an emulsion; if it is an organic material, it is called an organosol. When the liquid evaporates, a mixture of soft resin and pigment is left behind and fuses into a continuous film. The general types of resins used are listed below. Mixtures of two or more types may sometimes be used to improve certain properties.

(1) *Oils:* easy application; soluble in aliphatic solvents.

(2) *Alkyds:* all purpose; combined with other resins; most are soluble in aliphatic solvents.

(3) *Cellulosics:* (nitrate and acetate)–used in lacquers; fast dry.

(4) *Acrylics*: good colour and durability.

(5) *Vinyls:* good durability; abrasion resistant.

(6) *Phenolics:* good chemical resistance; yellow colour.

(7) *Epoxies:* good chemical resistance.

(8) *Polyurethanes:* good flexibility; abrasion resistance.

(9) *Silicones:* good heat resistance.

(10) *Amino resins:* (ureas and melamines)–blended with alkyds for baking finishes; tough, good colour.

(11) *Styrene-butadiene*: low cost; alkaii resistant.

(12) *Polyvinyl acetates*: low cost; good colour retention.

(13) *Acrylics*: good colour and durability.

Solvents

The type of binder determines the type of solvent or thinner used in a paint formulation. The preferred type of thinner is an odourless aliphatic hydrocarbon which can be used in all areas including the home. Unfortunately, aliphatic thinners do not dissolve all resins. In such cases, strong solvents such as aromatic hydrocarbons, esters, and ketones are used.

Solvents are generally classified as low boilers, medium boilers, and high boilers depending upon the evaporation rate. Various ranges are required depending on the specific application. Examples of solvents classified by their chemical composition are as follows:

(1) *Hydrocarbons*: naphtha and mineral spirits; and aromatics–benzene, toluene, and xylene.

(2) *Alcohols:* methyl alcohol, ethyl alcohol, and butyl alcohol.

(3) *Ethers*: dimethyl ether and ethylene glycol monoethyl ether.

(4) *Ketones:* acetone, methyl ethyl ketone, and methyl isobutyl ketone.

(5) *Esters:* ethyl acetate, butyl acetate, and butyl lactate.

(6) *Chlorinated:* tetrachloroethane.

(7) *Nitrated*: nitromethane, nitroethane, and l-nitropropane.

Latex or emulsion paints use water as the volatile component so that they may be used in the home.

General Properties of the Various Types of Paints

Acrylics

Acrylic resins are used in protective and decorative lacquers for paper, fabrics, leather, plastics, wood, and metal. White baking enamels having excellent resistance to chemicals and chemical fumes can be made from acrylic resins. Acrylic vehicles are used in luminescent paints. The present automobile finishes are based on acrylic resins.

Thermosetting types of acrylic resins have been developed which cross-link upon heating to form hard, insoluble, infusible coatings for appliance finishes that will withstand severe service on washing machines, stoves, and dishwashers.

Alkyds

Alkyds are oil-modified phthalic resins that dry by reacting with oxygen from the surrounding air. Alkyd finishes are usually of the general-purpose type and are available as clear or pigmented coatings. Paints are available in flat, semi-gloss, and high-gloss finishes in a wide range of colours. They are easy to apply and may be used on most surfaces with the exception of fresh concrete, masonry, and plaster which are alkaline. Alkyd finishes have good colour and gloss, and retain these characteristics in normal interior and exterior environments except under corrosive conditions.

Alkyd finishes are available in odourless formulations for use in hospitals, kitchens, sleeping quarters, and other areas where odour during painting might be objectional.

In trade sales applications, alkyds are used for interior walls and woodwork in both flat and gloss, and on the exterior as trim paints. In industrial finishes, alkyds are combined with amino resins to produce hard baking finishes for use on appliances, metal furniture, etc.

Cellulosics

The most widely used cellulosic derivatives are the esters, particularly cellulose nitrate and cellulose acetate. Cellulose nitrate has been combined with many different resins to produce useful coatings. Perhaps the most outstanding combinations have been those with alkyd and amino resins to produce tough, hard, durable lacquers capable of withstanding the severe service requirements of automotive, aircraft, and other industrial finishes.

Successful coatings have also been based on blends on nitrocellulose with natural resins and drying oils, resin derivatives, phenolic resins, acrylic resins, certain vinyl resins, and other materials. Applications include coatings for metal, wood, paper, fabrics, leather, and cellophane.

Epoxies

Epoxy binders are of two types: (i) the oil-modified compositions, which dry by oxidation; and (ii) the two-component materials, which comprise epoxies and amine or polyamide hardeners and are mixed just prior to use. In the latter type, when the two ingredients are mixed they react to form the cured coating. These coatings have a limited pot-life, usually a working day. Anything left at the end of the day must be discarded.

Epoxy paints can be used on any surface and can be applied with high-solids content, thus producing high film build per coat. The cured film has outstanding hardness, adhesion, flexibility, and resistance to abrasion, alkali, and solvents, as well as being highly corrosion resistant. Their major uses are as tile-like glaze coatings for concrete and masonry and for the protection of structural steel in corrosive environments. Their cost per gallon is high, but this is offset by the higher solids content and the reduced number of coats required to provide adequate film thickness. When used on exterior surfaces, epoxy paints tend to chalk to low-gloss levels and fade. Apart from this their durability is excellent.

Epoxy-coal tar coatings are made by adding coal tar as an ingredient to epoxy paints, thereby reducing the cost. They have outstanding corrosion resistance and are used for interior and submerged surfaces. Colour choice is limited to black.

Oils

Linseed oil is the major binder used in oil house paints. These paints are the oldest type of coatings in use. They are used primarily on exterior wood and metal since they dry too slowly for interior use. They are sensitive to alkaline masonry. Oil paints are easy to use and give high film build per coat. They also wet the surface very well so that surface preparation is less critical than with other types of paints for metal. Oil paints are not particularly hard or resistant to abrasion, chemicals, or strong solvents, but they are durable in normal environments.

Oleoresinous

These binders are made by processing drying oils with hard resins, such as natural resins, rosin esters, and hydrocarbon resins. They are generally used as spar varnishes or as mixing vehicles for aluminium paint.

Phenolics

Phenolic binders are made by processing a drying oil with a phenolic resin and, thus, are a type of oleoresinous binder. They may be used as flat (lustreless) or high gloss finishes which are clear or pigmented in a range of colours. The clear finishes may be used on exterior wood and as mixing vehicles for producing aluminium paints. The durability of the clear finish is very good for this class of material, i.e., one to two years; the durability of the aluminium paints is excellent.

Phenolic paints are used as top-coats on metal in extremely humid environments and as primers for fresh-water immersion. These paints require the same degree of surface preparation as alkyds, but they are slightly higher in cost than alkyds. Phenolic coatings have excellent resistance to abrasion, water, and mild chemical environments. They are not available in white or light tints because of the relatively dark colour of the binder.

Phenolic and alkyd binders are often blended to combine the hardness and resistance of phenolics with the colour and colour-retention of the alkyds. This may be done either by blending phenolic varnish with the alkyd vehicle or by the addition of phenolic resins during the processing of the alkyd resin.

Silicones

Silicones resins are used for heat-resistant finishes. They have good water-resistance and outstanding gloss-retention. When pigmented with aluminium, heat-resistant organic finishes containing a high concentration of silicone resins have the ability to withstand temperatures up to 1200°F. A combination of silicone and alkyd resins provides some heat resistance at a lower cost.

Urethanes

Urethane binders are of three types: (i) oil-modified, which are cured by oxygen from the air; (ii) moisture-curing, which are cured by moisture in the air; and (iii) two-part systems which are mixed just prior to use.

Oil-modified urethanes are similar to phenolic varnishes. Although somewhat more expensive, they have better initial colour and colour-retention, dry more rapidly, are harder, and have better abrasion-resistance. They can be used as exterior spar varnishes or as tough floor finishes. Oil-modified urethanes can be used on all surfaces. In common with all clear finishes, they have limited durability when used on exterior surfaces.

Moisture-curing urethanes are used in a manner similar to other one-package coatings except that the containers must be full to exclude moisture during storage. They have outstanding abrasion resistance and chemical resistance.

Vinyls

Lacquers based on modified polyvinyl chloride resins are used on steel where the ultimate in durability under abnormal environments is desired. They are moderate in cost, but they have low solids and require the most extensive degree of surface preparation to secure a firm bond. Because of their low solids, vinyl finishes require numerous coats to achieve adequate dry film thickness; thus the total cost of painting is higher than with most other paints. Since vinyl coatings are lacquers, they are best applied by spray, and they dry quickly, even at low temperatures. Recoating must be done with care to avoid lifting by the strong solvents which are present. In addition, these solvents present on odour problem. Vinyls can be used on metal or masonry, but they are not recommended for use on wood. They have exceptional resistance to water, chemicals, and corrosive environments, but they are not resistant to solvents.

Vinyl-alkyd combinations offer a compromise between the excellent durability and resistance of vinyls with the lower cost, higher film build, ease of handling, and adhesion of alkyds. Thcy can be applied by brush or spray, and they are widely used on structural steel in marine and moderately severe corrosive environments.

Rubber-base

So-called rubber-base binders are solvent-thinned and should not be confused with latex binders which are often called rubber-base emulsions. Four types are available: chlorinated rubber, styrene-butadiene, vinyl-toluene-butadiene, and styrene-acrylic. They are lacquer-type products which dry rapidly to form finishes which are highly resistant to water and mild chemicals. Recoating must be done carefully to avoid lifting by the strong solvents used. Rubber-base paints are available in a wide range of colours and levels of gloss. They are used for exterior masonry, also for areas which are wet, humid, or subject to frequent washing, e.g., swimming pools, wash rooms, shower rooms, kitchens, and laundry rooms.

Latices

Latex paints are based on aqueous emulsions of three basic types of polymers: polyvinyl acetate, polyacrylics, and polystyrene-butadiene. They dry by evaporation of the water, followed by coalescense of the polymer particles to form tough insoluble films. They have little odour, are easy to apply, and dry very rapidly. Interior latex paints are generally used either as a primer or finish coat on interior walls and ceilings made of plaster or wall board. Exterior latex paints are used directly on exterior masonry or on primed wood. They are non-inflammable, economical, and have excellent colour and colour-retention. Latex paint films are somewhat porous so that blistering due to moisture vapour is less of a problem than with solvent-thinned paints. They do not adhere readily to chalked or dirty surfaces nor to glossy surfaces under eaves. Therefore, careful surface preparation is required for their use.

Latex paints are very durable in normal environments, at least as durable as oil paints. The popularity of latex paints is due mainly to easy cleaning of brushes, etc., with water.

Inorganics

The major inorganic binders used in paints are sodium, potassium, lithium, and ethyl silicates. These binders are used in zinc-dust-pigmented primers in which they react with the fine zinc metal to form very hard films. These films are extremely resistant to corrosion in humid or marine environments. Many of these primers also contain substantial concentrations of lead oxides which react with the silicates in conjunction with the zinc to form an even more corrosion-resistant coating.

Another type of inorganic binder is Portland cement. The paint is supplied as a powder to which water is added before use. Cement paints are used on rough surfaces such as concrete, masonry, and stucco. They dry to form hard, flat, porous films which permit water vapour to pass through readily. Cement paints should not be used in arid regions. When properly cured, cement paints of good quality are quite durable; when improperly aired, they chalk excessively on exposure, and then they may present problems on repainting. A typical paint formula for an outside house paint is shown in Table 10.6.

Table 10.6. Analysis of a white outside house paint.

	Pounds	*Gallons*	*Per cent (by wt)*
Titanium dioxide (rutile)	105	3.05	7.4
White lead—basic carbonate	160	1.88	11.3
Zinc oxide (35 per cent leaded)	370	7.55	26.2
Talc	240	14.70	16.9
Linseed oil—bodied 3	85	10.54	6.0
Linseed oil alkali refined	350	45.40	24.8

(Contd...)

	Pounds	Gallons	Per cent (by wt)
Drier–manganese naphthenate 6 per cent	2	00.26	—
Drier–lead naphthenate 24 per cent	7	00.72	—
Mineral spirits	105	15.90	7.4
Total	1419	100.00	100.0
Constants			
Viscosity	90 KU		
PVC	30 per cent		
Vehicle solids	80 per cent		
Total solids	92.5 per cent		
Wt/gallon	13.1 lbs		

Manufacture of Paints

The manufacture of paint involves the following operations: mixing, grinding, thinning, adjusting, and filling (Fig. 10.1).

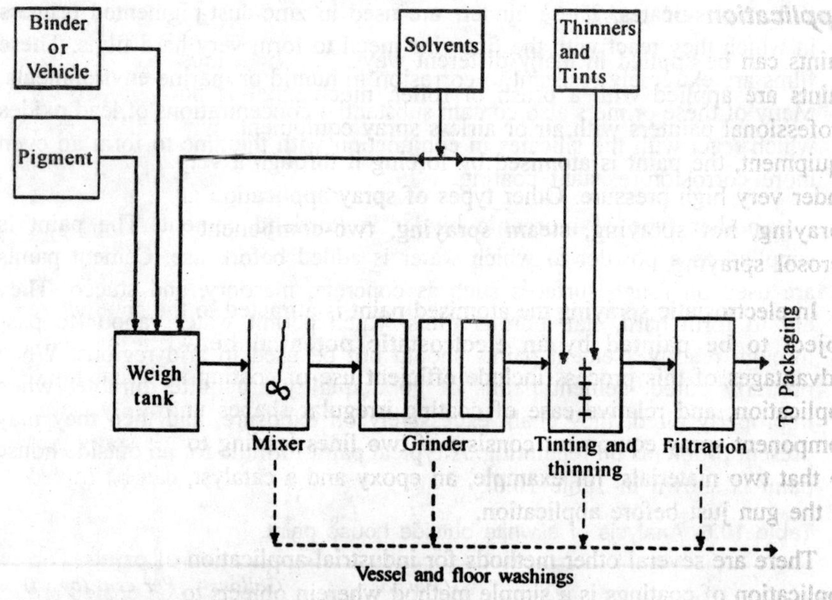

Fig. 10.1. Typical flow chart for paint production.

One of the older methods consisted of mixing all the pigment and part of the vehicle to make a paste of suitable consistency in a tub with rotating blades. This paste is then fed by means of a trough into a slow-speed stone mill which has two circular stones, one stationary and the other rotating above it. The pressure developed between the two stones is sufficient to disperse the pigments and liquids initimately, eliminating unwetted agglomerates and air pockets. This paste is fed in a continuous stream into a third piece of equipment, the thinning tank, which has rotating blades. At this point the remainder of the liquids is added, and the completed product was tested for viscosity, colour, and other physical properties pertinent to the formulation being prepared. The batch, having been approved, is then strained and filled into appropriate size containers. The term "grinding" is a misnomer, as little or no breakdown of pigment particles takes place.

Dispersion by old-fashioned low-speed stone mills is extremely slow and costly and is obsolete. Various types of more efficient equipment are now in use. The general principle, however, is the same that is, to wet each individual particle thoroughly with the vehicle and to eliminate flocculated aggregates. Mills in use today include steel roller, ball, pebble, sand, high-speed impeller, Morehouse (a high speed stone type), and Cowles disperser.

Application of paint

Paints can be applied in many different ways. Although most architectural paints are applied with a brush or roller, much paint is now applied by professional painters with air or airless spray equipment. With airless spray equipment, the paint is atomised by forcing it through a very small orifice under very high pressure. Other types of spray application are electrostatic spraying, hot spraying, steam spraying, two-component spraying, and aerosol spraying.

In electrostatic spraying the atomised paint is attracted to the conductive object to be painted by an electrostatic potential between the two. Advantages of this process include efficient use of coating material, rapid application, and relative ease of coating irregular shapes uniformly. Two-component spray equipment consists of two lines leading to the spray gun so that two materials, for example, an epoxy and a catalyst, can be mixed in the gun just before application.

There are several other methods for industrial application of paints. Dip application of coatings is a simple method wherein objects to be coated are suspended and dipped into a large tank containing the paint. This method is often used for undercoating objects where the uniformity and appearance of the paint are not important. Electrodeposition consists of depositing paint on a conductive surface from a water bath containing the paint. The negatively charged paint-component particles are attracted to the object being coated which becomes the anode when an electrical potential is applied. Paint can be applied to very irregular surfaces with very uniform thicknesses and little loss. The system is limited to one coat of limited film thicknesses and the equipment cost is high.

Yet, another method utilises roller-coating machines to apply paint to one or both sides of flat objects, such as metal or fibreboard. The thickness of the coating can be controlled by the clearance between a doctor blade and the applicator rolls. Decorative effects such as wood-grain patterns can be applied with these machines.

Other methods include flow coating, where the paint is allowed to flow over the object being coated, and powder coating, in which paint in dry powder form is fused on the surface of the object being coated.

VARNISHES

The term "varnish" is applied to clear, transparent coating materials that dry by a process comprising evaporation of the solvent, followed by oxidation and polymerisation of the drying oils and resins. Varnishes are a homogeneous mixture of resins, drying oils, driers, and solvents, and they contain no pigment.

Varnishes fall into three general classes, *spar varnishes* (exterior), *floor varnishes*, and *furniture finishes*. Typically flow chart for varnish manufacture is shown in Fig. 10.2.

The types of oils and resins and the ratio of oil to resin are the principal factors which determine the properties of a varnish. The selection depends on the compatibility of different oils and resins and on the intended use of the varnish. It is generally accepted that the oils in the finished coatings contribute to its elasticity, and the resin, to its hardness. In oleoresinous varnishes the ratio of oil to resin is expressed as the number of gallons of oil that are combined with 100 pounds of resin, and it is commonly referred to as the "length" of the varnish. Thus, where 50 gallons of oil are used with 100 pounds of resin, the varnish has a 50-gallon length. Varnishes containing less than 20 gallons of oil per 100 pounds of resin are usually classed as *short-oil* varnishes. A *medium-oil* varnish contains from 20 to 30 gallons of oil, and a *long-oil* varnish is one in which 30 gallons or more are used. Short-oil varnishes dry more rapidly than long-oil varnishes. They are used primarily where hardness, a high degree of impermeability, or resistance to alcohols, alkalies, or acids is desirable and where elasticity in the film is relatively unimportant. Short-oil varnishes are especially suitable where a "rubbed" finish is desired, e.g., on furniture, but they are too brittle for floors for which a medium-oil varnish is best suited.

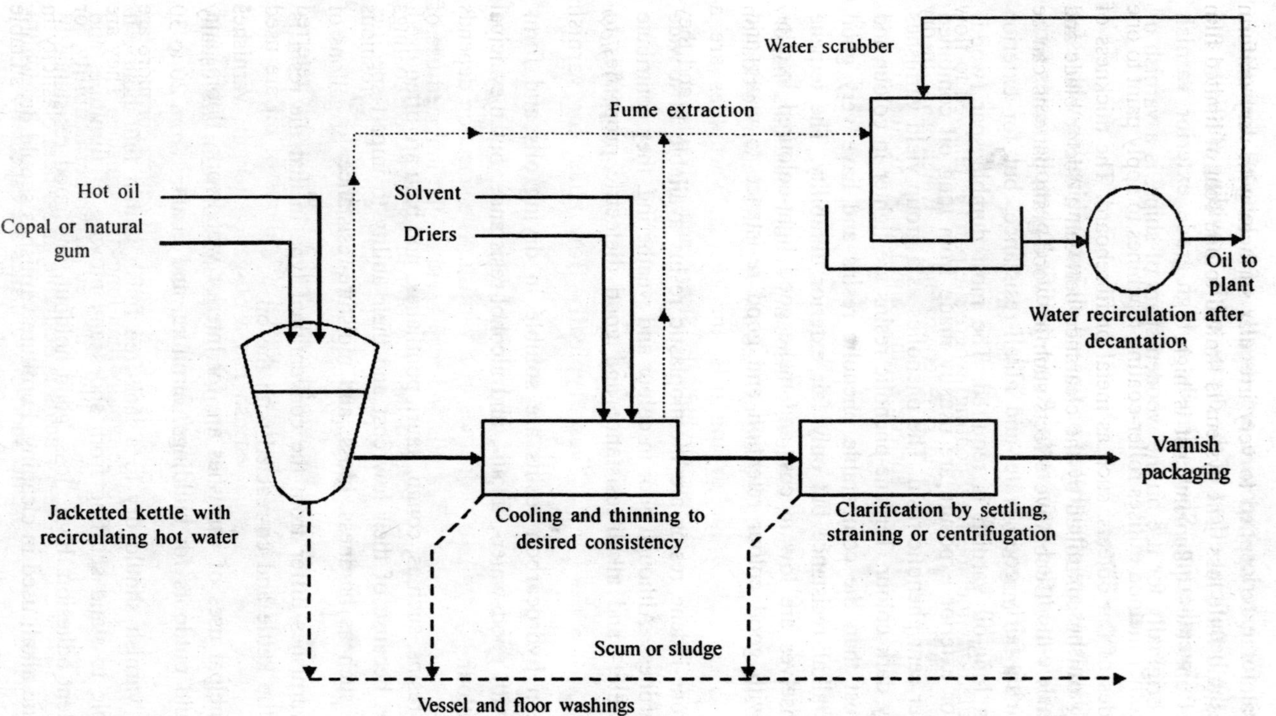

Fig. 10.2. Typical flow chart for varnish manufacture.

Varnishes for exterior exposure are usually of a long-oil formulation because of the beneficial effect of the drying oil in providing elasticity and resistance to weathering. Spar varnish, a high-grade exterior varnish formulated originally for use on the wooden spars of ships, is a varnish of this type.

Interior varnishes are often based on linseed oil in combination with ester gum or a maleic-modified-rosin resin. Rosin-modified phenolic resins form varnishes with fairly good water-and alkali-resistance, but for exterior durability a long-oil varnish is required. The most durable spar (wood) varnishes for exterior exposure are usually made from tung oil combined with 100 per cent phenolic resin. The phenolics as a group yield varnishes of relatively dark colour. Terpene-phenolic resins are lighter in colour and lower in cost than the comparable phenolic resins and have very good water- and alkali-resistance but only fair exterior durability. The terpene resins themselves are low in cost and make good light-coloured interior varnishes with good colour retention and good resistance to water and alkali.

Coumarone-indene resins are thermoplastic resins available in a wide range of hardness. Although poor in colour and weathering, they contribute excellent water- and alkali-resistance and good dielectric properties to varnishes.

Petroleum hydrocarbon resins are soluble in drying oils and form varnishes with good water-, alkali-, and alcohol-resistance, but they have only fair colour.

Natural resins, such as, congo, kauri, pontlianak, and batu are often used in varnishes because of their low cost and their ability to impart specific properties such as hardness, gloss, and moisture-resistance.

Alkyd varnishes differ from the conventional type in that the resin is formed in the kettle and co-reacted with the oil.

The principal uses of varnishes are for interior woodwork, floors, and furniture, and outdoors for buildings, furniture, and boats.

A floor varnish should dry to a tack-free finish within four hours. It must be able to stand scuffing from shoes and moving of furniture, and have excellent adhesion, high gloss, good holdout, and good resistance to the water and alkali used in cleaning. Furniture varnishes should dry within about four hours and must not soften from body heat, otherwise clothing would tend to stick when people sit on furniture such as chairs. They should have good sanding and polishing properties, resistance to water, acids (food), alkalies, alcohhol, etc.

Exterior varnishes should have good durability or weather-resistance. They must withstand the elements without failure by cracking, peeling, whitening, spotting, etc., and with minimum loss of gloss for a maximum period of time.

Table 10.7. Analysis of a phenolic marine spar varnish.

Lbs		*Gals*
75	*p*-Phenylphenol pure phenolic resin	7.5
30	High m. p. modified phenolic resin	3.3
180	Tung oil	23.1
125	Alkali-refined linseed oil	16.5
328	Mineral spirits	48.0
2	Cobalt naphthenate 6 per cent	.25
4.5	Lead naphthenate 24 per cent	.5
2.0	Antiskinning agent	.25
743.5	Total	100.4

Cooking directions: heat resins and linseed oil to 575°F; hold for one hour; add tung oil; reheat to 450°F; hold for body, then add drier and anti-skinning agent.

Oil length	37.5 gallon
Viscosity	E-F Gardner Holdt
NVM	55 per cent
Wt/gal	7.40 lbs

PAINT AND VARNISH REMOVERS

Paint and varnish removers are of two main types according to the method of use, i.e., application removers and immersion removers.

Organic

Methylene chloride

The most widely used removers are based on methylene chloride (dichloromethane), which is the most versatile stripping agent commonly available. Methylene chloride is the least toxic of chlorinated solvents and the fastest in lifting paint films.

Mechanism of removal

The effectiveness of methylene chloride results from the small molecular size. Its low molecular weight enables it to penetrate rapidly into a coating, and its intermediate solvency enables the coating not to be dissolved so that redeposition on the substrate is avoided. When the methylene chloride has reached the substrate, it swells the film to several times its original volume.

Functions of components

An application remover that is based on methylene chloride has several components, including solvents, e.g., methylene chloride for other chlorinated hydrocarbons; cosolvents; activators and corrosion inhibitors; evaporation retarders; thickeners; emulsifiers; and wetting agents.

Modifications

Methylene chloride removers are modified to increase stripping power for special purposes. The modifying chemicals include amines, alkalies, and organic acids.

Manufacture

Most formulae are made by simple mixing of components. Order of addition is important and care must be taken to incorporate the thickener properly.

Other organic paint and varnish removers

Some finishes are relatively easy to strip and can be dissolved by inexpensive solvents and blends of solvents which may be formulated with thickeners and waxes. Shellac is stripped by alcohols, and lacquers are removed by blends of alcohol and acetates, e.g., butyl acetate. One remover for exterior paints is a slowly evaporating oil-in-water (o/w) emulsion that is based on xylene and dimethylformamide.

Inorganic

Caustic soda

Caustic soda is one of the most common industrial strippers. It is used to clean paint-making and paint-application equipment, including jigs, hangers, and conveyors. Caustic soda baths also are used to salvage ferrous metal parts with defective finishes. Some alkali-resistant coatings are not removed in caustic systems. Caustic baths usually must be heated to ca 93°C to be effective.

Others

Paint can be removed by other alkali systems, inorganic acids, fused alkali, and molten-salt baths.

Choosing a Paint and Varnish Remover

Selection of the best compound or method for industrial or commercial paint and varnish removal can be complicated by many factors. Often the problems are best solved by companies who specialise in paint-stripper supplies and services. Selection of a paint stripper depends on the substrate, type of coating to the stripped, how much attack, if any, is allowed on the substrate, number of coats and primer to be stripped, available equipment, time and temperature limitations, odour and flammability restrictions, and disposal of spent stripper.

Health and safety factors

Many potent removers used in the past are no longer used because of carcinogenic and other toxicological effects. These include benzene, benzene derivatives, phenol, cresols, and some chlorinated and fluorinated hydrocarbons.

Handling and safety

Any chemical or formulation that blisters an organic coating film is likely to blister human skin. When working with ordinary methylene chloride removers, one should wear protective clothing and safety glasses. Good ventilation is necessary; established inhalation standards are associated with many of the chemicals and solvents.

Environment

Fume emission must be considered in industrial stripping operations. Emission standards have been established, for many of the solvents used in paint removers, by state and local governments. Disposal of liquid waste and sludge is a serious problem. Minimising quantities by reusing filtered or decanted stripper solutions is a partial solution.

LACQUERS

A lacquer is a protective coating which dries by evaporation of volatile components. The film-forming constituent is usually a cellulosic ester, i.e., nitrate, acetate, acetate-butyrate, or other high molecular weight polymer. Other types of lacquers are based on acrylics, polyurethanes, vinyls, etc. There are several differences between varnishes and lacquers. As stated previously, lacquers dry essentially by evaporation, while varnishes dry by a combination of oxidation and polymerisation. Lacquers are characterised by very rapid drying and distinctive odour. They are usually based on high molecular weight polymers which require low-boiling solvents of high solvency power such as alcohols, ketones, and esters. The film will redissolve in the original solvent. Lacquers are usually available in a solids range of 20–30 per cent, while varnishes have a solids range of 45–55 per cent. Fig. 10.3 highlights the manufacture of lacquers.

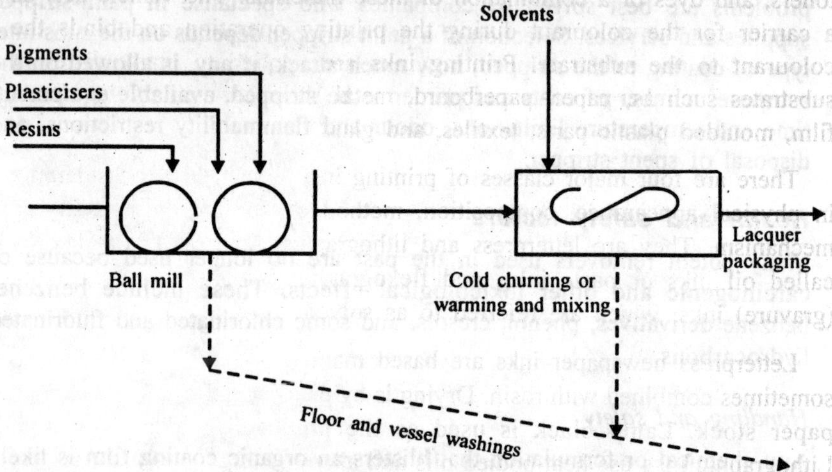

Fig. 10.3. Typical flow chart for lacquer manufacture.

Lacquers dry "tack-free" in 5 to 15 minutes and to a firm film in 30 minutes to 4 hours. They are usually applied by spraying, brush application often being impractical because of the very rapid drying. Pigmented lacquers predominate in use; clear lacquers are used where colourless, tough films are desired. Clear lacquers are not generally considered as durable as high grade varnishes on exposure to sunlight and moisture.

Nitrocellulose, which is one of the principal film-formers, will be considered in detail.

In addition to nitrated cotton, a nitrocellulose lacquer contains, for example: (i) a solvent mixture which usually contains a ketone, an alcohol, an ester, and frequently an ether-alcohol; (ii) a resin such as an alkyd, a phenolic, or ester gum to increase solids and improve adhesion; (iii) a plasticiser for flexibilising the film; (iv) an inexpensive volatile diluent such as toluene; and (v) a dye or pigment which is omitted if a clear lacquer is desired.

The main outlets for lacquers are automobile finishes, furniture finishes, metal finishes, and plastic, rubber, paper, and textile finishes.

PRINTING INKS

Printing inks are a mixture of colouring matter dispersed or dissolved in a vehicle or carrier, which forms a fluid or paste which can be printed on a substrate and then dried. The colourants used are generally pigments, toners, and dyes or a combination of these materials. The vehicle acts as a carrier for the colourant during the printing operation and binds the colourant to the substrate. Printing inks are applied to many different substrates such as, paper, paperboard, metal sheets, metallic foil, plastic film, moulded plastic parts, textiles, and glass.

There are four major classes of printing inks, which vary considerably in physical appearance, composition, method of application, and drying mechanism. They are letterpress and lithographic (litho) inks, commonly called oil inks or paste inks, and flexographic (flexo) and rotogravure (gravure) inks, which are referred to as solvent inks.

Letterpress newspaper inks are based mainly on mineral oil which is sometimes combined with rosin. Drying is by penetration into the absorbent paper stock. Lamp black is used as the pigment for the black ink. Lithographic inks use heat-bodied oils and alkyds as the vehicles. They are heat-dried in ovens.

Flexographic and rotogravure inks are low viscosity materials and contain highly volatile solvents. Flexographic inks require solvents such as alcohols that do not attack the rubber rollers and printing plates. In the case of gravure ink no contact with rubber is involved, and it is permissable to use solvents such as ketones and aromatic hydrocarbons. Flexographic inks use nitrocellulose, polyamides, and acrylates as vehicles.

INDUSTRIAL COATINGS

Coatings are available as liquid or fusible compositions. Table 10.8 gives a classification of coatings formulations. The liquids are usually aqueous or organic solutions. The coatings are applied by the user to the substrates, allowed to flow out smoothly, chiefly by forces of surface tension, and then cured to the final solid form. An industrial organic coating usually comprises an organic binder, pigments, a carrier liquid (sometimes omitted), and various additives.

The binder exists in the final film as a polymer of high molecular weight that may or may not be cross-linked. It is primarily responsible for the plastic quality of the film. Binders are grouped into certain overlapping classes such as acrylic, vinyl, alkyd, polyester, etc. The molecular structure of the binder and the forces operating between the molecules largely determine the mechanical properties.

The pigments, which may be organic or inorganic, contribute primarily to opacity and colour, in addition to durability, hardness, adhesion, and particular rheological properties of the coating in fluid form. Coatings are classified as primers that are applied directly to a substrate or as topcoats. The latter are applied over a primer and are usually the last coat. Intermediate coats are called sealers. In some uses, one coat satisfies all requirements.

Table 10.8. Formulation possibilities of synthetic resins in coatings formulation.

Vehicle system	*Uses*
Class 1: Vehicles containing oil-modified alkyds or other polymers containing drying oil	
(1) Oxidising alkyd resins (sometimes mixed with oleoresinous varnishes)	Architectural enamels, house paints, interior paints, flat wall paints, baking and air-drying undercoats and enamels for machinery, prefab housing structural units, and other factory products
(2) Alkyd and phenoplast Alkyd and nitrocellulose Alkyd and chlorinated rubber Alkyd and polystyrene Alkyd and diisocyanate Alkyd and vinyl and epoxy	Air-drying or low-temperature baking under coats and enamels for metals products) that have more plasticlike film properties than to possible to attain with alkyds alone
(3) Alkyd and aminoplast Alkyd and aminoplast and epoxy Alkyd and silicone	Similar uses as above, but where a high premium is placed on colour retention, and superior chemical and heat resistance
(4) Oil-modified epoxy resins and aminoplast.	Air-drying or baking-type undercoats or enamels; improved baking enamels and undercoats
Class 2: Vehicle systems containing no alkyd or drying oil	
(5) Vinyl acetals and/or phenolic	Chemically resistant baking under-coats and enamels

(Contd...)

Vehicle system	Uses
allylaminoplast, alkyd, epoxy } and 2,4,6-trimethylolphenyl ether	
(6) Phenoplasts (with or without epoxy, or vinylacetal, or aminoplast)	Thermosetting undercoats and/or enamels for high corrosion protection, especially in thin films (not resistant to discoloration); room-temperatu--re setting mastics for corrosion and abrasion protection
(7) Polyester and triazine resin allyl polyester silicone thermosetting acrylics complex amino resins some other polyesters	Chemical and discoloration resistant glossy, clear films and pigmen--ted baking enamels for metallic and non-metallic production goods (thermosetting film formers)
(8) Vinyl acetate–chloride, copolymers vinylidene or vinyl chloride-acryolonitrile copolymers butadiene copolymers acrylic copolymers poly(vinyl acetate)	Thermoplastic lacquers, baking or air-drying emulsion paints for production finishes on non-metallic goods such as acoustical board or molded plastics, fire-retardant and corrosion-protective mastics; exterior house paints and interior decorators' paints of emulsion type
(9) Nylons some cellulosic esters and ethers polyurethanes polytetrafluoroethylene poly(vinyl acetals) saturated polyesters unsaturated polyesters and styrene epoxy and polyamide copolymers of ethylene or propylene	Special type of coatings, potting compounds mastics, etc. for electrical insulation and corrosion production.

Industrial coatings are generally applied with specialised machinery representing a large investment. Coating properties must be controlled carefully to ensure continuous and satisfactory application. Properties to be controlled include rheological characteristics, stability, colour, smoothness or gloss, metallic lustre, durability, adhesion, permeability, hardness, flexibility, and protection of the substrate against deterioration.

MARINE COATINGS

Ships, offshore working platforms, and onshore waterfront structures are damaged by contact with the harsh marine environment. Control of this destructive action is best achieved through a programme of: (i) selection of the materials most resistant to deterioration; (ii) design to minimise conditions favourable to corrosion; and (iii) effective utilisation of protective coatings and/or cathodic protection (an electrical method of preventing metal corrosion in a conductive medium by placing a charge on the item to be protected) to deter corrosion. Protective (anti-corrosive) coatings impart protection to the substrate by forming a barrier to the water, salt, and oxygen which accelerate corrosion. Antifouling paints control the attachment and growth of marine fouling organisms on immersed areas.

The modern synthetic coatings have three common ingredients: solvent, binder, and pigment. Coatings that cure by chemical reaction of two component parts are the most widely used in submerged marine applications. For example, epoxies, coal-tar epoxies, urethanes, and polyesters are durable and resistant to water, solvents, and chemicals.

Modern synthetic marine coatings provide much longer protection than earlier ones, but require both complete cleanliness and surface profile (tooth) in order to obtain adequate bonding of prime coats. High speed abrasive (sand, grit, or shot) blasting by conventional air-pressured equipment or by newer equipment that utilises centrifugal force to propel the abrasive helps provide an appropriate surface. Actions taken by personnel to eliminate particulate emissions during abrasive blasting include the use of hard, sharp, and properly sized abrasives that produce minimum emission of particulates, blasting inside a building or under a temporary shroud, and using equipment that automatically moves across a regular surface and picks up and recycles the spent abrasive.

Manufacturers of marine coatings always have printed information available on the use of their products. This information includes recommendations on the equipment to be used, mixing of components, time and temperature requirements, coverage rate at a recommended dry-film thickness, and good application practices.

Protective (anticorrosive) coatings include vinyls, chlorinated rubbers, epoxides, coal-tar epoxies, urethanes, polyesters, inorganic zincs, zinc-rich organics, and specialised coatings (e.g., solvent-free epoxies, powder coatings, plastic-coated steel electrical conduits and fittings, and petrolatum-coated tapes).

Currently, the only two biocidal materials that are used extensively in antifouling paints are cuprous oxide and organotin compounds (usually tributyltin oxide or tributyltin fluoride). Most commercial antifouling paints use a vinyl binder, although products with other binders are also marketed. Rosin or some other leaching agent must be added to cuprous oxide formulations to permit its controlled release into sea-water where it is lethal to fouling larva forms. Organotins usually do not require leaching agents to dissolve slowly in seawater.

Recent research has lead to the development of organometallic polymers for use in antifouling paints. A sheet material of black neoprene rubber impregnated with tributyltin is currently marketed. Because the sheet is so much thicker than an antifouling coating system (usually about 100 μm) it has a larger reservoir of biocide that can result in longer-lasting fouling control.

RESISTANT COATINGS

Resistant or high performance coatings or linings are specialty products used to give long-term protection under difficult corrosive conditions to industrial structures. This contrasts with paint, which is used for general appearance and shorter-term protection against milder atmospheric conditions, and industrial coatings, i.e., coatings that are applied to manufactured products.

There are two basic methods by which coatings or linings protect the surface. The first is based on the principle of impermeability. The second method uses anodically active or inhibitive pigments in the primer or in the coating to regulate corrosion. The inert impervious system performs best as a lining where it is subject to continual moist, wet, or immersion conditions and where it is subject to little or no physical abrasion. Inhibitive coatings perform best in areas where the coating is subject to weathering, atmospheric conditions, high humidities, or chemical fumes.

Components of Resistant Coatings and Linings

The similarity to paint extends to the essential ingredients of the coating or linings which include the binder, colour-carrying pigments, inert and reinforcing pigments, inhibitive pigments for primer only, and solvents. In general, the same equipment and manufacturing procedures are used for paint or industrial coatings.

Application of Resistant Coating Systems

The characteristics provided by each part of a coating system are as follows :

(1) *Primer:* Good adhesion to surface, satisfactory bonding surface for next coat, ability to retard the spread of corrosion, enough chemical and weather resistance to protect the surface until application of next coat, chemical resistance equivalent to the remainder of the system.

(2) *Intermediate coats:* Adequate film thickness for the coating or lining system, uniform bond between the primer and the topcoats, a superior barrier to moisture vapour and aggressive chemicals.

(3) *Finish coats:* Must be pleasing in appearance as well as resistant, may serve to provide a non-skid surface, a matrix for antifouling agents or other specialised purposes, must have sufficient weather, chemical, and abrasion resistance in any environment to ensure its remaining intact and providing protection to the substrate. Proper surface preparation and coating or lining application are essential to the performance of the resistant coating.

Resistant coatings or linings represented about 9% of the overall paint or coatings market in 1983. Three health and safety factors are important to consider both during manufacture and application of protective coatings: air pollution, health hazards to the individual and fire and explosion hazards.

New Coating Developments

As a consequence of the new and increasing governmental environmental standards as well as health and safety requirements, there are three avenues of development that will have a strong influence on the future of protective coatings: the use of water-based coatings; the use of solvent-free coatings or 100% solids coatings; and the use of inorganic materials as coatings, e.g. organic zinc coatings.

Chapter 11
Pulp and Paper Industries

INTRODUCTION

Pulp is processed cellulosic fibre derived from hardwoods, softwoods, and other plants. There are two major types of pulp: (i) ground wood or mechanical pulp, which is merely finely divided wood without purification, and is made into newsprint, cheap manila papers, and non-permanent tissues; (ii) chemical pulp, of which there are three kinds: (a) soda process pulp, obtained from the digestion of wood chips (mostly poplar) by caustic soda; (b) sulphite process pulp (mostly spruce and other coniferous woods) obtained by digestion with a solution of magnesium, ammonium or calcium disulphite containing free sulphur dioxide; and (c) sulphate process (kraft) pulp, in which sodium sulphate is added to the caustic liquors but is reduced by the carbon present to the sulphide, which becomes a digesting agent. Sulphite and sulphate pulps (chiefly from softwoods), comprise the bulk of paper pulps. Sulphate pulps are known as kraft pulps because of their strength, and are used for wrapping, packaging, container board, etc. A relatively new process called holopulping replaces sodium sulphate with oxidants.

Paper is a semi-synthetic product made by chemically processing cellulosic fibres. Wide varieties of sources have been used for speciality paper (flax, bagasse, esparto, straw, papyrus, bamboo, jute, and others), but by far the largest quantity is made from softwoods (coniferous trees), such as spruce, hemlock, pine, etc.; some is also made from such hardwoods as poplar, oak, etc., as well as from synthetic fibres. Papermaking technology involves the following basic steps: (i) chipping or other subdivision of the logs; (ii) manufacture of chemical or semi-chemical pulp by digestion in acidic or alkaline solutions, which separates the cellulose from the lignin; (iii) beating the pulp to break down the fibres and permit proper bonding when the sheet is formed; (iv) addition of starches, resins, clays, and pigments to the liquid stock; (v) formation of the sheet continuously on a fourdrinier machine, where the water is screened out and the sheet dried by passing over a series of heated drums; (vi) high-speed calendering for brightness and finish; (vii) coating, either by machine applications or (for heavy finishes) by brushes.

MANUFACTURE OF PULP FOR PAPER

The raw materials for the preparation of pulp for ultimate conversion into paper are : (i) fibrous raw materials—(a) wood, (b) waste paper, (c) rags, (d) straw; (ii) non-fibrous raw materials—(a) sulphur, (b) NaOH, (c) Na_2CO_3, (d) lime, (e) chlorine, clay, alum, rosin.

To convert wood into pulp by any method used for pulp-making, it is first necessary to remove the bark, and in case of chemical pulp, to reduce the barked wood into small chips of fairly uniform size. The methods, however, differ depending on the quality and relative size of wood being pulped and the kind and quality of pulp being produced by the paper mill.

For removing bark from pulp wood, the barking drum is most frequently used. This machine consists of a cylinder-like drum, generally 12 ft in diameter and 45 ft long, placed horizontally, and sometimes incompletely partitioned into several compartments spaced in such a way as to permit the bark from the wood to pass between them. The drum rotates slowly, the wood cut to a uniform length enters continuously at one end and is discharged from it continuously at the other end, without the bark. Water is also used in the operation to free the barked wood of any adhering bark.

The barked wood, without any further preliminary treatment is ready for reduction to mechanical pulp. To make it ready for treatment by any of the chemical pulp-making process, it must further be reduced to chips approximately 1/2 to 1 inch long and from 1/8 to 3/16 inch thick.

The chipping is done by a machine known as a chipper which consists essentially of a hooded, heavy circular disc, made of cast iron or heavy steel plate, slotted for and equipped with three or four knives and placed in a vertical position on a horizontal shaft, together with a cast iron spout so placed in relation to the face of the disc that it directs the wood against it and to the cutting edge of the knives at the proper cutting angle. The chips produced by the chipper are screened by screens which may be revolving or rotatory shaker or vibrating or oscillatory.

Preparation of Chemical Pulp Liquor

The chemical pulp liquor is effected by the sulphite, soda or sulphate processes. In the sulphite process, the raw materials are sulphur (or source of sulphur as iron pyrites), lime or limestone and water. The principal equipments of the sulphite process are one or more sulphur burners, a combustion chamber arrangement, a gas cooler (for cooling the hot SO_2 gas produced by burning of sulphur), absorption apparatus and arrangement to permit the use of excess SO_2 gas relieved out from the digester for strengthening the acid solution. Absorption of the gas in water is carried out in the presence of milk of lime or limestone whereby a bisulphite of calcium, containing excess of sulphurous acid is formed. To this is added SO_2 gas released as excess from the digesters.

To prepare pulp liquor by the soda process, Na_2CO_3, lime and water are the principal raw materials used. The liquor produced, mainly a solution of NaOH with a strength of 8 to 15° Beaume at 60°F, is made by caustisising a solution of soda ash with quicklime.

The liquor of a charge following cooking is designated as 'black liquor' and is utilised for alkali recovery. A chemical recovery plant of a soda pulp installation, therefore, includes essentially an evaporator and an incinerator installation for making the alkali in the 'black liquor' available for use again as sodium carbonate.

The raw materials used in the production of sulphate cooking liquor are sodium sulphate, lime and water. In the preparation of this liquor, like that of the soda process, use is made of the cooking liquor from previous pulping operations. This liquor is also called 'black liquor' as in the soda process, and is concentrated by evaporation following the pulp-washing operation. The reactions in the sulphate process are

$$Na_2SO_4 + 2C = Na_2S + 2CO_2$$

$$Na_2CO_3 + Ca(OH)_2 = 2NaOH + CaCO_3$$

The $CaCO_3$ sludge is calcined to get CaO for re-use. The filtrate, then of clear colour, is used for cooking the cellulose fibres to produce pulp. Sulphate pulp is technically called 'kraft'.

Mechanical Wood Pulp

A grinder machine is used to reduce the barked wood to mechanical pulp. The grinder consists of a cylinder-like pulpstone, placed on a horizontal shaft. The wood is always placed length-wise across the pulpstone face, so that as the stone revolves, the wood fibres are separated from one another along their length. Sufficient water is used in the grinding operation both to prevent burning of the wood and to serve as a vehicle for carrying the pulp fibre away from the machine.

Pulp from Rags

An important source of fibre for the manufacture of fine paper is rags. New rags are mainly scraps from textile mills and garment factories and the old rags are collected from domestic wastes. The rags go through a thrasher where dust is removed before they are sorted. All are separated according to colour. Colours which are difficult to bleach are used in the manufacture of dark papers and those more bleached are used for lighter papers. After being sorted, the rags are cut into small pieces, the dust is removed and a magnet is passed over to remove any metallic foreign materials. After this they are ready for the cook. In this operation, the wax and resins in the fibre are removed, along with any additional dirt or grease and also to remove the dyes from the fibres. The cooking liquor contains either caustic lime, caustic soda or a mixture of caustic lime and soda ash. The digester is a horizontal rotary boiler. Cooking is done at 120°C for 10–12 hours. The pulp produced is washed, bleached with chlorine and rewashed.

Operation for the Chemical Pulp

After the cooking operations are completed, the contents from the digesters are discharged under pressure into specially designed tanks, which permits the rapid removal of steam, either to the atmosphere or to a condenser.

The following operations are then conducted :

(1) Washing the chemical pulp.

(2) Screening of the pulp.

(3) De-watering or thickening.

(4) Bleaching of the pulp by sodium hypochlorite or chlorine.

(5) Brightening the colour of the pulp (specially in case of groundwood pulp).

Two other processes of pulping, namely, semi-chemical pulping and chemico-mechanical pulping are also in use.

The pulp, by whichever process prepared, is then used for the manufacture of paper. The paper industry is a huge industry, requiring large space, machinery, chemical operations for sizing and finishing the paper, holding the stocks of paper and moving them to the market.

The operations leading to the preparation of finished paper including heating in a Hollander, refining, putting in fillers, sizing and colouring, adding requisite quantities of resins and forming the sheets in a Fourdrinier or a cylinder machine and then finishing. There are speciality papers like tissue papers, bank papers, hardboards, paper-based laminates and a variety of other papers.

PAPER

Paper consists of sheet materials that are compressed of bonded, small, discrete fibres. The fibres usually are cellulosic in nature and are held together by secondary bonds which, most probably, are hydrogen bonds. The fibres are formed into a sheet on a fine screen from a dilute water suspension. The word paper is derived from papyrus, a sheet made in ancient times by pressing together very thin strips of an Egyptian reed.

Paper is made in a wide variety of types and grades to serve many functions. Writing and printing papers constitute ca 30% of the total production. The balance, except for tissue and toweling, is used primarily for packaging. Paperboard differs from paper in that it is generally thicker, heavier, and less flexible than conventional paper.

Fibrous Raw Materials

The main components used in the manufacture of paper products are listed in Table 11.1. More than 95% of the base material is fibrous and more than 90% originates from wood. Many varieties of wood, e.g., hardwood and softwood, are used to produce pulp. In addition to the large number of wood types, there are many different manufacturing processes involved in the conversion of wood to pulp. These range from mechanical processes, by which only mechanical energy is used to separate the fibre from the wood matrix, to chemical processes, by which the bonding material, i.e., lignin, is removed chemically. Many combinations of mechanical and chemical methods also are employed. Pulp properties are determined by the raw material and manufacturing process, and must be matched to the needs of the final paper product.

Table 11.1. Main components of paper and paperboard.

Mechanical/semi-chemical wood pulp
Unbleached kraft chemical wood pulp
White chemical wood pulp[a]
Waste fibre
Non-wood fibres
Fillers/pigments

[a] Includes unbleached sulphite.

Physical properties

Most properties of paper depend upon direction. For example, tensile strength is greater if measured in the machine direction, i.e., the direction of manufacture, than in the cross-machine direction. For paper made on a Fourdrinier paper machine, the ratio of the two values varies ca 1.5-2.5. An even greater anisotropy is observed if either of the in-plane values is compared to the out-of-plane strength. Paper is quite weak in the thickness direction. Paper may be considered on orthotropic material, i.e., one possessing three mutually perpendicular symmetry planes. The three principal directions are the machine, cross-machine, and thickness directions.

Because the fibres generally are flat or ribbon-like, they tend to be deposited on the wire in layers. There is very little tendency for fibres to be oriented in an out-of-plane direction, except for small undulations where one fibre crosses or passes beneath another. The layered structure results in the different properties measured in the thickness direction as compared to those measured in the in-plane direction. The orthotropic behaviour of paper is observed in most paper properties and especially in the electrical and mechanical properties.

The basic weight is the mass in grams per square metre. It can also be expressed as pounds of a ream of 500 sheets of a given size, but the sheet sizes are not the same for all kinds of paper. Typical sizes are 43.2 × 55.9 cm for fine paper, 61.0 × 91.4 cm for newsprint, and 63.5 × 96.5 cm for several bookpapers. The most common designation is pounds per 3000 square feet (1.62 g/m^2) for paper. The basic weight of board usually is expressed as pounds per thousand square feet $(205\ kg/m)^2$. The thickness or caliper is the thickness of a single sheet measured under specified conditions. It is usually expressed in micrometers.

The tensile strength is the force per unit width parallel to the plane of the sheet that is required to produce failure in a specimen of specified width and length under specified conditions of loading. The strength of paper also is expressed in terms of a breaking length, i.e., the length of paper that can be supported by one end without breaking. Breaking lengths for typical papers are from ca 2 km for newsprint to 12 km for linerboards.

Chemical properties

The chemical composition of paper is determined by the types of fibres used and by any non-fibrous substances incorporated in or applied to the paper during the paper-making or subsequent converting operations. Paper usually is made from cellulose fibres obtained from the pulping of wood.

Occasionally, synthetic fibres and cellulose fibres from other plant sources are used. Paper properties that are affected directly by the fibres chemical composition include colour, opacity, strength, permanence, and electrical properties. Development of interfibre bonding during paper-making also is strongly influenced by the composition of the fibres. Because residual lignin in the fibres inhibits bonding, ground-wood pulp is used in newsprint and in some book and absorbent papers which do not require a highly bonded structure. Hemicelluloses in chemical pulps contribute to bonding; therefore, pulps containing hemicelluloses are used for wrapping papers and other grades which require bonding for strength, and in glassine, which requires bonding for transparency. In most papers, the chemical composition largely reflects those non-fibrous materials that were added to the paper to achieve the desired physical, optical, or electrical properties.

Manufacture and Processing

Stock preparation

Stock preparation denotes the several operations that must be undertaken in order to prepare the furnish from which paper is made. During the stock-preparation steps, paper-making pulps are most conveniently handled as aqueous slurries, so that they can be conveyed, measured, subjected to desired mechanical treatments, and mixed with non-fibrous additives before being delivered to the paper machine. In the case of adjacent pulping and paper-making operations, pulps usually are delivered to the paper mill in slush form directly from the pulping operation. Purchased pulps and wastepaper are received as dry sheets or laps and must be slushed before use. The objective of slushing is to separate the fibres and to disperse them in water with a minimum of mechanical work so as not to alter the fibre properties. Slushing is accomplished in several types of apparatus, e.g., the Hydrapulper.

Beating and refining

Virtually all pulps are subjected to certain mechanical actions before being formed into a paper sheet. Such treatments are used to improve the strength and other physical properties of the finished sheet and to influence the behaviour of the system during the sheet-forming and drying steps. Beating and refining may be considered synonymous.

Filling and loading

Materials, e.g., mineral pigments for filling and loading, are added to the pulp slurry to make the paper-making furnish. Pigments are used in varying amounts, depending upon the grade of paper, and may comprise 2–50% of the final sheet. Fillers can improve brightness, opacity, softness, smoothness, and ink receptivity. They almost invariably reduce the degree of sizing and the strength of the sheet. The brightness, particle size, and refractive index of fillers influence the optical properties of the finished sheet, and the particle size and specific gravity are important in regard to the filler retention during sheet formation. All commercial fillers are essentially insoluble in water under the conditions of use.

Sizing

Sizing is the process of adding materials to the paper in order to render the sheet more resistant to penetration by liquids, particularly water. Unsized or waterleaf paper freely absorbs liquids. Writing and wrapping papers are typical sized sheets, as contrasted with blotting paper and facial tissue which usually are unsized. Rosin, various hydrocarbon and natural waxes, starches, glues, casein, asphalt emulsions, synthetic resins, and cellulose derivatives are some of the materials that are used as sizing agents. The agents may be added directly to the stock as beater additives to produce internal or engine sizing, or the dry sheet may be passed through a size solution or over a roll that has been wetted with size solution; such sheets are said to be tub-sized or surface-sized.

Colouring

The colour of most papers and paperboards that are made from bleached pulps is achieved by the addition of dyes and other coloured chemicals. White papers frequently are treated with small amounts of blue materials to achieve a whiter visual appearance. By far the largest proportion of dyes is added to the stock during stock preparation, although a limited amount of dry paper is coloured by dip dyeing or by applying a dye solution during calendering. Water-soluble synthetic organic dyestuffs are the principal paper-colouring materials.

Other beater additives

Beater adhesives are employed widely to enhance fibre-to-fibre bonding. Starches probably are used in the greatest tonnage. Urea-formaldehyde and melamine-formaldehyde polymers provide wet strength to the finished paper sheet. Other natural and synthetic materials are used to alter the paper properties and to influence the behaviour of the system during sheet forming and drying.

Sheet Forming, Pressing and Drying

Sheet forming

The equipment is of two types: (i) the cylinder machine; and (ii) the fourdrinier machine. Continuous paper machines have undergone extensive mechanical developments in the past half century, although the principles employed have changed little. Cylinder machines still are operated and involve multiples of five to seven cylinders; they are used to produce heavy multi-ply boards. Fourdriniers are standard in the industry and are used to produce all grades of paper and paperboard. They vary 1–10 m in width and, including the press and dryer sections, may be > 200 m long.

Sheet pressing

The sheet leaving the wet end contains approximately four parts of water per part of fibre; however, it is possible to remove additional water mechanically without adversely affecting sheet properties. This is achieved in rotary presses, of which there may be one or several on a given paper machine.

Sheet drying

At a water content of ca 1.2–1.9 parts of water per part of fibre, additional water removal by mechanical means is not feasible and evaporative drying must be employed. This is at best an efficient but costly process and often is the production bottleneck of papermaking.

Converting

Almost all paper is converted by undergoing further treatment after manufacture. Among the many converting operations are embossing, impregnating, saturating, laminating, and the forming of special shapes and sizes, e.g., bags and boxes.

Pigment coatings

Pigment coatings are compositions of pigments and adhesives with small amounts of additives and are applied to one or both sides of a paper sheet. They generally are designed to mask or change the appearance of the base stock, improve opacity, impart a smooth and receptive surface for printing, or provide special properties for particular purposes.

Barrier coatings

In packaging applications, a barrier may be needed against water, water vapour, oxygen, carbon dioxide, hydrogen sulphide, greases, fats and oils, odours, or miscellaneous chemicals. A water barrier can be formed by changing the wettability of the paper surface with sizing agents. A grease or oil barrier can be provided by hydrating the cellulose fibres to form a pinhole-free sheet or by coating the paper with a continuous film of a material which is resistant to the particular grease. Gas or vapour barriers are formed by coating the paper with a continuous film of a suitable material.

Pollution Control

Methods of pollution control in pulp and paper mills include in-mill control measures, end-of-the-pipe control measures, and monitoring and assessing environmental quality. Control devices for gases may include stripping, scrubbing, condensation, incineration, and adsorption.

Control devices for particulate matter include cyclones, scrubbers, and electrostatic precipitators. Application of these devices individually or in combination has been quite effective in meeting the air-quality criteria.

The pulp and paper industry uses large amounts of water. Recently, water-quality requirements have forced mills to install end-of-pipe secondary treatments, which principally are biological. Means other than installation of expensive end-of-the-pipe treatments are being evaluated in terms of recent requirements. Mills are recycling and reusing more process water, which results in higher water temperatures and increased solids in the process water, thereby enhancing problems, e.g., corrosion, slime and other deposits. However, chemicals are used to control these problems. Dissolved solids may impart biochemical oxygen demand (BOD), colour, and toxicity to mill effluents. The suspended solids impart turbidity and long-term BOD. Directly or indirectly, all of these may affect aquatic life.

Sludge handling and disposal

Most waste-treatment processes generate solid wastes that must be disposed. Two kinds of sludges are generated by pulp and paper mills: primary sludges that contain fibres, clay filler materials, and other chemical additives, and secondary sludges that are largely biological in nature and are harder to handle and dewater. The disposal of sludges in landfills is being re-evaluated and alternative disposal approaches are being developed.

Uses

Paper is used in paper and paperboard such as folding cartons, fibre cans and tubes, corrugated and solid fibre boxes, paper bags, and multiwall shipping sacks.

PAPERMAKING ADDITIVES

In papermaking, chemicals can be added either to the pulp slurry prior to sheet formation, i.e., internal addition, or to the resulting sheet after complete or partial drying. The method used depends primarily on the desired effects. For example, strength additives usually are added internally if uniform strength throughout the sheet is wanted, but they are applied to the surface if increased surface strength is needed. If an additive cannot be retained efficiently from a dilute pulp slurry, then it is better to apply it to the surface of the sheet.

Functional internal additives improve specific properties of the paper, e.g., sizing, dry strength, wet strength, and optical properties, e.g., those that are affected by the presence of light-scattering particulate additives such as clay and titanium dioxide. Colour-modifying additives have been reviewed.

Functional surface treatments involve many of the same chemicals that are used internally. However, since retention of the chemical by pulp fibres is not important to surface treatment, many other materials may be used that are ineffective or, at best, inefficient when added internally.

Environmental constraints on the paper industry have resulted in drastic processing changes in recent years, primarily because very large amounts of water are used to produce paper. If chemical additives are not efficiently retained in the sheet, treatment of the large volumes of effluent is aggravated. Efficiency of various additives used in papermaking has increased and has permitted greater closure of white-water systems, i.e., water drained and pressed from the newly formed sheet. In alkaline sizing systems and some alkaline strengthening systems, the additives are retained so completely that effluents from these paper machines contain only very small amounts of dissolved materials.

Because many grades of paper and paper-board are used in direct contact with foods, most mills require paper chemicals that have been cleared for use are as per Food Development Act (FDA), so that it is not necessary to segregate machine broke and white water. Most of the chemicals discussed below are approved by the Food Development Act for use in paper and paper-board that are intended for applications in food packaging and processing. However, there are various restrictions and limitations on the amounts of chemicals that can be used and restrictions on some chemical additives in terms of specific functions uses.

Processing Aids

Processing aids improve the operation of the paper machine, whereas functional chemicals improve the properties of the paper. Processing aids include retention aids, drainage aids, formation aids, flocculants, defoamers, wet-web-strength additives, pitch-control agents, creping aids, and slimicides. Retention aids, flocculants, and drainage aids are phenomenologically related; all act by agglomerating filler particles, fines, or fibres with themselves or with each other.

Functional Internal Additives

Functional internal additives include sizing agents (rosin-based, cellulose-reactive, wax emulsions, and fluorochemical); dry-strength additives (natural gums, starches, cellulosics, and acrylamide polymers); wet-strength additives (urea-formaldehyde resins, melamine-formaldehyde resins, aminopolyamide-epichlorohydrin resins, polymeric amine-epichlorohydrin resins, aldehyde-modified resins, and polymeric amines); and fillers, most commonly mineral pigments, e.g., clay, calcium carbonate, silica, hydrated alumina, and talc. Kaolin clay is the least expensive and most used filler pigment.

In recent years, synthetic polymeric pigments have been promoted as fillers for paper. Pigments that are based on polystyrene latices and on highly cross-linked urea-formaldehyde resins have been evaluated for this application. These synthetic pigments are less dense than mineral fillers and, thus, can be used to produce lightweight grades of paper.

Functional Surface Treatments

Although many functional chemicals can be added to the wet end of the paper machine, some grades of paper require special properties that cannot be provided by the low levels of additives that are retained at the wet end. To achieve the properties required for these grades of paper, it is necessary to apply the chemicals to a preformed paper web.

Processes

The most common method for the application of chemicals to the surface of a paper web is by a size press. In the size press, dry paper, which usually is sized to prevent excess water penetration, is passed through a flooded nip and a solution or dispersion of the functional chemical contacts both sides of the paper. Excess liquid is squeezed out in the press and the paper is redried.

A process that is being developed involves the application of chemicals as a foam to the surface of a paper web. The chemical foams contain less water than their liquid equivalent; thus, savings in drying costs can be realised. Also, the foams can be applied to a wet web, e.g., in the press section, without breaking the sheet.

Sizing

The most commonly used materials for surface sizing are starches and modified starches, including oxidised, enzyme-converted, hydroxyethylated, and cationic starches. They are used not only for sizing but also to improve strength, especially surface strength, and to impart smoothness. Starches may be applied to the finished sheet by any of the previously discussed methods. Often starch is used with other surface-sizing agents, e.g., rosin-based emulsion sizes, wax and wax-rosin emulsions, alkylketone dimer emulsions, and various synthetic polymeric sizing agents. These combinations permit improved sizing against liquid penetrants and increased surface strength and better finish.

Fluorochemical emulsion sizing agents can be applied to the surface of paper or paperboard to provide good oil and grease repellency. If they are used with other sizing agents, e.g., alkylketone dimer emulsion, both oil and grease repellency and water repellency are obtained.

Other special sizing agents are the chromium complexes of long-chain fatty acids; these provide water-repellent surfaces and good release properties.

Application of dry-strength additives

The various water-soluble natural and synthetic polymers that are used for strength enhancement by internal addition can be applied to paper surfaces. This type of application usually is indicated when surface-strength properties are more important than increased internal strength. Starches and modified starches, especially cationic starches, are used in large quantities to improve the surface strength of paper; they also improve the printing quality of the paper as a result of increased surface strength and reduced linting.

Application of wet-strength resins

Wet-strength resins are seldom applied to the surface of paper for enhancing wet strength because the commercially available, cationic wet-strength resins are retained so effectively internally. However, wet-strength resins are applied frequently to the surface of towels and tissues as creping aids.

Creping

The products that are used as internally added creping aids can be applied to the surface of paper, usually by spraying an aqueous solution or emulsion in front of the Yankee dryer. One of the most commonly used resins for this purpose is an aminopolyamide-epichlorohydrin resin. Such resins yield coatings on the dryer with the required degree of adhesion for optimum creping.

Curl control

Many grades of paper tend to curl, especially as humidity varies, because of the stresses produced during the drying process. Judicious application of water to the opposite side of the dry sheet followed by redrying may correct the curling.

Pigmented coatings

Paper coatings are applied as coating colours which are aqueous slurries containing 35–65 wt% solids. There are three main components of the solids; pigments, binders, and minor additives. The pigment is the chief component of a paper coating and consists of small, white, particulate materials. Pigments usually are minerals, e.g., clay, calcium carbonate, or titanium dioxide.

Binders

Paper-coating binders are either polymers derived from natural sources or synthetic polymers. The largest-volume, naturally derived binder is starch.

AMMONIA PAPER

Practically all blue-print paper made today has as the basic light-sensitive compound ammonium oxalate, while the colour forming compound is a mixture of sodium and potassium ferricyanide. As a preservative, an alkali oxalate or citrate is generally used. The printing speed of the sensitised paper may be varied from slow to fast simply by varying the amount of redsol in the formulae.

At one time 100% rag stock was the only paper used for blue-prints and it was thought until fairly recently that if any filter other than rag were used, the resulting paper would not be suitable for the blue-printing process.

These have however changed and for the last 20 years excellent blue-print stock has been made with the rag content varying all the way from all rag to no rag.

It is available in every stationary shop, school, college, university and all other higher academic institutions. It is the only cheapest paper available in the market for getting copies, thus preferred by most of the collegiates, research fellows and scholars.

Manufacturing Process

Ferric salts when exposed to light in the presence of organic matter is reduced to the ferrous state. The experiment involved the use of a ferric salt coated paper which was then exposed and the paper treated with potassium ferricyanide to develop the image. The ferrous salt produced during the exposure period reacts with the ferricyanide to give blue ferrous-ferricyanide. This undergoes molecular rearrangement to form ferric-ferrocyanide or turnbulls blue and the soluble ferric-ferrocyanide which still remain on the exposed portious is removed by washing with water.

In actual practice there is not much divergence from this basic reaction. Ferric ammonium citrate and ferric ammonium oxalate are used as the light sensitive salts, either alone or in combination and are coated on the base sheet along with potassium ferricyanide. A typical formulae is as follows :

Ingredients	*Parts*
Ferric ammonium citrate	125
Potassium ferricyanide	45
Water	1000

Such a coated sheet must be dried in the darkness since the chemicals are photo reactive in the wet state. Such a paper is comparatively slow printing and will not age too satisfactorily. The addition of oxalic acid has been suggested but although it shortens the required exposure time, it also has a degrading effect on the paper and reduces the quality of the blue colour. Citric acid is added for helping the clarity of white.

ART PAPER

Art paper is a type of coated paper. Without any additional investment chromo paper, art card, black centred card can be produced in the same unit. The main raw material for these types of paper is white paper. This white paper is coated with certain types of chemicals (resin, china clay, casein, protein and minerals etc.).

It is generally coated to impart certain characteristics desired by the customer. Art paper is generally coated on both the sides making it possible to print it on either side by the end users whereas the chromo paper is coated on single side only.

A fraction of cotton, around 10% may significantly improve the quality. Cotton rags and linters and fluff from the cotton mills are also employed where available.

Manufacturing Process

The widely accepted process for these types of paper manufacturing is 'Soda process'. The maize cobs/stems and wheat or rice straw are all seasonal crops and are not available throughout the year. It is therefore necessary to store these materials for three to six months for consumption. These raw materials are stored in stacks covered with gunny bags or canvas sheets. The raw materials are brought to cutter house and fed to the cutter so that the material is uniformly cut. The cut material is then carried to the digester floor and fed into the rotary digester. The material is cooked with caustic soda under pressure at 165°C for about 4 hours. After the material is cooked, thc pressure is released and the cooked material is transformed to the washing section. The material is washed thoroughly to get rid of alkali. After ensuring that it contains no alkali then material is passed through battery of centri-cleaners.

The pulp is cleaned and transferred to the bleaching section, where the material is bleached thoroughly by chlorine gas in single or multi stage bleaching. In addition to this caustic soda and calcium hypochlorite solution is also used for the said purpose in a tower.

The bleached pulp is cleaned again by a centri-cleaner and is admixed with 10% stronger pulp of cotton wastes (cotton linters). The mixed pulp is refined continuously or batchwise refiners to impart desired property (strength) to the fibre. The refined pulp is mixed with adhesives like rosin, dye, alum, casein etc. Next step is to dilute the refined pulp by water. The material is pumped to the paper forming machine.

The diluted pulp slurry is passed over the fourdrinier machine to form an endless paper which is rolled on big rollers. The reel may be rerolled to tight reels and coated on one or both sides by chemicals. For further gloss the paper could then be passed through super calenders. The paper is then cut into different size sheets.

BITUMINISED WATER-PROOF PAPER

Water-proof paper are usually kraft papers that have been laminated, coated, or impregnated with a thermoplastic materials, capable of resisting the penetration of water in liquid form. These papers are used as box, bag, crates, barrel liners, and bale wrappers. Besides these if find application in linings for freight cars and trucks, covers, and shroulds for goods and machinery in transit, and interior wraps, especially, for export shipments. Other uses includes in agriculture as plant protectors, celery bleaching papers, mulch paper and in adhesive tapes.

Manufacturing Process

Water-proof papers are produced by coating the base paper with bitumen, and laminating if desired. Coating is performed, by a special type of coater machine. In plant the bitumen is held in storage tanks at the temperature required for the particular type of converting equipment and either pumped to the treating tanks of the machine with overflow provisions back to circulating tanks, or pumped to the machine bath. The molten bitumen is applied by pick-up and applicator type rolls of coating machine and metered automatically. As bitumen is solid material at normal temperature, it does not require drying. It solidify itself, after coming out of the roll system the coated rolls are rewinded in rolls and afterwards packed for despatch.

EMERY PAPER

Emery paper is a abrasive coated paper used in industry for various applications such as cutting, grinding and polishing etc. It consists of a sheet of good strength paper coated with an adhesive which holds and supports a coating of abrasive grains. Commonly used abrasives are aluminium oxide, silicon carbide, garnet, flint, emery and crocks etc. Due to the rapid industrialisation in India, the demand of emery paper is bound to increase at a great pace.

Manufacturing Process

In the manufacturing process a smooth film of adhesives like synthetic varnish, phenolic resin, hide glue or a combination of mineral filler and one of these adhesives is applied to the backing paper followed by final abrasive coating. A further coating of sizing material is applied to further increase the binding of emery powder. The size may be based on a water-proof resinous material. The paper is now ready for cutting to the required saleable size in the form of rolls or sheets.

NEWSPRINT PAPER

Newsprint paper is made mostly from mechanical pulp although it may contain upto 25-30% chemical pulp. The main aim of adding chemical pulp is to increase the strength of paper. The newsprint is relatively weak and made of non-bleached pulp. In turn it is a cheap grade of paper in view of the fact that it is not intended to last long.

Manufacturing Process

The paper is made by depositing fibres from dilute suspension onto a fine wire mesh by filteration. The water drains through while the fibres are retained back.

During the forming process the sheet is built up from the fibres or fibre aggregates generally in a random manner.

The sheet structure depends on distribution of the fibres in water suspension. The uniformity of the sheet formed improves with the reduction of size and density of the flocks and is at its optimum when each fibre is free to move independently.

TOILET PAPER PLANT

Development of raw materials and the invention of paper machines have ushered in an era of high quality and mass-production, and paper is now called the symbol of culture.

Toilet paper is in large and increasing demand. The mass-producing mills generally use wood pulp as the raw material for toilet paper. But smaller mills use wood pulp and waste paper together, or use waste paper 100 per cent. Waste paper is recovered at a high rate, and selection of quality is well controlled.

Manufacturing Process

Broadly speaking, there are two methods for processing waste paper, namely, continuous and batch systems. In general, the process is recommended when the daily production capacity of the mill is five tons.

The manufacturing process begins when waste paper is moved upto the sorting yard above the digester by elevator. The sorted waste paper is fed into the global digester. Simultaneously with the feeding, sodium sulphite, soda ash, etc. are sprinkled thoroughly over the waste paper. After stuffing of the digester has been completed, the gut for the digester is tightly closed.

Steam is blown into the digester as it is rotated. The next step is the charging of the material into the blow pocher. Circulation is performed by a propeller agitator, and the digesting chemicals are thoroughly washed in the washing drum. When the washing is over, the material is transferred into the dump chest, from where it is moved up into the head box by pump. Water is added for dilution in the head box. The material is then sent through the silver screen and Johnson screen for removal of dust. The material which comes out of the screen is transferred into the extractor for washing and clickering.

The material coming out of the extractor is stored in the bleaching chest. When a chest becomes full, bleach liquor is added. After the material has been left in the bleaching chest for a fixed time following the bleaching, it is flow-fed into the mixing chest, in which bleach liquor is thoroughly washed away. Washing takes place in the washing drum. After the washing, fluorescent dyestuff or rhodamine is applied for colouring as may be required by consumers. Now that this process has been completed, it is dropped into the stock chest, and sent into the machine chest. Out of the machine chest, the stock is fed into the head box for flow-fed into the mixing tank at a regulated feed level. In the mixing tank, the stock is diluted with white water coming out of the wire part, of which more hereafter, and then sent on to the vertical screen. On the vertical screen, stock with a comparatively larger specific gravity and non-digested stuff are removed. The material which has passed through the vertical screen goes into tank and, via the high pressure pump is fed into the cleaner, with only accepted material fed into the paper making machine.

The accepted material goes into the flow box of the paper machine, out of which a fixed amount flows out into the wire part in consistency of about 0.15–0.2 per cent. The stock which flows out is dehydrated by the table roll and suction box for transformation into a wet web which is sent into the press parts over a blanket. The wet web is dewatered under the pressure of the roll, and pushed against the dryer.

The wet web is then dried by the heat of steam inside the dryer. At the outlet of the dryer, the web is creped, after which it is wound into rolls. The base paper of the toilet paper, which has been ruled is cut into the fixed width by the slitter on the slitter napkins machine and wound into rolls in the required length.

WALL PAPER

Wall papers are essentially the type of resin-treated or coated paper specialities, used chiefly due to their enhanced decorative value. In the western countries the wall paper has been a commodity of common use for long. During the post World War II period the wall paper industry suffered a set-back. At present time, the momentum has been regenerated and the industry emerging today bids fair to make these wall coverings as popular again as it once was. This world-wide growing preference for wall papers is expected to make significant impact on an appreciable section of Indian community.

Manufacturing Process

The manufacture of wall papers include same phenomena such as plastic coating, printing, embossing and trimming. In the industry, by far the largest tonnage of raw-stock used today falls in basis weights ranging from 40–60 lbs/ream (24 × 36–500). The furnish usually varying between 70–75% ground wood with supplementary amounts of semi-bleached kraft-fibres. For greater wet-strength, the bulk of the tonnage consumed in wall-paper manufacture is treated with either melamine or urea-formaldehyde resins.

The principal requirements for a suitable hanging stock is that it should be as uniform as possible, especially in colour, sizing and texture. The finished products be free from shedding. It is one of the factors of paramount importance in the wall covering industry.

WAX COATED PAPER

Wax coated papers are widely used in packaging of various articles. In food packaging industries wax papers are widely used for packaging of food stuffs, because wax coated paper prevents moisture to penetrate in and does not allow the fragrance to escape out. Wax coated paper are used in metal goods industries (auto parts, razor blades etc.). In textile industry too waxed paper is used for wrapping thread, rayon and nylon etc. The waxed paper is also used for sanitary sheets, wraps for surgical instruments etc.

Wax coated paper also finds application in soap industries, cold drink straw industries, and many others such as cake glue, chewing gum, shrubs and rubber heels etc. can be wrapped in waxed papers. There are four general types of waxed paper: (i) dry waxed; (ii) wet waxed; (iii) one side waxed; and (iv) transparent papers.

STRAW BOARD

For the purpose of making straw board, pulping of the raw material is conducted slowly in stationary digester. The object of digestion is to remove dirt, grease and other accompanying impurities and to cause softening of the materials. It is followed by beating which is done slowly and carefully. The purpose of beating is to increase the sheet formation properties without sacrificing strength.

Manufacturing Process

Sheet formation

For the purposes of sheet formation, the pulp is held in vats. Vats may be of stone, copper or wood lined with metal. Extending the full length of the vat, at the bottom is a paddle wheel known as a 'hog', which runs at a low speed to keep the pulp in motion. The sheet is formed by immersing the mould in the vat containing the pulp which is diluted with the required amount of water.

As stated earlier, the mould essentially consists of a metal or wooden skeleton frame to suit the size of the sheet required; on the surface is stretched a fine copper wire gauze cloth which holds the pulp and simultaneously allows the water to drain through in a manner similar to the action of a sieve.

After the mould has been immersed in the vat containing the pulp of desired consistency, it is shaken horizontally to produce a well-felted sheet. A film of pulp is thus set by the draining away of the water through the wire gauze. The deckle is then removed and the mould is inverted on to a woollen felt. After the application of slight pressure, the mould is raised, leaving the film of board on the felt, this is known as the couching of the sheet.

A second woollen felt is 'pitched' on top of the sheet of board which has just been couched as above. The sequence is repeated until a pile of so-called "couch" is ready for the pressing in a hydraulic press where the sheets are squeezed to free the water and compact the sheets themselves.

The couching tray, holding the complete pile or the "couch" is rolled on a track to the hydraulic press, where it is subjected to high pressures which expel a large proportion of the water. The sheets are then subjected to the operations of drying, sizing and finishing.

Drying of the sheet

The pressed sheets of board are dried by handling up batches in drying lofts. The drying loft of a hand-making unit should be arranged so as to get circulation of air from all sides.

The length of time required to dry board depends upon the temperature of the loft, the size and the thickness of the sheets, the amount of parting and width of support on which the paper is hung. Drying results in considerable contraction of the sheet and thereby increasing its strength manifold.

After drying the sheets are still stuck together, and they are separated ('stripped') before the next operation, viz., sizing is undertaken.

Board finishing

The final finishing operation essentially involves plate-glazing of the sized sheets. The sheets of paper placed alternately between burnished copper plates a pile at a time being subjected to pressures between heavy rollers.

The finished sheets are then exercised while cutting, as the hand-made boards normally have a tendency to slip causing irregular slanting trims.

Chapter 12

Sugar and Starch Industry

INTRODUCTION

Sucrose (α-D-glucopyranosyl-β-D-fructofuranoside), $C_{12}H_{22}O_{11}$, is a disaccharide composed of D-glucosyl and D-fructosyl moieties.

Sucrose is extracted commercially from sugarcane, sorghum, and sugar maple. The cultivation of sugarcane is restricted to tropical and semi-tropical regions; the sugar beet is more suited to temperate zones as are sorghum and maple.

Sucrose crystallises as monoclinic, hemimorphic (sphenoidal) crystals. Its melting point has been reported between 160 and 200°C, with 188°C generally accepted. The structure is shown in Fig. 12.1.

Sucrose is soluble in water to the extent of 2.07 g/g H_2O at 25°C, and is readily soluble in aqueous protic solvents such as methanol and ethanol. It is insoluble in ether and chloroform and anhydrous ethanol and glycerol. It is moderately soluble in DMF, pyridine, and dimethyl sulphoxide. Sweetness is affected by total acidity, pH, viscosity, temperature, and other constituents present. A variety of sucrose esters has been prepared. They are used in baking, as emulsifiers and viscosity modifiers, and in drug and cosmetic formulations. Ethers have similar properties and uses. Detergents formulated with sucrose are biodegradable.

Sucrose reacts with alkali- and alkaline-earth metal salts to form complexes. Sucrose complexes with calcium form the basis for separating sucrose from beet molasses (Steffen process). Similar processes based on strontium and barium have been employed.

Under mild acid catalysis, usually dilute HCl, sucrose is easily hydrolysed and inversion occurs (change in the sign of polarisation from + 66.5 to a negative value, owing to the large negative rotation of fructose).

A small amount of the fructose formed is converted to D-fructose dianhydrides. A small degree of inversion is detectable under alkaline conditions.

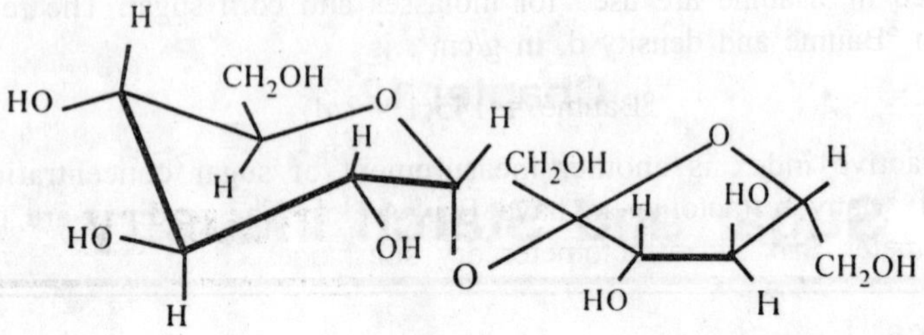

Fig. 12.1. Structure of sucrose.

Sucrose is hydrogenated with Raney nickel to a mixture of sorbitol and mannitol; under more drastic conditions, glycerol and propylene glycol are produced. Mild oxidation gives oxalic and tartaric acids, as well as acidic materials designated as saccharic acids.

Degradation of dry sucrose at temperatures of 90–200°C begins with cleavage of the glycosidic bond followed by condensation and formation of water. At 170–210°C, this process is referred to as caramelisation; the mixture of products formed is commercially called caramel. At high temperatures carbon-carbon cleavage occurs. Degradation reactions are sensitive to impurities. The thermal degradation solution produces 5-hydroxymethylfurfural and coloured condensation products.

SUGAR ANALYSIS

Physical Methods

The concentration of a pure sugar solution is determined by measurement of polarisation (optical rotation), refractive index, and density. The saccharimeter is a polarimeter reading directly in per cent sucrose. However, polarimeter scales are not inherently stable and must be calibrated frequently using quartz plates. Direct polarisation can give the value for sucrose only when no other optically active substances are present. When interfering substances are present, the correct sucrose value can be obtained by the Clerget method, which is a double polarisation, before and after inversion of the sucrose.

Density, measured by the standard hydrometer or spindle, is widely used to determine the sugar concentration of syrups, liquors, and juices.

Hydrometers graduated in sucrose concentration are called Brix hydrometers or Brix spindles; the readings are called spindle Brix. The degrees Brix equals the wt% of sucrose in the solution. Hydrometers graduated in °Baumé are used for molasses and corn sugar. The relation between °Baumé and density d, in g/cm^3, is

$$°\text{Baumé} = 145(1 - 1/d)$$

Refractive index is another measurement of sugar concentration in solution. Many refractometers have Brix scales. The readings are called refractometer Brix or refractometer dry substances (RDS).

Chemical Methods

Reducing action is the basis for methods for the determination of aldoses and ketoses. The alkaline copper tartrate reagent, Fehling's solution, is used in a number of methods. All have the problem that the reaction is not stoichiometric, and the procedure must be followed exactly. In the Lane and Eynon method, the boiling reagent is titrated with sugar solution and the point must be reached in exactly 3 min. The Munson and Walker method is convenient for an occasional analysis. The reduced cuprous oxide is determined gravimetrically.

The Ofner method is used for up to 10% invert sugar in sucrose. The reduced cuprous oxide is treated with excess standardised iodine, which is then back-titrated. Small amounts of reducing sugars in sucrose are determined by the Knight and Allen method. After the sugar reacts with the copper reagent, the excess unreduced copper is determined with EDTA. Very small amounts of reducing sugars are determined by the Emmerich method. It is based on the formation of a coloured complex of 3,6-dinitrophthalic acid and invert sugar.

Enzymatic methods can also be used. Glucose oxidase, for example, is specific for the oxidation of glucose. Sucrose content is also obtained by measuring reducing sugars before and after inversion.

Micromethods

These methods include reduction of ferricyanide to ferrocyanide; reduction of cupric sulphate to cuprous oxide; and the development of colour with phenol in strong sulphuric acid.

Chromatographic Methods

Paper chromatography gives good separation and, with highly specific sprays, the individual spots can be identified. Electrophoresis uses a high voltage gradient. The hydroxyl groups on sugars readily form complexes with borate and other ions. A charge is thus imparted and electrophoresis can be applied.

For gas chromatography, the sample is vapourised. Although sugars are not very volatile, their methyl esters and silyl derivatives are. Liquid chromatography offers the advantage that no derivatives are needed and the solvent can be water.

Purity

Purity denotes sucrose content as a percentage of total solids. It is calculated as pol/Brix.

Ash

Ash is determined by incineration. The residue is called carbonate ash. Bubbling and swelling of the incinerating mass can be avoided by adding concentrated sulphuric acid. The ash components are converted to sulphates and the ash is called sulphate ash. Electrical conductivity gives a conductivity ash that is simple and quick.

Colour

The standard conditions for colorimetry are 50 Brix concentration and pH 7.0. The light transmission T is measured at 420 nm. Sugar colour is expressed as:

$$a^* = -\log T/bc$$

where b is the cell depth in cm; c is the sugar concentration in g/mL; and a^* is the attenuation index. The number given by the above equation is multiplied by 1000 to give sugar-colour units. Raw sugars have a colour of 1000 or more, refined sugar, 35.

CANE SUGAR

Cane sugar is the name given to sucrose, a disaccharide produced from the sugarcane plant. The other principal source of sucrose is the sugar beet. Refined sugars from the two sources are practically indistinguishable. However, the trace constituents are different, notably raffinose and odour constituents. In addition, since the two plants have different photosynthetic pathways, the carbon isotope ratio in the sugars can be used to identify the source. The other important sweetener is corn sugar (glucose), derived from corn starch. The process flow diagram for manufacture of sugar cane is shown in Fig. 12.2.

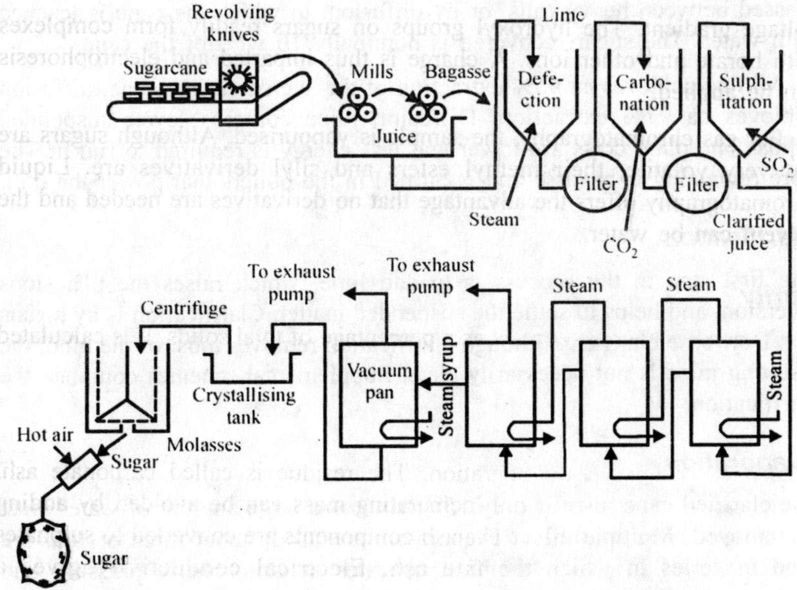

Fig. 12.2. Flow chart for the manufacture of cane sugar.

Cultivation

Sugar cane, a tropical grass, grows very tall, 3–5 or even 7 m, and has a thick stalk (2–3 cm) in which the sugar is stored in solution (in the juice). There are many varieties and all are hybrids. A continuing breeding programme is essential because all varieties suffer a gradual decline.

Cane is grown by planting stalks called seed cane or setts. The growing time is about one year, although in some areas it grows two or more years. Upon harvesting, the stalks are cut off even with the ground and the next day more sprouts start from the buds just below ground level. Good agricultural practice requires close attention to soils, fertilisers, weeds, pests, disease, and water supply. The sugarcane plant stores sugar at all times, and the harvesting season is therefore long. Sugarcane produces the heaviest yield of all crops in weight of both biomass and useful product per unit area of land. Table 12.1. gives sugarcane yields from various regions.

Raw Sugar Manufacture

Sugar cane cannot be stored more than a few hours after it is cut, so the raw sugar mills are located in the cane fields. The cane is washed and broken into pieces. The juice is extracted either by milling, in which the cane is pressed between heavy rolls, or by diffusion, in which the cane is leached with water. The sugar recovered is nominally 10 wt% of the cane.

Milling achieves ca 95% extraction of the sucrose in the cane, diffusion achieves ca 97% extraction. Diffusion juice contains fewer suspended solids. The plant costs are lower and less energy is required to run it. The fibre remaining is bagasse and is burned in the boilers that power the mill.

Clarification

The first step in the process is to add lime, which raises the pH, stops inversion, and helps to settle the suspended matter. Clarification is by means of a Dorr-type clarifier. Although clarification removes most of the mud, the resulting juice is not necessarily clear. Supplemental schemes complete the clarification.

Evaporation

The clarified cane juice is only 15° Brix and a large amount of water must be removed. Multiple effect evaporators are used; about three or four are used in series in which the first operates at atmospheric pressure and subsequent effects are at higher vacuum. Evaporation is carried to a final Brix of ca 65–68. The juice, after evaporation, is called syrup and is almost black and a little turbid.

Crystallisation

Crystallisation from the concentrated syrup is traditionally a batch process. After seeding, the evaporation and feeding of the syrup are balanced for the fastest possible rate of crystal growth.

Vacuum pans

Crystallisation is done under a vacuum such that the water boils at about 65°C in order to minimise thermal destruction of sucrose. The vacuum pan has a very large discharge opening, typically 1 m dia. At the end of a strike, the massecuite contains more crystals than syrup and is very viscous. The large opening is required in order to empty the pan in a reasonable time.

Centrifuging

The massecuites from the vacuum pans enter a holding tank called a mixer that has a slowly turning paddle to prevent the crystals from settling. The centrifuge is fed from the mixer. In batch-type centrifuges, the mother liquor is separated from the crystals in batches of ca 1 t. A centrifugal force of about one thousand *G* is applied. The starting and stopping of the centrifugal batch consumes a great amount of energy, much of which can be saved by the use of a continuous centrifuge consisting of a conical basket with a screen.

Boiling systems

In raw-sugar manufacture, the first strike is called the A strike, and the mother liquor obtained from the centrifuges is called A molasses. The so-called B sugar from the second strike is only half as pure as that obtained from the first strike. Usually, only three strikes are obtained from cane juice. Final strikes are not sent to the centrifuges directly but to crystallisers, which are holding tanks where the crystals grow, over a period of four d, and as much sucrose as possible is extracted from the final molasses.

Raw sugar

The A and B sugar from the centrifuges comprise the raw sugar of world commerce. Raw sugar can be stored for several years. If raw sugar is washed slightly in the centrifuge, and then dried and cooled, the resulting sugar has the very high pol (% sucrose) of 99.4%. This material is free flowing and keeps well.

Direct-consumption sugar

Many raw-sugar mills produce sugar for direct consumption (off-white) by performing a supplemental clarification on the cane juice, or by filtering it, followed by decolourisation. From these improved syrups one strike of sugar for direct consumption can be obtained; later strikes are treated as raw sugar.

Refining

Sugar refineries are located in large cities, near harbours. The first step in refining is to remove the molasses film from the raw sugar crystals by a washing process known as affination. The resulting syrup is called affination syrup. It is estimated that 90% of refining takes place in this first step.

In the recovery house, sugar is recovered from the affination syrup. The recovered sugar is called remelt and is sent back to processing. The residual syrup is refiners' molasses. The washed sugar is melted in hot water and the pH is adjusted with lime. The melted liquor is strained through a screen and then clarified.

The object of clarification is the complete removal of all particular matter by filtration, carbonation, or phosphatation. The most straightforward process is filtration, which requires a filter aid, such as diatomaceous earth or kieselguhr. With a tight, efficient filter aid, flow is very slow. Filtration, as a sole means of clarification, is being replaced by other methods.

To make liquor more suitable for clarification, substances that form precipitates and coagulate the impurities are added. Both processes used today, carbonation and phosphatation, employ lime. In the course of the precipitation of the calcium carbonate, insoluble lime salts are coprecipitated. The final step is filtration. In phosphate clarification, lime and phosphoric acid are added simultaneously. The calcium phosphate precipitate forms a floc. At the same time, air is injected. The precipitate floats to the surface as a scum and is scraped off without filtration. The clarified liquor from either process is brilliantly clear but dark.

The key process in sugar refining is decolourisation. Colour is the property that distinguishes refined sugar from raw sugar. The colour is measured by light transmission at 420 nm. Traditionally, carbon adsorbents have been used for decolourisation. However, they adsorb everything, including sugar, with little selectivity. Bone char and granular carbon behave similarly. Bone char is made by charring cattle bones. Granular carbon is made from selected mixtures of coal steam-activated by heating to ca 1000°C. Process variations include sending liquors of increasing colour to the adsorbent. Ion exchange is also used for decolourisation. Its advantages are *in situ* generation without heat, short contact time, and small equipment.

The Tafloc process takes advantage of the anionic properties of the sugar colourant as in ion exchange. The colourants are precipitated with a quaternary ammonium compound, and the precipitate is removed with a phosphate clarifier.

Crystallisation

The colour of the washed, clarified, and decolourised liquor proceeding to crystallisation ranges from water white to slightly yellow. The same vacuum pans are used as described above. They must be operated carefully to produce crystals of the desired size.

Packaging, storing and shipping

The present trend is away from consumer-sized (< 50 kg) packages and toward bulk shipment.

Products

Sugar marketing is primarily local. Each refinery serves its own area. Refined granulated sugar is the usual output of a refinery. By far, the largest production is in this class. Large-grain speciality sugars are used for candy and cookies. Fine-grain sugar consists of small crystals obtained by screening; it dissolves quickly. Powdered sugar is made by grinding granulated sugar and adding 3% cornstarch to prevent caking. Cubes are made by mixing a syrup with granulated sugar to form cubes. Reagent-grade sugar is sold by chemical supply houses but its quality is not higher than that of regular sugar. Brown sugars, known in the trade as soft sugars, can be made either by coating white sugar with syrup or by a special boiling technique that produces conglomerates. Liquid sugar is a solution of sucrose and invert sugar. Syrups have a high percentage of invert. High test molasses is partially inverted cane syrup. Molasses is the final syrup left after all the sugar that can be economically removed has been crystallised.

BEET SUGAR

Sugarbeets are often profitable in areas where few other crops, except potatoes, can be grown successfully. Sugar beets are an excellent crop to fit into a crop-rotation scheme, as their culture tends to improve the soil. Beets are deep-rooted and bring up plant food from a considerable depth, which is then made available to shallow-rooted crops. It is possible to plant them in populated areas.

Table 12.1. Composition of sugar beets.

Component	*Per cent*
Juice	ca 92
Insoluble matter[a]	ca 5
Water, chemically bound	ca 3
Soluble solids	11 – 25
sucrose in solids[b]	87.5
sucrose in beet	10 – 22
Non-sucrose substances soluble in juice	
organic nitrogen compounds	ca 44
nitrogen-free organic compounds	36
Inorganic compounds	20

[a]So-called marc.

[b]Expressed as purity of the juice.

Sugar Beet Composition

The composition of sugar beets is given in Table 12.1. Of the nitrogenous compounds, ca 58% is protein. Of the other nitrogen compounds, betaine constitutes 40%, which is carried through to the molasses. The remainder consists of amino acids, purines, amides, and ammonium salts. The principal acid is citric acid which, together with oxalates, is eliminated as insoluble calcium salts in the purification stage.

Manufacture

The various manufacturing processes are discussed as under.

Diffusion process

The sugar in the sugar beet is contained within the parenchyma cells. On heating to 70–80°C, diffusion of dissolved molecules takes place. The slow extraction of the small sodium and potassium ions is due to the electrostatic retarding effect of the high molecular weight polymeric anions.

The sugar beets are moved to the processing area in water flumes equipped with rock and stone removers, vegetation and trash catchers, and beet washers. At the factory, the beets are sliced into cossettes (long thin strips of V-shaped or square cross-section). Typical cossettes may be 2–3 mm thick and up to 15 cm long. The sucrose is extracted from the cossettes by passing them through a diffuser, counter-current to hot leaching water. The juice leaving the diffuser is screened and may contain ca 14.5 wt% total solids, which, with a sucrose content of 12 wt%, has a purity of ca 83 wt%.

The slope diffuser is a covered, sloping trough, ca 4–7.3 m wide and 15.8–19 m long, depending on the capacity. Today, two other types of diffusers are preferred. The RT-4 consists of a horizontal rotating cylinder built on the screw principle. A tower diffuser consists of a vertical tank with a rotating vertical central shaft equipped with conveying arms that move the cossettes upward against a downward flow of juice.

The dried pulp leaving the diffuser, having a water content of ca 95%, is pressed in screw-type presses to 74–80% moisture. The pressed pulp is enriched by the addition of molasses or concentrated Steffen filtrate, dried, cooled, and stored. Molasses-dried beet pulp is excellent cattle feed.

Purification with lime and carbon dioxide

Treatment with lime gives precipitates and soluble products; 60–70% of the non-sucrose products removed by precipitation are acids that form insoluble lime salts, e.g., oxalates, hydroxycitrates, citrates, tartrates, phosphates, and some sulphates. The other non-sucrose products are removed through coagulation.

Passing carbon dioxide through the limed mixture precipitates calcium carbonate on which non-sucrose materials are absorbed. The calcium carbonate crystals and the adsorbed impurities are removed by sedimentation and filtration.

Many modifications of liming and carbonation procedures have been devised, such as pre-defecation or pre-liming and main liming in advance of carbonation. In pre-liming, about one tenth of the total lime to be used is gradually added to the diffusion juice; many nitrogenous substances are precipitated at their isoelectric points and are not redispersed during main liming. In the Dorr-system purification, the hot juice is simultaneously limed and carbonated with the aid of recirculation, and high alkalinities are avoided.

Alternatives or supplements are treatments with activated carbon or ion-exchange resins. Deliming resins reduce the lime salts content of the thin juice.

A small quantity of sulphur dioxide is added to the thin juice to inhibit the Maillard or Browning reaction between reducing sugars and amino acids.

Evaporation and fuel economy

For fuel economy, all high pressure steam from the boilers is expanded to exhaust-steam pressure through turbines; all exhaust steam is used in the first effect of the evaporator; all possible evaporation takes place in the evaporator to reduce the load on the vacuum pans; and for heating of vacuum pans, heaters, or melters, vapours are employed from one of the evaporator effects at the lowest possible pressure. The beet-end juice is heated gradually, using portions of the later-effect vapours to start, and finishing with first-effect vapours.

Crystallisation

Sufficient feed liquor in the vacuum pan permits its concentration to supersaturation. Crystallisation may be induced by adding a small amount of fine crystals of sugar. The fully grown sugar crystals are spun free of syrup in centrifugal separators and washed briefly with hot water. The spun-off syrup is known as green syrup (although it is brown). It provides the feed syrup for the vacuum pan for the second or intermediate boiling. The spun syrup from the second boiling serves as feed syrup for the third (raw) boiling. The syrup spun from the third boiling is termed molasses. It may be used as cattle food, or for production of chemicals by fermentation, or may be treated for removal of more sucrose, as in the Steffen process. The sugar crystals produced in the second and third boilings are normally dissolved in the feed syrup for the first boiling.

Steffen process for desugaring molasses

In this process about 95% of the sucrose is recovered. Molasses, diluted with water to ca 6% sucrose, is treated with lime and the calcium saccharate is filtered off. The filter cake is slurried in thin juice, and the so-called saccharate milk is carbonated, yielding free sucrose and calcium hydroxide. Galactosidase may be used to hydrolyse the raffinose.

Granulated sugar is dried, screened, cooled, and bagged or stored in bulk. It is also used for the manufacture of liquid and powdered sugars.

Material Balance and Quality Control

Sucrose is relatively fragile in solution. Sucrose in sliced beets is determined first by weighing the beet cossettes, and then by analysing samples of cossettes for sucrose. The grower is paid on this basis. Sucrose in molasses is determined by analysing a sample.

Yields and losses statements are made for intervals of one day to one month. Losses of sucrose in the beets between the time of purchase and the time of slicing, so-called outside-factory losses, are partly due to metabolic changes. Total inside-factory losses are found by subtracting the weight of granulated sugar produced from the amount of sucrose in the beets sliced. Extraction denotes the percentage of granulated sucrose recovered of the total sucrose in the beets.

Analyses are based on the use of the refractometer or Brix hydrometer for measuring the total solids in solution; on the polariscope for the determination of sucrose; and on copper-reduction methods for invert sugar; pH is determined throughout.

SUGAR DERIVATIVES

Sucrose

The focus of interest on sucrose (1) as a chemical raw material has given rise to the name sucrochemistry for the field of carbohydrate chemistry concerned specifically with sucrose and its derivatives. Sucrose can be used as a starting material in degradative reactions, or in syntheses, in which case the carbon framework is retained and the chemistry of sucrose is that of a polyhydric alcohol.

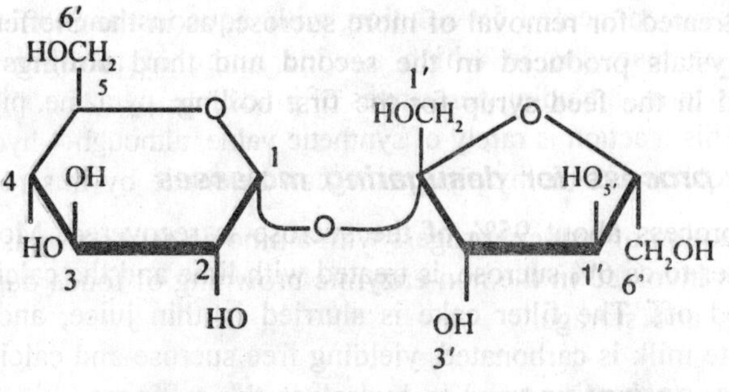

Fig. 12.3. Sucrose

Degradative reactions

Above its melting point (180°C), sucrose decomposes rapidly with the formation of a complex mixture of volatile compounds and an involatile residue. known as caramel, widely used as a colouring agent in foods and drinks.

In acidic aqueous solution, sucrose is hydrolysed to an equimolar mixture of glucose (2) and fructose (4) which, in alkaline solution, are interconvertible through their common intermediate enediol form (3) (Fig. 12.3). Under acidic conditions at moderate temperatures, hexoses undergo progressive loss of three molecules of water with the formation of 5-hydroxymethylfurfural. At high temperatures, in the presence of strong acids, sucrose gives a mixture of levulinic and formic acids. Under alkaline conditions, sucrose is relatively stable, but glucose and fructose decompose rapidly, forming highly coloured polymers and a complex mixture of decomposition products in which saccharinic .acids or lactic acid predominate according to the conditions

Sucrose heated in solution with an excess of lime gives predominantly lactic acid, which can be isolated in high yield. However, this is not an economically favourable route to lactic acid compared with fermentation or synthesis from acrylonitrile.

Sucrose is not readily oxidised by gaseous oxygen, unless it is first hydrolysed, as in strongly alkaline solution. Glucose is oxidised to arabonic acid when the alkaline solution is shaken with oxygen under pressure at ambient temperature, although in the presence of a palladium catalyst gluconic acid is obtained.

Sucrose is not reducible, but gives a mixture of sorbitol and mannitol upon hydrolysis and electrolytic reduction or hydrogenation in the presence of a nickel or palladium catalyst.

Sugars, when heated in solution with aqueous ammonia, undergo a sequence of reactions in which a wide range of nitrogen heterocyclic compounds form, including derivatives of imidazole, pyrazine, piperazine, and pyridine. This reaction is rarely of synthetic value, although 4-hydroxymethyl-1*H*-imidazole and 2-methylpiperazine can be made by this route.

The reaction of reducing sugars with amino acids, known as the Maillard reaction, is involved in the non-enzymic browning of foods during cooking, processing, and storage.

Synthetic derivatives

Sucrose undergoes reactions characteristic of alcohols, giving, for example, esters, ethers, acetals, and urethanes (Fig. 12.4).

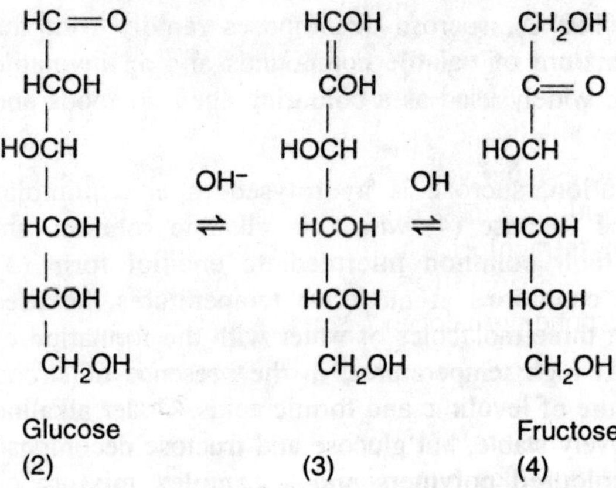

Fig. 12.4. Interconversion of glucose and fructose.

Sucrose is readily esterified to the octaester in high yield by reaction with excess of the anhydride or chloride of a sterically unhindered organic acid in the presence of pyridine or other suitable base.

An important alternative route to sucrose esters is by transesterification between sucrose and an alkyl ester or triglyceride, whereby an acyl group is transferred to sucrose in the presence of a basic catalyst, e.g., potassium carbonate, at temperatures > 100°C. Sucrose reacts with a reactive alkyl halide in the presence of a base to form an ether, e.g., the octabenzyl, trimethylsilyl and methyl ethers. Ethers o* ˙ucrose are also formed by addition to a reactive ethylenic bond, sucn as that of acrylonitrile or dihydropyran.

Reaction of sucrose with ethylene oxide gives the l-hydroxyethyl ethers. Sucrose under appropriate conditions will form cyclic acetals despite unfavourable steric factors. For example, benzylidene bromide in pyridine reacts with sucrose to give 4,6-*O*-benzylidenesucrose, which is isolated as the hexaacetate in 35% yield. Sucrose reacts readily with an isocyanate to give the corresponding urethane; a diisocyanate, e.g., toluene diisocyanate, giving a polyurethane. Such products are brittle, and ethoxylated or propoxylated sucrose are the preferred cross-linking polyols in the manufacture of polyurethane foams.

Sucrose can form internal ether bridges, creating multiple ring systems. The epoxide (oxirane), unlike the stable ether rings of sucrose anhydrides, is readily opened by nucleophilic displacement. Sucrose anhydrides are formed on treatment of a primary alkyl- or aryl-sulphonyl ester of sucrose with sodium methoxide in methanol, ring formation taking place with the appropriate secondary hydroxyl group at the 3, 3' or 4' position.

The alkyl- and arylsulphonyl esters of sucrose undergo nucleophilic displacement of the sulphonyl groups with the formation of the corresponding substituted deoxysucrose derivative, on treatment with the appropriate anion in a suitable solvent. This reaction provides a valuable synthetic route to deoxysucrose derivatives, e.g., azidosucrose. Sucrose can be chlorinated directly with an excess of sulphuryl chloride in pyridine, or preferably with carbon tetrachloride-triphenylphosphine reagent in pyridine. Deoxysucroses are usually obtained by the reductive dehalogenation of the corresponding halogen derivatives.

Enzymic conversion

Under the action of certain microbial enzymes, sucrose undergoes transglycosylation, a process by which a tri- or higher saccharide is formed by the transfer of a sugar residue. Simple enzymic rearrangement of the hexose linkages of sucrose may also occur as, for sample, in the formation of the isomeric reducing disaccharide palatinose (6-O-(α-D-glucopyranosyl)-D-fructose) by *Erwinia rhaptonici*.

Palatinose has only one-third the sweetness of sucrose, which it closely resembles in physical properties, and would be suitable for use in the manufacture of food products for human or animal consumption where the sweetness of sucrose is undesirable.

Sucrose monoesters of long-chain fatty acids have considerable potential as non-ionic surfactants in detergent formulations. They are particularly suited for use in cosmetics and shampoos, and also in food applications as emulsifying and dispersing agents. Trichlorogalactosucrose, which is 650 times sweeter than sucrose itself, is being developed as a non-caloric, high intensity sweetener.

A potentially large market for sucrose and its derivatives is in synthetic resins and plastics, e.g., polyurethane foams. Flame-retardant properties can be introduced into the resin by copolymerising with halogen or phosphorus derivatives of sucrose.

Fructose

D-Fructose (fruit sugar) is a monosaccharide constituting one half of the sucrose molecule. The sweetness of fructose is 1.3–1.8 times that of sucrose, which makes it an attractive alternative for sucrose. When stored in solution at high temperatures, it browns rapidly and polymerises to dianhydrides.

Because of its sweetness, fructose has been the object of commercial production for decades. Recent technologies involve ion-exchange separation from glucose in a mixture obtained by the isomerisation of glucose. In food formulations, fructose is used for special dietary purposes, e.g., to lower the calorie (kcal or 4184 J) intake. Fructose-based foods, then, may be helpful in the diets of diabetics and to control blood-sugar levels.

Maltose

Maltose (malt sugar) occurs occasionally in plants and fruits. It is a structural component of starch. Purification can be achieved by way of the β-maltose octaacetate. Commercial maltose typically contains 5–6 wt% of the trisaccharide maltotriose with traces of glucose. Maltose is about half as sweet as sucrose and has been used in intravenous feeding (Fig. 12.5). The maltose in malt syrups is important in brewing. High maltose syrups from starch typically contain ca 8–9 wt% glucose and 40–52 wt% maltose; the remainder is higher saccharides. Such syrups are used in the preparation of confections, preserves, and other foodstuffs. Hydrogenation gives maltitol.

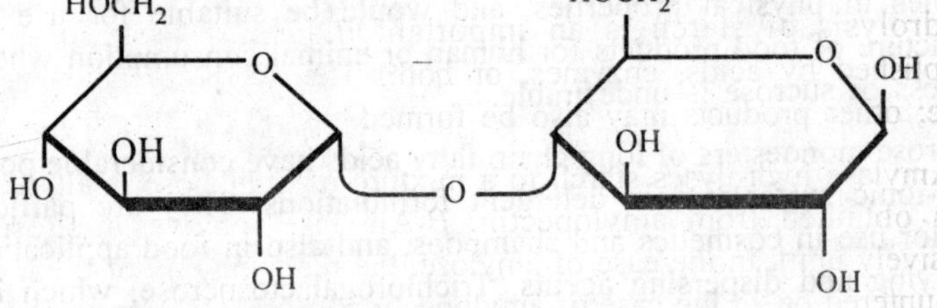

Fig. 12.5. 4-O-α-D-glucopyranosyl-D-glucose maltose.

Lactose

Lactose (milk sugar) is the only sugar available commercially that is derived from animal rather than plant sources. It is prepared commercially from whey and is available in a fermentation grade (98 wt% pure). Lactose, a reducing sugar, reacts with amines and undergoes browning. It has been used as a nutritional sugar, flavour enhancer, texture controller and colour retainer, and as a carrier for synthetic sweeteners.

STARCH

Starch, $(C_6H_{10}O_5)_n$, is a mixture of linear (amylose) and branched (amylopectin) polymers of α-D-glucopyranosyl units. It is the principal reserve polysaccharide in plants, and constitutes a substantial portion of the human diet.

Properties

Starch occurs in plants in the form of granules. Rice starch has the smallest granules and potato starch the largest. Undamaged starch granules are insoluble in cold water but imbibe water reversibly and swell slightly. In hot water, irreversible swelling occurs, producing gelatinisation. The property of forming thick pastes or gels is the basis of many starch uses.

Most common starches contain two types of D-glucopyranose polymers. Amylose is essentially a linear polymer of α-D-glucopyranosyl units linked (1 → 4). Amylopectin is a highly branched polymer of α-D-glucopyranosyl units containing 1 → 4 links with 1 → 6 links at branch points (Fig. 12.6).

Starch is heterogeneous not only in relation to polymer structure but also in relation to molecular weight. For amylose, osmotic pressure measurements give a range of 10,000–60,000. With an anaerobic isolation technique to prevent oxidative degradation, a range of $(1.6–7.0) \times 10^5$ was obtained. Amylopectin is a much larger molecule, ranging in mol. wt. between 50,000 and 10^6.

Hydrolysis of starch is an important industrial reaction which is accomplished by acids, enzymes, or both. The action of acids produces glucose; other products may also be formed.

α-Amylase hydrolyses starch to a mixture of D-glucose, maltose, and a dextrin obtained from amylopectin. β-Amylase removes maltose units successively until, in the case of amylose, the reducing end of the molecule is encountered or, in the case of amylopectin, an α-(1 → 6) branch point is met. Glucoamylase hydrolyses both amylose and amylopectin, yielding D-glucose.

Oxidation of the hydroxyl groups of starch, e.g., with hypochlorite, gives aldehydes, ketones, or carboxylic acids. Starch gives a characteristic blue complex with iodine which is used as a test for starch in various systems.

Manufacture

Milling of corn, *Zea mays*, provides corn starch, which is extensively used in food and non-food applications. Steeped corn is coarsely ground in an attrition mill to break loose the germ. The mill gap during this step must be adjusted to maximise the amount of germ freed, but minimise rupture of the germ. Germ is removed from the aqueous slurry in a cyclone separator. The germ fraction is processed for corn oil. The cyclone underflow is milled a second time for complete release of the starch granules. Following the second milling, the kernel suspension contains starch, gluten, and fibre. The collected fibre is combined with gluten for feed use.

The starch-gluten suspension, commonly known as mill starch, is concentrated by centrifugation. The concentrated starch is passed through hydroclones to remove the last traces of protein. Starch suspension may be processed dry and marketed as unmodified corn starch, modified by chemical or physical means, gelatinised and dried, or hydrolysed to corn syrup.

Chemical modification

Acid-treated starches exhibit decreased intrinsic viscosity, decreased gel strength, and increased gelatinisation temperature. In industrial production, a 40% slurry of corn or waxy maize starch is acidified with hydrochloric or sulphuric acid at 25–55°C. When the desired degree of thinning is attained, the mixture is neutralised.

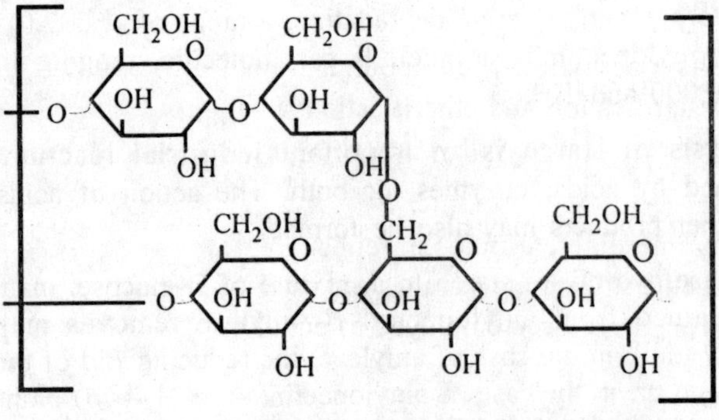

Fig. 12.6. Branch point structure in amylopectin.

In the manufacture of oxidised starch, a slurry of starch granules is treated with alkaline hypochlorite, neutralised, washed, and dried. Oxidation results in lower pasting temperatures, decreased thickening power, and lower paste setback.

In the manufacture of pyrodextrins, dried starch is sprayed with an acid and dried to 1–5% water content. The acidified starch is hydrolysed and reverted by heating. Chemical changes produced by acid-heat treatment lead to starches with lower water content, decreased solution viscosity, and greater solubility in hot water.

Uses

Unmodified corn starch is used in non-food applications in the mining, adhesives, and paper industries. Acid-modified or thin-boiling starches are employed mainly in the textile industry as warp sizes and fabric finishes. Oxidised starches are used in paper coatings and adhesives. Starch pyrodextrins and British gums have extensive use as adhesives for envelopes, postage stamps, and other products.

Ethanol, isopropyl alcohol, *n*-butanol, acetone, 2,3-butylene glycol, glycerol, and fumaric acid are produced from starch by fermentation.

Food uses

Unmodified starch is used in food preparations requiring thickening, gelling, or similar properties. Pre-gelatinised starch is used where thickening is required but cooking is to be avoided. Acid-modified starches are used for gum candies.

Derivatives

Starches, as polyhydroxy compounds, undergo reactions of alcohols, including esterification and etherification. Such modifications may produce useful polymers. Hydroxyethyl starches are used as paper coating sand sizes.

The commercial cationic starches are the tertiary and quaternary aminoalkyl ethers. Cationic starches are used on paper for fibre and pigment retention. They have been employed as emulsifiers for water-repellent paper sizes.

Starch monophosphates are useful in food applications because of their freeze-thaw stability. Degree of substitution is generally low (DS < 0.15).

Starch phosphate diesters contain ester cross-links and are used as thickeners and stabilisers in baby foods, salad dressings, fruit pie filling, and cream-style corn.

Starch acetates are used in food because of their clarity and stability. Applications include frozen fruit pies and gravies, baked goods, instant puddings, and pie fillings.

Chapter 13

Fermentation and Distillery

INTRODUCTION

Although the word fermentation referred originally to the anaerobic metabolism of organic compounds by micro-organisms or their enzymes to produce products simpler than the starting material, the modern definition is that of any microbial action controlled by man to make useful products. Some of the substances produced from carbohydrates on a commercial scale by anaerobic microbial metabolism include ethanol and lactic acid and products containing these materials such as beer, wine, vinegar, pickles, and sauerkraut (Food processing). Products from aerobic metabolism include antibiotics, organic acids, enzymes and vitamins.

Moulds, yeasts, bacteria, and streptomycetes are among the micro-organisms used in fermentation processes today. Most of the successful fermentation operations have some degree of simplicity, a rather uniform series of techniques, and the possibility of using the same manufacturing facilities to make many chemically unrelated substances.

Microbiological Aspects

The most important consideration in industrial fermentations is the selection of the proper micro-organism. This choice must provide for production of product at the highest possible concentration and yield. Stock cultures of micro-organisms useful in industrial fermentations are usually maintained by the manufacturer, and less frequently, in commercial collections or collections at academic or government laboratories.

For antibiotic-producing fermentations, the careful selection of a micro-organism to produce high titers of drugs appears to be the most important method for increasing productivity of the process. Other important factors are formulation of the medium used for growth of the micro-organism, and changes in bio-engineering technology. New developments in recombinant DNA techniques and microbial genetics may lead to technology that can be generally applied to all micro-organisms of interest for industrial processes.

Fermentation Operations

The micro-organisms used in industrial fermentations require various nutrients for growth, including carbohydrates, nitrogen-containing compounds (or ammonium salts), growth factors, vitamins, and minerals. In most of the fermentations, these nutritional requirements are met by including in the medium peptones, yeast products, agricultural materials, e.g., cornsteep liquor (steepwater), soybean meal, cottonseed flour, and buffer salts.

Specific precursors are sometimes added to fermentations to increase the rate of formation and amount of product formed. These include substances such as cobalt salts for the vitamin B_{12} fermentation, and phenylacetic acid and phenoxyacetic acid for fermentations producing penicillin G (benzylpenicillin) and penicillin V (phenoxymethyl penicillin) respectively.

In most fermentation processes, the nutrient medium is first sterilised to inactivate or kill extraneous micro-organisms which may be present. This sterilisation is usually accomplished by heating at high temperatures (e.g., 150°C) for a few minutes or at lower temperatures (e.g., 126°C) for longer periods. Any material that enters the fermenter during the process must be sterilised. Air may be sterilised by filtration through hydrophobic sterilisable filters, and nutrients are often sterilised by continuous or batch sterilisation at 126°C.

It has been found that the desired metabolic changes taking place in the fermentation frequently have a rather narrow temperature optimum. Product yields can often be increased through control of the pH of the growing culture or the level of dissolved oxygen and this can be programmed and controlled by using computerised process control to optimise product yield.

Vessels for fermentation operations range in size from 380-L (100-gal) for production of specialised products, including enzymes and vaccines, to 900 m^3 (237,000 gal). A flow sheet of a typical fermentation process for production of penicillin G is shown in Fig. 13.1.

The production of antibiotics and the use of micro-organisms as replacements for chemical operations in complicated chemical syntheses are becoming increasingly common. Uses of micro-organisms are being considered for preparation of peptide hormones, e.g., somatostatin, insulin, and growth hormone, substances that are quite foreign to the microbiologists's domain. This potential application of genetics to fermentations, is so large that there may be a complete revolution in the fermentation industries in the coming years in that previously scarce human proteins will be produced at a low cost by large-scale fermentation procedures.

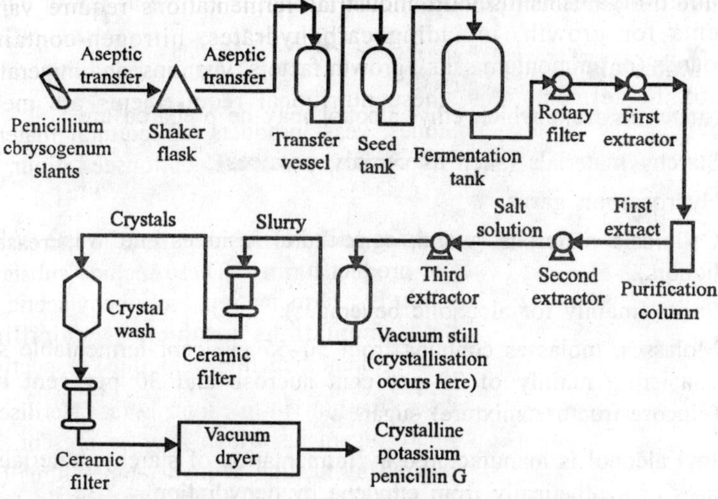

Fig. 13.1. Flow sheet for penicillin G process.

INDUSTRIAL ALCOHOL

Ethyl alcohol, C_2H_5OH (simply called alcohol) has been known as a beverage since antiquity. Industrial alcohol and power alcohol are outgrowths of alcoholic beverages and have now become extremely important by virtue of their economically useful properties as solvents and for the synthesis of other chemicals. Alcohol, for medicinal, pharmaceutical, flavouring and beverage purposes is duty or tax paid, while denatured alcohol is duty free. There are two classes of denatured alcohol, completely denatured and specially denatured alcohol. Denaturing agents are chemicals approved by authorities and the proportions to be mixed with ethyl alcohol are also fixed.

Uses of Alcohols

Duty paid pure ethyl alcohol is used only for medicinal, pharmaceutical, flavouring and beverage purposes. Duty free alcohol is low priced and hence its consumption has increased enormously and the field of application has also widened. Alcohol is second only to water in solvent value and is employed nearly in all industries. It is also the raw material for making numerous chemicals, such as acetaldehyde, ethyl acetate, acetic acid, ethylene dibromide, glycols, ethyl chloride, etc. It may also be used as a raw material for butadiene and styrene. In countries with poor petroleum supply, it is economical to mix absolute alcohol with the gasoline in order to conserve the petroleum products. Such alcohol is known as power alcohol.

Raw Materials

Raw materials from which ethyl alcohol may be prepared are :

(1) Starchy materials (such as cereals, potatoes).

(2) Hydrocarbon gases.

(3) Cellulosic materials (wood, agricultural residues and waste sulphite liquors).

(4) Fruits (mainly for alcoholic beverages).

(5) Molasses, molasses contains from 50–55 per cent fermentable sugar, consisting mainly of 70 per cent sucrose and 30 per cent invert (glucose-fructose mixture) sugar.

Ethyl alcohol is manufactured by fermentation of starchy materials and molasses or synthetically from ethylene by dehydration.

Manufacture of Alcohol by Fermentation

Ethyl alcohol is made from :

(1) Starchy materials such as corn, wheat, rice, rye, barley potatoes, etc.

(2) Molasses of cane and beet sugar factories.

(3) Waste sulphite liquors from wood pulp industry (Fig. 13.2).

Alcohol from starchy materials

The starchy material is ground, cooked in water at 180°C (the process is known as mashing) and then converted into fermentable sugars with dilute HCl or H_2SO_4 (also by using 'diastase' or germinating barley) at 60°C, starch thus forms maltose and in about 30 minutes, some dextrin $(C_6H_{10}O_5)_n$ is also formed. Certain moulds such as some species of 'Aspergillus' and Mucor which contain diastase are also employed for hydrolysis. The sugar solution is cooked to about 150°C and fermented with yeast. Maltase, present in the yeast, first hydrolyses maltose to glucose which is then fermented by zymase to ethyl alcohol.

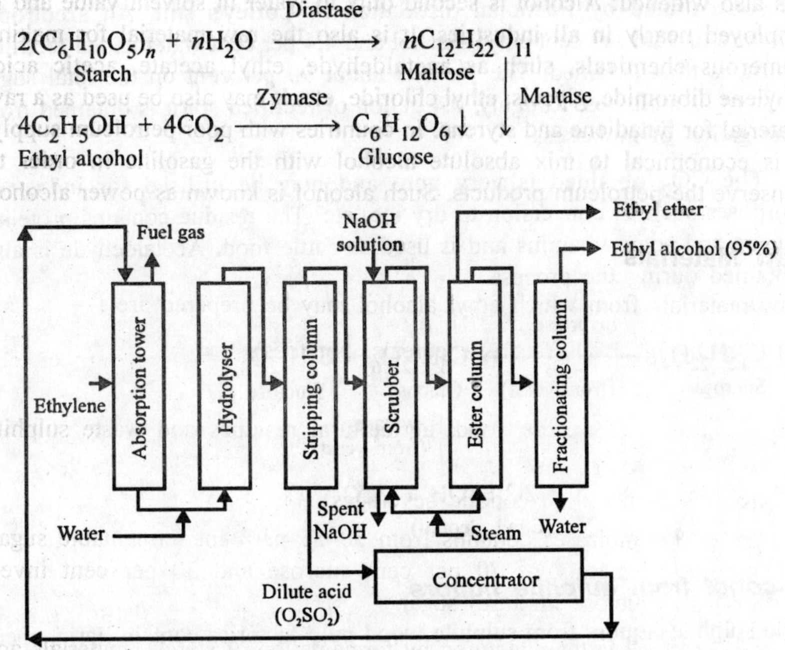

Fig. 13.2. Flow diagram for manufacture of industrial alcohol.

Alcohol from molasses

Molasses is the mother liquor left after the recovery of crystalline cane sugar and is a dark, thick liquid. It has 50–55 per cent total sugar content, of which 30–40 per cent is cane sugar or sucrose.

The molasses is diluted to 10–15 per cent sugar content because yeast cannot ferment concentrated sugar solutions. The pH of the solution is adjusted at 4.0–5.0 with H_2SO_4 and some nutrients for yeast such as ammonium sulphate and phosphate are added, if required. The mixture, called 'mash', is run into wooden or steel tanks to which 5 per cent by weight of yeast is added and the temperature maintained at 30°C. As the reaction is exothermic, cooling coils are necessary to maintain this temperature, 8–10 per cent alcohol is produced, fermentation is arrested by passing air through the liquid. The enzyme invertase present in the yeast, converts sucrose to glucose and fructose, which is then fermented to ethyl alcohol and CO_2 by the enzyme 'zymase' also present in the yeast. Alcohol is concentrated by fractional distillation in Coffey's still. An azeotropic mixture of 95.6 per cent ethyl alcohol and 4.4 per cent water called rectified spirit, is thus obtained. The yield is about 90 per cent on the total sugar content of molasses. Roughly, 0.4 gallon of rectified spirit is obtained from one gallon of molasses.

The CO_2 obtained is very pure and may be utilised for industrial purposes such as conversion to dry ice, etc. The residue contains proteins, sugars and some vitamins and is used as cattle food. Acetaldehyde is also obtained during the process.

$$\underset{\text{Sucrose}}{C_{12}H_{22}O_{11}} \xrightarrow[\text{(from yeast)}]{\text{Invertase}} \underset{\text{Glucose}}{C_6H_{12}O_6} + \underset{\text{Fructose}}{C_6H_{12}O_6}$$

$$\xrightarrow[\text{(from yeast)}]{\text{Zymase}} \underset{\text{(Ethyl alcohol)}}{2C_6H_5OH} + 2CO_2$$

Alcohol from sulphite liquors

The sulphite liquors from sulphite wood pulp factories contain ligrin, spent chemicals such as $CaHSO_3$, and H_2SO_3 and about 2 per cent fermentable sugars (derived from cellulose) mainly as glucose. SO_2 is boiled off, the liquid filtered through screens to remove fibres, cooled to about 30°C and pH adjusted to 6.0 with milk of lime. Urea is added as yeast food and the liquid fermented with the addition of 10 per cent yeast. When the liquid contains 10 per cent alcohol, the fermentation is arrested. The yeast is removed by centrifuging and ethyl alcohol fractionated in the usual way.

$$C_6H_{12}O_6 \xrightarrow[\text{(from yeast)}]{\text{Zymase}} 2C_2H_5OH + 2CO_2$$

Synthetic Preparation of Ethyl Alcohol

Ethyl alcohol is now-a-days synthesised from ethylene by conversion to ethyl hydrogen sulphate and subsequent hydrolysis, or by catalytic dehydration. (Fig. 13.3).

Alcohol from ethylene and sulphuric acid

Ethylene (35–95%) under 150–200 P.S.I. Pressure is led to an absorption tower, containing 90–95 per cent H_2SO_4 at 50–80°C. Catalysts, such as oxides of iron or copper, may be added for better absorption of the gas and ethyl hydrogen sulphate and diethyl sulphate are thus formed. The reaction being exothermic the requisite temperature is maintained by controlling the flow of ethylene or by employing cooling coils. The mixture of ethyl hydrogen sulphate and diethyl sulphate is pumped to a hydrolyser and diluted with 2 volumes of water. It is kept at about 70°C for 2 hrs. when diethyl ether (5–10%) is also formed. The mixture passes through a stripping column into which steam is led. Alcohol, ether and a little acid are carried over by the steam. This is treated with concentrated NaOH solution to fix up the acid. Ethyl alcohol is purified by fractional distillation. One gallon of 95 per cent ethyl alcohol is obtained from 4 lb of ethylene. The H_2SO_4 (50–60%) is concentrated for reuse. The reactions are :

$$\underset{\text{Ethylene}}{CH_2 = CH_2} + H_2SO_4 \longrightarrow \underset{\text{Ethyl hydrogen sulphate}}{CH_3CH_2OSO_3H}$$

$$2CH_2 = CH_2 + H_2SO_4 \longrightarrow \underset{\text{Diethyl sulphate}}{(CH_3CH_2)_2SO_4}$$

$$CH_3CH_2OSO_3H + H_2O \longrightarrow \underset{\text{Ethyl alcohol}}{CH_3CH_2OH} + H_2SO_4$$

$$(CH_3CH_2)_2SO_4 + 2H_2O \longrightarrow \underset{\text{Ethyl alcohol}}{2CH_3CH_2OH} + H_2SO_4$$

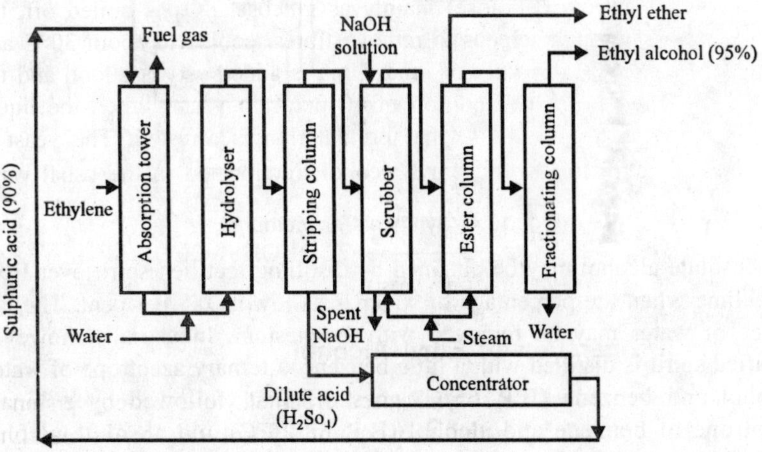

Fig. 13.3. Synthesis of ethyl alcohol from ethylene.

Alcohol from ethylene by catalytic hydration

Almost pure ethylene (97%) under 65 atmosphere pressure and water (0.6 mole per mole of ethylene) are heated to 300°C and led over the catalyst, which is inert diatomaceous earth soaked with 85–90 per cent H_3PO_4, in a reactor. Conversion per passage is 4 per cent. The gases from the reactor are passed through a dilute NaOH solution to remove any adhering acid. The reaction is exothermic. The unreacted gas is recycled. Some acetaldehyde is present due to hydration of acetylene present in ethylene. This is reduced to ethyl alcohol by catalytic hydrogenation over nickel. The yield is 95-97 per cent of ethyl alcohol.

$$\underset{\text{Ethylene}}{CH_2{=}CH_2} + H_2O = \underset{\text{Ethyl alcohol}}{CH_3CH_2OH}$$

ABSOLUTE ALCOHOL

Alcohol is concentrated by fractional distillation in Coffey's still, and an azeotropic mixture of 95.6 per cent alcohol and 4.4 per cent water called rectified spirit, is obtained (Fig. 13.4).

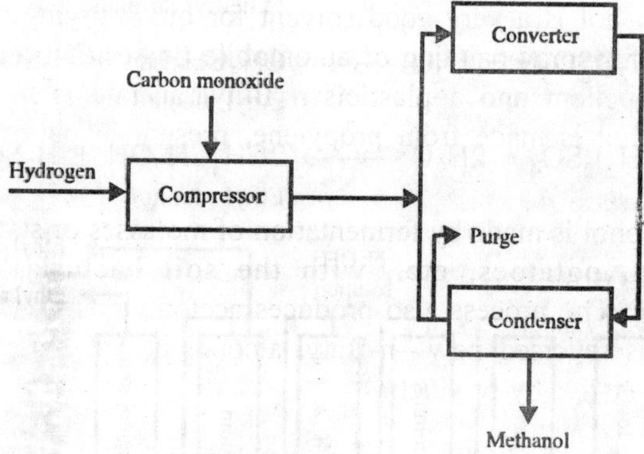

Fig. 13.4. Synthesis of methanol.

Absolute alcohol may be obtained by distilling rectified spirit over fresh quicklime when the percentage of water is as low as 0.3 per cent. The last traces of water may be removed with magnesium. Industrially, however, rectified spirit is distilled with a little benzene–a ternary azeotrope of water, alcohol and benzene (B.P. 65°C) goes off first, followed by a binary azeotrope of benzene and alcohol (B.P. 68.25°C) and absolute alcohol (B.P. 78°C) distills last.

Methylated Spirit

This is rectified spirit to which methanol (10%), pyridine (5%) and light caotchoucine (0.5%) obtained by distillation of vulcanised rubber, have been added to make it unfit for drinking. It is duty free and so much cheaper than rectified spirit. It is used mainly for making varnish and for lighting stoves.

Proof Spirit

It is 57.1 per cent ethyl alcohol (by volume), specific gravity 0.91976 at 60°F (by definition). Strength of alcohol for excise duty is determined by density measurement e.g., 10 over proof means that 100 ml of it, diluted with water, will make 110 ml proof spirit. Similarly, 100 ml of 10 under proof is equivalent to 90 ml of proof spirit. By-products of alcoholic fermentation are :

Carbon dioxide, acid potassium tartrate (cream of tartar), fusel oil, some esters, etc., isopropyl and n-butyl alcohols. These two alcohols are also commercially important now-a-days. Isopropyl alcohol is a duty free solvent, an anti-freeze and a component of shaving lotions, nail polish, etc. It is mostly converted into acetone.

n-Butyl alcohol is a very good solvent for quick drying nitro-cellulose lacquers used for spray painting of automobile bodies. Dibutyl phthalate is a mosquito repellent and a plasticiser. Butyl acetate is a fruit essence. Isopropyl alcohol is made from propylene, present in petroleum cracking gases.

n-Butyl alcohol is made by fermentation of molasses or starchy materials such as corn, potatoes, etc., with the soil bacteria, *Clostridium acetobutylicum*. The process also produces acetone along with the alcohol. The position is reversed now—n-Butyl alcohol is the main demand now, with acetone as the by-product.

METHYL ALCOHOL

Methyl alcohol (CH_3OH) was first obtained from the products of wood distillation. Till recently this substance was a solely obtained from pyroligneous acid, a product of wood distillation. But it is obtained in large quantities by the synthetic method.

Synthetic Method

In this method the raw material is a mixture of CO and H_2 in the proportion of 1 : 2. This gas mixture is compressed under a tremendous pressure of 3000–5000 psi at a temperature of 300°C, in copper lined converter over a catalyst, a mixture of ZnO and Cr_2O_3. The reaction being highly exothermic, the temperature is maintained at 300°C by means of cooling coils once the reaction has started. The issuing gases are cooled to 0.20°C, and the methyl alcohol formed is drawn off. The residual gases are recycled. The yield is about 60 per cent.

$$CO + 2H_2 = CH_3OH$$

Methyl alcohol can also be directly made from methane by passing a mixture of CH_4 and O_2 under 100 atmospheric pressure over a copper catalyst at 200°C. The conversion per passage is 17 per cent.

$$CH_4 + 1/2O_2 = CH_3OH$$

Alcohol from Wood Distillation

Here hardwood chips are used. Air-dried chips are heated, in the absence of air, in cast iron retorts at 450°–560°C for about 30 hrs. This process is destructive distillation. Wood gas consisting of methane, CO and CO_2 is utilised in heating the retort. Pyroligneous acid (45%) containing 2–6 per cent methyl alcohol, 0.5 per cent acetone and about 8 per cent acetic acid are obtained in a red coloured watery solution. Wood tar (5%) and wood charcoal (25%) and other products obtained. Pyroligneous acid vapour is led into milk of lime to fix up the acetic acid as calcium acetate while methyl alcohol and acetone pass over. By fractional distillation over quicklime methyl alcohol with 1–2 per cent acetone is obtained. The yield of methyl alcohol is about one per cent of the wood distilled.

BEER

Beer is generally defined as an alcoholic beverage made by fermentation of a farinaceous extract that is obtained from a starchy raw material, barley, in the form of malt. Although it is possible to replace some part of the barley with other starchy materials (e.g., corn, rice, wheat, oats, or potatoes). barley is usually the main constituent of brewing materials.

Properties of Beer

The properties of the finished beer vary with the type of beer and place of origin. Most beers have an alcohol content of 3–5 wt%, a protein content of 0.3–0.6%, and a pH of ca 4. The quality of a beer is judged by the following attributes: aroma, taste, appearai ., formation and stability of foam.

The caloric content of beer is significant, but not especially high. A 355 mL (12 oz) bottle of average beer yields ca 600 kJ (143 kcal). The calories are provided by the unfermented residues and by the alcohol. The metabolic role of the latter is not fully understood, but it replaces carbohydrates, fats, and proteins, so that there may be a gain in body weight. Besides its caloric value, beer also contributes to the mineral requirements of the body and supplies useful quantities of B-complex vitamins.

A special use of beer is for the control of sodium intake in the treatment of diseases such as congestive heart failure, high blood pressure, and certain kidney and liver ailments. Beer cannot, because of its low pH, harbour any pathogenic germs. The content of nourishing components is all in dissolved form. Beer is free from fat, it acts as a diuretic, and it promotes the formation of gastric acid, thus acting as an appetite inducer. The alcohol in beer is effective to the amount of the concentration (i.e., the speed at which it is drunk).

Beer defects

Among beer defects, turbidity is of primary importance. It may be of either biological or physio-chemical nature. The former occurs only in unpasteurised beer and is caused by growth of micro-organisms, such as brewers yeast, wild yeast, or certain bacteria. The latter may be induced by oxalic acid (from the barley), metals, chilling, and heating (pasteurisation). Chill haze is rich in sulphur (1.8–2.0%) indicating a rather high content of active suphydril groups (—SH) which are readily subject to oxidation. When oxidation occurs, the —SH groups combine to form large molecules with sulphur to sulphur linkage (—S—S—). The oxidised compounds are insoluble, thus producing an oxidation haze with its consequent deleterious effect on the flavour and shelf life of the beer. The effect of sunlight in the sunstruck flavour in beer is caused by formation of mercaptans. Wild or gushing beer is a defect observed as a rather violent over-foaming from the bottle immediately after opening; this defect, however, does not affect the taste of the beer. The fundamental cause of gushing is attributed to the formation of microbubbles, which are caused by certain substances in barley or malt that has been moulding.

Besides the culture yeasts, there are many wild yeasts, moulds, and bacteria that may contaminate the beer. By constant cleaning, brewers try to keep to a minimum the number of harmful micro-organisms. The infecting organisms are non-pathogenic, but they increase the acidity which makes the beer less palatable, though not toxic. Foreign organisms capable of growing in beer are most often introduced through contamination of innoculation yeast.

Brewing Materials

Barley has technical advantages that make it superior for malting and brewing. Barley differs from the other common cereals in that the husk adheres to the kernel after threshing. This makes the process of malting, and subsequent extraction of the wort, much easier than with wheat or other grains.

During the malting process, the raw, hard, flat-tasting barley is changed into a crisp, mellow, sweet-tasting malt. The aim of malting is to bring forth substances, i.e., enzymes, that will break down (hydrolyse) starch and proteins to less complex, water-soluble compounds, i.e., amino acids, fermentable sugars, and small peptides. When these compounds are dissolved in water, the resulting liquid is known as wort. Brewers are especially interested in the starch-splitting amylases. They are formed in barley during germination, but they react during mashing. The protein-splitting enzymes, proteases, are formed and react during the germination so that some of the proteins of the barley are split into compounds soluble in water. The hemicellulases, another group of enzymes, split the cell tissue, thereby leaving the starch open to attack by amylases.

The character of the malt plays a large role in the resulting beer. The two extremes in malt character are reflected in Pilsner and Munich beers. For Pilsner, a pale malt with no pronounced aroma must be used; Munich beer demands malts with a pronounced aroma. Although the water content of the final malt is 3–4%, the interplay of water content and temperature during the drying is of fundamental importance for the character of the malt. The pale malt must be quickly dried to reach the final temperature; the dark malt must have a suitable high water content as the temperature rises. Special malts such as caramel and black are used for the darkest beers such as porter and stout. These malts have pronounced colour and aroma but little or no diastatic power.

Adjuncts in the form of cereals (especially corn) other than barley are often part of the brewing materials when the demand is for a stable, non-satiating, sparkling beer. Sugar can be added during various steps of the brewing instead of forming fermentable sugar through the splitting of starch during mashing. Since adjuncts are essentially starch with very little protein, they are a source of additional alcohol, but contribute little to the colour, taste, aroma, or protein content of the beer.

Originally, the principal aim of adding hops was to compensate for the insipid, sweet taste of the unhopped beer with the characteristic bitter taste and aroma of hops. Other benefits of adding hops include increasing the biological stability of the beer and improving the head retention and body of the beer. The amount of hops added varies from 0.4 to 4.0 g/L.

Yeast for brewing is propagated from single-cell cultures and is recovered after fermentation for use throughout many generations. There are two types of yeast in brewing: top yeast, which forms spores, ferments vigorously at elevated temperatures, and tends to float on top of the beer; and bottom yeast, which does not usually form spores, is well adapted to slow fermentation at low temperatures, and settles to the bottom of the tank at the end of fermentation. *Saccharomyces carlsbergensis* is considered the type of species for bottom fermenters, and certain strains of *S. cervisiae* for the top fermenter.

Perhaps the greatest preoccupation in relation to brewing yeasts is with stability, i.e., whether a strain or mixture of strains remains unchanged and uncontaminated with prolonged use and whether the range of performance of the constant strain or mixture of strains is small enough to provide adequate constancy of beer quality despite inevitable minor variation in wort composition. Various strains of yeast have individual flavour characteristics; thus, brewers yeasts are not selected solely on the basis of fermenting power, but decidedly on the flavour they give beer.

Water has a great influence on the character of the beer, and the hardness of the water manifests itself by the extent of its reaction (alkalinity) with the weak acids of the mash. Certain ions are harmful to brewing; nitrates slow down fermentation, iron destroys the colloidal stability of beer, and calcium ions give beer a purer flavour than magnesium or sodium ions.

If the production of a pale beer with a fine hop taste is desired and the water is rich in carbonates, one of the following methods of treatment may be used: addition of gypsum (burtonising); addition of acid; decarbonation; and demineralising by ion exchange. These methods are easily employed, but there is a chance that salt may remain in the water, which is undesirable. Today, it is possible to treat brewing water through double ion exchange, a process in which all salts are completely removed.

Production

The most important phases during brewing are mashing, wort boiling, fermenting, lagering, filtering, and bottling. During mashing, a mixture of finely crushed malt and warm water is exposed to enzymatic activity, thus converting starch into miscellaneous sugars and protein into peptides and amino acids. Adjuncts must be pretreated to be accessible to attack by the malt enzymes, usually by boiling before addition to the mash. The dissolved product from mashing is called wort and the insoluble remainder (mostly husks of malt) is called spent grain, which is sold as cattle fodder. The wort is boiled with hops, and during boiling, the enzymes are destroyed while bitter substances are extracted from the hops. Boiling of the wort also causes a certain amount of unconverted protein to coagulate and flocculate. This flocculation appears in the boiled wort as flakes; the brewer says that the wort has a fine break. After separation from the wort, it is called sludge.

After cooling of the wort to ca 10°C, the yeast is added in order to convert the sugar into alcohol and carbon dioxide during fermentation. After fermentation, most of the yeast is separated from the beer followed by a slow after-fermentation and maturing of the green beer in lagering tanks at ca 0°C. By the end of the lagering there is a sediment in the tank, called draff, consisting of the remaining yeast together with precipitated proteins and tannins. The beer is then filtered into pressure tanks and transferred to the bottling machine. It is then pasteurised to keep its clear colour for longer periods of time. The finished product is completed in 2 mo for Pilsner beer and 4–6 mo for the stronger beers.

Much has changed in the technology of beer making; however, the methods of transforming barley into beer have not undergone profound modifications. Today, attempts are being made to avoid the malting process by brewing directly from barley with enzymes extracted from micro-organisms. Industrial enzymes, such as bacterial or fungal α-amylases, amyloglucosidases, proteases, and pullulanase, mainly of microbial origin, are being considered for use. Using these enzymes will yield hitherto unknown degrees of freedom in selecting conditions of the mashing and, consequently, the control of wort composition.

Although there is controversy over the principles of continuous brewing, with a large setback in the United States and a success of Spain, more cautious approaches in other parts of the world, particularly the UK, are proving their worth. Continuous fermentation is the most recently studied development in breweries. Principally, the primary fermentation is continuous, and the secondary fermentation and lagering is conventionally done. Because great demands are put on the operative yeast, the right yeast must be found for the system to work at an optimum. Use of the right yeast eliminates the need for the following lagering step.

Two systems have been described, and both are in commercial operation with good results. The open system depends on the wort being continually supplied to the system and the beer removed simultaneously. The partially closed system is built on the principle that a high concentration of yeast is artificially held back in the system.

The degree of automation varies widely. Newer plants have the ultimate in automatic wort production where one man per shift, per brewhouse unit, makes twelve brews per day. An important consideration in current brewhouse operations is to increase the utilisation of all vessels by the rapid batch process. A rapid filtration of grains is required and achieved with the use of a lauter tun, mash filter, or strain master. The separation of hops from the boiled wort has been accelerated by the use of hop powder pellets or omitted when hop extract is used. High density brewing involves the production of worts with high initial density and the introduction of a substantial part of water at as late a stage in the conventional brewing process as possible, usually after preliminary beer filtration. A further development in the process is the use of universal refrigerant-cooled tanks, designed to ferment, age, and finish beer in a single tank without the usual transfers.

The production of concentrated beer is basically a crystallisation process in which water, constituting ca 90% of a normal beer, forms pure ice crystals at temperatures below the freezing point of water which are separated from the beer concentrate. After filtering, the beer concentrate is ready for reconstitution by the addition of water and carbon dioxide. In the concentration and subsequent filtering step, some of the constituents that contribute to the instability of finished beer are removed. Since these processes can be accomplished in a matter of hours, the resulting beer can be ready for consumption shortly after fermentation is completed, rather than weeks later as is the case with present lagering techniques. Also, in recent years, automatic regulation of processes has steadily gained a foothold in the brewing industry. The aim has been to produce beer with a better and more even quality and at lower costs.

Cleaning and disinfection of brewery equipment at every stage of the process is necessary. It has been the object of much research, and more effective chemicals are constantly marketed. Physiological, chemical, and bacteriological cleanliness all are important, though bacteriological cleanliness through disinfection is the most important. In the breweries, both chemical and bacteriological degrees of cleanliness are nearly always obtained.

Environmental Problems

The awakening of public interest toward environmental pollution means that brewing companies must anticipate future legislation. Recent changes in state laws have placed severe restrictions on breweries in terms of amounts of suspended solids and biological oxygen demand (BOD) that can be discharged with waste-waters. Most breweries have an average BOD of

1.0–1.5 g/l; however, with strict controls and collection of waste streams, this can be decreased to as low as 0.35 g/l BOD.

The impact on the environment is not caused solely by effluents; it is also a matter of disposal of one-way bottles and cans.

WINE

The word wine was possibly first applied to the fermentation product of the sugars in the juice of grapes; this is its primary meaning. However, the fermented juices of many fruits are now called wine, and the term is also sometimes incorrectly applied to the alcoholic fermented juice of various plant materials containing sugars, e.g., rice wine. It is thus, used in contrast to fermented liquids made from starch-containing materials, such as beer and related beverages.

Classification

Wines may be classified in many ways: by alcohol (ethanol) content, place of production, colour, method of production, or variety of grape or fruit. Wines with ≤ 14% alcohol are defined as table wines since they are normally consumed with meals, especially if they do not taste sweet. Wines with > 14% are called dessert or aperitif wines since they are usually consumed after meals if sweet or before meals if not sweet.

Production

The special character of the fermentation of the various types of wines depends to a considerable extent on the composition of the fruit juice fermented.

Composition of grapes

Grapes contain ca 15–25% sugar; partially dried grapes contain 30–40%. The percentage of sugar in the grapes, the extent of the fermentation, and the losses or additions of alcohol during treatment and storage determine the percentage of alcohol in the finished product.

Only a small amount of nitrogenous material (0.3–1.0%) is found in grapes. However, this material is of considerable significance for yeast nutrition, bacterial stability, and flavour development, presumably because of the many amino acids present.

The pigments of grapes and fruits are usually located in the epidermal cells. A few varieties have red juice as well. During alcoholic fermentation, the epidermal cells are killed and the coloured pigments are released. By separating the skin of red grapes from the juice before fermentation, it is possible to produce a white or nearly white wine.

The pectins of some fruits and grapes are a source of difficulty in juice clarification. They are rather insoluble in alcohol and precipitate during alcoholic fermentation. The inorganic constituents are not of critical importance as they are usually present in sufficient amounts to catalyse yeast or enzyme functioning; excess iron or copper may cause turbidity.

Micro-organisms

Wines are normally produced by fermentation with the yeast *Saccharomyces cerevisiae*, sometimes with *S. bayanus* or *S. oviformis*. Taxonomists are still reclassifying the genus Saccharomyces. These and other yeasts, found in grapes or fruit, multiply rapidly in the sweet juice, eventually causing fermentation. To prevent growth and competition or undesirable organisms, 50–150 mg sulphur dioxide are usually added per litre ca 2 hour before the pure yeast culture is added.

Equipment

Wine production can be very simple, but for large-scale operation, specialised equipment has been developed. The grapes are crushed in combined stemmers and crushers.

Fermentation tanks may be of wood, concrete, stainless steel, or iron-lined with epoxy resins or a thin layer of stainless steel; they may be open or closed, and possibly, temperature-controlled.

After fermentation, the residue, i.e., stems and skins, often called pomace or marc, must be transferred from the fermenter to the press. The oldest type of press still in use is the screw-type basket press, usually operated vertically, either from above or below. It has been largely replaced by the horizontally operated hydraulic press or other continuous-type presses. These presses are more expensive to operate but produce a relatively clear juice from both red pomaces or white musts.

White table wines

White wines of < 14% alcohol are designated as white table wines. They are usually made from white grapes. Occasionally, a white wine is made from red grapes by separating the skins from the juice immediately after crushing. They include dry and sweet wines.

Red table wines

Most of the world's wine is red table wine of < 14% alcohol. In France and Italy, particularly, they constitute an important part of the daily calorie intake. Red wines are relatively easy to produce compared to white table wines, and they are less subject to spoilage or clouding during aging.

The grapes should be hárvested when they are sufficiently ripe to produce 11–13% alcohol. A sugar content of ca 21.5–23.5% is best. The grapes should be transported from the vineyard to the crusher without delay to prevent development of spoilage bacteria. Prompt crushing and destemming are essential.

Sparkling wines

Wines containing a permanent visible excess of carbon dioxide are called sparkling wines. The nomenclature of the sparkling wines is somewhat complicated. The most famous name is champagne, originally produced only in the region of that name in France where this appellation may be used only for wines produced and fermented in that region. In the United States, the term champagne is used for most sparkling wines produced by a secondary fermentation of sugar in closed containers. If fermented in tanks, it must be so stated on the label.

Dessert wines

Less than 15% of all the Californian wines are fortified as dessert wines at an alcohol content of 17–21%. Both red and white dessert wines are produced; the white may be either dry or sweet. Worldwide, probably less than 10% of the wines are dessert types.

White dessert wines

Muscatel, Angelica, white port, and sweet sherry are the most important sweet dessert wines produced in the United States; dry sherry is the only dry type and it is often at least slightly sweet. Famous European dessert wines include Malaga (Spain), Marsala (Sicily), and Madeira (Madeira Islands).

Red desert wines

The production of fruit wines is similar to that of grape wines, except that the sugar must always be added because of the low sugar content of most berries and fruit.

Fruit and berry wines

The production of fruit wines is similar to that of grape wines, except that the sugar must always be added because of the low sugar content of most berries and fruit.

Naturally flavoured wines

Natural herbs, spices, fruit juices, aromatics, essences, and other natural flavourings may provide the base, and sugar and caramel may be added; apple wine may also be added. These wines are made both as table (< 14% alcohol) and dessert wines.

Vermouth

Vermouths are nearly dry or sweet fortified wines to which herbs or herb extracts are added. The nearly dry of French type is used straight or for martini cocktails. The herbs in vermouth should be easily detectable, but the odour of no single herb should be allowed to predominate.

Finishing

Finishing steps include filtration, fining, refrigeration, and, occasionally, pasteurisation.

Packaging

Automatic bottling, corking or capping, labeling and casing lines are employed in large wineries. These are less labour-intensive and give better results than manual procedures. The bottles are packed in cardboard boxes for shipment.

Spoilage

Wines are relatively immune to spoilage because of their alcohol content and low pH. Contact with air may result in acetification, which can be avoided by keeping the containers full or by pasteurisation. Low acid wines may be spoiled by lactic acid bacteria which can be prevented by low concentrations of sulphur dioxide. Yeast infection may be a problem. Iron and copper originating from production equipment occasionally cause cloudiness in white table wines.

BEVERAGE SPIRITS (DISTILLED)

The early methods of distillation utilised the alembic, a simple closed container to which heat was applied. The vapours were transferred through a tube to a cooling chamber in which they were condensed. Alembics made of copper, iron, or tin were preferred. From this developed stills consisting of clay or brick fireboxes into which the copper pots of the still were fitted. Direct heat was applied to the body or cucurbit containing the fermented mixture, and the vapours rose into the head, passed through a pipe (the cane neck) to the worm tub, and condensed by a copper coil immersed in a barrel of cool water. Variations and improvements of this technique resulted in pot stills, some of which are still in use in the production of malt whiskies, notably in Scotland, and of brandies in France.

Within each category of product, wide variations in flavour can be caused by types of materials and their proportions, methods of preparation, selection of yeast types, fermentation conditions, distillation processes, maturation techniques, and blending.

Since the alcohol and water components are insignificant factors in flavour intensity or palatability, distillers are primarily interested in the more flavourful constituents, i.e., the so-called congeners, substances that are generated with the alcohol during the fermentation process and in the course of maturation. In order to produce a palatable product, it is therefore necessary to select the proper configuration of these constituents (congeners). This cannot be accomplished by production techniques such as listed above, and therefore, the majority of beverage spirits are blended to provide uniformity, balanced bouquet, and palatability.

Whiskies

Whiskies are by far the leading distilled beverage spirits. Whisky is manufactured in compliance with the laws regulating the manufacture of whisky for consumption in and containing no distilled spirits less than three years old. They are light-bodied, and though delicate in flavour, they nevertheless retain a distinct positive character.

The principal cereal grains, i.e., corn, rye, and barley malt, are used, and their proportions in the mashing formula remain a distiller's trade secret; otherwise, the process is substantially the same as is found in the main distilleries of the India.

Scotch

Scotch whisky is a distinctive product of Scotland, manufactured in Scotland in compliance with the laws of the UK regulating the manufacture of Scotch whisky for consumption in the UK, and containing no distilled spirits less than three years old.

The outstanding taste characteristic of Scotch, its subtle smoky flavour, results from the techniques used in the production of malt whiskies. Malted barley, dried over peat fires, is the only grain ingredient in the mash. The kind and the amount of peat used in the fires determines the intensity of flavour in the final product. The aroma of the burning peat, known as peat reek, is absorbed by the barley malt. This smoky flavour is carried through to the distillate and becomes a characteristic of single-malt whisky. Peat is heather, fern, and evergreen that have been subjected to aging and compression processes over the centuries.

The dried malted barley is ground to a grist, gelatinised in a mash tub, and after conversion is completed, the liquid portion or wort is drained, cooled, and placed in a fermenter.

After fermentation is complete, the separation of malt whisky from the fermented wort takes place in a batch distillation system, a copper kettle with a "worm" or spiral tube leading from its head. The size and shape of these pot stills exert a definite influence on the character of the whiskies. Another critical factor is the selection of the product during that portion of the distillation cycle that will produce the desired flavours. The first portion in the cycle is referred to as foreshots and the last as feints. The middle portion, after further distillation, becomes high wines and is subsequently reduced to maturation proof for storage in oak casks. The final distillation proof is in the 140–160°C (70–80%) range.

Gin

Gin is a product obtained by original distillation from mash or by the redistillation of distilled spirits, by mixing neutral spirits with or over juniper and other aromatic berries, or with or over extracts derived from infusion, percolations, or maceration of such materials, and includes mixtures of gin and neutral spirits. Gin derives its main characteristic flavour from juniper berries. In addition to juniper berries, other botanicals may be used, including angelica root, anise, coriander, caraway seeds, lime, lemon and orange peel, licorice, calamus, cardamon, cassia bark, orris root, and bitter almonds. The use and proportion of any of these botanicals in the gin formula is left to the producer, and the character and quality of the gin depends to a great extent on the skill of the craftsman in formulating his recipe. The more skilled producers formulate their aromatic ingredients on the basis of the essential oil content in the raw materials to assure a greater degree of product uniformity.

Brandy

Brandy is a distillate or a mixture of distillates obtained solely from the fermented juice, mash, or wine of fruit, or from the residue thereof, distilled at < 190° proof (95%) in such a manner as to produce the taste, aroma, and characteristics generally attributed to it. The most important category of brandy is fruit brandy. Cognac is a brandy produced in the Cognac region of France, and it enjoys an exclusive identity under which no other brandy may be labelled. Brandy derived exclusively from one variety of fruit is so designated. However, a fruit brandy derived exclusively from grapes may be designated as brandy without further qualifications. A minimum of two years' maturation in oak casks is required; otherwise, the term immature must be included in the designation. Blended applejack is accorded a special classification as a mixture that contains at least 20% apple brandy (applejack) on a proof basis, stored in oak containers for not less than two years, and not more than 80% neutral spirits on a proof gallon basis and bottled at not less than 80% proof (40%). Another class of beverage spirits, flavoured brandy, is a brandy to which natural flavouring materials have been added with or without addition of sugar; it is bottled at not less than 70° proof (35%).

Rum

Rum is produced from the fermented juice of sugar cane, its syrup, molasses, or other by-products distilled at less than 190°C proof (95%) and bottled at not less than 80° proof (40%). Blackstrap molasses is the most common raw material for the manufacture of rum. Rums are characterised as light-bodied, of which the Puerto Ricans are the best known, and full-bodied, which come from Jamaica and certain other islands of the West Indies.

Tequila

Tequila, a distinctive product of Mexico, is produced in Mexico from the fermented juice of the heads of Agave Tequilana Weber (blue variety), with or without additional fermentable substances. This is Mexico's most popular distilled spirits drink. Tequila, as consumed in Mexico, is unaged and usually bottled at 80–86° proof (40–43%). Tequila aged one year is identified as Anejo; aged as much as 2–4 years, it is identified as Muy Anejo.

Cordials and Liquers

Cordials and liqueurs are obtained by mixing or redistilling neutral spirits, brandy, gin, or other distilled spirits with or over fruits, flowers, plants, or pure juices there from, or other natural flavouring materials, or with extracts derived from infusions, percolations, or maceration of such materials. Cordials must contain a minimum of 2.5 wt% of sugar or dextrose, or a combination of both. If the added sugar and dextrose are < 10 wt%, then it may be designated dry. Three basic processes, namely, maceration, percolation, and distillation, or any combination thereof, are used to make cordials.

Vodka

Vodka, produced on a multicolumn distillation system from fermented mash of grain at or above 190° proof (95%), must be further treated with charcoal or activated carbon or further refined by distillation in such a manner as to be without character, aroma, or taste. If any flavouring material is added to the distillate, the vodka is so characterised.

Manufacturing process

Any material rich in carbohydrates is a potential source of ethyl alcohol, which for industrial purposes is obtained by the fermentation of materials containing sugar (molasses), or a substance convertible into sugar, such as the starches. In the production of distilled spirits for beverage purposes, however, cereal grains are the principal types of raw material used. Any reference to alcohol in beverages is always to ethyl alcohol, C_2H_5OH.

The chemical composition of grain varies considerably and depends to a large extent on environmental factors such as climatic conditions and the nature of the soil. Another variable is the malt (sprouted or germinated grain) used. Malt is generally understood to be germinated barley, unless it is further qualified as rye malt, wheat malt, etc. The purpose of malting is to develop the amylases, the active ingredients in malt. Amylases are enzymes of organic origin that change grain starch into sugar (maltose). Besides providing the means of converting the grain starch into sugar, the malt also contributes to the final flavour and aroma of the distillate.

Milling breaks the outer cellulose protective wall around the kernel and exposes the starch to the cooking and conversion processes. Distillers require an even, coarse meal without flour. The mashing process consists of cooking, i.e., gelatinisation of starch, and conversion (saccharification), i.e., changing starch to grain sugar (maltose). Distillers vary mashing procedures, but generally conform to basic principles, especially in the maintenance of sanitary conditions. The cooking and conversion equipment is provided with direct or indirect steam, propeller or rake-type agitation, and cooling coils or a barometric condenser.

In fermentation, the grain sugars (largely maltose), produced by the action of malt enzymes (amylases) on gelatinised starch are converted into nearly equal parts of ethyl alcohol and carbon dioxide. This is accomplished by zymase, which is produced by yeasts. Although yeasts of several genera are capable of some degree of fermentation, *Saccharomyces cerevisiae* is almost exclusively used by the distilling industry. A great variety of strains exist, and the characteristics of each strain are evidenced by the type and amount of congeners the yeast is capable of producing. Alcohol fermentation is represented by the following reactions :

$$\underset{\text{maltase}}{C_{12}H_{22}O_{11}} \xrightarrow[H_2O]{\text{maltase}} \underset{\text{dextrose}}{C_6H_{12}O_6} \xrightarrow{\text{zymase}} \underset{\text{ethyl alcohol}}{4C_2H_5OH} + 4CO_2$$

Distillation separates, selects, and concentrates the alcohol and flavour constituents produced by yeast fermentation from the fermented grain mash, sometimes referred to as fermented wort or distillers beer. In addition to the alcohol and the desirable secondary products (congeners), the fermented mash contains solid grain particles, yeast cells, water-soluble proteins, mineral salts, lactic acid, fatty acids, and traces of glycerol and succinic acid. The most common distillation systems used are the continuous whisky separating column, the continuous multicolumn system used for the production of grain neutral spirits, and the batch column and kettle unit used primarily in the production of grain neutral spirits. The discharge from the base of the whisky column, called stillage, contains substances derived from grain and is processed into distillers feeds which are used to fortify dairy, poultry, and swine formula feeds.

The final phase of the whisky production process, called maturation, is the storage of beverage spirits in new, white oak barrels whose staves and headings are charred. The duration of storage in the barrel depends on the time it takes a particular whisky to attain the desirable ripeness or maturity. The thickness of the stave, depth of the char, temperature and humidity, entry proof, and length of storage impart definite and intended changes in the aromatic and taste characteristics of a whisky. These changes are caused by three types of reactions occurring in the barrel: extraction of complex wood constituents by the liquid; oxidation of components originally in the liquid and material extracted from the wood; and reaction between the various organic substances present in the liquid, leading to the formation of new congeners.

BUTYL ALCOHOL (BUTANOL)

From Carbohydrates by Fermentation

$$\underset{\text{Starch}}{(C_6H_{10}O_5)_x} \xrightarrow{H_2O} \underset{\text{Glucose}}{C_6H_{12}O_6} \longrightarrow$$

$$CH_3COCH_3 + CH_3CH_2CH_2CH_2OH + C_2H_2OH + CO_2 + H_2$$

Process

Butyl alcohol (butanol), acetone, and ethyl alcohol are produced by the selective bacterial fermentation of carbohydrate-containing materials such as molasses and grain.

Molasses is diluted with water to a concentration of approximately 5% sugar, sterilised, cooled to 30°C, and pumped to a fermenter. A culture of bacteria (*Clostridium saccharobutyl acetonicum liquefaciens*) grown in sterile molasses is added to start the fermentation. Protein nutrients and an alkaline buffer to control the pH are also added to improve yields. (Fig. 13.5).

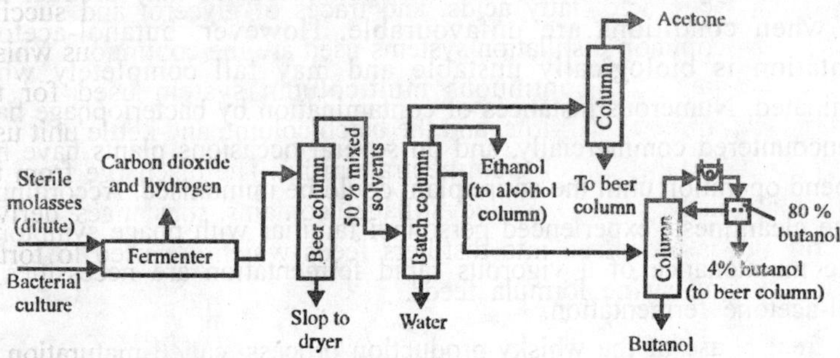

Fig. 13.5. Flow diagram for manufacture of butyl alcohol from carbohydrates by fermentation.

After fermentation has proceeded for 36 to 48 hours, the beer (fermentation mixture) containing 1.5 to 2.5% mixed solvents is pumped to a column whence a 50% solvent mixture is taken off overhead and distillers' slop is removed as bottoms. This slop may be dried and sold as animal feed or utilised as a source of riboflavin and other components of the vitamin B complex, since it has been discovered that the bacteria used in the fermentation synthesise vitamins. Another by-product is a mixture of carbon dioxide and hydrogen evolved during the fermentation.

The mixed-solvent vapours from the beer column are led to a batch fractionating column from which three fractions (acetone, ethanol, and butanol) are removed overhead, leaving water as bottoms. The acetone and ethanol fractions are purified by conventional fractionation. The butanol fraction, containing about 15% water, is led to a column from which vapours containing 70% butanol and 30% water are removed overhead. On condensation, two layers are formed. The top layer (80% butanol and 20% water) is returned to the butanol column, and the bottom layer (4% butanol and 96% water) is returned to the beer column.

The yield of mixed solvents is approximately 30% by weight based on the sugar charged. Solvent yield ratios are 70% butanol, 25% acetone, and 5% ethanol by weight.

Sterilised degermed corn starch mash may be substituted for the molasses, in which case the fermentation is carried out at 37°C using bacteria of the *Clostridium* genus such as *Clostridium acetobutylicum.*

Considerably more care must be exercised in carrying out butanol-acetone fermentation than either ethyl alcohol or acetic acid fermentations. The latter are biologically stable fermentations which usually proceed readily and easily. Ethyl alcohol fermentation of course may product low yields when conditions are unfavourable. However, butanol-acetone fermentation is biologically unstable and may fail completely when contaminated. Numerous instances of contamination by bacteriophage have been encountered commercially, and on several occasions plants have had to suspend operation until the entire plant could be immunised. Accordingly, absolute cleanliness, experienced personnel familiar with phage symptoms, and the maintenance of a vigorous rapid fermentation are necessities in butanol-acetone fermentation.

From Butyraldehyde by Hydrogenation

$$C_3H_7CHO + H_2 \rightarrow C_3H_7CH_2OH$$

95% yield

Process

Normal butyl alcohol is manufactured by .e catalytic hydrogenation of *n*-butyraldehyde in the liquid phase.

The *n*-butyraldehyde is produced by the oxonation of propylene. Isobutyraldehyde is also produced during the process. The two aldehydes may be hydrogenated and the corresponding alcohols separated by distillation, or the aldehydes may be separated and hydrogenated individually. (Fig. 13.6).

In either case the hydrogenation is carried out under pressure (up to 100 atm) in the liquid phase. From 1 to 10% water is added to the hydrogenator to minimise ether formation. A variety of catalysts may be used, among them chromium and copper oxides on silica, cobalt (6 to 38%) on kieselguhr, silica or alumina, or a supported nickel catalyst.

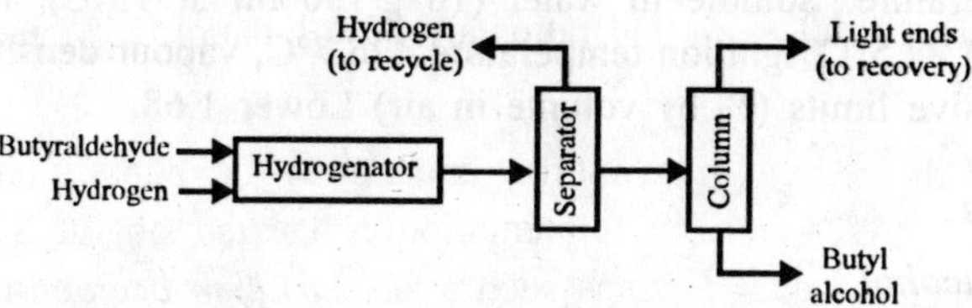

Fig. 13.6. Flow diagram for manufacture of butyl alcohol from butyraldehyde by hydrogenation.

From crotonaldehyde

n-Butyl alcohol is produced synthetically by the condensation of acetaldehyde to form aldol, which is then dehydrated to crotonaldehyde. The latter is hydrogenated (30 psi, 207 kPa, 180°C, nickel-chromium catalyst) to yield *n*-butyl alcohol. Some *n*-butyl alcohol is obtained in the high-pressure oxidation of propane and butane. A small amount is produced via the Ziegler route.

Secondary and tertiary butyl alcohols are produced by the absorption of butylene and isobutylene in sulphuric acid, similar to the production of isopropyl alcohol from propylene. Isobutyl alcohol is produced by the oxo process.

Uses

Solvent, *n*-Butyl acetate, glycol ethers and esters, plasticisers, amine resins, butyl acrylate and butyl methacrylate, miscellaneous and export.

Properties

n-Butyl alcohol

Colourless liquid possessing a vinous odour. Mol. wt 74.12, mp. –89.53°C, Sp gr 0.811 20°C/4, bp 117.25°C, weight per gallon 6.75 lb (20°C). Soluble in ethanol and ether in all proportions at room temperature. Slightly soluble in water (9 g/100 ml at 15°C).

Flash point (closed cup) 35 to 35.5°C, ignition temperature 343.3°C, vapour density (air = 1) 2.55, explosive limits (% by volume in air) lower 1.7, upper 11.2, threshold limit value (ppm) 100.

Properties

Isobutyl alcohol

Mol. wt 74.12, mp –108°C, Sp gr 0.798 25°C/4, bp 107.9°C, weight per gallon 6.65 lb (25°C). Soluble in ethanol and ether in all proportions at room temperature. Soluble in water (10 g/100 ml at 15°C). Flash point (closed cup) 27.5°C, ignition temperature 426.7°C, vapour density (air = 1) 2.55, explosive limits (% by volume in air) Lower 1.68.

Properties

sec-Butyl Alcohol

Colourless liquid with a strong, pleasant odour. Mol. wt 74.12, mp –114.7°C, Sp gr 0.808 20°C/4, bp 99.5°C, weight per gallon (6.74 lb (20°C). Soluble in water (12.5 g/100 ml at 20°C), ethanol, and ether. Flash point (closed cup) 24.4°C

Properties

tert-Butyl Alcohol

White crystals with a camphor like odour. Mol. wt 74.12, mp 25.55°C, Sp gr 0.789 20°C/4, bp 82.2°C. Soluble in water, ethanol, and ether in all proportions at room temperature.

Flash point (closed cup) 8.9°C, ignition temperature 477.8°C, vapour density (air = 1) 2.55, explosive limits (% by volume in air) lower 2.35, upper 8.0.

ACETONE (CH_3COCH_3)

From Isopropyl Alcohol

Process

Acetone is produced by the catalytic dehydrogenation of isopropyl alcohol. Isopropyl alcohol vapours are preheated (by heat exchange with hot effluent gases from the reactor) and passed into a reactor containing a brass or copper catalyst. Reaction conditions are usually 500°C and 40 to 50 psi (275 to 350 kPa). The hot reaction gases containing acetone, isopropyl alcohol, and hydrogen pass through a water-cooled condenser and hence into a water scrubber, where final traces of isopropyl alcohol and acetone are removed from the non-condensable hydrogen. The water solution from the bottom of the scrubber is mixed with condenser product and fed to a fractionating column. Concentrated acetone is taken overhead, and an isopropyl alcohol-water mixture is removed from the bottom of the column. To recover isopropyl alcohol, the bottoms are led to a column, where a binary constant-boiling mixture containing 91% isopropyl alcohol is taken off the top and water is discharged at the bottom. The 91% isopropyl alcohol may be recycled without further purification. The water from the bottom of the alcohol concentrating column is reused in the scrubber. (Fig. 13.7.)

Reaction

$$(CH_3)_2CHOH + 1/2O_2 \text{ (air)} \rightarrow CH_3C'\ CH_3 + H_2O \text{ or}$$

$$\underset{\text{85–90\% yield}}{(CH_3)_2CHOH \rightarrow CH_3COCH_3 + H_2}$$

Other methods

Acetone is also produced by: (i) catalytic dehydrogenation of isopropyl alcohol; or (ii) as a coproduct with phenol in the oxidation of cumene. Other less significant sources include: (i) as a by-product in the production of hydrogen peroxide by the oxidation of isopropyl alcohol; (ii) as a coproduct in the manufacture of hydroquinone from p-diisopropylbenzene; (iii) as a coproduct with methyl ethyl ketone in the dehydrogenation of butenes; (iv) as a coproduct with butyl alcohol in the fermentation of carbohydrate-containing materials; (v) as a by-product of the propylene oxide process; and (vi) as a by-product of propene oxidation.

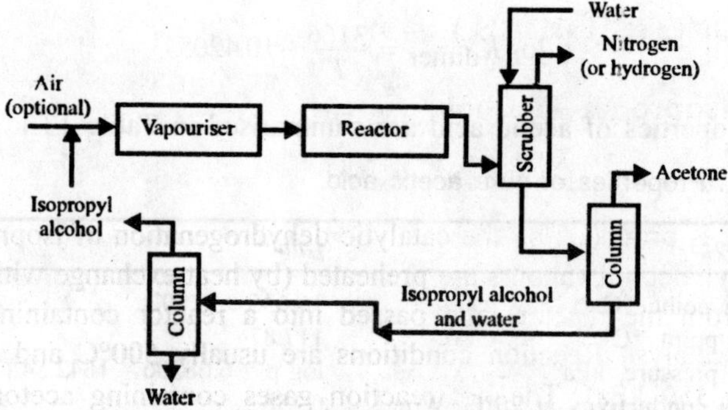

Fig. 13.7. Flow diagram for manufacture of acetone from isopropyl alcohol.

Uses

It is used in methyl methacrylate, methyl isobutyl ketone, methacrylic acid and higher methacrylates, methyl isobutyl carbinol, bisphenol A, miscellaneous chemicals (isophorone, diacetone alcohol, hexylene glycol, mesityl oxide), pharmaceuticals, solvent—cellulose acetate, spinning solvent, solvent-protective coatings and miscellaneous.

Properties

Colourless, volatile, highly flammable liquid with an ethereal odour. Mol. wt 58.08, mp –95.35°C, Sp gr 0.791 20°C/20, bp 56.1°C, weight per gallon 6.58 lb (20°C). Soluble in all proportions in water, ethanol, and ether at room temperature. Slightly soluble in chloroform. Flash point (open cup) –9°C, ignition temperature 560°C, vapour density (air = 1) 2.0, explosive limits (% by volume in air) lower 2.6-upper 12.8, threshold limit value (ppm) 1000.

ACETIC ACID

Acetic acid, CH_3COOH, is a colourless, waterlike liquid that has a piercingly sharp, vinegary odour and a burning taste. Although the formula weight of acetic acid is 60.06, its apparent molecular weight varies both with the temperature and the other associating substances present. At pressures below saturation in the temperature range 25–120°C, the vapour is well described as a mixture of the dimer and the monomer in equilibrium. Dimerisation, represented by the equation below, accounts for many of the apparent anomalies found in its equation of state, thermodynamic functions, transport properties, etc. At the atmospheric boiling point, acetic acid dehydrates to give traces of acetic anhydride and, at high temperatures in the presence of a catalyst, yields commercially useful amounts of anhydride.

$$\log K_{\text{dimer}} = \frac{3166}{T} - 10.4205$$

The properties of acetic acid are summarised in Table 13.1.

Table 13.1. Properties of pure acetic acid.

Property	*Value*
Melting point, °C	16.635 ± 0.002
Boiling point, °C	117.87
Vapour pressure, kPa[a]	log p = 6.68206 – 1642.54/(233.386 + t)
Thermal conductivity at 20°C, W/m.K	0.158
Heat of vapourisation at bp. J/g[a]	

(Contd...)

Property	Value
Specific heat, J/(g . K)[b]	394.5
of vapour at 124°C	5.029
of solid at 100 K	0.837
Density at 20.0°C, g/cm^3	1.0493
Refractive index, n_D^{25}	1.36965
Viscosity of liquid, mPa/s (= cP)	
at 20°C	11.83
at 25°C	10.97
at 40°C	8.18
Dielectric constant	
solid at 10.0°C	2.665
liquid at 20°C	6.170
Surface tension at 20.1°C, mN/m	
(= dyne cm)	27.57
Flash point, open cup, °C	57

[a]To convert kPa to mm Hg, multiply by 7.5.
[b]To convert J to cal. multiply by 0.239.

Acetic acid has important properties as a solvent for acids and bases; it protonates bases too weak to detect in water and is often used in their analysis. Other acids also exhibit very powerful acidity properties in acetic acid solutions. Most of the esters of acetic acid are prepared to direct reaction with the corresponding alcohol. A notable exception is ethyl acetate, prepared by the Tischenko reaction of acetaldehyde.

$$2CH_3CHO \rightarrow CH_3C(O)OC_2H_5$$

Some ethyl acetate is isolated as a by-product of acetic acid manufacture from liquid-phase hydrocarbon oxidation. Non-catalytic esterification of alcohols is accomplished with a two- to fivefold excess of acetic acid; water, which substantially slows the reaction, is usually removed azeotropically. Catalytic esterification of alcohols uses a mineral acid. Olefins react with anhydrous acetic acid to give esters of secondary or tertiary alcohols. Ethylene, oxygen, and acetic acid passed over a palladium-lithium acetate (with copper) catalyst produce vinyl acetate, $CH_3C(O)CH{=}CH_2$.

Ethylene oxidation in acetic acid gives glycol diacetate, which may be thermally decomposed to vinyl acetate. Acetylene adds to give vinyl acetate. Many metals, their oxides, or their carbonates, dissolve in acetic acid to give simple salts. Alkali metal acetates are prepared from corresponding hydroxides or carbonates. Transition-metal acetates can be made from free metal.

Manufacture

Virtually all acetic acid commercially produced is made by one or three routes: (i) acetaldehyde oxidation, involving direct air or oxygen oxidation of liquid acetaldehyde in the presence of manganese acetate, cobalt acetate, or copper acetate; (ii) liquid-phase oxidation of butane or naphtha, with butane yielding more acetic acid than the higher paraffins, but naphtha having the chief advantages of greater accessibility, initial lower costs, and more ready oxidation; and (iii) methanol carbonylation using a variety of techniques. Cobalt and several other carbonyl-forming metals were studied but found less active than nickel iodide catalysts at 215°C and 14 MPa (13-atm) pressure.

From Acetaldehyde

Process

A highly concentrated acetaldehyde solution (99 to 99.8%) is introduced into a stainless-steel reaction kettle in which air at 70 to 75 psi (480 to 520 kPa) bubbles through the liquid (55 to 65°C) which contains 0.1 to 0.5% manganous acetate used to decompose the explosive intermediate-peracetic acid. In batches, 4085 kg acetaldehyde can be completely oxidised in 12 hours. The gases leaving the reactor are scrubbed with water, releasing nitrogen and decomposition products to the atmosphere. The dilute acetaldehyde solution leaving the bottom of the scrubber is sent to a column for recovery. The crude acetic acid (94 to 96%) is discharged from the reactor and continuously rectified to yield 99% glacial acetic acid. (Fig. 13.8).

$$CH_3CHO + 1/2O_2 \xrightarrow[\text{88–95\% yield}]{(CH_3COO)_2Mn} CH_3COOH$$

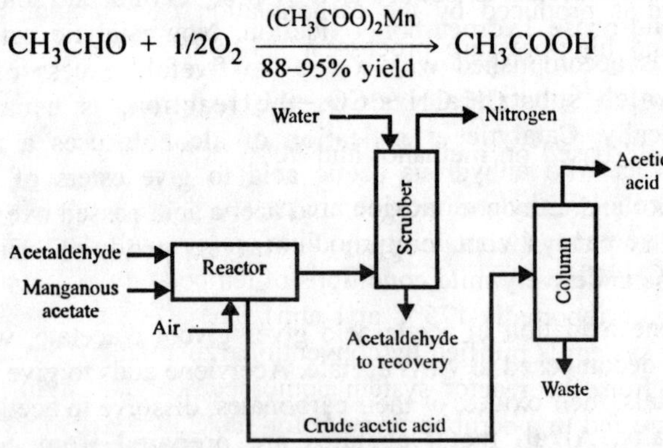

Fig. 13.8. Flow diagram for manufacture of acetic acid from acetaldehyde.

In some modern plants the oxidation is carried out continuously rather than in batches. The catalyst in this case is cobalt acetate dissolved in acetic acid. The reactor is operated at 15 psi (100 kPa) and 70 to 80°C. Approximately 4 moles of air enter the converter per mole of acetaldehyde. Conversions of 20 to 30% per pass and yields of 95 to 96% of theory may be obtained. If oxygen is used in place of air, the pressure in the system may be reduced to 30 psi (210 kPa) and good recovery obtained.

An alternative vapour-phase oxidation process currently in use employs a mixture of alcohol and acetaldehyde, which on catalytic air oxidation yields both acetaldehyde and acetic acid as products. By recycling the acetaldehyde, adding only alcohol to the system as makeup, and removing only acetic acid, the process becomes essentially a catalytic vapour-phase alcohol oxidation to acetic acid.

In a liquid-phase version of the process, the reaction is carried out in an acetic acid solution, using a mixture of cobalt and chromium acetates as the catalyst. Feed to the reactor is in the ratio of 11 moles of ethyl alcohol to 89 moles of acetaldehyde. The acetic acid yield is 95%.

In still another version a two-step process is used, consisting of an alcohol dehydrogenation unit followed by intermediate fractionation units from which acetaldehyde is delivered continuously to an oxidation unit where acetic acid is formed. Unconverted alcohol and by-product ethyl acetate from the dehydrogenation unit are separated and the alcohol recycled.

From Carbonylation of Methanol (Monsanto Process)

Process

Acetic acid is produced by the carbonylation of methanol in both low-pressure and high-pressure processes.

$$CH_3OH + CO \rightarrow CH_3COOH$$

99 + % yield based on methanol and 90 + % based on carbon monoxide.

Methanol and carbon monoxide are passed through a continuous reactor system with a catalyst containing rhodium and iodine. The reaction is in the liquid phase under very mild conditions of temperature and pressure. (These conditions are reportedly 175°C at 1 atm). The acetic acid withdrawn from the reactor system is purified by conventional distillation. Small amounts of light ends from the reactor system combined with the light ends column overhead are fed to a scrubber system for subsequent recycle to the reactor system. Vent gases from the scrubber to the flare and to the atmosphere at specific locations are continuously monitored by on-stream process analysers. (Fig. 13.9).

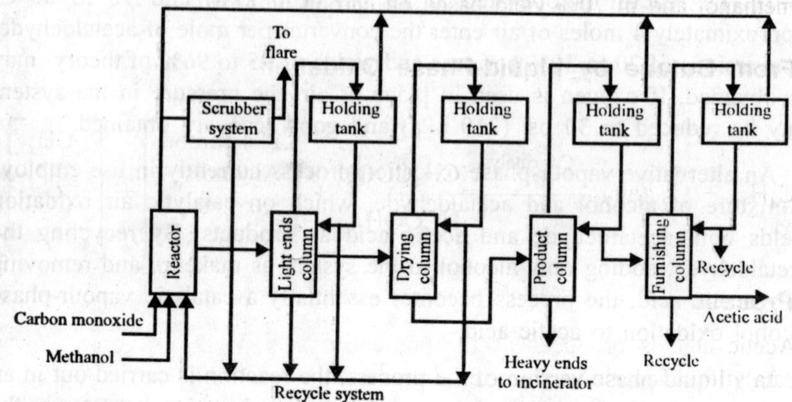

Fig. 13.9. Flow diagram for manufacture of acetic acid from carbonylation of methanol.

The wet acetic acid, taken as a side stream from the light ends column, is sent to the drying column. From the drying column dry acetic acid is removed as bottoms and fed to the product column. Small quantities of propionic and acetic acids are removed as bottoms. The overhead from the product column is fed to the finishing column, where acetic acid of high purity is recovered as a side draw. Overhead and bottoms from the finishing column, as well as the wet acid taken as overhead from the drying column, are recycled to the reaction system. The acetic acid product has a purity of 99.9%. The yield is 99.5% based on the methanol feed.

The advantages of this process are the extremely high yield of acetic acid, the reduction in capital and operating cost for the reaction and purification systems as compared with other processes, and the absence of significant amounts of by-products even in the presence of hydrogen.

Another process used commercially for the direct carbonylation of methanol is the BASF process. In this Reppe-based process, carbon monoxide is bubbled through liquid methanol in a reactor maintained at about 250°C and 7500 psi (52 MPa) in the presence of a cobalt iodide catalyst. Unreacted gas and crude acid are taken from the top of the reactor, cooled, and reduced in pressure at 10 atm. Methyl iodide in the unreacted gas is recovered by washing the vent gas with feed methanol. Methanol is recycled, and the washed gas goes to the fuel supply.

After degassing, crude acid is first distilled to separate low boilers, then separated from the catalyst, and finally purified by azeotropic distillation. Acetic acid of over 99.8% purity is recovered, in 90% yield based on methanol and in 70% yield based on carbon monoxide.

From Butane by Liquid-Phase Oxidation

$$C_4H_{10} + O_2 \text{ (air)} \xrightarrow{\text{Co catalyst}} \begin{matrix} CH_3COOH \\ + \\ HCOOH \end{matrix} + \begin{bmatrix} \text{2–4 carbon} \\ \text{Alcohols} \\ \text{Ketones} \\ \text{Acids} \end{bmatrix} + \begin{bmatrix} CH_4 \\ CO_2 \\ CO \\ N_2 \end{bmatrix}$$

Process

Acetic acid is produced as the predominant product by the liquid-phase catalytic oxidation of butane. Commercial butane (95% *n*-butane, 2.5% isobutane, and 2.5% pentane) is charged to a reactor at 800 psi (5.5 MPa) which also contains an acetic acid solution of cobalt, manganese, or chromium acetate. Air is bubbled through the reactant solution at 150 to 250°C. About 5 parts of air (by weight) are required for 1 part of butane. Unreacted nitrogen leaving the reactor carries various oxidation products (formic, acetic, and propionic acids; acetone; methyl ethyl ketone; methanol; etc.) and some unreacted butane with it. Unlike vapour-phase oxidation processes, aldehydes are not formed. The vapours leaving the reactor are condensed and then separated by methods similar to those described for the vapour-phase oxidation of butane (Fig. 13.10). Under proper reaction conditions the principal product is acetic acid; another set of conditions yields methyl ethyl ketone as the principal product.

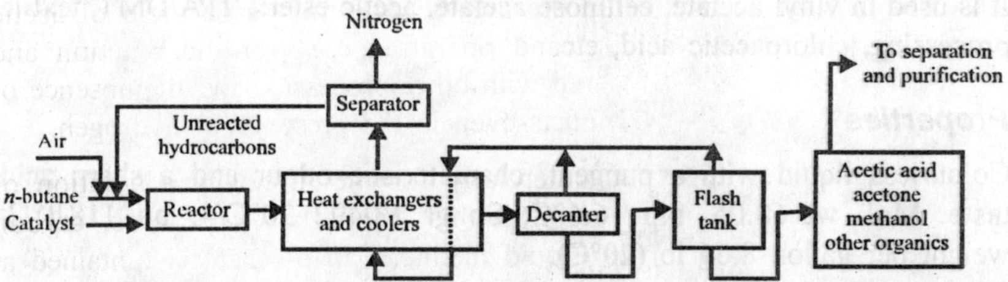

Fig. 13.10. Flow diagram for manufacture of acetic acid from butane by liquid-phase oxidation.

When a cobalt catalyst is used, the product distribution in weight percent is; acetic acid-76; formic acid-6; ethanol-6; methanol-4; other organics-8. Under the acidic conditions used, the alcohols show up principally as acetates. When a manganese catalyst is used, the weight percentage of formic acid is 23, largely at the expense of acetic acid production.

From Wood Distillation

Pyroligneous acid liquor is one of the products obtained by the destructive distillation of hardwood. The wood is charred in either continuous or batchwise retorts in the absence of air. Processes vary in procedure in such details as size and type of wood used and length of predrying periods (2 days to 18 months), but all produce about the same quality and quantity of charcoal and raw pyroligneous liquor. One cord (1814 kg) of hardwood yields about 950 litres (1025 kg) of pyroligneous acid liquor (containing 7% acetic acid or equivalent, 4% crude methanol and acetone, 9% tar and oils, and 80% water), 454 kg of charcoal, and 212 m^3 of fuel gas (53% carbon dioxide, 27% carbon monoxide, and 15% methane) having an average heating value of 2225 to 2670 kcal/m^3.

The charcoal is cooled, graded, and packaged. The vapours are condensed, and the non-condensable gases are used for fuel. The resulting liquor is allowed to settle to remove insoluble tar, and the clear decanted liquor containing 6 to 7% acetic acid is subjected to extraction or distillation or both to separate the acid from methanol and tar. Recovery of acetic acid from the pyroligneous liquor varies from 95 to 98%. For each metric ton of acetic acid (100%) produced, 625 to 800 litres of crude methanol is recovered.

Uses

It is used in vinyl acetate, cellulose acetate, acetic esters, TPA/DMT, textile processing, chloroacetic acid, etc.

Properties

Colourless liquid with a pungent, characteristic odour and a sharp acid taste. Mol. wt 60.05, mp 16.6°C, Sp gr 1.0492 20°C/4, bp 118.1°C, weight per gallon 8.64 lb (20°C).

Soluble in all proportions with water, ethanol, glycerine, and ether at room temperature. Insoluble in carbon disulphide.

Flash point (closed cup) 39°C, ignition temperature 516°C, vapour density (air = 1) 2.07, explosive limits (% by volume in air) lower 6.6, higher 19.9, threshold limit value (ppm) 10.

Economic Aspects

Acetic acid has a place in the organic chemical industry comparable to sulphuric acid in the inorganic chemical industry. Acetic acid is used for the production of vinyl acetate, an important monomer. Acetic acid is used in the fermentation industry in large-scale amino acids production and as the substrate of choice for the production of many dietary supplements, e.g., glutamic acid, citric acid and lysine. Cellulose acetate was formerly the foremost of synthetic polymers and is still an important thermoplastic and fibre constituents used in the manufacture of plastic sheeting film and in the formulation of lacquers.

Acetic acid is used as the solvent in the liquid-phase oxidation of p-xylene to terephthalic acid. The principal use for acetate esters, from methyl through amyl, is as solvents for lacquers and coating. These solvents are attractive because of their great convenience and low price. All acetate solvents may be made obsolete by air-pollution legislation, although these laws have not been promulgated as rapidly as first expected. Acetic acid is an effective fungicide used to control the fungi that cause ropiness in bread; to preserve high moisture grain, and in combination with propionic acid, to combat.

Health and Safety Factors

Any solution containing more than half acetic acid in water or organic solvents is considered corrosive. These solutions cause damage to skin, eyes, nose, and mouth. There is no burning sensation upon application to unbroken skin, although symptoms appear as blisters within 30 minutes to 4 hours. There is little or no pain experienced, unless sensory nerve receptors are attacked. Medical care should be sought immediately. Particular care is needed when handling acetic acid to avoid spilling it on the skin or clothing and to avoid breathing the vapours. When exposed, wash acetic acid with large amounts of water or dilute sodium bicarbonate solution. Washing blisters that have already appeared will not alleviate the pain. Leave areas where the odour of acetic acid vapours is noticeable. Glacial acetic acid causes severe gastric difficulties if swallowed, but the precise toxic dose for humans is not known.

VINEGAR

Vinegar is used as a preservative in pickling and to give a sharp or sour taste to foods. Most table vinegars derive from the acetic acid-bacterial fermentation of wine or cider. These are produced by alcoholic fermentation of dilute sugar solutions, such as grape juice, cider, or malt. (Table 13.2).

A number of factors govern the composition of vinegar, including the nature of the raw material, the substances added to promote alcoholic fermentation and the growth and activity of *Acetobacter*, the procedure used for the acetification, and finally the aging, stabilisation, and bottling operations.

Table. 13.2. Raw materials and fermentation products.

Raw materials	*Products*
Tropical fruits	Wines vinegars
Mango waste	Syrups, wines vinegars
Whey	Vinegar
Rejected bananas	Vinegar
Palm sap	Vinegar
Dates	Vinegar
White soy sauce	Rice vinegar
Enzyme preparation	Rice vinegar

Manufacture

Vinegar is produced by surface or submerged-culture oxidation. Production is accelerated by increasing thc surface of the liquid and bacteria exposed to oxygen by passing the liquid through a column, tank, or vat packed with inert material on which the bacteria are absorbed.

In starch hydrolysis, yeasts cannot utilise starch as carbon source for growth, and the starch must be hydrolysed to sugar. Malt vinegars are made from malted barley which may be mixed with other starches or grains. The enzymes present hydrolyse the starch to glucose and maltose, which are readily fermented by *Saccharomyces* yeast. In Japan, vinegars are made from rice.

In alcoholic fermentation, glucose from hydrolysed starch or the glucose and fructose from fruits and honey are converted by yeast enzymes to ethyl alcohol and carbon dioxide. Both the fermentation of the sugars and the oxidation of ethanol are exothermic processes. The excess heat must be removed during the oxidation. Practical yields of acetic acid are 77–85%.

In the Orleans process, the wine oxidises slowly in a barrel where it is covered with a film of *Acetobacter*. Holes in each barrel head, covered with screens, permit access to atmospheric oxygen. Wines with 10–12% ethanol give vinegars of 8–10% acetic acid concentration. Orleans vinegars have a relatively high concentration of ethyl acetate, detected by its characteristic odour.

In generator processes, generators are filled with beech wood shavings which provide a loose packing that allows open spaces for the free flow of liquid and air; cooling coils are provided. In submerged-culture generators, a mechanical system keeps the bacteria in suspension in the liquid in the tank and in intimate contact with fine bubbles of air. The Frings acetator uses a bottom-driven hollow rotor turning in a field of stationary vanes; the air which is drawn in is intimately mixed with the liquid throughout the bottom area of the tank. Submerged-culture oxidisers are smaller and more flexible in operation. They are usually operated on a semi-continuous basis. Foam production impedes bacterial growth and is minimised by keeping nutrient, ethanol, and acetic acid concentrations in the optimum ranges. Dead cells promote foaming, which is usually broken by centrifugation.

In the tower fermenter the liquid is held without packing on a porous plate by the pressure of air introduced below.

Concentrated vinegars

The regulations require at least 4 g acetic acid per 100 cm^3 vinegar. Submerged-culture oxidisers can give concentrations of 10–13 g/100 cm^3. Continuous aeration and carefully stepwise addition of ethanol seem to be the keys to successful operation.

Although aesthetically undesirable, nematodes, known as vinegar eels, consume dead bacteria, prolong the operation, and make nutrients more readily available to *Acetobacter*. "Mother of vinegar" is the cellulosic slime coating the bacterial cells. It may block passageways in packed-tank generators.

Processing

Raw vinegars vary widely in stability and may contain materials that form cloudiness or deposits. Vinegars that carry a high and cloudy suspension of bacterial cells are clarified and stabilised with bentonite and similar agents or activated carbon. Membrane filtration can be combined with aseptic bottling to provide a product free of all micro-organisms. Wine, cider, and malt vinegars benefit from aging, which improves the character of the product. Vinegars bottled for table use or pickling usually are pasteurised at 77–78°C.

CITRIC ACID

Citric acid (2-hydroxy-1,2,3-propanetricarboxylic acid), $C_6H_8O_7$, a natural constituent and common metabolite of plants and animals, is the most versatile and widely used organic acid in the field of foods and pharmaceuticals.

Citric acid is also used in many industrial applications to sequester ions, neutralise bases, and act as a buffer. In cosmetics it is used as a buffer to control pH in shampoos, hair rinses, and setting lotions. Many citrates, especially the neutral sodium salt, are used extensively in food and pharmaceutical products and in detergents. Esters of citric acid are used commercially as plasticisers in the preparation of polymer compositions, protective coatings and adhesives.

Physical Properties

Anhydrous citric acid, mol. wt. 192.13, crystallises from hot concentrated aqueous solutions. The crystals are translucent and colourless; they belong to the holohedral class of the monoclinic system. The melting point of the anhydrous form is 153°C; the density is 1.665. A monohydrate may also be formed. Citric acid is optically inactive and manifests no piezoelectric effect.

Citric acid is a relatively strong organic acid as indicated by the first dissociation constant, 8.2×10^{-4} at 18°C. Second and third dissociation constants are 1.77×10^{-5} and 3.9×10^{-7}, respectively. pK values at 25°C are pK_1 3.128, pK_2 4.761, and pK_3 6.396. Citric acid is readily soluble in water and moderately soluble in alcohol.

Chemical Properties

As a tribasic acid, citric acid manifests the usual properties of a polybasic acid. It forms a variety of salts, including those of the alkali metal and alkaline earth metal families. In addition, citric acid forms a variety of esters, amides, and acyl chlorides. Mixed compounds such as the salts of the acid esters can also be formed. The anhydride itself cannot be formed, but acyl derivatives of the acid can be dehydrated to form acyl citric anhydrides. The hydroxyl group may form acyl derivatives, ethers, etc. A wide variety of such mixed compounds is possible, and many have been prepared and studied. In aqueous solution, citric acid can be mildly corrosive toward carbon steels and, therefore, should be used with an appropriate inhibitor. It is not corrosive to stainless steels which are most often employed as the material of construction for processes involving citric acid.

At proper pH in aqueous media, citric acid's hydroxyl and carboxylic acid groups act as multidentate ligands forming complexes or chelates with metal ions. These chelating reactions are the basis for many of today's industrial processes, including elimination or control of metal-ion catalysis, lowering of metal oxidation potentials, removal of corrosion products (i.e., Fe^{3+}), regeneration of ion-exchange resins, recovery of valuable metals by precipitation of insoluble chelates, decontamination of radioactive materials, quenching reactions, and driving reactions to completion.

Occurrence

Citric acid occurs widely in both the plant and animal kingdoms. It is found most abundantly in the fruits of the citrus species such as lemons (4.0–8.0%), grapefruit (1.2–2.1%), tangerines (0.9–1.2%), oranges (0.6–1.0%), and limes (ca 7.0%). The citrate ion occurs in all animal tissues and fluids, e.g., in human whole blood (15 ppm). The total circulating citric acid in the serum of humans is approximately 1 mg/kg body weight. Citric acid occurs in the terminal oxidative metabolic system of all but a few organisms. This system, variously referred to as the Krebs cycle, the tricarboxylic acid cycle, or the citric acid cycle, is a metabolic intermediate cycle involving the terminal steps in the conversion of carbohydrates, fats, or proteins to carbon dioxide and water with concommitant release of energy necessary for growth, movement, luminescence, chemosynthesis, and reproduction. This cycle also provides the carbonaceous materials from which amino acids and fats are synthesised by the cell.

From Molasses by Fermentation

Process

Beet molasses (about 41°Bé, sp gr 1.40) containing 48 to 52% sugar by weight or cane molasses (about 40° Bé, sp gr 1.39) of the blackstrap (52 to 57% sugar) or high-test (70 to 80% sugar) variety is introduced into a mixer. Dilute sulphuric acid is added to adjust the pH of the molasses between 5.5 and 6.5. Phosphorus, potassium, and nitrogen in the form of acids or salts are added as nutrients for proper mould growth in amounts depending on the nature of the raw material. The mix is then sterilised with live steam for about 0.5 hour and is finally diluted with water to a 15 to 20% sugar concentration. (Fig. 13.11)

$$\underset{\text{Sucrose or dextrose}}{C_{12}H_{22}O_{11} \text{ or } C_6H_{12}O_6} + H_2O \xrightarrow[\text{Air}]{\text{Organism}} \underset{\text{Citric acid}}{C_6H_8O_7} + CO_2 + H_2O$$

35–65% weight yield

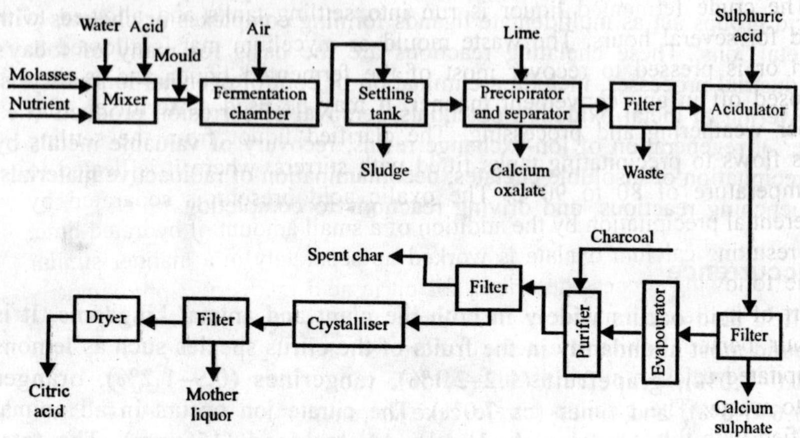

Fig. 13.11. Flow diagram for manufacture of citric acid from molasses by fermentation.

This ferment medium flows by gravity into shallow aluminium pans or trays, which are arranged in tiers in sterile fermentation chambers. The size and number of trays and chambers vary from plant to plant, but most chambers have provision for regulation and control of temperature, relative humidity, and air circulation. One plant has 80 trays per chamber with each tray holding about 400 litres of solution at a depth of 76 mm.

The culture is prepared in a laboratory, using a special strain of *Aspergillus niger*. When the fermentation medium has cooled to about 30°C, it is inoculated with spores of this organism.

The fermentation requires 8 to 12 days, during which time evaporation reduces the volume of solution in the each tray. The medium becomes more acid as the fermentation continues, and the course of the reaction may be followed by analyses; the pH of the solution drops to about 2 at the end of the fermentation, and the acid content varies from 10 to 20%. The exact mechanism of the fermentation process is not well known, even though many hypotheses have been proposed. The temperature of the medium during fermentation is maintained at 28 to 32°C. Even with careful control of the temperature and the selection of high citric acid-producing organisms, some oxalic and gluconic acids are formed. Purified air is generally circulated through the fermentation chamber. and the relative humidity is controlled between 40 and 60%.

After the completion of each run, the fermentation chamber and trays are thoroughly sterilised, using water, dilute formaldehyde solution, and sulphur dioxide gas.

The crude fermented liquor is run into settling tanks and allowed to stand for several hours. The waste mould or mycelium mat is allowed to drain or is pressed to recover most of the fermented liquor and is then disposed off in any convenient manner. It may be used as fertiliser after proper weathering and processing. The clarified liquor from the settling tanks flows to precipitating tanks fitted with stirrers, where it is heated to a temperature of 80 to 90°C. The oxalic acid present is separated by preferential precipitation by the addition of a small amount of hydrated lime. The resulting calcium oxalate is worked up separately in a manner similar to the following process described for citric acid recovery. Approximately 1 part of hydrated lime for every 2 parts of liquor is slowly added over a 1 hour period, while the temperature is raised to about 95°C. The precipitated calcium citrate is filtered on a vacuum filter, and the filtrate is run to waste. The calcium citrate cake is run to acidulators, where it is acidified with dilute sulphuric acid. It is then filtered (the cake consists of calcium sulphate), and the citric acid mother liquor is decolourised by a charcoal treatment. From the purifier the liquor goes to a vacuum evaporator, where it is concentrated. It is then run into a crystalliser where, on cooling, citric acid crystallises, generally in the form of the monohydrate. The resulting acid is of I.P. grade, and the overall yield from the process varies from 35 to 65% by weight, based on the sugar content of the raw material.

By Submerged Fermentation

Most newly constructed plants have adopted a submerged or deep-fermentation process rather than the shallow or tray fermentation described in the foregoing. Generally, submerged fermentation is a two-stage process. There is a "growth" stage and a "fermentation" stage, each employing a different nutrient medium. After 3 or 4 days of growth the mycelia are separated from solution and added to the fermentation solution which may then consist of sucrose (200 g/litre), dry mycelia (5 g/litre), ammonium nitrate (1.1 g/litre), potassium chloride (0.15 g/litre), magnesium sulphate heptahydrate (0.25 g/litre), and sufficient hydrochloric acid for pH adjustment.

The fermentation is carried out at 25°C while oxygen is dispersed through the liquor. After 4 days about 70% of the sugar has been converted to citric acid.

As with the tray process, each manufacturer uses certain modifications of his own. These may be in the choice of the mould (although *Aspergillus niger* is still common) or the addition of a fermentation accelerator (such as activated carbon or ascorbic acid). Recovery of citric acid from the fermented liquor is essentially the same as for the tray process.

The first submerged fermentation processes gave poor yields on a commercial scale, but apparently this has been overcome by adding methanol to the fermentation solution.

Uses

Beverages, food and candy, pharmaceuticals, sodium salts and esters, cosmetics, miscellaneous and export.

Properties

Citric acid is generally produced in the anhydrous form by direct crystallisation. It is also commercially available as the monohydrate in the form of colourless, translucent crystals or white, crystalline powder. It is odourless, has an acid taste, and is efflorescent in dry air.

Anhydrous ($C_6H_8O_7$)		Monohydrate ($C_6H_8O_7.H_2O$)
Mol. wt	192.12,	210.14
Sp. gr	1.665 20°C/4	1.542
mp	153°C	Loses H_2O at 70 to 75°C
bp		Decomposes about 175°C

Soluble in cold water (133 g/100 g), cold (15°C) absolute ethanol (76 g/100 g), and cold (15°C) ether (2.2 g/100 g).

Chapter 14

Food Processing Industries

INTRODUCTION

Food processing, regardless of the type of food, can be divided into three classes: separation (e.g, rendering, skimming, boning, coring, defeathering, husking, peeling, shelling, etc.); assembly (e.g, coating, enrobing, baking, homogenisation, roasting, pelleting, stuffing, etc.); and preservation. These can occur at harvest, at the food-processing plant, or even at the point of retail sales.

Foods can be categorised as living-tissue or raw foods, and as non-living tissue. Living-tissue foods include fresh fruits, vegetables, meats, and grains.

The genetically controlled taste, colour, nutritive value, cell structure, and conformation of fresh foods usually are best at harvest. Protection of these qualities is achieved by retarding or inhibiting the detrimental action of microbes and enzymes, insects, bruising, and chemical degradation, while maintaining the integrity of the membranes, enzyme systems, and gross structure of the food.

The intensity of processing operations is limited by the tissue itself in terms of temperature range, water activity, respiratory-gas composition, mechanical stresses, and concentrations of chemicals. Except for certain fruits and vegetables, storage should be to 0°C without ice formation.

Most processed foods, e.g., canned, frozen, and dried products, are marketed as non-living tissue or manufactured foods. Separation, assembly, and preservation operations determine the final quality of the product. Non-living tissue products must be defined in terms of composition, absence of defects, nutritional value, fill of container, etc, rather than of genetic quality.

Table 14.1. Preservation unit operations of the food industry arranged by mode of preservation and spoilage vectors.

	Mode of preservation				
	Physical				
Spoilage vector	*Heat*	*Cold*	*Drying*	*Chemical*	*Mechanical*
Microbes	Frying, boiling pasteurising retorting, pasteurisation	Cooling, freezing	Dehydration, desiccation, evaporation, lyophilisation	Acidification, brining, fermentation, fumigating, irradiation, isomerisation, pickling, smoking	Cleaning, centrifugation, compressing, washing
Enzymes	Blanching, boiling, scalding,	Hydrocooling		Acidification	
Chemical	Exhausting, deodorising	Chilling		Hydrogenation, esterification, smoking	Cleaning, degassing, washing
Mechanical damage		Freezing			Aspiration, cleaning, centrifugation, sorting, washing

Table 14.1 lists modes of preservation used to treat various spoilage vectors by the food industry. Typical processing temperatures, pressures, and pH values are shown in Table 14.2.

Table 14.2. Typical temperatures, pressures, and pH values in food-processing operations.

Temperature	*Example of use*	*Limits of application*
Heat-transfer media		
Cryogenic freezing liquid nitrogen, - 196°C	Rapid freezing to minimise cell damage, moisture loss	Cost, stress cracking during freezing
Solid carbon dioxide, –78.5°C	Rapid freezing	cost
air, plate, and aqueous base freezants, –40 to –5°C	Commercial freezing parasite and insect destruction, commercial freezing, storage; freeze drying	Foods not suitable for freezing
Refrigerated storage, 0–10°C	Commercial refrigerated storage	Microbial growth possible
Room temperature storage, air, 20–40°C	Canned and dry food storage	Suitable only for preserved and packaged foods
Water; air; atmospheric steam, 50–100°C	Pasteurisation, milk, eggs, blanching, sterilisation of acid foods (pH 4.5); air drying (product temperature): "cooking"	Excessive times will cause poor colour, flavour, and structure
Steam, 110–130°C	Thermal sterilisation of non-acid foods; destruction of anti-nutritive factors	Rapid thermal degradation of nutrients, pigments, structure, flavours
Oil, steam, infrared radiation, 180°C	Frying, roasting, baking, generation of browning reaction products	Short duration surface treatments
Various infrared radiation sources, 180°C	Surface heating for flash drying, peeling	Charring, pyrolysis without precision control
Pressure[a]		
<0.5 kPa	Freeze-drying	Cost
0.5-4 kPa	Hypobaric storage, deaeration, vacuum concentration, vacuum cooling	Cost
101 kPa	Most processing operations	
0.1-1 MPa	Steam sterilisation, over	Cost

(Contd...)

Temperature	*Example of use*	*Limits of application*
	pressure for glass and pouch packs; carbonated and aerosol packaged foods	
> 1 MPa	Extruders, hydraulic pressing, homogenisation, potential microbial and enzyme inactivation at 1 GPa (100 atm) and higher	Cost, specialised products
Hydrogen-ion concentration		
pH 1	Acid hydrolysis	Neutralise to pH range of 7
pH 2.5	Lemons, limes, vinegar, organic acids	Taste
pH 3-4.5	Fruits, acidified foods	Non-compatability of foods; protein denaturation
pH 7.35	Normal pH of most foods	
pH 8-9	Solubilisation of certain proteins for extraction, alkali process cocoa	Neutralise to normal pH for use
pH 12	Limed corn	Colour, flavour, nutrient loss
33% sodium hydroxide	Peeling of fruits and vegetables	Surface treatment only, neutralise to normal pH

[a] To convert kPa to mm Hg, multiply by 7.5; to convert MPa to atm, divide by 0.101.

Toxicological Implications of Food Deterioration and Spoilage

A primary objective of food processing is to ensure a safe food supply. Certain harvested foods can contain naturally occurring poisonous substances, compounds with pharmacological effects, or compounds that can interfere with the utilisation of nutrients present in other foods when they are consumed together. Foods just harvested also may contain or become contaminated with spore or vegetative forms of microbes, parasites, or the wastes of higher life forms, as well as pesticides and other chemicals used for crop protection.

A third source of potential health hazards in food supplies is toxic compounds formed during the chemical degradation of processed and stored foods.

Food-Processing Facilities

Factors important in the selection of sites for food-processing operations include: availability of raw materials; abundant potable water supply; low cost waste-disposal facilities; adequate low cost energy supply; adequate seasonal and non-seasonal labour supply; ease of access to rail and truck transportation; proximity to consumers; and adequate storage areas.

Food Packaging

Food packaging includes rigid glass or metal containers, easily moulded paper, plastics, and aluminium, and pouches formed by heat-sealing paper-foil-plastic laminates.

Because of the potential migration of packaging materials into foods, current food and drug regulations should be consulted prior to marketing foods packaged in non-standard materials.

FOODS (NON-CONVENTIONAL)

Non-conventional food differ from the usual materials of plant and animal origin that are used for human food or animal feed in that they can be produced from chemicals such as carbohydrates, hydrocarbons, or industrial organic chemicals by processes such as microbiological, enzymatic, or chemical synthesis, or from existing natural products containing carbohydrates, proteins, and fats and by physical, chemical, microbiological, or enzymatic modification.

Single-Cell Protein (SCP)

Two broad classes of micro-organisms are of interest for single-cell protein (SCP) production: photosynthetic organisms including the algae and certain bacteria; and non-photosynthetic organisms, including bacteria, actinomycetes, yeasts, moulds, and higher fungi. In addition, two different uses of SCP are distinguished: food for humans and feed for animals.

Photosynthetic organisms

Mass cultivation of algae in ponds or tanks under photosynthetic conditions, using incident sunlight as the energy sources and CO_2 as the carbon source, has been investigated in Japan, Mexico, Algeria, India, and in California in the United States. Artificial-illumination systems have been used for experimental mass cultivation of algae in Japan and erstwhile Czechoslovakia and in bioregenerative systems for converting CO_2 and human wastes into breathable oxygen and food as part of life-support systems for long duration, space-exploration missions.

Key factors influencing growth include temperatures, pH, availability of CO_2, nitrogen, phosphorus, and other inorganic nutrients, and availability of sunlight as influenced by latitude, cloud cover, and depth of the culture pond or tank. Slow, erratic growth results from wide fluctuations between day and night temperatures in outdoor ponds, and from season to season.

Table 14.3 lists proximate analyses of selected algae as determined by animal feeding studies. It is important to note that the N × 6.25 factor used to calculate the crude protein contents of the algae does not reflect accurately the true protein content because algal cells may contain non-protein nitrogen substances.

Algae tend to have lower contents of methionine than is desirable in human and animal nutrition and supplementation with this amino acid is necessary with many species. *Spirulina maxima* apparently has a higher protein quality than other species since its protein efficiency ration (PER) is nearly equivalent to that of casein (2.5%).

Dried algae are not accepted readily by livestock. Palatability of dried algae for swine was improved by pelletising with steam-rolled barley. In general, many species of algae have cell walls that are resistant to digestive enzymes, have dark colours, and bitter flavours; all of which must be altered to make an acceptable food or feed product.

Non-photosynthetic organisms

Non-photosynthetic micro-organisms of interest in SCP production include bacteria, actinomycetes, yeasts, molds, and higher fungi. Carbon and energy sources that have been considered for growing these organisms include carbohydrates such as simple sugars, starches, cellulose, and agricultural, forestry, pulp, and paper, or food processing wastes containing these carbohydrates, hydrocarbons, and chemicals derived from them including alcohols and organic acids. Commercial-scale operations are conducted in batch, semi-continuous or continuous culture systems.

The product quality considerations for non-photosynthetic micro-organisms are similar to those mentioned previously for algae. Table 14.4 presents typical proximate analyses for selected bacteria, yeasts, moulds, and higher fungi that have been produced on a large pilot-plant or commercial scale.

Table 14.3. Proximate analyses of selected algae grown photosynthetically.

Organism	Content, % (dry wt basis) Nitrogen	Crude protein N × 6.25	Fat	Ash
Chlorella pyrenoidosa 71105 (Sorokiniana)	9.6	60	8.1	8.9
Chlorella regularis	9.3	58	16	6.7
Chlorella ellipsoidea and Scenedesmus obliquus	9.7	61	19.5	13.4
Scenedesmus acutus 276-3A	8-8.8	50-55	12-19	6-8
Spirulina maxima	9.9	62	2-3	

Table 14.4. Proximate analyses of selected non-photosynthetic micro-organisms grown on various substrates.

Organism	Substrate	Composition, g/100 g dry wt Nitrogen	Protein	Fat	Crude fibre	Ash
Bacteria						
Acinetobacter cerificans	Hexadecane	11	72			
Cellulomonas sp	Bagasse	14	87	8		7
Methylophilus methylotrophus	Methanol	13	83	7	< 0.05	8.6
Yeast						
Candida lipolytica (Toprina)	*n*-alkanes	10	65	8.1		6
Candida lipolytica (Toprina)	Gas oil	11	69	1.5		8
Candida utilis	Ethanol	8.3	52	7	5	8
	sulphite liquor	9	55	5		8
Kluyveromyces fragilis	Cheese whey	9	54	1		9
Saccharomyces cerevisae	Molasses	8.4	53	6.3		7.3
Moulds and higher fungi						
Morchella hortensis	Glucose	5.4	34	1.4		
Paecilomyces variboti (Pekilo)	Sulphite waste liquor	9.1–10.11	57–63			

It is apparent that most of the bacteria, yeasts, moulds, and higher fungi of interest for SCP production are deficient in methionine and must be supplemented with this amino acid to be suitable for animal feeding or human food applications. Also, lysine-arginine ratios should be adjusted in poultry ratios in which yeast SCP is used. Human feeding studies have shown that only limited quantities of yeast such as *Candida utilis* can be added to food products without adverse effects on flavour.

Derived Plant and Animal Products

Plant protein products—leaf protein concentrates

Leaf protein concentrates (LPC) are prepared by crushing plant material, extracting the juices, using the juice *per se* or recovering protein from the juice by heating or chemical precipitation.

Table 14.5 gives proximate analyses of several LPC products. In general, the composition of LPCs varies widely with the raw materials and processes used. Typical protein-efficiency-ratio values for LPCs derived from alfalfa range from 1.87 without supplementation to 2.57 when 0.4% methionine was added as compared with cascin adjusted to a PER of 2.50.

It is apparent that there are definite limits to the use of LPC product in human diets and that raw materials used for LPC production should be evaluated for possible allergenic problems.

Plant protein products—seed-meal concentrates and isolates

Seed-meal protein concentrates and isolates, particularly soy protein products, can be used as extenders for meat, seafood, poultry, eggs, or cheese.

Animal proteins—fish-protein concentrates

Fish-protein concentrates (FPC) are distinguished from fish meal and fish solubles in that the former are produced for human food use from whole edible species of fish using sanitary processing methods rather than for animal feed.

Synthetic Protein Products

Plastein synthesis

Plasteins are mixtures of high molecular weight proteinaceous peptides. They are synthesised by enzyme-catalysed growth of peptide chains from lower molecular weight peptides. The process by which plasteins are formed is called the plastein reaction and is the reverse of the proteolytic enzyme hydrolysis of peptide bonds of proteins. At present, plasteins are in the experimental stage of development.

Table 14.5. Proximate analysis of leaf protein products.

	Composition, % (dry wt)			
Leaf proteins	*Protein N × 6.25*	*Fat*	*Fibre*	*Ash*
Pro-Xan II (alfalfa)	47	14	4	16
Cauliflower (pressed leaves)	26	2.9	12.5	10.7
Pea vine (spray-dried juice)	18.5	2.5	1.4	12.8

Synthetic proteins

Apparently, these protenoids are digestible by mammalian proteinases and can serve as sources of nutrients for *Lactobacillus arabinosus* and *Proteus vulgaris.* The possibility exists that products could be nutritionally imbalanced, have mammalian toxicity and undesirable tastes, odours and stability. These problems must be investigated further before any assessment of the utility of polyamino acids or protenoids can be made.

Textured and Formulated Foods

The availability of protein concentrates and isolates prepared from soyabean protein has made it possible to produce a wide range of textured protein analogue foods in which the soy protein is substituted for animal protein.

FOOD TOXICANTS (NATURALLY OCCURRING)

Toxicants are substances that, upon ingestion, produce changes in homeostasis that are threatening to the normal function of the organism. Some naturally occurring food toxicants are shown in Table 14.6.

Toxic Proteins, Peptides, Amides and Amino Acids

Nitrogenous compounds are present in every living cell, and are the most frequently implicated natural toxicants in foods. Their mode of action is often obscure but they may be loosely grouped according to either gross manifestations or specific structural characteristics. Accordingly, vitamin-destroying enzymes, hemaglutenins, enzyme inhibitors, and many hepatotoxins (many of which are carcinogens) are of protein, peptide, or amino acid composition.

Thiocyanates and Related S-Containing Compounds

Goitrogens are compounds that produce goiter by interfering with thyroxine synthesis in the thyroid gland. Food-borne goitrogens are often characterised by the presence of sulphur and most are thiocyanates or closely related compounds. Because of their widespread occurrence in *Cruciferae*, e.g., cabbage, kale, onions, cress, broccoli, cauliflower, rhubarb, turnip and radish, goitrogens are among the most common and longest-recognised substances of toxic nature in the human food supply.

Table 14.6. Some common naturally occurring food toxicants.

Compound	*Toxin classification*	*Typical food sources*
Aflatoxin B_1	Mycotoxin	Mouldy grains, nuts, oilseeds
Amygdalin	Cyanogenic glycoside	Apricot pits, peach pits
Avidin (chicken), mol wt ca 60,000	Forms insoluble complex with biotin	Raw egg white
Caffeic acid	Destroys thiamine	Bracken fern
Caffeine	Alkaloid, stimulant	Coffee, tea, cola nuts
Capsaicin	Amide	Capsicum, peppers
Goitrin	Goitrogen	Cabbage, kale, onions, cress, cauliflower, broccoli, turnips
Myristicin	Alkaloid, psychoactive	Nutmeg, mace, carrots
Oxalic acid	Reacts with calcium to reduce availability	Rhubarb
Phytic acid	Reacts with calcium to reduce availability	Oats
Solanine	Glycoalkaloid, antiacetylcholinesterase	Potatoes, tomatoes, apples, eggplant
Thiaminase, mol wt ca $(75–100) \times 10^3$	Enzyme, inactivates thiamine	Raw fish
Tyramine	Vasoactive amine	Cheeses, bananas, plantains, tomatoes, pineapple

Oxalates, Phytates, and Other Chelates

Of nutrient chelates in the human diet, oxalates and phytates are the most common. Oxalic acid, found principally in spinach, rhubarb leaves, beet leaves, and mushrooms, is a primary chelator of calcium. Phytic acid, found mainly in grain products, complexes a broader spectrum of minerals than does oxalic acid. Decreased availability of P is probably the most widely recognised result of excessive intakes of phytic acid, yet Ca, Cu, Zn, Fe, and Mn are also complexed and rendered unavailable by this compound.

Vasoactive and Psychoactive Amines and Alkaloids

Most compounds producing hypertensive eposides are classified as amines and are found in greatest concentration in banana, plantain, tomato, avocado, pineapple, broad beans, and various cheeses. Amines that are vasoactive include dopamine, tyramine, histamine, tryptamine, noradrenaline, and dihydroxyphenylalanine (DOPA).

Caffeine, a xanthine derivative, is perhaps the best known of the natural stimulants and is found in coffee beans, tea leaves, and cola nuts.

Depressant symptoms, which include burning abdominal pain, decreased excitability, convulsions, nausea, and coma, become the general syndrome for all oral alkaloid poisoning. Dioscorine, alkaloid derivatives of lysergic acid produced by a parasitic fungus, and myristicin are some common alkaloids.

Antinutrients

Any substance that destroys, inactivates, or in other ways renders unavailable an essential dietary constituent can be termed an antinutrient. The most widely studied are antivitamins. Some antivitamins are enzyme thiaminase, niacin inhibitors, biotin antagonist, avidin, linatine, a pyridoxine antagonist, and pantothenic acid antagonists.

Vitamin Toxicity

Because fat-soluble vitamins (e.g, A and D) tend to accumulate in the body with a relatively inactive mechanism for excretion, they cause greater toxicological difficulties than do water-soluble vitamins.

Of the water-soluble vitamins, overdoses of nicotinic acid and folic acid may cause toxic effects as may ascorbic acid in doses over 9 g/d.

Essential Minerals and Heavy Trace Elements

Toxic ingestion of essential minerals from naturally occurring foods is almost beyond comprehension. Cases involving human toxicity from heavy trace elements, such as Pb, Hg, As, and Cd, are not uncommon but almost exclusively traced to accidental contamination rather than true natural occurrences.

Cyanogenic Glycosides

Complex glycosides, which upon hydrolysis yield hydrocyanic acid, are found commonly among plant materials. The toxicity of this class of compounds is directly related to their liberation of HCN upon digestive hydrolysis. They are found in the bitter almond, pits of stone fruits, sorghum, and lima beans.

Nitrates, Nitrites and Nitrosamines

The carcinogenicity of nitrosamines has created widespread concern over the safety of food products that are significant sources of nitrates and nitrites. Nitrates are found in fairly high concentrations in beets, spinach, kale, collards, eggplant, celery, and lettuce. Additionally, nitrates and nitrites are commonly used in the curing solutions of bacon, ham, and other cured meats.

Sodium Chloride

The fact that excessive intake of NaCl contributes to increased fluid retention has led to the conclusion that there may be a relationship between NaCl intake and hypertension.

Mycotoxins

The condition produced by the consumption of mouldy foods containing toxic material is referred to as mycotoxicosis. Moulds and fungi fall into this category, and several derive their toxicity from the production of oxalic acid, although the majority of mycotoxins are much more complex.

Seafood Toxins

Several species of the *Moray eel* have caused toxic reactions. The toxic principle appears to be proteinaceous, and is found predominantly in the blood but it may occur also in the flesh as well. Pufferfish toxin has been identified as having the empirical formula $C_{11}H_{17}N_3O_8$. The liver of sharks, and other oily fishes sometime accumulate toxic levels of vitamin A.

FOOD ADDITIVES

A food additive as a substance or mixture of substances, other than a basic foodstuff, that is present in food as a result of any aspect of production, processing, storage, and packaging. The term does not include chance contaminants.

In most nations, direct food-additive legislation limits the use of a substance to maximum doses in specified foods and are subject to premarketing approvals.

In the United States, a new substance gains approval for use in food through the successful submission of a Food Additive Petition (FAP) in which the following information is documented: safety, including negative carcinogenic studies; declaration of intended use(s); efficacy data at specific levels in specified food systems; manufacturing details including intermediates and final product specifications backed by analytical methods; methods for analysing the substance in foods; and an environmental impact statement.

It is important to note that, in the United States, colourants are handled in a special way. A new colourant intended for use in foods is the subject of a Colour Additive Petition (CAP). This follows the basic format of an Food Regulation Act, but does not require specific proof of efficacy, other than imparting colour, and is not stringently limited to specified use levels. Moreover, colourants are classified as either certified or uncertified. Substances that are not intentionally added to foods but nonetheless find their way into foods from packaging materials, agriculture chemicals residues, etc., are also regulated.

Categories

Depending on the nation, food ingredients and additives are described in the following 52 food additive categories: acidulants, amino acids, anticaking agents (drying agents), antimicrobial agents (fumigants), antioxidants, baking acids (yeast foods), binders (plasticisers, fillers), bleaching agents (oxidising agents), buffering agents, bulking agents, caramelisation aids, carriers (disintegrating agents, dispersing agents), clarifying agents, clouding agents (density adjusters for essential oils), colourants (pigments, lakes), colour fixatives (colour retention aids), cooking media, defoamers, dough conditioners, emulsifiers (foaming agents, whipping agents), enzymes, filter aids, firming agents, flavour enhancers (flavour intensifiers), flavours, foam stabilisers, food starch modifiers (starch modifiers), gums, humectants (moisturising agents), hydrocolloids, leavening agents, lubricants (antistick agents, release agents), masticatory substances (chewing gum base) maturing agents, micro-organisms (moulds, yeasts), neutralisers (acids, bases, alkalies), non-nutritive sweeteners, nutrients and dietary supplements, preservatives, propellants, protein sources, refrigerants, salt substitutes, sequestrants (chelating agents), solvents (solubilisers, vehicles, diluents, extractants), stabilisers (suspending agents), surface-active agents (wetting agents), surface-finishing agents (glazes, coating agents, film forms, micro-encapsulating agents), sweeteners (carbohydrates), texturisers (texture modifiers), thickeners (gelling agents), and vitamins.

The technology, specific uses, and effects of food additives in various food systems have been described in detail and kept current by updates. Table 14.7 is a list of a few chemical additives and their estimated levels of use. The compound Anoxomer is an antioxidant and represents the first in a class of non-absorbable, polymeric food additives.

Table 14.7. Selected chemical additives and estimated level of use.

Type of additive	*Examples*
Antioxidants	Butylated hydroxyanisole, erythorbic acid, nitrogen gas
Colourants	Caramel, certified colours, beet powder
Emulsifiers and surface-active agents	Mono- and diglycerides propylene glycol sorbitol, glycerol monostearate
Enzymes	Papain, rennet, glucose-isomerase, pectinmethylesterase
Flavour enhancers	Mono sodium glutamate
Flavours and spices	Mustard, pepper, ethylacetate, carbon dioxide, vanillin
Leavening agents and baking aids	Benzoyl, peroxide, sodium acid pyrophosphate, various carbonates
Non-nutritive sweeteners	Sodium saccharin
Nutrient supplements	Ascorbic acid, niacin, ferrous sulphate, thiamin hydrochloride, zinc sulphate
pH control agents	Acetic acid, calcium carbonate, calcium phosphates. citric acid, phosphoric acid, sodium bicarbonate, malic acid
Preservatives	Sodium benzoate, sulphur dioxide, sorbic acid, Propionic acid
Stabilisers, thickeners	Modified food starch, various gums, carboxymethyl cellulose, cellulose
Anti-caking agents	Calcium silicate, magnesium stearate
Stimulants	Caffeine
Protein hydrolysate	
Sequestrants	Citric acid, disodium ethylene-diaminetetraacetate

MILK AND MILK PRODUCTS

Composition and Properties

Milk consists of 85–89 wt% water and 11–15 wt% total solids (Table 14.8). The latter comprises solids-not-fat (SNF) and fat. Milk with a higher fat content also has higher SNF content with an increase of SNF of 0.4% for each 1% fat increase. The principal components of SNF are protein, lactose, and minerals (ash). The fat content and other constituents of the milk vary with the species, and for the milk, cow with the breed. Likewise, the composition of milk varies with feed, stage of lactation, health of animals, udder position of withdrawal, and seasonal and environmental conditions.

Table 14.8. Constituents of milk from various mammals, average, wt%.

Species	*Water*	*Fat*	*Protein*	*Lactose*	*Ash*	*Non-fat solids (SNF)*	*Total solids*
Human	87.4	3.75	1.63	6.98	0.21	8.82	12.57
Cows							
Holstein	88.1	3.44	3.11	4.61	0.71	8.43	11.87
Ayrshire	87.4	3.93	3.47	4.48	0.73	8.68	12.61
Brown Swiss	87.3	3.97	3.37	4.63	0.72	8.72	12.69
Guernsey	86.4	4.5	3.6	4.79	0.75	9.14	13.64
Jersey	85.6	5.15	3.7	4.75	0.74	9.19	14.34
Goat	87.0	4.25	3.52	4.27	0.86	865	12.90
Buffalo (India)	82.76	7.38	3.6	5.48	0.78	9.86	17.24
Camel	87.61	5.38	2.98	3.26	0.70	6.94	12.32
Mare	89.04	1.59	2.69	6.14	0.51	9.34	10.93
Ass	89.03	2.53	2.01	6.07	0.41	8.49	11.02
Reindeer	63.3	22.46	10.3	2.50	1.44	14.24	36.70

The non-fat solids, fat solids, and moisture relationships are well-established and can be used as a basis for detecting adulteration with water. The physical properties of milk are given in Table 14.9.

Table 14.9. Physical properties of milk.

Property	*Value*
Density at 20°C of milk with 3–5% fat, average, g/cm^3	1.032
Weight at 20°C, kg/L[a]	1.03
Density at 20°C of milk serum, 0.025% fat, g/cm^3	1.035
Weight at 20°C of milk serum, 0.025% fat, kg/L[a]	1.03
Freezing point, °C	–0.540
Boiling point, °C	100.17
Maximum density at °C	–5.2
Electrical conductivity, S (=Ω^{-1})	(45–48) × 10^{-8}
Specific heat at 15°C, kJ/(kg.K)[b]	
skim milk	3.94

(Contd...)

Property	Value
whole milk	3.92
40% cream	3.22
fat	1.95
Relative volume 4% milk at 20°C=1, volume at 25°C	1.002
40% cream 20°C=1.0010 at 25°C	1.0065
Viscosity at 20°C, mPa.8(=cP)	
skim milk	1.5
whole milk	2.0
whey	1.2
Surface tension of whole milk at 20°C, mN/m (=dyn/cm)	50
Acidity, pH	6.3 to 6.9
Titratable acid, %	0.12 to 0.15
Refractive index at 20°C	1.3440–1.3485

[a] To convert kg/L to lb/gal, multiply by 8.34.

[b] To convert kJ/(kg./K) to Btu/(lb.°F), divide by 4.183.

Nutritional content

Milk fat is a mixture of triglycerides and diglycerides. The triglycerides are short chain, C_{24}–C_{46}; medium chain, C_{34}–C_{54}; and long chain, C_{40}–C_{60}. Milk fat contains more fatty acids than vegetable fats. In addition to being classified according to the number of carbon atoms, the fatty acids in milk may be classified as saturated and unsaturated as well as soluble and insoluble. The fat carries numerous lipids (Table 14.10), and vitamins A, D, E and K, which are fat-soluble.

Table 14.10. Composition of lipids in cow milk.

Class of lipid	Range of occurrence
Triglycerides of fatty acids, %	97.0–98.0
Diglycerides, %	0.25–0.48
Monoglycerides, %	0.016–0.038
Keto acid glycerides, %	0.85–1.28
Aldehydrogenic glycerides, %	0.011–0.015
Glyceryl ethers, %	0.011–0.023
Free fatty acids, %	0.10–0.44
Phospholipids, %	0.2–1.0
Cerebrosides, %	0.013–0.066
Sterols, %	0.22–0.41

(Contd...)

Class of lipid	*Range of occurrence*
Free neutral carbonyls, ppm	0.1–0.8
Squalene, ppm	70
Carotenoids, ppm	7–9
Vitamin A, ppm	6–9
Vitamin D, ppm	0.0085–0.021
Vitamin E, ppm	24
Vitamin K, ppm	1

Milk is an emulsion of fat in water (serum) stabilised by phospholipids which are absorbed on the fat globules. In treatments such as homogenisation and churning, the emulsion is broken.

Processing

The processing operations for fluid milk or manufactured milk products include centrifugal sediment removal and cream separation, pasteurisation and sterilisation, homogenisation, and packaging, handling, and storing.

Cooling

After removal from the cow by a mechanical milking machine, usually at 35°C, the milk should be cooled as rapidly as possible to 4.4°C or below to maintain quality. At this low temperature, enzyme activity and growth of micro-organisms are minimised. Commercial dairy operations usually consist of a milking machine, a pipeline to convey the milk directly to the tank, and a refrigerated bulk milk tank in which the milk is cooled and stored. A meter may be in the line to measure the quantity of milk from each cow. Development of rancidity must be avoided by preventing air from passing through the warm milk as a result of excessive air flow, air leaks, and long risers in the pipeline. The pipelines, made of glass or stainless steel, are usually cleaned by a CIP (cleaning-in-place) process.

Centrifugation

Centrifugal devices include clarifiers for removal of sediment and extraneous particulates, and separators for removal of fat from milk. Modifications include a standardising clarifier that removes fat to provide a certain fat content of the product as sediment is removed; a clarifixator that partially homogenises while separating the fat; and a high speed clarifier that removes bacterial cells in a bactofuge process. Bactofugation is a specialised process of clarification in which two high velocity centrifugal devices (bactofuges) operate at 20,000 rpm in series. The first device removes 90% of the bacteria. The second removes 90% of the remaining bacteria providing a 99% bacteria-free product.

Homogenisation

Homogenisation is the process by which a mixture of components is treated mechanically to give a uniform product that does not separate. In milk, the fat globules are broken up into small particles that form a more stable emulsion in the milk. In homogenised milk, the fat globules do not rise by gravity to form a creamline. The fat globules in raw milk are 1–15 μm in diameter; they are reduced to 1–2 μm by homogenisation. The homogenised milk can be defined as "milk that has been treated to insure the breakup of fat globules to such an extent that, after 48 hours or quiescent storage at 7.2°C, no visible cream separation occurs in the milk..." Today, most fluid milk is homogenised.

Milk is homogenised in a homogeniser or viscoliser. The milk is forced at high pressure through small openings of a homogenising valve formed by a valve, or a seat, or a disposable compressed stainless-steel conical valve in the flowstream (Fig. 14.1). The globules are broken up as a result of shearing, impingement on the wall adjacent to the valve, and perhaps to some extent by the effects of cavitation and explosion after the product passes through the valve.

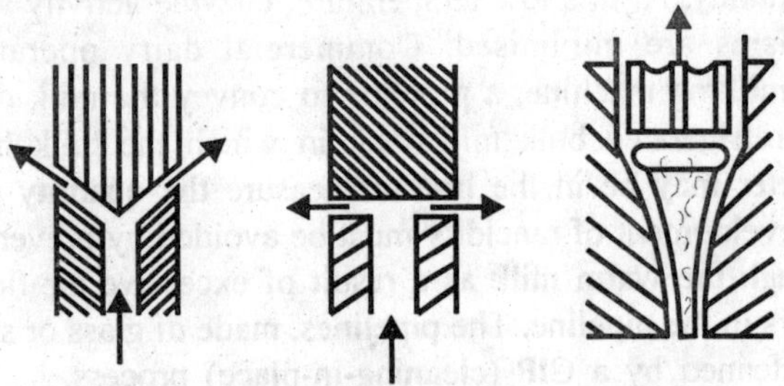

Fig. 14.1. Types of homogeniser valves.

Pasteurisation

Pasteurisation is the process of heating milk to kill yeasts, moulds, and pathogenic bacteria, and most other bacteria, and to inactivate certain enzymes, without greatly altering the flavour.

Pasteurisation may be carried out by batch or continuous-flow processes. In the batch process, each particle of milk must be heated to at least 62.8°C and held continuously at or above this temperature for at 71.7°C for at least 15 seconds. This is known as the HTST (high temperature short-time) pasteurisation. For milk products with a fat content above that of milk or with added sweeteners, 65.5°C is required for the batch process and 74.4°C for the HTST process. For either method, following pasteurisation, the product should be cooled quickly to 7.2°C or less. Other time-temperature relationships have been established for other products. Ice-cream mix is pasteurised at 79–81°C and held for 25 seconds.

Another continuous pasteurisation process is known as ultrahigh temperature (UHT), in which shorter time, 1–2 seconds and higher temperatures, 87–132°C, are employed.

Batch holding

The milk in the batch holding tanks is heated by a hot-water spray on the tank liner, a large-diameter coil which circulates in the milk through which the hot water is pumped, a flooded tank around which hot water or steam is circulated, or by coils surrounding the liner through which the heating medium is pumped at a high velocity.

High temperature short-time pasteurisers

The principal continuous-flow process is the high temperature short-time method. The product is heated to at least 72°C and held at that temperature for not less than 15 seconds. Other features are nearly the same as in the holding method. The following equipment is needed: balance tank, regenerative heating unit, positive pump, plates for heating to pasteurisation temperature, tube or plates for holding for the specified time, flow-diversion valve (FDV), regenerative cooling unit, and cooling unit. Other devices often incorporated in the HTST circuit are clarifier, standardising-clarifier, homogeniser, and booster pump.

Cleaning systems

Both manual and automatic methods of cleaning food-processing equipment are used. Even in a plant with advanced cleaning equipment, some manual cleaning is involved. Today, in most plants, the equipment surfaces are cleaned in place, generally at least once every 24 hours. Continuously operated equipment may be cleaned every few days, depending upon the product and the potential build-up of residue on the surfaces.

Cleaning in-place (CIP) systems

CIP systems evolved from recirculating of cleaning solutions in pipelines and equipment of a highly automatic system with valves, controls, and timers. In the early circulation systems, considerable manual operation was required in the assembly and disassembly of units. Homogenisers and heat exchangers were cleaned manually or by a system of circulating solutions.

In the CIP procedure, a cold or tempered aqueous prerinse is followed by the circulation of a cleaning solution for 10 mins. to 1 hour at 54–82°C. Hot water rinses may harden the food product on the surface to be cleaned. The temperature of the cleaning solutions should be as low as possible but high enough to avoid an excess of cleaning chemicals. A wide variety of cleaning solution may be used, depending upon the food product, the hardness of the water, and the equipment. Alkali or chlorinated acid cleaners are preferred. A chlorinated alkaline cleaner may be used separately or in combination with an acid detergent.

Manufactured Products

Most of the fluid milk production is used for manufactured products, mainly cheese, evaporated and sweetened condensed milk, non-fat dry milk, and ice cream. Evaporated and condensed milk and dry milk are made from milk only; other ingredients are added to ice cream and sweetened condensed milk.

Evaporated and condensed milk

Evaporated and condensed milk are produced by removing moisture from milk under a vacuum followed by packaging and sterilising in cans. The milk is condensed to half its volume in single- or multiple-effect evaporators. The product has a fat to SNF ratio of 1:2.2785, and is standardised before and after evaporation. It must have at least 7.9% fat and 25.9% of total milk solids, including fat.

Sweetened condensed milk

Sweetened condensed milk, unlike evaporated milk, is not sterilised. Sugar is added as a preservative which replaces sterilisation as a means of maintaining keeping quality. The equipment is similar to that used for evaporated milk, except that sugar is added in a hot well before condensing (evaporating) the liquid. According to standards, sweetened condensed milk must contain 8.5% fat min and 28.0% min total milk solids, including fat (fat to SNF ratio = 1:2.294). The product contains 43-45% sugar.

Cream

Cream is a high fat product which is secured by gravity or mechanical separation through differential density of the fat and the serum. Fat content may range from 10 to 40% depending on use and as per food laws. Whipping cream has a fat content of 34–40%, and table coffee or light cream has a fat content of 20–25%. Half-and-half, suggesting a mixture of cream and milk, has not less than 10.5% milk fat, and in some states, up to 12%. Normally about 10 wt% of the edible fat used is butter, which is defined as a product that contains 80% milk fat with not more than 16% moisture. It is made of cream with 25–40% milk fat.

Cheese

Cheese production is based on the coagulation of casein from milk, or to a minor extent, of the proteins of whey. The casein is precipitated by acidification, which can be accomplished by natural souring of milk. The procedures for making cheese vary greatly and cheese products are countless.

Yogurt

Yogurt is a fermented milk product that is rapidly increasing in consumption. Milk is fermented with *Lactobacillus bulgaricus* and *Streptococcus thermophilus* organisms producing lactic acid. Usually some cream or non-fat dried milk is added to the milk in order to obtain a heavy-bodied product.

Non-Dairy Products

Many so-called non-dairy products such as coffee cream, topping, and icings utilise caseinates. In addition to fulfilling a nutritional role, the caseinates impart creaminess, firmness, smoothness, and consistency of products.

POULTRY MEAT PROCESSING

The poultry production is also increased due to intensive agriculture system and increased grain production. Poultry is the most efficient converter of grain and by-products into meat with added advantages of high productive safe.

Poultry meat consist of several desirable properties, compared to other meats. It provides important class of nutrients, proteins and relatively little fat. Carcasses from young birds have higher proportion of moisture than old ones. Raw poultry meat consist of about 75% water 19% protein and 2.5% fat, species is perhaps the most important factor affecting the composition of poultry meat. Poultry meat is very popular due to excellent flavour, ease of digestion, low fat content, excellent source of protein and for fair source of vitamins and minerals.

Uses and Application

Poultry meat is one of best meats; rich in protein contain and is pallatable to the tongue. As it is very much relishing in taste. It is soft in texture, which is easily digestible. It does not require much quantity of vegetable oil for cooking purposes. It is directly consumed by restaurants, hotels, motels, dhabas, guest houses etc.

Manufacturing Process

A typical poultry meat processing plant consists of two main processing zones. First processing zone consist of receiving, killing and defeathering units and second processing zone contain evisceration final processing and packing unit. Separation of the two zones is intended to facilitate sanitation by separating the inherently dirtier phases of processing from those involved in preparation of the final product. Other aspects of poultry meat processing plant provide office site and personnel need area. The major steps involved in poultry meat processing are: (i) receiving; (ii) killing; (iii) bleeding; (iv) scalding; (v) defeathering; (vi) singeing; (vii) washing; (viii) pinning; (ix) eviscerating; (x) removal of various parts (lungs, kidney, heart etc.); (xi) washing; and (xii) chilling cutting and packing (Fig. 14.2).

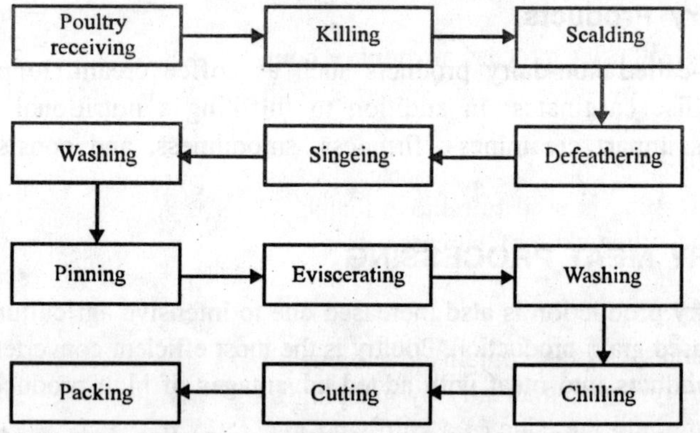

Fig. 14.2. Process flow sheet for poultry meat processing plant.

Receiving

Generally poultry meat processing plants are installed near poultry farms, poultry arrives at plant and convey to the shacking area. In shacking area poultry are shacked by their feets than it is transported to killing section.

Killing

Killing is usually performed manually. Many plants employ electric stunning prior to killing. This serves to reduce stuggling and also relaxes the feather papillac, thereby facilitating feather removal.

Bleeding

After killing, blood comes approximately for 1.5 minutes. Incomplete bleeding cause discolouration of the carcasses and consequent lowgrading.

Scalding

Scalding is done at 123 to 128°F for 1.5 to 2 minutes commonly, it is sufficient to allow feather removal, yet the outer layer of skin on the carcass remain intact. Finished carcass scalded in this manner have an attractive natural appearance. Under scalding cause incomplete feather removal by the automated pickers. Over scalding may lead to irregular removal of the outer skin, causing an unattractive splotchy appearance.

Defeathering

Defeathering of carcasses is completed in rotating drum. In the rotating drum carcasses are brought in contact with rubber fingers, the feathers loosened by scalding, are stripped away with little damage to the skin.

Singeing

After defeathering process all poultry contain some hair like feather which is removed by passing the carcass quickly through a flame.

Washing

Spray washing is used to remove feather, blood and other foreign materials from the outer skin of carcass.

Pinning

Some times carcass contain developing feather that have just pierced the skin, are removed manually by grasping the protuding pin feather between the thumb and the edge of a knife and giving a strong tugging motion. The carcass is transported to eviscerating room.

Eviscerating

In the eviscerating process, first of all a vertical cut is made from the tip of the sternum to the vent, and with the vent still attached viscera and drawn out, but remain attached viscera is used for the postmortem inspection.

Removal of various parts

After inspection the liver and gizzard are removed and remaining viscera discarded. Then lungs and kidney removed mechanically or by use of a strong vacuum aspirator. Head of bird is removed by using v-shape knife. The skin is slit along the dorsal surface of neck to the shoulders then skin is cut on the dorsal side to the shoulders and neck is cut or broken off at the shoulders manually.

Washing

Carcasses are washed with cold water to remove attached tissues, foreign materials, blood etc.

Chilling

Chilling of carcasses is done to remove body heat and to protect from microbial action. Normally chilling is done below 40°F for 30 minutes.

Packing

Sometimes carcasses are cut into parts according to market demand, otherwise then are packed in frozen form.

MEAT PRODUCTS

Meat, an excellent source of protein, iron, and B vitamins, was processed as early as prehistoric times, probably by drying under the sun and later by smoking and drying over wood fires. The purpose of meat processing was to prepare products that could be stored for considerable time periods at ambient temperatures. Today, meat is processed with salt, colour-fixing ingredients, and seasonings in order to impart desired palatability traits to intact meat products (e.g., bacon, corned beef, ham, smoked butt, and pork hocks) and comminuted meat products (e.g, all types of sausage items). Products intermediate to these categories are sectioned, or chunked and formed meats.

Curing

Meat-curing agents include sodium chloride, nitrite, ascorbate or erythorbate, and possibly sodium phosphate, sucrose, dextrose, or corn syrup and seasonings. The salt content of processed meats varies from 1–12% according to the type of product. Salt is used for flavour, reservation, and extraction of myofibrillar protein, whereas nitrite promotes colour development, flavour, and preservation by inhibiting the growth of micro-organisms and fat oxidation, and ascorbate or erythorbate inhibits undesirable nitrite reactions. Phosphates facilitate myofibrillar protein extraction, inhibit fat oxidation, and improve colour development. Sugars and seasonings are used principally for flavour.

Comminution

Comminuted meat may be cured or a fresh product. The degree of comminution varies considerably from one product to another. Sectioned or chunked and formed products may be composed of particles that weigh more than 450 g each, whereas finely comminuted meats are chopped to a paste-like texture of very small particles. Comminution equipment includes grinders, bowl choppers, emulsion mills, and flaking machines. In addition to comminution, the meat is blended with other ingredients. Blenders, mixers, tumblers, and massagers are used to subject the meat protein to mechanical action in the presence of salt. This causes the salt to extract the principal myofibrillar protein, myosin, from the muscle. The extracted myosin gels when the comminuted meat is heated to form a matrix which entraps water and fat and binds the meat particles to each other.

Smoking

Many intact and comminuted, cured meat products are smoked to impart a desirable smoked flavour and colour. The smoking process may also include a drying or cooking cycle.

Canning

Canned meats may be processed to be commercially sterile or semipreserved. The objective of commercial sterilisation is the destruction of all harmful bacteria, or bacteria that may cause spoilage of the product under normal unrefrigerated storage. However, the process does not kill the spores of all heat-resistant bacteria. Therefore, it is essential to cool the cans rapidly after processing and avoid storage above 35°C. The most persistent type of bacteria are sporeforming organisms. The common vegetative non-sporeforming pathogenic bacteria are killed with adequate processing and are of little or no importance in spoilage.

Dehydration

Freeze-drying

Freeze-drying of meat results in a product with a spongelike appearance, practically devoid of moisture but resembling the original meat. The composition and type of meat influence the acceptability and stability of freeze-dried meat. Freeze-drying extends shelf life and reduces weight. The meat is rapidly defrosted by immersing in water before cooking. Under optimal processing and storage conditions, reconstituted meats have acceptable flavour, colour, texture, and nutrient retention.

Air-drying

Precooked comminuted lean meat is dried in a forced-air rotary or tunnel dryer to less than 10% moisture for use as an ingredient in dried soups and stews. The dried product is often compressed to reduce its volume to about one third of the fresh meat volume for shipment.

By-Products

By-products in the meat-packing industry represent a substantial part of the sales value of the production derived from the slaughter of animals. By-products include variety meats, edible and inedible fats, and hides and other inedibles. The value of meat or by-products depends upon the species and age of the animal, degree of finish, and price.

Approved portions of the carcasses of clean, sound meat animals yield edible fats including lard (from swine), edible tallow (from cattle and sheep), and related products such as oleo stock. These may be consumed in the form of shortening for baking and similar food applications, or as frying fats. Inedible fats are used widely in animal feeds and for other industrial uses. The principal products of this type are inedible tallow and grease.

Chapter 15

Leather and Tannery

INTRODUCTION

The skins and bones of animals and their products such as leather, gelatin and glue are mostly colloidal materials having indefinite properties. The raw skin consists of number of different complex proteins having different structures in different animals and even vary in different parts of the same hide. A large number of active substances such as enzymes, bacteria, alkalies, acids, tannin, tannin substitutes, oils, fats and salts have been found to affect the skin. Leather is most historic of useful materials. The art of leather manufacturing from hides and skins antedates by centuries, any of man's scientific knowledge of chemistry. The original primitive methods of preservation of skin probably consisted of simply drying them in air and sunlight. Later on, oils were used for preservation. Recent preservations methods also include tanning effects of leaves, twigs, and barks of certain trees soaked in water.

Animal skins are received by the tanner may be divided into three layers: The *epidermis*, the *derma*, or *corium*, and the *flesh*. The epidermis, constituting about 1% of the total skin, is the outer layer and consists chiefly of the protein keratin. The hair, which grows through both the derma and epidermis is also largely keratin. This layer is dense and chemically resistant, thus allowing the epidermis to be readily removed by chemical means. Hot water, however, causes slow solubilisation by hydrolysis of the collagen to produce gelatin. The elastin is not affected by this treatment. The flesh is an extraneous layer, mostly adipose tissue, which must be removed to ensure tannin penetration on both sides of the corium.

LEATHER

Despite efforts, particularly during the last 50 years, to produce leather-like materials from other fibres, sheet materials, and petrochemicals and to substitute them in the marketplace for leather products, leather still is the product of choice by the consumer. The demand for leather far exceeds the supply, and the principal function of the substitutes has been to fill those needs for which the supply of leather is inadequate.

Physical Properties

Leather has unique properties that make it ideally suited for use in the manufacture of a variety of products, the most notable being footwear. Even if leather is considered for a single use, e.g., shoe uppers, a large variation in properties is advisable to allow for different shoe types, e.g., casual, military, work, or dress shoes. A range of values for various physical properties of cattlehide shoe-upper leather and the representative value are listed in Table 15.1.

One of the most important properties of leather, and one for which there is a paucity of quantitative data, is the ability of leather to conform to the shape of the foot. Plastic flow imparts the necessary give to provide foot comfort; elastic flow is responsible for the maintenance of shape by ensuring proper recovery during flexing action. Leather substitutes have not achieved this combination of properties. Nor are they able to remove moisture from the foot as leather does. The mechanism of the transmission of water vapour through leather has been demonstrated to be a function of the material rather than a linear diffusion process. The evidence indicates that water is sorbed at polar groups of the protein molecules and is conducted by an activated diffusion process along the fibres, even against an air-pressure head.

Manufacture

Raw material

Fresh cattlehides, like most biological materials, contain 65–70% water, 30–35% dry substance, and less than 1%, ash. The dry substance is largely made up of the fibrous proteins, collagen, keratin, elastin, and reticulin. The main components of the ash, listed in decreasing concentration, are phosphorus, potassium, sodium, arsenic, magnesium, and calcium.

Defects

Defects in leather include those that result from poor flaying technique in the slaughterhouse, inadequate preservation and handling methods, or improper use of chemicals and heat in the tannery. The scar produced by

heat-branding, those left by parasitic skin diseases, and biological defects also mar the surface appearance and impair the physical properties of leather.

Table 15.1. Physical properties of shoe upper leather.

Property	*Range*	*Representative value*
Tensile strength, MPa[a]	15.26–37.48	27.6
Elongation at break, %	29.5–73.0	40
Stitch tear strength, N/cm[b]	1280–2275	1751
Tongue tear strength, N/cm[b]	226–961	525
Thickness, mm	1.5–2.4	1.8
Bursting strength, kN/cm[b]	1.10–24.5	17.5
Grain-cracking, N/cm[b]	525–1489	1051
Wet shrinkage temperature,[c] °C	96–120	100
Apparent density, g/cm³	0.6–0.9	0.75
Real density, g/cm³	1.4–1.6	1.5
Flexibility (Flexometer)		
Bending length, cm	6–9	7
Flexural rigidity, mg/cm	10,000–50,000	20,000
Bending modulus, MPa[a]	19.7–68.9	34.5
Compression modulus, MPa[a]		0.345
Cold resistance,[d] K		
(Without finish or fat liquor)		
Workable without cracking at		92
Heat resistance	shrinks depending on moisture content, anhydrous decomposition at 160–165°C	

[a] To convert MPa to psi, multiply by 145.
[b] To convert N/cm to lbf/ft, divide by 14.6.
[c] Chrome-tanned leather values.
[d] Ordinarily limited by finish-crack or lubricant hardening properties.

The necessary steps in the manufacture of leather, may be broken down into the following sequences:

(1) The skins are opened, examined, trimmed, and graded.

(2) The skins are water-soaked for 3 to 24 hours (for dried hides, longer time may be necessary), and the flesh is removed by a machine equipped with a spiral-bladed cylinder (beaming). Soaking assists are used; sodium tetrasulphide, 1/5 to 2/5 of 1%, hastens rehydration and produces more uniform hydration.

(3) Then the skins are treated with a saturated lime solution plus certain accelerators, such as sodium sulphide or sulphydrate and dimethylamine, for 3 to 7 days. Liming may be reduced to 24 hour on light hides.

(4) Any hair remaining is removed by machine and hand-scraped.

(5) The clean, dehaired hides are treated with certain enzymatic preparations for several hours to soften the skin and remove the lime (bating).

(6) After rinsing, the hides are ready for tanning.

(7) The vegetable-tanning process consists of the following procedures:

   (a) The prepared skins are treated with organic acids to neutralise any residual lime.

   (b) The tanning is carried out in several tanning baths, starting very weak to prevent case-hardening and becoming progressively stronger. This requires about 1 month.

   (c) The skins are then allowed to soak in a vat containing strong tanning extract.

(8) The chrome-tanning process is somewhat different from the tanning process. The following pertains to one bath processes:

   (a) The prepared skins are pickled in a bath of salt and sulphuric acid.

   (b) The pickled hides are soaked in a solution of sodium dichromate.

   (c) The dichromate-saturated skins are treated with sodium thiosulphate in the reduction drum.

   (d) The chrome salt is set on the fibres by adding borax to the skins in the settling drum.

After washing, the leather is ready for the finishing operations. Finishing operations are somewhat similar for both methods of tanning:

(1) The tanned hides may be dyed.

(2) The leather is stuffed with oil.

(3) The leather is split to produce a thinner, more supple product.

(4) Gloss is developed on the surface of the leather by ironing and glazing it with a glass cylinder.

(5) The jacked leather is dried, embossed and measured.

In industrial application these chemical conversions and operations are complicated and require experienced workers to carry them out. The manufacture of leather is an engineering process requiring careful chemical control in order to ensure uniform quality (Table 15.2).

Table 15.2. Comparison of skins and leather.

	Skins or hides	*Leather*
Pliability	Soon lose pliability and become hard and brittle	Retains pliability
Permanence	Putresce very quickly	Extremely durable, not attacked by bacteria
Water resistance	Absorb water, and are permeated easily by it	Possesses great resistance to water.
Boiling water	Are converted to gelatin by hydration	Attacked with great difficulty
Mechanical strength	Good	Good

Preservation and disinfection of skins

Since raw skins decay rapidly, some method of preservation must be used to arrest bacterial action and consequent hide disintegration. Tannery experience with domestic hides has demonstrated the superiority of brine curing, in whioh a bactericide is employed, over curing by the conventional green-salt method. Curing is more rapid and uniform, affords greater protection prior to tanning, practically eliminates "salt stains," requires no washing of hides before "soaking back," and produces leather of plumper grain. In salt curing, standard domestic practice piles the hides flesh side up in alternate layers with salt. These stacks are arranged for proper drainage of the brine. This method of curing requires a minimum of 3 or 4 weeks at about 13°C, during which the hide loses part of its moisture by dehydration and gains weight through salt absorption. The loss is the greater, and the overall difference is termed the "shrink." Many irregularities in leather may be traced to careless curing and storing.

Preparation of hides for tanning

The first step in the tanning process is inspection of the hides for defects as they come into the tannery, and the cutting off of ears and ends. The next step is the cleansing of the hides to remove dirt, manure, and salt, in order to restore them to a natural soft hydrated state, by paddle washing, drum washing, or soaking in vats (Fig. 15.1).

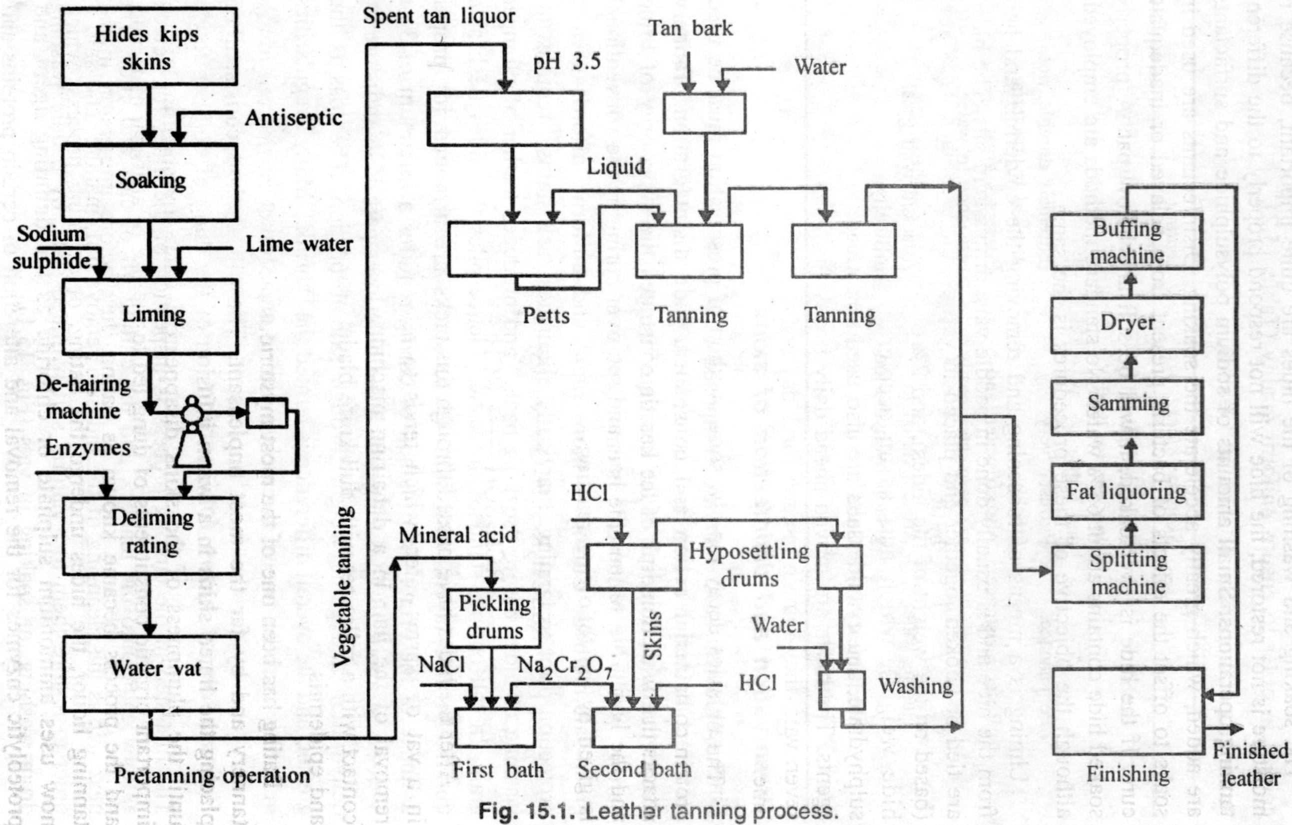

Fig. 15.1. Leather tanning process.

The soaking and washing of the hides are quite important, because if moisture is not restored, the hide will not respond properly to the different tanning operations. Small amounts of sodium polysulphide and surfactants are added, which greatly accelerate the soaking. Disinfectants are used in soaks to offset the effects of bacteria present from delayed or inadequate cure. If the hide is oversoaked, it will be soft and below par. A properly soaked hide contains about 65% water. No standard methods are employed, although the objective of these procedures is the same.

Liming is a means of loosening and removing the epidermis and hair from the hide and is usually done in paddle vats, drums, and pits. The hides are tied or hooked together and placed in vats containing water, 10% lime (based on the weight of the hides), and 2% sodium sulphide (based on the hide weight), which acts as an accelerating agent. Dimethylamine, sulphydrate, and cyanide salts are also used as accelerating, or sharpening, agents. The hides are moved ahead daily in a ser' s consisting of three to seven vats; they remain in each for 1 day and then enter a vat with fresher lime.

The epidermis and hair are composed chiefly of keratin. Keratin is a protein containing a cystine residue which is easily attacked by alkali. Lime attacks the disulphide link of the keratin of the epidermis, the cortical layer, and the hair. This softens the hair and removes the epidermis. This action is given by the following reaction :

$$RSSCH_2R + H_2O \longrightarrow RSH + RCH_2SOH$$

$$RCH_2SOH \longrightarrow RCHO + H_2S$$

After the hides have passed through this process, they are usually placed in a vat of warm water, which tends to relax them and permits easier removal of the hair in a dehairing machine. The skins are brought into contact with a roller set with dull knife blades which rub off the loose hair and epidermis.

Bating has been one of the most mysterious processes carried out in the tannery and by far the most unpleasant. The old method consisted of placing the limed skins in a warm infusion of the dung of dogs or fowls until the plumpness of the skins disappeared. Later investigation of the important organic constituents of dung led to the use of artificial materials, and the process became known as bating. Just before immersion in the tanning liquor, the hides undergo the bating, or deliming process, which now uses ammonium sulphate or chloride as the deliming agent and proteolytic enzymes for the removal and alteration of certain proteins and for the improvement of the colour of the grain. Another important reaction in bating is called falling—a reduction in the degree of swelling of the protein constituents of the limed skin previously swollen. The principal function of bating is to regulate the degree to which the skin structure is prepared to make it accessible to uniform absorption of the tanning agent.

Vegetable tanning

Vegetable tanning is chiefly used for heavy leathers, for soles and belting. The wedding of the tanner's art and the chemist's knowledge has resulted in the modern technology of leathermaking. Vegetable tannins may be divided into two broad groups, catechols and pyrogallols, depending on which substance is yielded on fusion with alkali.

Tannins are complex mixtures of glucosides of various polyphenols. Their action is to combine with and between the collagen fibres of the skins. During the tanning process tannins liberate sugars. The acidity, or pH, is controlled by the use of sulphuric acid.

$$\text{skin} + \text{tannin} \longrightarrow \text{leather} + \text{sugar}$$

The protein in the aggregate has the capacity to absorb large quantities of tannin. Skins may increase in weight as much as 350% during the tanning operation. This combined and absorbed material fills up the holes and stiffens the leather.

Vegetable tanning is usually done in 8 × 7 ft vats, 5 or 6 ft deep, which contain the tan liquors and hides. Since the hides remain in the liquor for one or more months, many vats are required, and they must therefore be cheap and durable. Hides first enter the oldest, weakest tanning liquors in order to prevent complete plugging of the surface pores and subsequent poor penetration of the tannins into the skins by a strong liquor. They pass successively through liquors of increasing strength until they leave the freshest and strongest liquor as fully tanned leather. The ratio of liquor to hide is kept as low as possible and is usually 6 to 4 lb of liquor to 1 lb of hide. Modern processing now speeds up the tanning time and shortens the expensive and tedious aging and mellowing in tan liquors. Much research has been devoted to expediting the process and reducing the hand labour required.

Since approximately 250 gal of iron-free water is required for each hide produced, water supply and disposal are important factors in locating a tannery. As in most other fields of industrial endeavour, there is no standard treatment of waste disposal in the tanning industry. Because of the relatively high amount of solids found in tannery wastes, some form of precipitation and sedimentation is usually included. Tannery waste disposal has proved to be expensive, but plants have been able to meet the strict antipollution standards by careful operation.

Synthetic tanning agents, or syntans

Synthetic tanning agents or syntans, are gaining in use. Syntans are condensation products of sulphonated phenols (or higher homologs) and formaldehyde capable of converting animal skin into leather. Comparatively few syntans lack a sulphonic acid group. Syntans may be classified according to use as auxiliary, complementary, or replacement tannins. Introduction of the first domestic replacement tannin, all syntans belonged to the first two classes. Their function is adjuncts and dye mordants and are used for bleaching chrome leather, for solubilisation of vegetable tans, and for increasing the speed of penetration of natural tannins into leather. Research has been done on a potential source in the petroleum industry, and lignosulphonate by-products of the paper industry remain attractive.

Chrome tanning

The uppers of a pair of chrome-tanned leather shoes will outwear two or three pairs made from a vegetable-tanned product. For this reason more than 90% of the world's production of light leather is chrome-tanned by the one-bath process. To condition hides and skins for subsequent mineral tannage, a process called pickling is utilised, wherein bated hides and skins are treated with solutions of acid, usually sulphuric, and salt. In the chrome process, combination of the chromium salts and hide fibres is much more rapid and takes place without the degree of swelling that occurs in vegetable tanning. Therefore chrome-tanned leather is more pliable and looser in structure. In sole leather, fillers are usually added to give water-resistant properties. Chrome-tanned leather is characterised by a high content of original hide proteins and mineral matter and a low content of water-soluble materials. The main mechanism of chrome tanning is the coordination of chromium to the carboxyl groups of the protein. Chrome tanning is usually subdivided into two processes: (i) the one-bath process, using basic chromic sulphate, and (ii) the two-bath process, using sodium dichromate. The chrome liquor for the one-bath process is usually made by reducing an acidified solution of sodium dichromate by slowly adding a solution of glucose to it or by bubbling sulphur dioxide through it until the reduction is complete.

$$Na_2Cr_2O_7 + 3SO_2 + H_2O \longrightarrow 2Cr(OH)SO_4 + Na_2SO_4$$

Various modifications of the two-bath process are in use, and the method is shown in Fig. 15.1 serves as an illustration. After pickling, the hides are placed, successively, in two drums or paddle vats, containing about 5% sodium dichromate solution (based on the weight of the hides), with hydrochloric acid, and salt. The skins are allowed to remain in each liquor overnight and are then pulled out and allowed to drain. After an hour or two, they are placed in another paddle vat containing approximately 15% of the weight of the hides in sodium thiosulphate. The hides are left in this liquor overnight, pulled out in the morning, and washed in a drum containing water. Borax is added to reduce the acidity of the leather to the desired extent.

Finishing operations

After the tanning operation is completed, the leather, if dried rapidly, is quite stiff, usually breaks if bent sharply, and is not yet fit for use. The fibres must be lubricated to give them pliability and softness. Actually, more work is required in the operations following tannage than in all the preceding operations and processes put together. Since each of the numerous kinds of leather requires a special series of operations.

If the tanned leather is not smooth, it is shaved by a machine with sharp blades. Very often it is split by means of a bandlike knife to obtain the desired thickness. Vegetable-tanned leathers are sometimes washed, bleached, or scoured. The bleaching or washing is usually done in a revolving drum containing sodium carbonate, sodium hydroxide, or sulphuric acid and water. Other bleaching agents, such as sodium bisulphite, sulphurous acid, sulphite cellulose solutions, and various syntans, are also used.

The next operation involves the incorporation of oils and greases and is called stuffing, or fat liquoring. Sulphonated oils blended with raw oils are the most important fat liquors and have mostly replaced raw oils, natural greases, emulsions of raw oil, soaps, waxes, and resins. The leather, either dry or wet, may be stuffed by hand or in a rotating drum. A more uniform job is achieved by hand. Sole leathers are usually "loaded" with some material, such as magnesium sulphate and cellulose, to give them desired properties.

The colour and shade of the finished leather are produced by a combination of dyeing and finishing operations. Dyeing may be combined with the fat-liquoring process, or it may be done before or after it. Most of the dyestuffs used now are synthetic coal-tar derivatives. The most common method of dyeing employs revolving drums or vats equipped with rotating paddles.

The finishing series of operations in the production of leather includes both mechanical treatments and applications of finishing materials to the leather. The following classes represent the more common materials used in leather finishing : (i) cellulose ethers, (ii) waxes, (iii) resins, (iv) dyes, (v) pigments, (vi) lacquer materials, (vii) antiseptics, and (viii) miscellaneous materials such as solvents, perfumes, soaps, sulphonated oils, metal salts, plasticisers, acids, and alkalies. Mechanical operations carried out in the finishing of leathers include staking (flexing the leather over an edge to loosen fibres and soften it), glazing, buffing, graining, trimming, rolling, brushing, and plating. The finishing operations alone embrace a technique requiring great skill, wide experience, and close attention.

Oil tanning

From very early times skins have been tanned or preserved by the use of certain drying oils which have been worked into the skins and allowed to oxidise and react, thus becoming fixed. An example of this type of leather is *chamois*. Sheep or lambskins are subjected to the usual rinsing, liming, trimming, fleshing, and bating. They are then swollen preparatory to splitting, frequently with the aid of an acid such as sulphuric acid at a pH of about 2.0. The pickled skins in the plumped condition can be split easily, separating the grain from the flesh layer. The former is the skiver, and the flesh layer is processed to make chamois leather. After this, depickling is carried out with an alkali such as borax, or even soda ash or ordinary chalk, with the pH at about 7.0. The damp skins are treated with about 10% of their weight of fish oil, often in a drum in which the revolving action lasts over 4 hours, so that the oil can distribute itself and penetrate the skin. The skins are removed and piled in layers, during which fermentation and oxidation proceed, resulting in a reaction of collagen with liberated aldehydes formed by the oxidation of fish oil during tannage. The skins are hung or hooked while the reactions proceed. The temperature rises about 20°F during this maturing process. The oiling and hooking are repeated until 40 to 50% of the oil (based on the weight of the moist skin) has been incorporated. After the oil-tanning process is completed, the chamois leather is washed with warm, alkalised water and hydraulically pressed to remove excess oil. Further washing, buffing, and drying complete the operations, except for bleaching or dyeing.

This classical method of tanning just described has been largely superseded by the "new-chamois" tanning process, wherein pre-tanning with formaldehyde precedes the actual fish-oil tannage; however, the latter is not superfluous and must be carefully executed. A modification which shortens the process involves the addition of water-soluble fatty agents to the fish oil.

By-products

As raw material for tanneries, skins of cattle, hogs, and sheep are themselves by-products of the packing industry; in fact, sheep skins also come from the wool industry. Since it has been estimated that, for each 100 lb of green-salted hides, only about 68 lb of leather will be produced, it is obvious that by-products, of which scrap leather is only one, must also yield revenue. For example, fleshings and trimmings are sold for glue or low-grade gelatin. From them also recent research has developed a protein supplement for use as chicken and dog feed. Beamhouse waste, which includes lime, hair, and shavings, can be converted into tiles that have many desirable features, such as tensile strength and water resistance. Spent vegetable tans, so troublesome from a standpoint of waste disposal, can be concentrated by evaporation and the concentrates sold as water-conditioning agents. Hide house trimmings and beam-house splits are used in the production of gelatin. Although gelatin itself is a by-product both of the packing and tanning industries, its manufacture produces several more by-products, such as calcium phosphate for the pottery industry, bone meal, and bone black.

Other tannages

Most of these tannages are pre-tannages or post-tannages for chrome or vegetable processes and are employed to give such qualities as filling, lighter shade for dyeing, and reduction of tannery effluent. Only occasionally are they used alone. They include mineral tannages, polyphenolic syntans, resin tannages, oil tannage, sulphonyl chloride tannage, and aldehyde tannage.

Post-tanning

During the early aging period, when the leather is stacked, there is a gradual decline in the pH of the stock as increased chrome fixation, olation, and oxolation occur. Consequently, after transfer to a drum, the stock is neutralised to a desired pH, frequently is retanned to develop certain properties for different end uses, and then is dyed and fatliquored in rapid succession in the smae drum. Fat-liquoring is the application of oil-in-water emulsions to the leather.

Once all of the wet operations have been completed, the leather can be dried, which not only involves removal of excess water, bringing the moisture content close to that of the finished leather, but completes the reactions of some of the materials (e.g., tanning agents, fat-liquors, and dyes) with which the hide has been treated.

It is the finishing process with its application of natural or synthetic polymers and of colourants within and on the surface of the leather that produces the uniformity, appearance characteristics, and resistance to scuffing and abrasion that are required for a commercial product. Coatings for leather function as decoration and protection.

Energy consumption

On the basis of quantity consumed per unit of product produced, the leather-manufacturing industry would be categorised with aluminium, paper, steel, cement, and petroleum-manufacturing industries as a gross consumer of energy. An energy wastage in the amount of 70% of energy input offers considerable opportunity for improvements.

Health and Safety Factors

The effluent streams that must be treated can be separated into two areas of the tanning operation, the beamhouse and the tanyard. The main pollutants found in beamhouse effluent are sulphide and BOD, and the pH is ca 12. The second, and even more difficult, stream to treat is that from the tanning operation. This effluent contains chromium as well as conventional pollutants, e.g., BOD and suspended solids.

Uses

The principal domestic markets for leather are leather footwear, luggage, handbags, small personal leather goods, leather apparel, and gloves. Hide collagen, when it can be separated from the hide in a native state, has use in food products, cosmetics, pet treats, and biomedical products.

LEATHERLIKE MATERIALS

Leatherlike materials normally refers to synthetic materials that are used as substitutes for leather in those applications where the latter traditionally has been used. Periodic shortages caused by the cyclical nature of leather supplies cause continued inroads by new products into traditional leather markets. These include the use of urethane-coated fabrics in shoes and apparel and the development of poromerics for men's shoes. Each of these products is designed to simulate certain properties of leather that are relevant to a specific application. The suitability of synthetics as leather substitutes depends upon their ability to duplicate certain desirable characteristics of leather. These characteristics arise from the physico-chemical structure of the leather. The high permeability for air and moisture vapour is one of leather's most desirable properties in its main use area, i.e., footwear.

Types

Leatherlike materials have one common characteristic: they look like leather in the goods in which they are used. For purposes of technological description, however, they may be divided into two general categories: coated fabrics and poromerics. Poromerics are manufactured so as to resemble leather closely in both its barrier and transition properties, e.g., for heat or moisture, and its workability and machinability. The barrier or permeability properties normally are obtained by manufacturing a controlled microporous structure. The modulus characteristics of synthetics should match those of leather at 80°C rather than at room temperature. Although the latter temperature commonly is used as the standard in the industry, the former more closely matches the lasting conditions.

Coated fabrics

All commercial non-poromerics are coated fabrics. Several types of fabrics have been used, and the coating may be either vinyl or urethane, depending upon the desired balance between physical properties and economic considerations.

Poromerics

Poromerics have been developed in order to match the moisture-vapour permeability, skivability, and non-fraying characteristics of leather; the products have microscopic, open-celled structures. The structure of a poromeric is more homogeneous than that of a coated fabric, since it is either all polymer or polymer reinforced with non-woven fibres. The size and configuration of the cells is such that the passage of moisture vapour but not liquid water can occur. Although the vapour transmission rates of commercial poromerics vary considerably, the better ones can equal that of leather.

Manufacture and Processing

Coating of fabrics

The type of fabric used as a coating substrate is determined by the intended application of the product. Woven fabrics are used for most footwear products, whereas knits are used for upholstery, apparel, or accessories where greater extensibility is required.

The weight of solid vinyl coatings varies from 200–270 g/m^2 (6–8 oz/yd^2) and the thickness of the vinyl from 0.30–0.75 mm, depending upon the specific use. The vinyl usually is a plasticised vinyl chloride homopolymer that is compounded with various stabilisers and pigments as desired. The formulated vinyl is applied to the continuously moving textile web by coating, calendering, or extrusion.

Expanded vinyl, coatings are prepared by either the transfer-coating or calender-coating technique. Foaming or blowing the interlayers is accomplished by including a blowing agent in the formulation of the vinyl. Upon being heated, the blowing agent decomposes to yield a gas, e.g., N_2 or CO_2. The gas, in turn, produces a closed-cell foam structure.

Urethanes have an advantage over vinyl in that their molecular structure renders them inherently flexible. The commercial products are stock, or segmented, ester-urethane or ether-urethane copolymers. Urethanes also have tensile strengths that are as much as ten times greater than those of vinyls. Urethanes do not require the stabilisation against thermal degradation that vinyl does, but they do require stabilisation against hydrolytic decomposition. One of the main disadvantages of urethanes, with respect to vinyls, is that urethane solutions require strong solvents, e.g., dimethylformamide (DMF) or tetrahydrofuran, both of which are expensive and toxic.

Formation of poromeric structures

The manufacture of poromeric structures, particularly those with sufficient permeability for footwear applications, has presented the largest technological problem in the leather-substitute field.

The simplest poromeric products are those that contain no reinforcing. The most direct method of forming such a homogeneous microporous web is by polymer coagulation, which involves excluding a web of urethane solution onto a suitable support. The web is immersed in a non-solvent (for the polymer) which is miscible with the solvent used. The usual solvent/non-solvent combination is DMF/water. When the urethane web is immersed in the water, DMF migrates to the aqueous phase and water begins to migrate into the web; simultaneously, the polymer precipitates from solution. The net effect of this precipitation/migration is the formation of a solid urethane that contains microscopic channels or pores.

The various techniques for producing fibre-reinforced poromerics result in a large variety of structures. The main variations include the type of fibre structure used and the degree to which it is impregnated with polymer. Although some poromerics, e.g., Corfam, include woven-fabric reinforcement, recent practice has tended toward using only non-woven materials; the non-woven-based products tend to match leather more closely in modulus.

The Japanese have developed an elegant non-woven approach in the production of Clarino by producing fibres that are close in properties to the collagen fibres in leather. Two immiscible molten polymers are extruded to live a fibre with a cross-section, as shown in Fig. 15.2. By subsequent selective solvent extraction of either the island or the sea phase, a hollow tripple, compressible fibre or a bundle of very fine fibres is obtained.

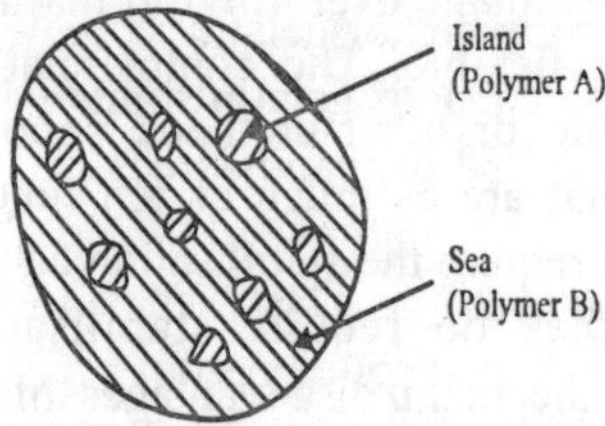

Fig. 15.2. Cross-sectional view of islands-in-the-sea fibre.

The best overall properties in laminated structures have been obtained with a combination woven/non-woven structure. The structure is prepared by heavily napping a sateen fabric, shearing the nap, and then re-pregnating the fabric with a reinforcing resin.

Finishing

In addition to having suitable physical properties, it is necessary for synthetic leather to have a leatherlike appearance. The appearance of a piece of leather results principally from two factors: colour and texture. The finished surface of the material must have some resistance to abrasion and appropriate tactile properties. Three types of surfaces are commercially important: grained, patent, and suede, and such requires different finishing techniques.

GELATIN

A mixture of proteins obtained by hydrolysis of collagen by boiling skin, ligaments, tendons, etc. Its production differs from that of animal glue in that the raw materials are selected, cleaned and treated with special care so that the product is cleaner and purer than glue. Gelatin is obtained from acid-treated raw materials, or alkali-treated raw materials. Gelatin is strongly hydrophilic, absorbing up to ten times its weight of water and forming reversible gels of high strength and viscosity. It can be chemically modified to make it insoluble in water for such special applications as microencapsulation of fish nutrients for fish culture.

Physical and Chemical Properties

Commercial gelatin produced from coarse granules to fine powder. Dry commercial gelatin contains about 9–12% moisture and is an essentially tasteless, odourless, brittle solid with specific gravity of 1.3–1.4. The physical and chemical properties of gelatin, such as solubility, stability, and swelling, are measured on aqueous gelatin solutions and are functions of the source of collagen, the method of manufacture, conditions during extraction and concentration, thermal history, pH, and the chemical nature of impurities or additives. Gelatin is classified as derived protein since it is obtained from collagen by a controlled partial hydrolysis. This makes gelatin a heterogeneous mixture of polypeptides; properties desired are obtained by blending products from selected extracts.

Gelatin is not a single chemical substance. The main constituents of gelatin are large and complex polypeptide molecules of the same amino acid composition as the parent collagen, covering a broad molecular weight distribution range. Most commercial gelatin contains molecular species from 15,000 to ca 250,000 mol wt, the average being between 50,000 and 70,000 mol wt.

The amphoteric character of gelatin is due to the functional groups of the amino acids and the terminal amino and carboxyl groups created during hydrolysis. The distribution of these groups and the resulting isoionic point is determined by the manufacturing process.

A most useful and unique property of gelatin solution is its capability to form reversible gel-sols. A gelatin solution gels at temperatures below 35°C. This conversion temperature is determined as the setting point (sol to gel) or melting point (gel to sol). Commercial gelatins melt between 23 and 30°C, the setting point being lower by 2–5°C. Gelatin is used mainly by the food, pharmaceutical, and photographic industries.

Types of Gelatin

Derivatised gelatins

Commercially successful derivatised gelatins are made mostly for the photographic gelatin and microencapsulation markets. In both instances, the amino groups are acylated.

Gelatin dynamite

A high explosive which contains nitrocellulose in addition to nitroglyceride. The product is a gelatinised mass, less sensitive to shock and friction than straight dynamite.

ADHESIVES

Adhesives, often referred to as cements, glues, or pastes, are defined as substances "capable of holding materials together by surface attachment". Adhesive technologists do not all agree as to what is and what is not an adhesive. Substances may attach to surfaces and develop the internal or cohesive strength necessary to hold materials together while cooling from liquid to solid state, while losing solvent, or during chemical reaction. Pressure-sensitive adhesives do not undergo a phase change in order to hold materials together.

Many of the substances designated as adhesives may be called paints, finishes, or coatings when applied in thin films to only one surface, or may be called caulking, potting, casting, or encapsulating compounds when employed in thick masses.

To be termed an adhesive, a substance must be a liquid or tacky semisolid, at least for an instant to contact and wet a surface, and be used in relatively thin layer to form a useful joint capable of transmitting stresses from one substrate to another. Adhesives and sealants include cement, glue, paste and waxes.

Theory

Adhesion is an interfacial phenomenon: wetting of substrates is essential. The electrical theory presumes that the adhesive and the substrates are like the plates of a capacitor that become charged due to the contact of two substances. The theory fails to predict the strong joints that result when a layer of water is frozen to join two blocks of ice, or when an epoxy adhesive is used to join two previously cured blocks of cast epoxy.

Diffusion theory presumes the penetration of the substrates by the adhesive prior to its solidification; this is easily applied to many porous plastics; however, not to metal, glass, or glazed ceramic. Adsorption theory specifies the concept of forces, such as van der Waals forces, acting across the space between molecules in a material. Rheological theory suggests that the removal of weak boundary layers in plastics leaves the joint's mechanical properties determined by the mechanical properties of the materials making up the joining and local stresses. In the absence of weak boundary layers, joint failure must be cohesive within the bondline of the substrates.

Types and Application

Adhesives have been categorised by suitability for bonding various substrates by physical form, by method of application, and by temperature-resistance recommendations for many substrates when using a particular type of bonding agent.

Much of the equipment used to apply adhesives is the same as used to apply surface coatings such as roller, flow, curtain, and knife coaters. A brief classification by type is as follows :

Inorganic

(1) Soluble silicates (water glass).

(2) Phosphate cements.

(3) Portland cement (calcium oxide-silica).

(4) Other hydraulic cements (mortar, gypsum).

(5) Ceramic (silica-boric acid).

(6) Thermosetting powdered glasses ("Pyroceram").

Organic

(1) Natural

(a) Animal

Hide and bone glue; fish glue

Blood and casein glues

(b) Vegetable

Soyabean, starch, cellulosics, rubber latex and rubber-solvent (pressure-sensitive). Gums, terpene resins (rosin), mucilages.

(c) Mineral

Asphalt, pitches, hydrocarbon resins.

(2) Synthetic
 - (a) Elastomer-solvent cements
 - (b) Polysulphide sealants
 - (c) Thermoplastic resins (for hot-melts)

   Polyethylene, isobutylene, polyamides, polyvinyl acetate
 - (d) Thermosetting resins

   Epoxy, phenolformaldehyde, polyvinyl butyral, cyanoacrylates.
 - (e) Silicone polymers and cements.

High-temperature Adhesive

(1) Organic polymers, e.g., polybenzimidazoles, that retain bonding strength up to 260°C for a relatively long time (500–1000 hours); above 260°C strength drops rapidly, 80% being lost after 10 minutes at 535°C.

(2) Inorganic (ceramic), e.g., silica-boric acid mixtures or cermets produce bonds having high strength above 2000°F; adhesive lap-bond strengths can be over 2000 psi at 1000°F. These adhesives are used largely for aerospace service, and metal/metal and glass/metal seals.

Hot-melt Adhesive

A solid, thermoplastic material which quickly melts upon heating, and then sets to a firm bond on cooling. Most other types of adhesives set by evaporation of solvent. Hot-melt types offer the possibility of almost instantaneous bonding, making them well-suited to automated operation. In general, they are low-cost, low-strength products, but are entirely adequate for bonding cellulosic materials. Ingredients of hot-melts are polyethylene, polyvinyl acetate, polyamides, hydrocarbon resins, as well as natural asphalts, bitumens, resinous materials, and waxes.

Rubber-based (Cement, Rubber) Adhesive

(1) A solution of natural or synthetic rubber in a suitable organic solvent, without sulphur or other curing agent.

(2) A mixture of rubber (often reclaimed), filler, and tackifier (pine tar, liquid asphalt) applied to fabric backing (pressure-sensitive friction tape).

(3) A room-temperature curing rubber-solvent-curative mixture, often made up in two parts, which are blended just before use.

(4) Rubber latex, especially for on-the-job repairing, such as conveyor belts.

(5) Silicone rubber cement.

Hazard

Those containing organic solvents, (1) and (3) above, are flammable.

Setting and Curing

Only the reactive adhesives truly cure. For example, acrylics, unsaturated polyesters, and other monomer adhesives containing ethylenic unsaturation cure by formation of free radicals when catalysed by peroxides and accelerated by metal-ion donors. Epoxies cure by addition mechanisms. Polysulphides react with active oxidising agents to form rubbery polymers particularly useful in joining low expansion glasses to metals. Many urethane and silicone adhesive cure in the presence of moisture. Most of the reactive adhesive cures are accelerated by heat. Cyanoacrylate adhesives cure only in thin bondlines, i.e. 1.0 μm to 0.1 mm, without the addition of a catalyst or hardener. The basicity present on many substrates initiates cure when inhibitors present in the adhesive are overcome.

Uses

Adhesives are used widely in a variety of industries including automotive; construction; electronic bonding, with special applications to computer memories and both active and inactive micro- and macro-electronic elements; packaging; plastics; and in the textile products and apparel industry.

Chapter 16

Dye and Dye Intermediates

INTRODUCTION

Intermediate products is the name given to chemical compounds obtained at an intermediate stage in the process of conversion of simple hydrocarbons to complex synthetic organic dyes, i.e. they are compounds which occupy an intermediate position between the raw materials and the end products. The most common of the intermediates are aniline, dimethyl aniline, phenol, nitrobenzene, chlorobenzene, etc. Dye intermediates also often represent valuable industrial products in their own right or intermediate products for manufacturing pharmaceutical and photographic materials, plastics, and other organic chemicals.

METHODS USED FOR MANUFACTURING DYE INTERMEDIATES

The raw materials employed in manufacturing intermediate compounds belong mainly to the cyclic hydrocarbons of the aromatic series: they are benzene, toluene, xylene, naphthalene, anthracene, and other primary products obtained in the by-product coke and petrochemical industries. By a series of operations and processes these compounds are converted into intermediate products of various degree of complexity.

The principal processes used in converting the raw materials into intermediate products are sulphonation, nitration, reduction, halogenation, substitution of a hydroxyl group for a sulpho group or a haloid, alkylation, oxidation, and condensation. The processes used in converting the raw materials into an intermediate product vary widely, but they can be reduced to three classes. To the first class belong processes by which a hydrogen atom of an aromatic compound is replaced by a substitute, for instance a sulpho group, a nitro group or a haloid; to the second class, processes of replacement or conversion of the hydrogen substitute, for instance, in the transformation of chlorobenzene into phenol; to the third class, processes in which the carbon skeleton of the molecule of the initial compound is changed, for instance in the oxidation of naphthalene to phthalic anhydride. A brief review of some of the most common processes used in manufacturing dye intermediates follows.

Sulphonation

One of the most important reactions in the manufacture of intermediates and dyes is that of sulphonation, i.e. the reaction of an aromatic compound with sulphuric acid resulting in the formation of a sulphonic acid. In the general form it can be described by the following equation :

$$R\text{—H} + \text{HO} — SO_3H \rightleftarrows \underset{\text{sulphonic acid}}{R\text{—}SO_3H} + H_2O$$

where R is an aromatic radical, for instance C_6H_5— (phenyl ring with H atoms), etc.

The above equation shows only the initial and final products of the reaction. In actual fact the process is much more complicated. The sulphonation reaction is reversible. When sulphonic acids are heated with dilute sulphuric acid, the sulphonic acids are hydrolysed (a desulphonation reaction occurs), i.e. the process is reversed.

Let us now consider the effect of three factors—the concentration of the sulphonating agent (sulphuric acid, oleum, or monohydrate), the temperature and the use of a catalyst—on the course of the sulphonation reaction. As can be seen from the above reaction, sulphonation results in the formation of water, which dilutes the sulphuric acid, reduces the rate of the sulphonation reaction and increases the rate of the desulphonation reaction. As a result of these opposing processes equilibrium is established, i.e. the sulphonation process ends. For each organic chemical a definite limiting value of the sulphuric acid concentration exists below which sulphonation at a given temperature will not take place. The concentration of the sulphuric acid cannot be increased excessively, however, since this will lead to side reactions, such as the formation of tarry substances.

The best concentration of sulphuric acid for sulphonation of a particular substance is different for each substance and depends on the reaction temperature. When the reaction is run at a low temperature there should be an excess of the acid to prevent the dilution of the reaction mixture with the water which is formed. When sulphonation is carried out at a high temperature, no excess of the acid is necessary because the water formed evaporates.

It should also be borne in mind that there is a certain optimum temperature for each reaction above which undesirable transformations may take place—either oxidation by sulphur trioxide of the compound obtained or a shift of the sulpho group (—SO_3H). By changing the process conditions—the temperature, the concentration of the sulphonating agent and the reaction time, different end products can be obtained. Occasionally the sulphonation reaction is conducted in the presence of a catalyst. For instance, mercury is sometimes used as a catalyst in sulphonation of anthraquinone.

The sulphonation process is carried out in iron or cast iron kettles, or sulphonators, provided with an agitator, a cooling jacket and a coil for indirect heating with steam. As an example production of benzene-sulphonic acid will be considered.

Manufacture of Benzene-Sulphonic Acid

Benzene-sulphonic acid, C_6H_5—SO_3H, is a solid hygroscopic substance readily soluble in water. Benzene-sulphonic acid is a strong acid. Its sodium salt is used in the synthesis of phenol.

The sulphonic acid is obtained by sulphonation of benzene. Both batch processes and continuous processes are used for making benzene-sulphonic acid. The best of the batch procedures is that of benzene sulphonation in the vapour phase. This method is schematically represented in Fig. 16.1.

From metering tank 1, the benzene enters tube-in-tube vapouriser 2, which is heated by steam. The benzene vapours go from the vapouriser through superheater 3, where they are heated to 160-170°C, to sulphonator 4, to which 80 to 100% - concentration sulphuric acid is fed from metering tank 5. The benzene vapour bubbles through the sulphuric acid and is partially sulphonated (about 10% of it); the main part (90%) leaves the sulphonator unconverted, is separated and re-used. The mixture obtained, containing 92 to 94% benzene-sulphonic acid, has residual benzene removed and is re-processed into the sodium salt of benzene-sulphonic acid, for which purpose the mixture is diluted with water and treated with a solution of sodium sulphite (Na_2SO_3). Formation of the sodium salt of the benzene-sulphonic acid (the sulphonate) proceeds by the reaction:

$$2\,C_6H_5{-}SO_3H + Na_2SO_3 \rightarrow 2\,C_6H_5{-}SO_3Na + H_2O + \uparrow SO_2$$

The excess sulphuric acid used for sulphonation is neutralised by the sodium sulphite :

$$H_2SO_4 + Na_2SO_3 \rightarrow Na_2SO_4 + H_2O + \uparrow SO_2$$

The sulphur dioxide released is used in the process of production of phenol. The amount of the impurity, sodium sulphate, Na_2SO_4, formed is small and the product can be employed without purification for making phenol.

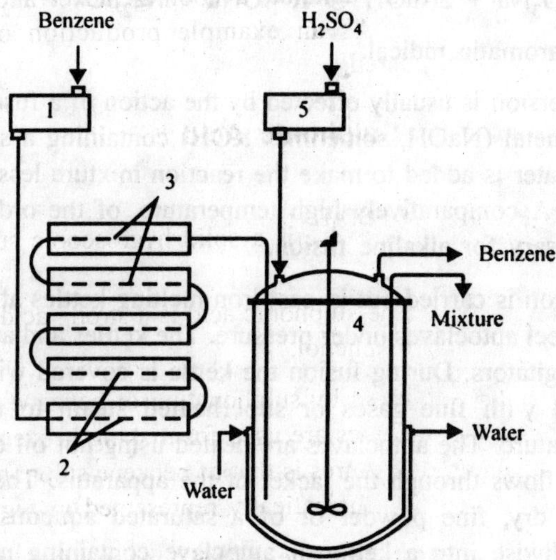

Fig. 16.1. Manufacture of benzene-sulphonic acid.

Recently a continuous sulphonation procedure has been developed. In this procedure benzene reacts with a countercurrent of sulphuric acid in a battery of four sulphonators. Continuous sulphonation is also carried out in cast-iron tray columns in which there is a countercurrent of the sulphuric acid and benzene vapours.

Alkaline fusion

Alkaline fusion is a process in which salts of aromatic sulphonic acids and alkalies react with substitution of a hydroxyl (—OH) group for the sulpho (—SO_3H) group.

This method is used to obtain various intermediates employed in the production of dyes, plastics, drugs, and perfumes.

The most important products obtained by alkaline fusion of salts of the aromatic sulphonic acids are phenol, [benzene ring]—OH, resorcinol [benzene ring with OH]—OH, and β-naphthol, [naphthalene ring]—OH

The conversion of salts of aromatic sulphonic acids (sulphonates) to the corresponding phenolates (*R*—ONa) by alkaline fusion can be represented schematically by the following equation :

$$R\text{—}SO_3Na + 2NaOH \rightarrow R\text{—}ONa + Na_2SO_3 + H_2O,$$

where *R* is an aromatic radical.

Such a conversion is usually effected by the action of a fused hydroxide of an alkaline metal (NaOH, sometimes KOH) containing a small amount of water. The water is added to make the reaction mixture less viscous and better miscible. A comparatively high temperature, of the order of 160 to 330°C, is necessary for alkaline fusion.

Alkaline fusion is carried out in cast-iron melting kettles at atmospheric pressure or in steel autoclaves under pressure. The kettles and autoclaves are provided with agitators. During fusion the kettle is covered with a lid. The kettle is heated with flue gases or superheated steam to maintain the required temperature. The autoclaves are heated using hot oil or a diphenyl mixture, which flows through the jacket of the apparatus. The sulphonate, in the form of dry, fine powder or of a saturated aqueous solution, is introduced batchwise into a kettle or autoclave containing molten alkali. When all the sulphonate has been introduced the reaction mixture is heated for 0.5 to 1 hour with stirring and the temperature is raised to the required value. The melt obtained is quenched (diluted with water), and the phenolates decompose :

$$2R\text{—}ONa + SO_2 + H_2O \rightarrow 2R\text{—}OH + Na_2SO_3$$

The phenol compounds thus produced are separated and purified.

Nitration

In the nitration reaction usually one (or more) of the hydrogen atoms of an aromatic ring is replaced with a nitro (—NO_2) group as a result of treatment of the aromatic compound with nitric acid. The overall equation of the nitration reaction is

$$R\text{—H} + \text{HO—NO}_2 \xrightarrow{H_2SO_4} R\text{—NO}_2 + H_2O,$$

where R is an aromatic radical.

Nitration is widely used in manufacturing various dye intermediates, dyes, explosives, etc. In the nitration process water is formed and it gradually lowers the concentration of the nitric acid. This leads to a reduction in the reaction rate, equilibrium may set in, and the nitration process will end. Therefore substances which bind the water must be introduced into the reaction mixture. For this purpose concentrated sulphuric acid can be used, either as oil of vitriol (about 94% H_2SO_4) or as monohydrate (100% H_2SO_4). Oleum is also used, but less often.

A so-called nitrating mixture—a mixture of concentrated nitric and sulphuric acids in the ratio of 1 : 3 to 1 : 4—is used for nitrating the aromatic compounds. The nitrating agent in the mixture is concentrated nitric acid. Dilute nitric acid, which dissociates into H^+ and NO_3^- ions, cannot be used for nitration. The NO_3^- anion has no nitrating effect, but it oxidises the aromatic compounds, converting them into tarry substances. The nitrating mixture for nitration of benzene to nitrobenzene contains nitric acid in an amount only slightly exceeding the theoretical value; the amount of concentrated sulphuric acid used is such that in the nitration process it is diluted to about 70% by the water released. This corresponds to the following composition of the nitrating mixture (in % by weight): 60% H_2SO_4, 32% HNO_3, 8% H_2O. The nitration temperature varies widely and lies between −10 and +120°C for different substances.

Nitration is accompanied by a considerable release of heat, and therefore the reaction mixture must be cooled to maintain a fixed temperature. The nitrating mixture must be added slowly and with good mechanical agitation to prevent the formation of hot spots.

Manufacture of Nitrobenzene

Nitrobenzene, C_6H_5—NO_2 (benzene ring) or $C_6H_5NO_2$, is a yellowish poisonous liquid with an odour of bitter almonds. It is used in the production of dyes and explosives. Nitrobenzene is made by nitration of benzene :

$$C_6H_5H + \text{HO—NO}_2 \rightarrow C_6H_5NO_2 + H_2O$$

Fig. 16.2 depicts the nitrobenzene production process. Benzene and the nitrating mixture are delivered from metering vessels 2 and 3 to the bottom of nitrator 1. Nitration occurs mainly in the lower part of the nitrator, where a temperature of 40 to 50°C is maintained. The reaction mixture near the top of the nitrator contains nitrobenzene, sulphuric acid and the remainder of the nitric acid. The reaction mixture remains in the nitrator for 6 to 12 min, after which it is discharged to separator 4, where the benzene is separated from the acids. From the separator the nitrobenzene is removed to tank 5; the crude product is subjected to further treatment: the acid residues are removed by washing with water and the remainder neutralised with soda ash. The spent acid (about 70% H_2SO_4) gathers in tank 6. For return to the nitration process it is denitrated and concentrated by evaporation to oil of vitriol (about 94% H_2SO_4).

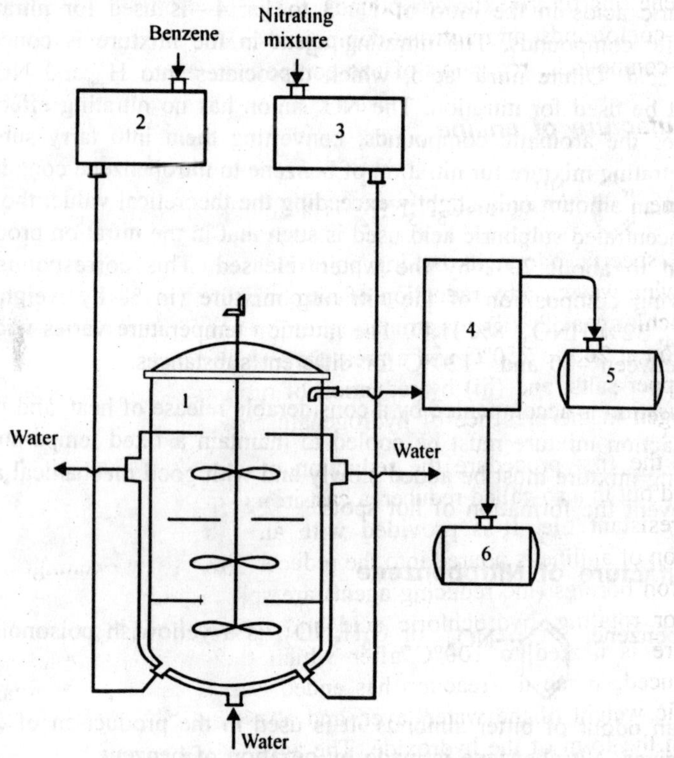

Fig. 16.2. Nitrobenzene production.

The nitrator, metering vessels, separator, tanks, and all the pipes are made either of high-silicon acid-resistant cast iron, or of stainless steel with an acid-resistant ceramic lining. The nitrator is provided with a water jacket for controlling the temperature.

Reduction of nitro-compounds

The process of reduction of nitro-compounds is used in making various intermediates necessary mainly for the manufacture of dyes. The products of the reduction of nitro-derivatives of benzene and its homologues are aromatic amines (aromatic compounds containing a $—NH_2$ group). Thus, aniline, or aminobenzene, $C_6H_5NH_2$, is obtained by reduction of nitrobenzene :

$$C_6H_5NO_2 + 3H_2 \rightarrow C_6H_5NH_2 + 2H_2O$$

Many different methods are used for introducing an amino group into the benzene ring of aromatic compounds, the principal ones being reduction of nitro-compounds and nitroso-compounds, electrochemical reduction of nitro-compounds, reduction of azo-compounds.

Manufacture of aniline

Aniline, (benzene ring)$-NH_2$, or $C_6H_5NH_2$, is a colourless poisonous liquid with a weak, specific odour. Its boiling point is 184.4°C. Aniline is made in the following ways: (i) by reduction of nitrobenzene by iron in the presence of hydrochloric acid; (ii) by amination of a chlorobenzene by an ammonia solution at 200 to 250°C and a pressure as high as 100 atm in the presence of copper salts; and (iii) by reduction of nitrobenzene in the gas phase by hydrogen in the presence of hydrogenating catalysts.

In the first procedure the reduction of the nitrobenzene to aniline is carried out in a so-called reducer, a cast-iron cylindrical apparatus lined with acid-resistant tile. It is provided with an agitator. First a 3% aqueous solution of aniline is poured into the reducer (through an opening in the lid), cast-iron borings (the reducing agent) are spilled into it and then, with the agitator rotating, hydrochloric acid and live steam are introduced. The mixture is heated to 100°C after which the nitrobenzene is gradually introduced. When the reaction has ended, salt is added for increasing the specific weight of the water layer, and also slaked lime to precipitate the iron in the form of the hydroxide. The agitator is stopped and the reaction mixture is left to stand; it divides into three layers; the upper one—crude aniline, the middle one—an aqueous salt solution, the lower one—iron oxides.

The upper aniline layer is decanted and subjected to further processing, first it goes to the condenser-cooler to separate the water from the aniline, and then to a vacuum distillation unit.

This method requires the use of cumbersome equipment; aniline losses are considerable. In recent years a catalytic method of making aniline has come to be widely used. Nitrobenzene is evaporated in a current of high-temperature hydrogen to produce a vapour-gas mixture of nitrobenzene and hydrogen at a temperature of 170°C. This mixture containing a large excess of hydrogen is passed through two converters. The mixture leaves the second converter at a temperature of 350 to 370°C. The mixture is cooled, and the reaction products condense; the excess hydrogen is returned to the process. The aniline yield is 98% of the theoretical value. The catalyst used in the process is obtained by depositing copper salts on pumice from their suspension in an aqueous solution of sodium silicate, drying and reducing with hydrogen.

Diazotisation

Azo compounds are the numerous organic compounds which contain an azo (—N=N—) group. In the most important of them the azo group is linked with two aromatic radicals (aryls).

Diazo compounds are the low-chemical-stability products of reactions between nitrous acid, HNO_2, and primary amines of the aromatic series. These compounds are highly reactive and they are widely used for making intermediates and synthetic dyes (azo dyes). *Diazotisation* consists in reacting an aromatic (more seldom a heterocyclic) amine with nitrous acid obtained in the diazotisation process by reaction between an aqueous solution of hydrochloric (or sulphuric) acid and sodium nitrite ($NaNO_2 + HCl \rightarrow HNO_2 + NaCl$). For instance, the diazotisation of aniline, $C_6H_5NH_2$, in the presence of hydrochloric acid produces the diazo compound of aniline in the form of its hydrochloric acid salt, $C_6H_5N_2Cl$, called benzene diazonium chloride:

$$C_6H_5NH_2 + HNO_2 + HCl \rightarrow [C_6H_5N \equiv N]^+Cl^- + 2H_2O$$

Aniline — Benzene diazonium chloride

Diazotisation is carried out as when amine is dissolved in dilute mineral acid and the solution of the nitrous acid salt is added to the solution obtained at a fixed temperature. Usually the process is run at a temperature of 0 to 10°C. Such low temperatures are necessary because of the instability of nitrous acid, and also of the diazonium salts formed at higher temperatures. The end of diazotisation is determined by testing with an iodine-starch paper (dark-blue colour). The reaction rate depends on the solubility of the amine (higher solubility results in a higher rate).

Azo coupling

The *azo coupling reaction*, used in producing intermediates and dyes, consists in the interaction of the aromatic diazo compounds with amines and phenols. The aromatic diazo compound is called the *diazo component of the reaction*. The amine or phenol which reacts with the diazo compound is called the *azo component of the reaction*. Coupling with amines is usually conducted in an acid medium (in the presence of dilute hydrochloric acid or acetic acid):

$$C_6H_5{-}N{=}N{-}\boxed{OH + H}{-}C_6H_3(NH_2){-}NH_2 \rightarrow$$

$$\rightarrow C_6H_5{-}N{=}N{-}C_6H_3(NH_2){-}NH_2 + H_2O$$

Coupling with phenols takes place in an alkaline solution:

$$O_2N{-}C_6H_4{-}N{=}N{-}\boxed{OH + H}{-}C_{10}H_7 \rightarrow$$

$$\rightarrow O_2N{-}C_6H_4{-}N{=}N{-}C_{10}H_6(OH) + H_2O$$

The rate of the azo coupling reaction depends on the reaction temperature, the concentration of the solutions, the structure of the reacting molecules and the concentration of hydrogen ions in the solution.

The reaction rate increases with the temperature. But at higher temperatures the diazo compounds begin to decompose, and undesirable by-products are formed. The reaction is usually begun, therefore, at 2 to 5°C and terminated at 25 to 30°C. As the reaction proceeds, the concentration of the reactants decreases and the rate of the azo coupling drops.

Chlorination of aromatic compounds

The *chlorination reaction* is widely used in manufacturing of intermediates, dyes and insecticides. The chlorination of benzene to chlorobenzene, C_6H_5Cl, can be represented by the equation

$$C_6H_6 + Cl_2 \rightarrow C_6H_5Cl + HCl$$

If a compound being chlorinated has a side chain, chlorination in two ways is possible: chlorine can enter into the aromatic nucleus or into the side chain. The conditions differ radically for these reactions. Chlorination with substitution in the aromatic nucleus is carried out at moderate temperatures in the presence of a catalyst. The catalysts used are metallic iron, ferric chloride or aluminium chloride. Chlorination of the side chain of an aromatic hydrocarbon proceeds without a catalyst, under heating and in the presence of light.

Manufacture of Chlorobenzene

Chlorobenzene, C_6H_5Cl, is a colourless liquid with an odour resembling that of benzene. Chlorobenzene is used in making various benzene derivatives. It is manufactured by passing chlorine through liquid benzene in the presence of a catalyst—ferric chloride or, more often, metallic iron. Fig. 16.3 is a flowsheet of the production of chlorobenzene by continuous chlorination of benzene.

The industrial unit consists of three (or more) series-connected chlorinators 1, 2, and 3 arranged in tiers. The chlorinator is a cylindrical cast-iron tower filled with ceramic rings. It is provided with a water jacket. The first, upstream chlorinator, in addition to the ceramic rings, contains iron rings, which act as a catalyst. The benzene is fed to the lower part of the first chlorinator, chlorine is passed through each chlorinator. From the upper part of the first chlorinator the benzene flows to the lower part of chlorinator 2, and from it to chlorinator 3. The hydrogen chloride formed in the chlorination reaction is removed from the top of each chlorinator. The chlorinated benzene is withdrawn from the third chlorinator and then subjected to further treatment. The hydrogen chloride and ferric chloride impurities are removed from the chlorobenzene by washing it with water and a soda ash solution or by mixing it with a concentrated solution of caustic soda. The purified chlorinated benzene is allowed to settle and is separated, after which it is sent for fractional distillation.

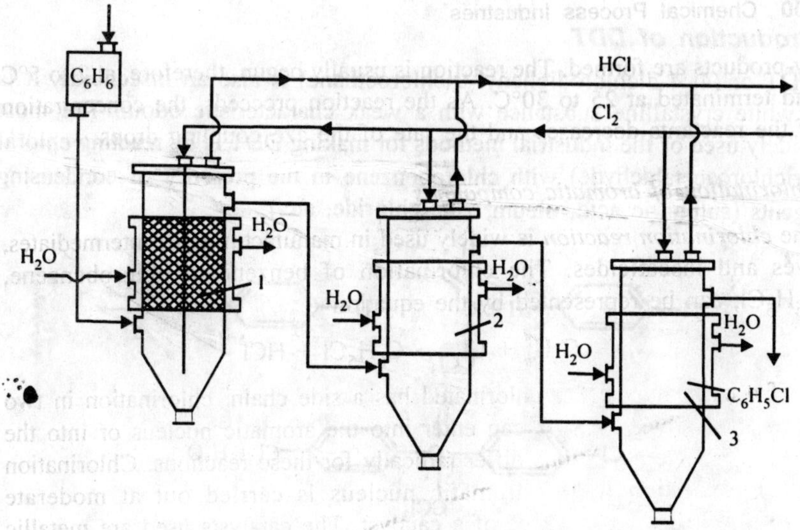

Fig. 16.3. Production of chlorobenzene by continuous chlorination of benzene.

In recent years continuous chlorination of benzene in the gas-vapour phase (with the reaction mixture boiling) has become an important procedure.

Production of hexachlorane

Hexachlorane (or hexachlorocyclohexane), $C_6H_6Cl_6$, is an insecticide. It is used in the form of powders (dusts) in mixtures with various fillers (talc, kaolin, etc.). Hexachlorane is obtained by adding chlorine to benzene:

$$C_6H_6 + 3Cl_2 \rightarrow C_6H_6Cl_6 + Q$$

Gaseous chlorine practically will not react with pure benzene in the dark and without the use of a catalyst. Production of hexachlorane is an example of the photochemical processes, i.e. reactions proceeding under the effect of light. The chlorine is reacted with the benzene in a continuous-action column. The benzene is fed to the column at the top and the gaseous chlorine at the bottom. Quartz-bulb mercury-vapour electric lamps are mounted inside the columns. A temperature of 30°C is maintained in the upper part of the column and 50°C in the lower part. The chlorinated benzene leaving the column is subjected to further treatment.

Hexachlorane is also obtained by a catalytic method. Peroxides, bases (alkalies, calcium hydroxide) and some other substances can be used as catalysts. The process is run at low temperatures and without the presence of light.

Production of DDT

DDT or (4, 4-dichlorodiphenyl-trichloroethane) is also an insecticide. It is a white crystalline substance with a weak characteristic odour. The most widely used of the industrial methods for making DDT is by reacting chloral (trichloroacetaldehyde) with chlorobenzene in the presence of condensing agents (sulphuric acid, oleum, zinc chloride, etc.):

$$Cl-C_6H_4-H + H-C(=O)-CCl_3 + H-C_6H_4-Cl \xrightarrow{H_2SO_4}$$

$$\rightarrow Cl-C_6H_4-CH(CCl_3)-C_6H_4-Cl + H_2O$$

The reaction is conducted at a moderate temperature and with an excess of the chlorobenzene to avoid side processes, in particular that leading to the formation of *n*-chlorobenzene-sulphonic acid.

Production of phenol

Phenol, C_6H_5OH, is one of the most important synthetic organic products. Phenol is used in huge amounts in mañufacturing plastics and drugs, in the synehtsis of dyes, and for making explosives. Phenol is obtained in the following ways: by fusion of the sodium salt of benzene-sulphonic acid with an alkali, by hydrolysis of chlorobenzene with an alkali, by catalytic hydrolysis of chlorobenzene with water vapour, and by decomposition of the hydroperoxide of isopropyl-benzene.

In making phenol from benzene-sulphonic acid the starting material is benzene. The production process consists of the following stages: (i) sulphonation of the benzene (preparation of benzene-sulphonic acid); (ii) dormation of the sodium salt of the benzene-sulphonic acid; fusion with caustic soda; decomposition of the phenolate to form phenol. Schematically the process can be represented as follows:

$$C_6H_6 \xrightarrow{H_2SO_4} C_6H_5SO_3H \xrightarrow{Na_2SO_3} C_6H_5SO_3Na \xrightarrow{NaOH}$$

$$C_6H_5ONa \xrightarrow{SO_2 + H_2O} C_6H_5OH$$

A flowsheet of the production of phenol from benzene-sulphonic acid is presented in Fig. 16.4. The benzene is first sulphonated in column 1. The benzene-sulphonic acid formed is fed to a neutralisation column, 2, where it is treated with sodium sulphite to convert it to the sodium salt of benzene-sulphonic acid. From column 2 the sodium salt is transferred to cast-iron kettle 3, supplied with an agitator, for fusion with the alkali.

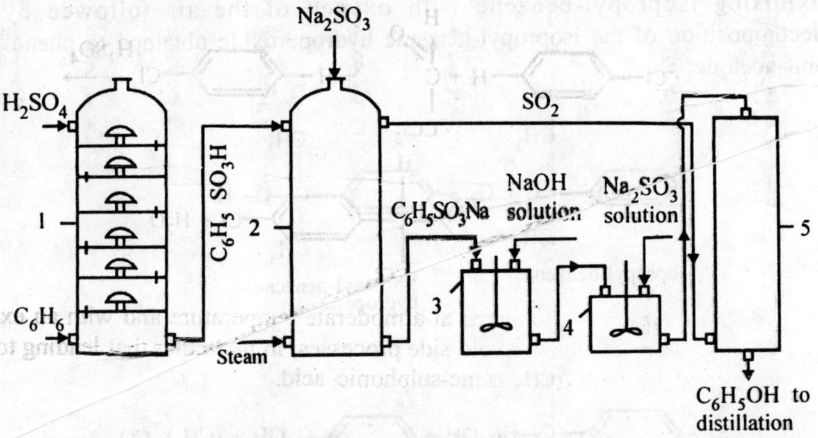

Fig. 16.4. Production of phenol from benzene-sulphonic acid.

Crushed caustic soda is loaded into the kettle together with a small amount of water, and the alkali is fused. The solution of the sodium salt of benzene-sulphonic acid is added while stirring, and the mixture is boiled until all water is evaporated. The temperature is then raised to 300°C. The entire process lasts 6 to 9 hr. The finished melt is discharged into quencher 4, which was previously filled with a solution of sodium sulphite. The sodium phenolate formed in the quencher is in solution, whereas the sodium sulphite crystallises. From the quencher the sodium phenolate solution goes to a continuous-action packed column, 5, to the lower part of which SO_2 is fed. The phenol produced in the column is removed at the bottom and then distilled.

Two methods for making phenol from chlorobenzene have been developed.

(1) In the Raschig method phenol is obtained by hydrolysis of chlorobenzene with steam at 450 to 500°C over a catalyst (e.g. silica gel):

$$C_6H_2Cl + \underset{\text{vapour}}{H_2O} \rightarrow C_6H_5OH + HCl$$

(2) In the hydrolysis of chlorobenzene it is treated with a 10% aqueous solution of caustic soda at 250 to 300°C; the process is run under a pressure of 200 to 250 atm in a tubular reactor:

$$C_6H_5Cl + 2NaOH \rightarrow C_6H_5ONa + NaCl + H_2O$$

$$C_6H_5ONa + H_2O \rightarrow C_6H_5OH + NaOH$$

The method for making phenol from isopropyl-benzene consists in oxidising isopropyl-benzene with oxygen of the air followed by decomposition of the isopropyl-benzene hydroperoxide obtained to phenol and acetone:

$$C_6H_5\text{—}CH(CH_3)_2 + O_2 \rightarrow C_6H_5\text{—}C(CH_3)_2\text{—}O\text{—}H$$

isopropyl-benzene → isopropyl-benzene hydroperoxide

$$C_6H_5\text{—}C(CH_3)_2\text{—}O\text{—}OH \rightarrow C_6H_5\text{—}OH + (CH_3)_2CO$$

phenol, acetone

The first reaction is accelerated by catalysts—salts of iron, copper, manganese, etc. Sometimes the reaction is conducted in an alkaline aqueous emulsion at a temperature of about 130°C. The decomposition of the hydroperoxide (the second reaction) is carried out under pressure by boiling with dilute sulphuric acid. This method is the most attractive, because acetone, a valuable product, is obtained along with phenol.

ORGANIC DYES

Dyes are coloured organic compounds capable of colouring various materials are articles. Dyes are used for colouring wool, silk, and flax fabrics, chemical fibres, leather, fur, paper, wood, soaps, oils, some foodstuffs, and other materials. Other fields where dyes are used are in the motion picture and photographic industries, analytical chemistry, bacteriological research, the printing industry, etc.

Organic dyes can be of *natural* (vegetable or animal) or *synthetic* origin. Examples of natural dyes are red alizarin and indigo, which can be obtained from plants, and red carmine, made from dried *cochineal* insects.

In connection with the rapid development of the textile industry there was a sharp increase in the demand for dyes in the middle of the 19th century. Natural dyes could no longer satisfy the requirements of the industry. It became necessary to develop methods for the synthetic production of dyes, and extensive research was started in this field. Since the artificial dyes were cheaper, faster, gave brighter colours and could be made in a considerably wider assortment, they had almost completely supplanted natural dyes by the beginning of the 20th century.

There are a number of theories of the colour of dyes. The colour of a dye is due to the presence of specific chromophore groups in its molecule. Among such groups are the azo group (—N=N—), the nitro group (—NO), the carbonyl group (> C=O), etc. Dyes also contain auxochrome groups (—NH_2, —OH, etc.), which strengthen and deepen the colour of the dyes and also strengthen their bond with the textile fibres.

Contemporary science bases the explanation of colour on the ability of the outer electrons of an atom in a molecule to absorb light of a specific wavelength. Ultraviolet and visible lights are mainly absorbed by the outer (valency) electrons of the molecules. After having absorbed energy an electron becomes "excited", i.e. it is transferred to a more removed orbit. This is an unstable electron state and the electron spontaneously returns to the previous orbit, emitting a photon of a definite wavelength. It is the wavelength of radiation or a combination of different wavelengths of radiation that determines the colour of a compound. Colourless substances may absorb ultraviolet rays, but they do not absorb visible light, whereas coloured substances absorb visible light of corresponding wavelength.

Classification of Dyes

Two systems are used for classifying artificial organic dyes: a chemical system and a commercial one. The chemical classification is based on the chemical structure of the dyes; the dyes are divided into the following groups: nitro dyes, nitroso dyes, azo dyes, arylmethane, quinonimine, sulphur, indigoid, phthalocyanine, anthraquinone (mordant and acid), polycyclic (vat), and polymethine dyes.

The commercial classification is based on the applications of the dyes:

(1) *Acid dyes (salts of organic acids):* They colour materials of animal origin (wool, natural silk, leather) and polyamide fibres in an acid medium, but do not colour cellulose fibres.

(2) *Basic dyes (organic bases or salts):* They colour protein fibres; they also colour cellulose fibres with the aid of special substances (mordants).

(3) *Direct dyes (salts of organic acids):* They colour protein fibres and cellulose fibres (without mordants).

(4) *Mordant dyes:* The material is preliminarily treated with salts of metals (Al, Cr, Fe); the dye forms a fast and insoluble compound called a lake.

(5) *Vat dyes:* They have a wide range of application.

(6) *Sulphur dyes:* They colour vegetable fibres.

(7) *Oxidative dyes (aniline black):* They colour furs.

(8) Azo dyes formed directly on the fibre (azoic dyes).

(9) Oil-soluble and spirit-soluble dyes.

(10) *Pigments and lakes:* Water-insoluble dyes.

(11) Acetate rayon dyes.

(12) Active dyes, which form a chemical bond with fibres.

Azo dyes

Azo dyes make up one of the most important dye groups. It includes dyes which are used practically in all branches of industry. More than half of the brands of dyes in use belong to this group. The molecule of an azo dye contains one or more azo (—N=N—) groups. Azo dyes containing one azo group are called *monoazo dyes*, those containing two azo groups—*diazo dyes*, three groups—*triazo dyes*, more than three—*polyazo dyes*. Azo dyes are obtained as follows. The intermediates for making the dyes are obtained by the diazotisation reaction. The dye is synthesised from the intermediate by azo coupling reaction. By way of illustration let us consider the manufacture of one of the simplest representatives of the azo dye class—acid orange light fast:

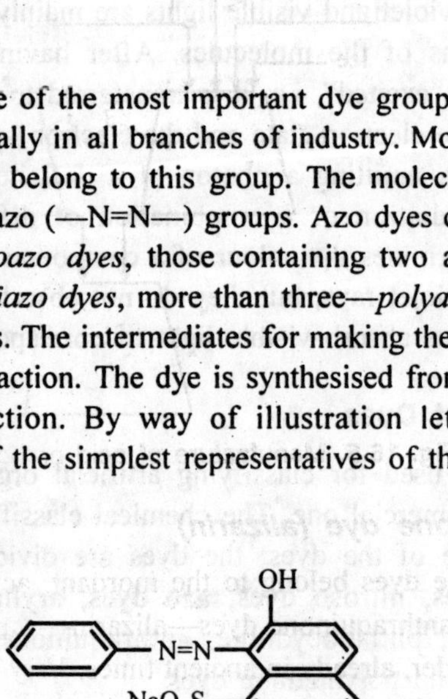

OH
N=N
NaO_3S
SO_3Na

The synthesis is carried out in a vat provided with an agitator (Fig. 16.5). Water is poured into vat 3, ice is introduced and sulphuric acid and aniline are added from measuring tank 1. Then, at a temperature of 0 to 2°C, with stirring, a sodium nitrite solution is added from metering tank 2, and diazotisation of the aniline takes place. The diazotisation reaction continues for about one hour. G-salt (disodium salt of 2-naphthol-6,8-disulphonic acid) and soda ash are dissolved in water in vat 4. The solution from vat 4 is transferred to vat 5, ice is added, and the diazo compound is gradually added from vat 3 with stirring. The azo coupling reaction terminates within an hour. After that the dye is precipitated by adding salt, and the suspension is sent to a filter press. After it is filtered off the dye is dried and ground to a fine powder, which is mixed with salt to produce a dye of standard concentration.

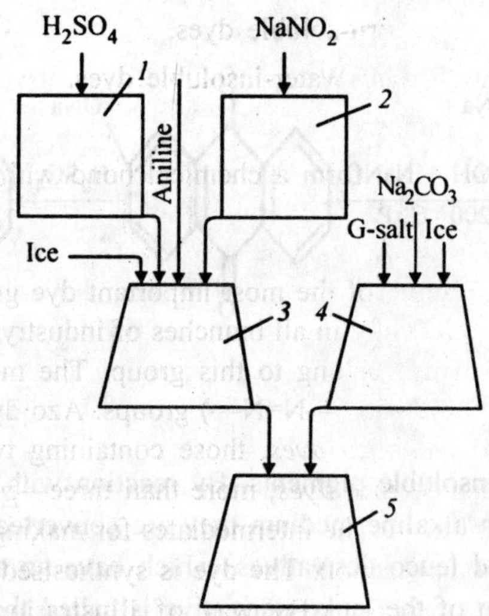

Fig. 16.5. Manufacture of acid orange light-fast dye.

Anthraquinone dye (alizarin)

Anthraquinone dyes belong to the mordant, acid and vat dye classes. The oldest of the anthraquinone dyes—alizarin—was obtained from the roots of a plant—madder, already in ancient times. Very early, it was established that alizarin is 1,2-dioxy-anthraquinone, and it was synthesised.

The starting material from which anthraquinone dyes are made is anthraquinone, which is sulphonated with oleum at a temperature of 125 to 140°C. Then the 2-sulphonic acid anthraquinone obtained is converted to its sodium salt with the aid of sodium sulphate.

$$\text{anthraquinone} \xrightarrow{H_2SO_4 + SO_3} \text{anthraquinone-}SO_3H \xrightarrow{Na_2SO_4} \text{anthraquinone-}SO_3Na$$

anthraquinone

The salt formed is heated at 200°C under pressure (in an autoclave) together with a concentrated solution of caustic soda in the presence of sodium acetate, which is an oxidising agent, this results in the formation of a disodium salt of alizarin, which is converted to alizarin by dilute sulphuric acid.

$$\text{anthraquinone-}SO_3Na \xrightarrow[200° \text{ C, P}]{NaOH + NaNO_3} \text{anthraquinone-}ONa \xrightarrow{H_2SO_4} \text{anthraquinone-}OH$$

alizarin red

Vat dyes (indigo)

Vat dyes are water-insoluble pigments. By reaction with a reducing agent (hydrosulphite) in an alkaline medium they are converted to water-soluble compounds, so-called leuco derivatives, which have an affinity for fibres. The alkaline solution of the leuko derivative is called the vat solution. At one time reduction to the leuco derivative was done in vats—hence the name, vat dyes. They have a wide range of application. They are used for uniform dyeing and printing on various types of fabrics, for dyeing artificial and synthetic fibres, plastics, etc.

The vat dyes are divided into the two large groups of *indigoid* and *anthraquinone* dyes. The first group owes its name to the typical and oldest of all dyes indigo blue.

More than 30 different methods of making indigo are known. Two of them are commonly used in industry. The raw material for manufacturing indigo is aniline. The manufacture of indigo consists of three basic stages:

making the intermediate—the sodium salt of phenylglycine; making indoxyl and oxidation of the latter to blue indigo; processing the waste.

A reaction between aniline and monochloroacetic acid is used to produce phenylglycine. Then the process is continued in the presence of ferrous hydrate; a sparingly water-soluble salt of phenylglycine of ferrous is the result and it is converted into the soluble phenylglycine potassium salt by reacting it with caustic potash (KOH):

$$2\,C_6H_5NH_2 \xrightarrow{ClCH_2\text{—}COOH} 2\,C_6H_5NHCH_2COOH \xrightarrow{Fe(OH)_2}$$

aniline — phenylglycine

$$\rightarrow Fe\left[C_6H_5NHCH_2COO^-\right]_2 \xrightarrow{KOH} 2\,C_6H_5NHCH_2COOK$$

ferrous oxide salt precipitate — soluble potassium salt of phenylglycine

The phenylglycine potassium salt is filtered off, evaporated and ground; after this the potassium salt of indoxyl is made from it. For this purpose sodium amide, $NaNH_2$, is introduced into a molten mixture of sodium hydroxide and potassium hydroxide, and the phenylglycine potassium salt is gradually added at 230°C:

The indoxyl obtained is oxidised by air to form the indigo:

$$C_6H_5NHCH_2COOK \xrightarrow{NaNH_2,\ KOH,\ NaOH} C_6H_4(NH)(CH)C\text{—}OK$$

indoxyl

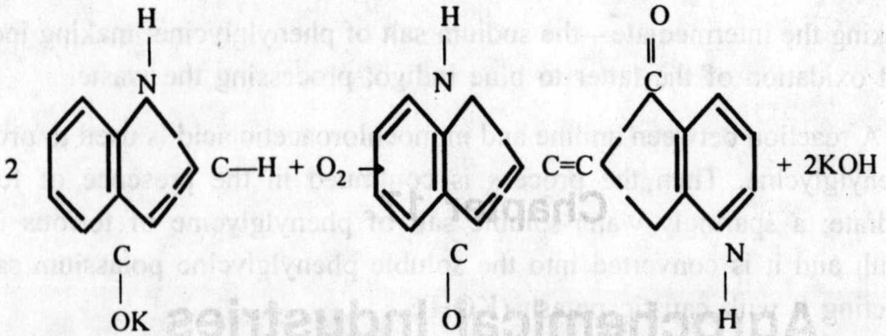

$$2\,[\text{C}_6\text{H}_4(\text{NH})\text{C(OK)}{=}]\text{C—H} + O_2 \rightarrow \text{indigo} + 2KOH$$

indigo

The production of chemical fibres and plastics is constantly increasing. Many of the dyes which were successfully used for dyeing natural fibres are unsuitable for dyeing synthetic fibres. Therefore the main new trend in the dye industry is to develop new products for colouring synthetic fibres. The assortment of dyes is widening and dyes with new properties are being manufactured.

Chapter 17

Agrochemical Industries

INTRODUCTION

India has one of the most dynamic generic pesticide industries in the world. Exports of generic agrochemicals will hold the key to profitable operations for players in the Indian agrochemical industry. Industry is in the process of developing and manufacturing environment-friendly formulations such as flowable; water dispersable granules and suspension concentrates with better bio-efficiency and shelf life. These products are having a huge world market of approximately US$ 1 billion and in the next three years India will be exporting these products worth Rs. 300 crore if sincere efforts are made.

The Indian industry is fortunate to have the expertise of organisations such as the Institute of Pesticide. Formulation Technology which is aided by UNIDO. Also the credit goes to various CSIR laboratories particularly National Chemical Laboratory (NCL), Pune and the Indian Institute of Chemical Technology (IICT), Hyderabad. These laboratories worked on processes for a number of technicals, and transferred this is the industry as a complete engineering package, which led to the development of the industry. Thus the agrochemical industry, in particular, is seeing dramatic developments, involving the constant phase-out of old molecules, and their replacement by environ-friendly, low dosage, more potent moieties. Fluorinated pyrethroids, including, Tefluthrin, Bifenthrin and Acrinathrin which have good activity when applied to soil, are now on the anvil to replace organophosphates, and aryl heterocyclics have been developed as a new class of compounds.

A new class of respiration inhibitors which kill pests by preventing oxidative phosphorylation are also now available, and these include pyridaben and tebufenpyrad. Aryl heterocyclics are a new class of neutrotoxic insecticides, of which Fipronil, the first compound, along with some others, have shown greater activity than synthetic pyrethroids. New avenues should be explore, so that industry is in a position to offer broad-based product for use for the national and international markets.

Backward integration is a must to make production cost effective and, diversification into fine chemicals segments is necessary. Both the pharmaceuticals and the agrochemical industries need various chemicals and many of these products are still imported.

Structurally, pesticide molecules will undergo significant changes in the future to become more specific and eco-compatible. From the process point of view, shortened and high performance synthesis enriched through novel chemistry and biotechnology will provide immense opportunities and challenges to the researchers in the Asia-Pacific region. A number of enabling sciences have to be employed to provide the cutting edge to these technologies. These include process science and engineering, computational and combinatorial chemistry, molecular modelling to relate structure with the mode of pesticidal action, analytical measurements and environmental sciences. Biotechnology will play an increasing role in process technologies, as will assymmetric synthesis and chiral separations. However, the integrated pest management is the likely trump card for the future growth process. The product processes waste paradigm will undergo significant transformation to ensure sustainability of agrochemical sector in this region.

Changing customer priorities in USA, Europe and domestic markets will create excellent opportunities for agrochemical sector in Asia-Pacific region to create new business designs. The health risks posed by chemical pesticides which arise during manufacture, transportation and application are still much lower than for many harmless looking operations. This is mainly due to abundant precautions taken during pesticides manufacture. Notwithstanding these facts it has become increasingly difficult for the industry to convince the public at large, that not every pesticide plant is a 'Bhopal' in the making.

INORGANIC INSECTICIDES

Some of the inorganic insecticides are discussed below.

Lead Arsenate

Lead arsenate is a very poisonous, colourless and crystalline compound and is usually employed for the potato beetle and for the codling moth in apple orchards.

It is soluble in nitric acid and insoluble in water, sp. gr. 5.8 m.p. 1024°C. It is usually employed for the potato beetle and for the coding moth, in apple orchards. It is used as an insecticide and herbicide. Due to its high toxicity 1 gm. is lethal to man, may cause skin irritation.

Calcium Arsenate

It can be prepared by interaction of calcium chloride, calcium hydroxide, and sodium arsenate. It is a white powder, slightly soluble in water, soluble in dilute acids. Decomposes on heating. It is used as an insecticide and germicide. It is toxic by ingestion and inhalation. Tolerance, 1 mg per cubic metre of air (as As). It is applied to leaves in the form of dust.

ORGANIC INSECTICIDES

Dinitrophenol

It can be prepared: (i) by heating phenol with dilute sulphuric acid, cooling the product, and then nitrating, keeping the temperature below 50°C; (ii) by nitration with mixed acid with careful temperature control. It can be purified by crystallisation. It is in the form of yellow crystals, sp. gr. 1.681; m.p. 144°C (2,4 dinitrophenol), sp.gr. 1.683; m.p. 114–115°C. m.p. 63°C (2,6 dinitrophenol). Soluble in alcohol, ether, benzene and chloroform; slightly soluble in water. It is used in dyes, especially sulphur colours; picric acid; picramic acid; preservation of lumber; manufacture the photographic developer diaminophenol hydrochloride; explosives manufacture; indicator and can be used as a reagent for K and NH_4 ions. It is highly toxic; absorbed by skin. Dust inhalation may be fatal. Severe explosion hazard when dry.

Dichloro Diphenyl Trichloroethane (DDT)

It can be prepared by condensing chloral or chloral hydrate with chlorobenzene in presence of sulphuric acid. It is in the form of colourless crystals or white to slightly off-white powder. Odourless or with slight aromatic odour. Insoluble in water; soluble in acetone, ether, benzene, carbon tetrachloride, kerosene, dioxane, and pyridine. Not comparable with alkaline materials.

It is toxic by ingestion, inhalation, and skin absorption, especially in solution. Acute oral toxicity for man said to be 0.25 g/kg of body weight. Since DDT is not biodegradable and is ecologically damaging. Claims of human carcinogenicity have not been proved. DDT can be used for a few specialised purposes, e.g., to combat the tussock moth. Tolerance, 1 mg per cubic metre of air; 5 ppm in foods. May accumulate in the soil incompatible with alkaline materials. It should not be stored in iron containers. Toxic to bees. Accumulates in the body fat of animals. It is used as an insecticides, especially for tobacco and cotton; pesticides (tussock moth).

Methoxychlor (Methoxy DDT; DMDT)

It can be prepared by reaction of methyl phenyl ether and chloral hydrate. It is in the form of white, crystalline solid; m.p. 89°C; insoluble in water; soluble in alcohol. Not compatible with alkaline materials. It is used as an insecticide, effective against mosquito larvae and house flies. Especially recommended for use in dairy barns. It is moderately toxic. Tolerance, 10 mg per cubic metre in air.

BHC or Benzene Hexachloride

BHC or benzene hexachloride or more correctly 1,2,3,4,5,6 hexachlorocyclohexane is another important insecticide, which exists in a number of stereoisomers, the gamma being the most toxic.

The commercial BHC possesses 5 main geometrical isomers. The percentage of each of the isomer in the commercial BHC is alpha (55–70%), beta (5–14%), gamma (10–13%), delta (6–8%), epsilon (3–4%). Later on various isomers were separately prepared or separated from the mixture and it was found that BHC has insecticidal property of gamma-isomer.

The gamma BHC is usually known as gammexane or lindane. It is refined material containing 99% of gamma isomer of BHC. The BHC is prepared by the chlorination of benzene in presence of sunlight.

The product is a mixture of five out of eight possible isomers of which the physiologically active isomer does not exceed 18%. The amount and formation of isomer has been found to increase in presence of some substances such as sulphur, selenium, tellurium, nitrotoluenes, dialkyl peroxy dicarbonates etc.

Excess of benzene is used in order to keep the products in solution and to continue the reaction under homogeneous-conditions. The reaction can also be carried out in presence of γ-radiation or fluorine (elemental) instead of uv radiation.

The manufacture can be carried out by batch process or by continuous process. In batch process technical grade benzene and chlorine are reacted in a lead lined vessel provided with cooling arrangements and mercury vapour lamp is used as a uv source. The temperature is kept between 15–20°C and pressure is maintained slightly above atmospheric pressure, in order to minimise the loss of benzene. The reaction is continued till the BHC concentration attains 12–15%. The excess of chlorine is air blown and the mother liquor is concentrated and finally fractionally crystallised to isolate the biologically active isomer.

In continuous process chlorination is carried out in a glass lined pipe chlorinator in the central tube of which benzene is taken and uv source is two 40 watt fluorescent lamps. The temperature is 40–60°C and pressure is atmospheric. The chlorination is continued till the concentration of BHC reaches 13–14.5%. The chlorinated liquor is concentrated at 85–88°C till it becomes saturated in isomer and almost all the γ- and β-isomers are crystallised out. The crystals are separated by filtration and the filtrate or mother liquor is subjected to steam distillation. The molten residue in the vessel is cooled to get lumps of crude BHC containing 25% of the isomer. The isomer is then extracted by solvent extraction using hexane-nitromethane solvent mixture. The ground solid is then deodourised with treatment with oleum or agitation in air with kaolin and manganese linoleate. Isomer of BHC is more soluble than other in most organic solvents. It kills or controls insects, grass hoppers and common pests and used in the form of dust containing 5–10% of the ingredients in emulsion containing 25% of the γ-isomer and also as granules having 6–10% of the active principal.

Gammexane

Crude BHC is generally not suitable for food crops, because of its strong musty odour. Gammexane or lindane, a refined material containing 99% of the BHC, has largely overcome the odour. Benzene hexachloride is a colourless, practically odourless crystalline solid, insoluble in water, but soluble in organic solvents.

Gammexane has exceptional insecticidal activity, comparable to that of DDT and may be used in food crops. Insects that are not killed by DDT are killed by gammexane. Gammexane has also been applied successfully against the common disease carrying insects, household stock, horticulture, stored food and agriculture pests. Gammexane has widely been used also as an agricultural pesticide. 99% pure BHC, called lindane or gammexane is prepared by its selective crystallisation followed by repeated solvent extraction. Lindane or gammexane is sparingly soluble in water but soluble in acetone. It is stable to air, heat and light. It is also stable to strong acids but decomposes in contact with hot alkalies. It has no contact action on human system and so it is safe insecticide for domestic use. It is applied as dust, emulsion, granules and as water dispersion.

Chlordane

Chlordane is a chlorinated hydrocarbon, insecticide showing contact, stomach poison, and fumigant action with long residual effects. It is colourless, odourless, viscous liquid; sp.gr. 1.57–1.67 (60/60°F); viscosity SSU 100 seconds (38°C); organic chlorine 64–67% by wt, purity 98%; b.p. 175°C (2 mm); refractive index (n 25/D) 1.56–1.57; soluble in many organic solvents; insoluble in water; miscible in deodorised kerosene; decomposes in weak alkalies. It is used as an insecticide and fumigant. It is toxic by ingestion, inhalation, and skin absorption. Tolerance, 0.5 mg per cubic metre of air.

Heptachlor

Heptachlor is a chlorinated hydrocarbon, insecticide with contact and stomach poison action, especially effective on soil insects. It is a white, waxy solid; m.p. 95–96°C; sp. gr. 1.57–1.59. Insoluble in water; soluble in xylene, and alcohol. It is used as an insecticide, in soil treatment, seed treatment and on the plant foliage. It is toxic by ingestion, inhalation and skin absorption. Tolerance, 0.5 mg per cubic metre of air.

Aldrin

Aldrin is a chlorinated hydrocarbon insecticide with contact and stomach poison action. It is in the form of brown to white crystalline solid, m.p. 104–105.5°C; insoluble in water; soluble in most organic solvents. Not affected by alkalies or dilute acids; compatible with most fertilisers, herbicides, fungicides, and insecticides. It is valuable as a soil insecticide for termites. It is highly toxic by ingestion and inhalation; absorbed by skin. Tolerance, 0.25 mg per cubic metre of air. It is a suspected carcinogen. It's use restricted to non-agricultural applications.

Endrin

Endrin is isomeric to dieldrin and can be prepared by first preparing hexachloro bicylochlorinated by an inorganic base. Hexachloro bicycloheptane can also be prepared by adding Hex to acetylene in a tubular nickel alloy reactor at 150°C and 300 psi pressure. The hexachloro bicycloheptadine is then heated with cyclopentadiene at 70–100°C to obtain isodrin, the endoendo isomer of aldrin. The isodrin on epoxidation with paracetic acid produces endrin, the endoendo isomer of aldrin. Endrin is also used as contact cum stomach poison and employed as dust or emulsion.

Toxaphene

Toxaphene is a chlorinated hydrocarbon insecticide showing contact with stomach poison activity. It is amber, waxy solid with a mild odour of chlorine and camphor; melting range 65–90°C; sp. gr. 1.66 (27°C). Soluble in common organic solvents. It is used as an insecticide (primarily for cotton and early growth stage of vegetables). It is toxic by ingestion, tolerance 0.5 mg per cubic metre of air inhalation, skin absorption. Causes cancer in mice. It is corrosive to iron. Lactating dairy animals should not be treated. Birds feeding on treated areas may be killed. Use may be restricted.

Malathion

Malathion is an organic phosphate insecticide-acaricide. It can be prepared from diethyl maleate and dimethyl di-thiophosphoric acid. It is in the form of yellow, high-boiling liquid; (b.p. 156–157°C, under 0.7 mm with slight decomposition); m.p. 3.0°C; refractive index (n 25/D) 1.4985; sp. gr. 1.2315 (25°C); vapour pressure (20°C) approximately 0.00004 mm. Miscible with most polar organic solvents. Slightly soluble in water. It is toxic by ingestion and inhalation. Absorbed by skin. Tolerance, 10 mg per cubic metre of air. Chloinesterase inhibitor.

Parathion

Parathion is an organic phosphate insecticide, a caricide with both contact and stomach poison activity. It can be prepared from sodium ethylate, thiophosphoryl chloride and sodium para-nitrophenate. It is deep brown to yellow liquid; usually has faint odour. Refractive index (n 25/D) 1.5367; sp. gr. 1.26 (25/4°C); b.p. 375°C; f.p. 6°C; vapour pressure 0.003 mm (24°C); very slightly soluble in water (20 ppm); completely soluble in esters, alcohols, ketones, ethers, aromatic hydrocarbons, animal and vegetable oils; insoluble in petroleum ether, kerosene, spray oils. Stable in distilled water and in acid solution. Hydrolysed in the presence of alkaline materials; slowly decomposes in air. It is highly toxic by skin contact, inhalation or ingestion. Cholinesterase inhibitor. Repeated exposure may, without symptoms, be increasingly hazardous. Fatalities have resulted from its accidental use. Tolerance, 0.1 mg per cubic metre of air. Use has been restricted.

FUMIGANTS

A toxic agent in vapour form that destroys rodents, insects, and infectious organisms; a type of pesticide. The most effective temperature for their use is about 70°F. They are used chiefly in enclosed or limited areas (barns, greenhouses, ships' holds, and the like); also are applied locally to soils, grains, fruits and garments. Some commonly used fumigants are formaldehyde, sulphur dioxide and other sulphur compounds, hydrogen cyanide, methyl bromide, carbon tetrachloride, p-dichlorobenzene, and ethylene oxide. Care must be used when handling and applying fumigants in view of their toxicity. Some of the fumigants are discussed below:

Ethylene Dibromide

It can be prepared by the action of bromine on ethylene. It is colourless, nonflammable liquid having sweetish odour and emulsifiable. Miscible with most solvents and thinners; slightly soluble in water. Sp. gr. 2.17–2.18 (20°C); wt/gal 18.1 lb; b.p. 131°C; vapour pressure 17.4 mm (30°C); f.p. 9°C; refractive index 1.5337 (25°C); flash point, none. It can be used as scavenger for lead in gasoline; grain fumigant; general solvent; waterproofing preparations; organic synthesis; fumigant for tree crops. It is toxic by inhalation, ingestion, and skin absorption. Carcinogen in test animals. Strong irritant to eyes and skin.

Ethylene Dichloride

It can be prepared by the action of chlorine on ethylene with subsequent distillation, with metallic catalyst; also by reaction of acetylene and HCl. It is colourless, oily liquid; chloroform-like odour; sweet taste. Stable to water, alkalies, acids, or active chemicals. Resistant to oxidation and will not corrode metals. Miscible with most common solvents; slightly soluble in water. B.p. 83.5°C; f.p. -35.5°C; sp.gr. 1.2554 (20/4°C); wt/gal 10.4 lb; refractive index 1.444; flash point 56°F (13.3°C).

It is used in vinyl chloride solvent; lead scavenger in anti-knock gasoline; paint, varnish and finish removers; metal degreasing; soaps and scouring compounds; wetting and penetrating agents; organic synthesis; ore flotation. It is toxic by ingestion, inhalation, skin absorption. Strong irritant to eyes and skin. Tolerance, 10 ppm in air. Flammable, dangerous fire risk; explosive limits in air 6 to 16%. and it may be carcinogenic.

Para-dichlorobenzene

Para-dichlorobenzene is prepared by the chlorination of monochlorobenzene. It is in the form of white crystals; volatile (sublimes readily); penetrating odour and it is combustible. Sp.gr. 1.458; b.p. 173.7°C; m.p. 53°C; flash point 150°F (65.5°C) (closed cup). Soluble in alcohol, benzene, and ether; insoluble in water. It is used as moth repellent; general insecticide; germicide; space odorant; manufacture of 2,5-dichloroaniline; dyes; intermediates; pharmacy; agriculture (fumigating soil). It is toxic by ingestion; irritant to eyes. Tolerance, 75 ppm in air.

MITICIDES

Miticide generally is a pesticide which kills mites, small animals of the spider class, among them the European red mite and the common red spider which infest fruit trees.

RODENTICIDES

Both inorganic (zinc phosphide) and organic compounds are used as rodenticides. They all kill rodents when they penetrate through the gastro-intestinal tract, although the mechanism of action of each pesticide is different. Rodenticides are mainly used for preparing poisoned baits. Grains of wheat, rye, corn, etc. are used as the bait material for controlling gophers, voles, midday gerbils and mice in field conditions. Grain bait can be used to control voles only in cold weather. In summer, they do not eat such bait. To eradicate them in warm weather, and also the great gerbil in spring and autumn, green bait is used such as alfalfa, clover, and other succulent food.

To exterminate rats and mice in premises (granaries, livestock, buildings, greenhouses, etc.), bread crumbs, porridge, grain, minced meat and fish, etc. are used as the bait. At stock-raising farms, bait is prepared on the fodder forming the ration of the livestock and is placed in the feed bunks in stalls freed of animals. Baits must include adhesive (stickers). Vegetable oils are the best for this purpose. In controlling gophers, motor oil is mainly used as the adhesive.

Water-soluble poisons are used in the form of aqueous solutions and the bait is soaked in them. A very small assortment of substances is used in practice for poisoned baits. This is associated with the difficulty of preparing substances with a narrow selective toxicity.

Zinc Phosphide (Phosvin)

The active ingredient is Zn_3P_2. The pure substance is a dark grey odourless powder with a high melting point. It is insoluble in water and ethanol, and soluble in benzene and carbon disulphide. It is thermally stable. When reacted with acids, even weak organic ones, zinc phosphide decomposes with the evolution of the very poisonous phosphine, capable of spontaneous flammation :

$$Zn_3P_2 + 6HCl \rightarrow 2PH_3 + 3ZnCl_2$$

Such a process also occurs in an animal's stomach upon the souring of baits consisting of bread, dough, porridge, or greens. Such baits lose their toxic properties in a few days. Zinc phosphide decomposes comparatively slowly in other poisoned baits, and their toxicity persists during a month and more. Consequently, when zinc phosphide is used, domestic and wild animals and birds may be poisoned.

The toxic action of zinc phosphide on rodents is due to the fact that when it is ingested into the stomach, it decomposes into phosphine, which is the toxic substance. Phosphine penetrates with the blood stream into the animal's nervous system and results in violation of its functions. The death of animals is preceded by vomitting, weakening of the heart's activity, emphysema, and convulsions.

The toxicity of zinc phosphide to rodents depends on the quality of the poison and on the natural resistance of the animals to its action. The quality of the rodenticide is determined by the fineness of grinding, which underlies the dimensions of the poison's reactive surface coming into contact with the acid in an animal's stomach. Finer particles accelerate and amplify the action of the poison. The fineness of grinding is determined by the relevant USSR State Standard. The resistance of rodents to zinc phosphide depends on their mass, physiological state, age, and specific features. Animals with a high vitality (in the period of intensive reproduction) are more resistant to poisons.

The specific resistance of rodents to zinc phosphide is associated with the features of their feeding. The high resistance of gophers is apparently explained by the lower acidity of their gastric juice. The amount of acid evolved in the stomach depends on the composition and quality of the food and on the time when it enters the stomach. It has been established that the more the food correpsonds to the needs of rodents in the given conditions of their existence, the higher is the acidity of their gastric juice and the more rapidly does poisoning with zinc phosphide occur. Baits with zinc phosphide have no repelling taste or odour and are readily eaten by rodents.

As regards its toxicity to humans and warm-blooded animals, zinc phosphide belongs to strong poisons with selective toxicity; it has no chronic toxicity. It completely decomposes in the stomachs of killed animals, therefore the secondary poisoning of predators is excluded.

FUNGICIDES

Any substance which kills or inhibits the growth of fungi is known as fungicide. Older types include mixtures of lime and sulphur, copper oxychloride, and Bordeaux mixture. Copper naphthenate has been used to impregnate textile fabrics such as tenting and military clothing. Mercury compounds are also effective, but have been discontinued because of their toxicity to humans. Hypochlorite solutions are used in swimming pools and water-cooled heat exchangers. Some types of fungi that infect the human body are extremely hard to eradicate and require highly specific medical treatment.

Eptam

It is trade-mark for a selective herbicide containing ethyl-N,N-di-n-propylthiolcarbamate.

2,3-dichloroallyl diisopropylthiolcarbamate (Diallate)

It is in the form of brown liquid; b.p. 159°C (9 mm). Almost insoluble in water; soluble in acetone, benzene, chloroform, kerosene, and xylene. It is generally used as herbicide. It is toxic and absorbed by skin.

Ferbam

It can be prepared by addition of carbon disulphide to an alcoholic solution of dimethylamine and precipitation with a ferric salt. It is in black or dark coloured fluffy powder; decomposes above 180°C; usually readily dispersible but very slightly soluble in water; pH of saturated solution 5.0. It is used as a fungicide. It is irritant to eyes and mucous membranes. Tolerance, 10 mg per cubic metre of air.

2,4,6-trichlorophenol

It is in the form of yellow flakes with strong phenolic odour; sp. gr. (25/4°C) 1.675; f.p. 61°C; b.p. 248–249°C; no flash point. Soluble in acetone, alcohol, and ether. It is non-flammable. It is used as fungicide. herbicide, defoliant. It may cause skin irritation.

Para-chlorophenol

It can be prepared by chlorination of phenol; from chloroaniline through the diazonium salt. It is the form of white crystals (yellow or pink when impure) with unpleasant penetrating odour. Slightly soluble in water; soluble in benzene, alcohol, and ether. Volatile with steam; b.p. 217°C; m.p. 42–43°C; sp. gr. 1.306; refractive index (n40/D) 1.5579. A 1% solution is acid to litmus. Flash point 250°F (121°C). It is combustible. It is used as an intermediate in synthesis of dyes and drugs; denaturant for alcohol; selective solvent in refining minerals oils; antiseptic. It is toxic by skin absorption, inhalation or ingestion. Strong irritant to tissue.

Glyodin

It can be prepared by the action of ethylenediamine on stearic acid. It is in the form of light orange crystals; m.p. 62°C; sp. gr. 1.03; insoluble in water. Toxicity probably low. It is used as fungicide (fruits and vegetables).

HERBICIDES

A pesticide, either organic or inorganic, used to destroy unwanted vegetation, especially various types of weeds, grasses, and woody plants. Until 1924 inorganics such as sodium chlorate, sodium chloride, ammonium sulphamate, arsenic, and boron compounds were used. At that time the more specific organics were introduced, typified by 2,4-dichlorophenoxyacetic acid (2,4-D). Herbicides may be of two major types: (i) selective, such as 2,4-D; 2,4,5-T; phenols; carbamates, and urea derivatives, permitting elimination or weeds without injury to the crop; and (ii) non-selective, comprising soil sterilants (sodium compounds) and silvicides (ammonium sulphamate). The latter kill wood plants and trees. Some types act as overstimulating growth hormones. Many herbicides are highly toxic and should be handled and applied with care; use of chlorinated types may be restricted.

2,4-dichlorophenol

It can be prepared by chlorination of phenol. It is in the form of white, low melting solid; b.p. 210°C; m.p. 45°C; flash point 237°F (113°C); combustible. Soluble in alcohol and carbon tetrachloride; slightly soluble in water. It is used in organic synthesis. It is strong irritant to tissue; moderately toxic by ingestion.

Sodium Trichloroacetate

It is toxic by ingestion. Irritant to skin and eyes. It is used as herbicide and pesticide.

Chloro-IPC

It is a light tan powder; m.p. 41.4°C; vapour pressure (149°C) 2 mm; sp. gr. 1.18 (30°C); very slightly soluble in water. It is a pre-emergence herbicide; prevents sprouting of potatoes. It is toxic by ingestion.

Sesone

It is a crystalline solid; m.p. 245°C; soluble in water. It is toxic and is used as pre-emergence herbicide.

MCPA

It is a white crystalline solid; m.p. 118–119°C. Free acid insoluble in water but sodium and amine salts are soluble. It is used as selective herbicide.

SYNTHETIC INSECTICIDES

These insecticides, unlike other types, are absorbed by the plant roots and foliage and translocated through the plants. In this manner, they render plants toxic to aphides, red spider mites and other sucking insects that are otherwise difficult to be killed. These insecticides remain within the plants and so remain effective for much longer time than the other insecticides sprayed on their surface. If synthetic insecticides are applied to seeds, they are capable of protecting the young plants for a considerable period of time. *Schradan, dimefox, systox* and *dimetholate* etc. are well known and important synthetic insectides. These are described briefly as under :

(1) Schradan is octamethyl pyrophosphoramide and can be prepared by the reaction of tetramethyl diamido chlorophosphate with water in the presence of tertiary amine. Treatment of ethyl tetramethyl diamidophosphate with tetramethyl amidochlorophosphate at 40°C gives schradan in a good yield. It is soluble in water as well as in most organic solvents. It provides long term protection from aphides and mites after absorption into plants.

(2) Dimefox is bis (dimethylamino) phosphoryl fluoride and can be prepared by the following important methods :

(a) By the reaction of dimethylamine with dimethylamido difluorophosphate.

(b) By the action of sodium fluoride on bis (dimethyl amino phsophoryl chloride).

It is a colourless, mobile liquid, which is soluble in water as well as in common organic solvents. It is most suitable for foilage application. It is an insecticide for sucking pests of hops and vactors of a virus disease of Cocao trees.

(3) Systox or Demeton can be prepared by the action of diethyl chlorothiophosphate with 2-hydroxy diethyl sulphide in presence of NaOH. It is in the form of yellow colour liquid which soluble in organic solvents. It is long-lasting, rapidly absorbed by roots, stems or foliage.

(4) Dimethoate can be prepared by the reaction of potassium salt of dimethyl dithio phosphoric acid with N-methyl chloroacetamide. It is in the form of white crystalline solid, which is soluble in water as well as in common organic solvents. It is used for fruit larvae and for side-dressing of soil.

Chapter 18

Coal and Coal Chemicals

INTRODUCTION

Coal is a dark burnable solid, usually layered, that resulted from the accumulation and burial of partially decayed plant matter over earlier geological ages. These deposits were converted to coal through biological changes and later effects of pressure and temperature. Composition as well as chemical and physical properties vary widely for the different kinds of coal. Differences in type are caused by variations in the amounts of different plant parts exemplified by common-banded, splint, cannel, and boghead coals. The degree of coalification is referred to as rank. Brown coal and lignite, sub-bituminous coal, bituminous coal, and anthracite make up a natural series with increasing carbon content. The impurities in coal cause differences in grade.

Coal is red-brown black or brownish-black and is sold in a range of sizes. The colour, lustre, texture, and fracture vary with the type, rank, and grade. Coal is composed chiefly of carbon, hydrogen, and oxygen with small amounts of nitrogen and sulphur, and varying amounts of moisture and ash or mineral impurities.

The formation of coal, the variation in its composition, its microstructure, and its chemical reactions indicate that coal is a mixture of compounds. The organic composition depends on the degree of biochemical change of the original plant material, and on the later coalification by pressure and temperature effects in the deposit. The major organic compounds have resulted from the formation and condensation of polynuclear carbocyclic and heterocyclic ring compounds containing carbon, hydrogen, oxygen, nitrogen, and sulphur. The amount of carbon in aromatic ring structures increases with rank.

Nearly all coal is utilised in combustion and coking. Over 80% is burned directly for generation of electricity or steam generation for industrial uses, transportation or residential heating, and metallurgical processes, in the firing of ceramic products, etc. The balance is mostly carbonised to produce coke, coal gas, ammonia, coal tar, and light oil products from which many chemicals are derived. Combustible gases can also be produced by the gasification of coal. Carbon products are produced by heat treatment. A relatively small amount of coal is used for miscellaneous purposes such as fillers, pigments, foundry material, and water filtration.

Most organic chemicals can be synthesised from coal carbonisation products or by different processes utilising gases derived from coal. Some aliphatic and aromatic carbocyclic acids can be produced by the oxidation of coal. Processes to produce gaseous and liquid fuels, as well as chemicals, have been extensively developed and are being improved in the world. Development has been limited by the ability of these products to compete economically with petroleum and natural gas.

The reserves of petroleum and natural gas are being depleted more rapidly than coal. Coal consumption has been cyclic and, as energy consumption increases, coal will be used in continually larger amounts. Many new processes for coal conversion and utilisation are being developed and will be of significant interest to chemists and chemical engineers.

Microscopic examination of coal has been carried out using transmitted light through thin sections and reflected light from polished samples.

The mineral matter in coal is the result of several separate processes. Part is caused by the inherent material present in all living matter; another type is caused by the detrital minerals that were deposited during and after the time of peat formation; and a third type is caused by secondary minerals that have crystallised from water that has percolated through the coal seams. Clay is the most common detrital mineral; however, other common ones include quartz, feldspar, garnet, apatite, zircon, muscovite, epidote, biotite, augite, kyanite, rutile, staurolite, topaz, and tourmaline. The secondary minerals are generally kaolinite, calcite, and pyrite. Analyses have shown the presence of almost all elements in at least trace quantities in the minerals matter. Certain elements, i.e., germanium, beryllium, boron, and antimony are found primarily with the organic matter in coal. Zinc, cadmium, manganese, arsenic, molybdenum, and iron are found with the inorganic material in coal. The primary mineral constituents in coal are aluminium, silicon, iron, calcium, magnesium, sodium, and sulphur. The relative concentrations will depend primarily on the geographical location of the coal seam, and will vary from place to place within a given field.

Coal petrology is used to identify the macerals and microlitho types present in coals. The behaviour of coal depends on the properties of the individual constituents, the relative amounts in the coal, and the rank of the coal. The differences in resistance of macerals to breakage together with other properties is used in the Longwy-Burstlein or Sovaco process in the selection of coals for blending in the optimum proportions for producing high quality coke. By careful control of thc petrological composition and the rank of a coal blend, its behaviour during carbonisation can be controlled. The petrological composition of a coal is quantitatively determined on polished surfaces by measuring the volume percentage of the different macerals with a point count method. The rank of the coal is determined microscopically by measuring reflectance of vitrinite with a microphotometer. The coking behaviour of a coal or a blend of coals of different ranks can be reliably predicted with these two measurements.

Two types of classification may be identified, i.e., scientific and commercial. In the scientific category, the Seyler chart has found the widest acceptance. The commercial classification uses scientific measurements to make the distinctions that occur in the classifications. Both methods are complementary and are used in research. In industry the commercial classification is essential.

In the ASTM classification system, the higher rank coals are specified by fixed carbon (for volatile matter ≤ 31%) on a dry, mineral-matter-free basis and lower rank coals are classified by calorific value on the moist, mineral-matter-free basis. The latter parameter depends on two properties: the moisture absorbing capacity and the calorific value of the pure coal matter. Some overlap between bituminous and sub-bituminous coals occurs, and is resolved on the basis of the agglomerating properties.

Similar to the other methods, volatile matter and gross calorific value on a moist, ash-free basis are the parameters used. Coals with volatile matter up to 33% are divided into classes 1–5. Coals with volatile matter greater than 33% are divided into classes 6–9. The calorific values are given for a moisture content obtained after equilibrating at 30°C and 96% rh.

The nine classes are divided into four groups determined by caking properties as measured through either the free swelling index or the Roga index. These tests indicate properties observed when the coal is heated rapidly.

The brown coals and lignites have been classified separately and have been defined as those coals with heating values less than 23.85 MJ/kg (10,260 Btu/lb, 5700 kcal/kg).

Constitution

The functional groups of interest contain O, N, or S. The significant oxygen-containing groups found in coals are those of carbonyl, hydroxyl, carboxylic acid, and methoxy. The nitrogen-containing groups include aromatic nitriles, pyridines, and pyrroles. The sulphur is largely found in dialkyl, aryl alkyl thioether, and thiophene groups.

The relative and absolute amounts of these groups vary with coal rank and maceral type. The major oxygen-containing functional groups in vitrinites of mature coals are phenolic hydroxyl and conjugated carbonyl as in quinones. Evidence exists for hydrogen bonding of hydroxyl and carbonyl groups. There are unconjugated carbonyl groups such as ketones in exinites. The infrared absorption bands are displaced from the normal range for simple ketones by the conjugation in vitrinites. The interactions between the carbonyl and hydroxyl groups affect the normal reactions.

The anthracites approach graphite in composition. As such they have higher rank than bituminous coals and have less oxygen and hydrogen. They are less reactive and insoluble in organic solvents. These characteristics are more pronounced as rank increases in the anthracite group. The anthracites have greater optical and mechanical anisotropy than lower rank coals. The internal pore volume and surface increase with rank after the minimum below about 90 wt% C.

Properties

Most of the physical properties depend on the direction of measurement compared to the bedding plane of the coal. Additionally, the properties vary according to the history of the piece of coal. They also vary between pieces because of the brittle nature of coal and of the crack and pore structure. The specific electrical conductivity of dry coals is very low (specific resistance 10^{10}–10^{14} Ω.cm), although it increases with rank. Coal has semi-conducting properties. The conductivity tends to increase exponentially with increasing temperatures.

The dielectric constant (ε) is also affected by structural changes on strong heating. Also it is very rank dependent, exhibiting a minimum at about 88 wt% C, and rises rapidly for carbon contents over 90 wt%. The dielectric constant (in Debye units) equals the square of the refractive index only at the minimum value.

Density values of coals differ considerably, even after correcting for the mineral matter, depending on the method of determination. The true density of coal matter is most accurately obtained from measurement of the displacement of helium after the absorbed gases have been removed from the coal sample. Density values increase with carbon content or rank for vitrinites. They are 1.4–1.6 g/cm^3 above 85 wt% carbon where there is a shallow minimum.

The thermal conductivity and thermal diffusivity are also dependent on the pore and crack structure. Thermal conductivities for coals of different ranks at room temperature are in the range of 0.23–0.35 W/(m.K).

The specific heat of coal can be determined by direct measurement or from the ratio of separate measurements of thermal conductivity and thermal diffusivity. The latter method gives values decreasing from 1.25 J/(g.K) [(0.3 cal/(g.K))] at 20°C to $\leq$ 0.4 J/(g.K) [($\leq$ 0.1 cal/(g.K))] at 800°C.

Coal contains an extensive network of ultrafine capillaries that pass in all directions through any particle. The smallest and most extensive passages are caused by the voids from imperfect packing of the large organic molecules. Vapours pass through these passages during adsorption, chemical reaction, or thermal decomposition. The rates of these processes depend on the diameters of the capillaries and any restrictions in them. Most of the inherent moisture in the coal is contained in these capillaries. The porous structure of the coal and products derived from it will have a significant effect on the absorptive properties of these materials. Effectiveness of coal conversion processes depends on rapid contact of gases with the surface. Large internal surfaces are required for satisfactory rates.

Mechanical properties are important for a number of steps in coal preparation from mining through handling, crushing, and grinding. The properties include elasticity and strength as measured by standard laboratory tests and empirical tests for grindability and friability; and indirect measurements based on particle size distributions. A number of efforts to correlate grinding energy input to size reduction have been made.

The Young's modulus for medium rank coals has been found to be ca 4 GPa (4 × 10^{10} dyne/cm^2). Sharp increases in the Young's and shear moduli have been found in vitrinites with increase in carbon content over 92%.

There are indications that compressive strength (measured by compression of a disk) may give useful correlations with the ease of cutting with different kinds of equipment. Studies of the probability of survival of pieces of different size suggest that the breaking stress should be most closely related to the linear size rather than the area or volume of the piece.

The effects of rank on both compressive and impact strength have been studied, and usual minima were found at 20–25% dry, ash-free volatile matter (88–90 wt% carbon). Accordingly, the Hardgrove grindability index exhibits maximum values in this area.

Solvent extraction

A wide range of organic solvents can dissolve part of coal samples but dissolution is never complete and usually requires heating to temperatures sufficient for some thermal degradation or reaction with the solvent to take place (ca 400°C). At room temperature the best solvents are primary aliphatic amines, pyridine, and some higher ketones, especially when used with dimethylformamide. Above 300°C large amounts can be dissolved with phenanthrene, 1-naphthol, and some coal-derived high boiling fraction. Dissolution of up to 40 wt% can be achieved near room temperature, and up to 90% near 400°C. Coals with 80–85% carbon in the vitrinite give the largest yields of extract. Very little coal above 90 wt% C dissolves.

Properties involving utilisation

Coal rank is the most important single property for almost every application of coal. The rank sets limits on many properties such as volatile matter, calorific value, and swelling and coking characteristics. Other properties of significance include grindability, ash content and composition, sulphur content, and size.

Most of the mined coal is burned in boilers to produce steam for electric-power generation. The calorific value determines the amount of steam that can be generated. The design and operation of a boiler requires consideration of a number of properties.

In general, high rank coals are more difficult to ignite, requiring supplemental oil firing and slower burning, with large furnaces to complete combustion. The greater reactivity of lower rank coals makes them better suited for cyclone burners that carry out rapid, intense combustion to maximise carbon utilisation and minimise smoke emission.

Ash content is important. Ash discharge at high temperature, as molten ash from a slagging boiler, involves substantial amounts of sensible heat. The higher cost of washed coal of lower ash content will not always merit its use. Ash disposal and extra freight costs for high ash coals enter the selection of coal. The current use of continuous mining equipment produces coal with about 25% ash content. The average ash content of steam coal is about 15%. For some applications, such as chain-grate stokers, a minimum ash content of about 7–10% is needed to protect the metal parts.

For pulverised coal firing, a high Hardgrove index or grindability index is desired. This implies a relatively low energy cost for pulverising since the coal is easier to grind. The abrasiveness of the coal is also important since this determines the wear rate on pulveriser elements.

Moisture content affects handling characteristics. It is most important for fines smaller than 0.5 mm. The moisture content of peat or brown coal that is briquetted for fuel must be reduced to about 15% for satisfactory briquetting. Mechanical or natural means are used because of the cost of thermal drying.

The sulphur content is important in meeting air quality standards. Volatile matter is important for ease of ignition. High rank coals have low volatile matter and burn more slowly.

Coke production

Coking coals are mainly selected on the basis of the quality and amount of coke that they will produce. Gas yield is also considered. About 65-70% of the coal charged is produced as coke. The gas quality depends on the coal rank and is a maximum (in energy in gas per mass of coal) for coals of about 89% carbon (dry, mineral-matter-free) or 30% volatile matter.

A Clean Coke process is being developed with partial Department of Energy support, to convert coals not normally used for coke production, including high sulphur coal, to a low ash, low sulphur metallurgical coke, chemical feedstocks, and other fuels. The coal is initially carbonised, then pelletised using a binder derived from another part of the process before heating in a continuous coking kiln.

Gasification

A number of gasifiers are either available commercially or in various stages of development. The range of coals that may be used vary from one gasifier type to another. Many of the coal selection criteria for combustion apply to gasification, which is a form of partial oxidation.

Reactions

Mature coals (> 75% C) tend to be built of assemblages of polynuclear ring systems connected by a variety of functional groups. The ring systems themselves contain many functional groups. These so-called molecules differ from each other to some extent in the coal matter. For bituminous coal, a tar-like material fills some of the interstices between the molecules. Generally, the initial volatile constituents of coal are moisture and light hydrocarbons. The volatile matter produced on carbonisation reflects decomposition of parts of the molecule and the release of moisture. Rate of heating affects the volatile-matter content. Faster rates give higher volatile-matter yields. Coal is not carbon or a hydrocarbon. The composition will depend on the rank. The empirical composition of a vitrinite of ca 84 wt% carbon content is approx $C_{13}H_{10}O$ with small quantities of N and S.

The surface of coal particles undergoes air oxidation. This process may initiate spontaneous combustion in storage piles or weathering with a loss of heating value and coking value during storage.

Many of the products made by hydrogenation, oxidation, hydrolysis, or fluorination are of industrial interest. The increasing cost of petroleum and natural gas is increasing interest in some of these materials primarily as upgraded fuels to meet air-pollution control requirements as well as to take advantage of the greater ease of handling of the liquid of gaseous material, and to utilise existing facilities such as pipelines and furnaces. Demonstration plants are planned for conversion of coal to methane (substitute natural gas or SNG), industrial fuel gas, ammonia, coal-derived refinery feedstock, and boiler fuel. A chemistry based on synthesis-gas conversion to other materials is being developed and may have more extensive application in another decade. At present, the major production of chemicals from coal involves the by-products of coke manufacturing.

Plasticity of heated coals

Coals with a certain range of composition associated with the bend and having 88–90 wt% carbon will soften to a liquid condition when heated. These are known as prime coking coals. The soft condition is somewhat reversible for a time, but does not persist for many hours at 400°C, and is not be observed above ca 550°C if the sample is continuously heated as in a coking process. This is caused by degradation of coal matter, releasing vapours and resulting in polymerisation of the remaining material.

Several laboratory tests are used to determine the desirability of a coal or blend of coals for making coke. These are empirical and are carried out under conditions that approach the coking process. The three properties that have been studied are swelling, plasticity, and agglomeration.

Pyrolysis of coal

Most coals decompose below temperatures of about 400°C, characteristic of the onset of plasticity. Moisture is released near 100°C, and traces of oil and gases appear between 100–400°C depending on the coal rank. As the temperature is raised in an inert atmosphere at a rate of 1–2°C/min, the evolution of decomposition products reaches a maximum rate near 450°C, and most of the tar is produced in the range of 400–500°C. Gas evolution begins in the same range but most evolves about 500°C. If the coal temperature in a single reactor exceeds 900°C, the tars can be cracked, the yields are reduced, and the products are more aromatic. Heating beyond 900°C results in minor additional weight losses but the solid matter changes its structure. The tests for volatile matter indicate the loss in weight at a specified temperature in the range of 875–1050°C from a covered crucible. The weight loss represents the loss of volatile decomposition products rather than volatile components.

The loss of fusible material can be accelerated by reducing the pressure, and can also reduce coking properties. Further reactions of the volatile products are also reduced. Increasing the pressure has little effect. Mild oxidation is used to destroy the coking properties and can eliminate the production of tars.

Mining and Preparation

Coal is obtained by either surface mining of outcrops or seams near the surface, or by underground mining. The method chosen depends on the geological conditions which may vary from thick, flat seams to thin, inclined seams that are folded and need special methods of mining.

Coal preparation normally involves some size reduction of coal particles and the systematic removal of some ash-forming material and very fine coal.

Storage of coal may be necessary at any of the major steps in production or consumption. For utilities, two types of storage are used. A small amount of coal in storage meets daily needs and will be continually turned over. This coal is loaded into storage bins or bunkers. Long-term reserves are carefully piled and left undisturbed except as necessary to sustain production. Coal storage results in some deterioration of the fuel owing to air oxidation. If inadequate care is taken, spontaneous heating and combustion will result. As the rank of coal decreases, it oxidises more easily and must be piled more carefully. Anthracite does not usually present a problem.

The usual means of transporting coal are railroad, water shipment, truck, and in some instances by conveyor belt from mine to plant.

Health and Safety Factors

Coarse-coal cleaning

The most popular concentrating processes are jigging and dense-medium washing.

Jigging

Separation of coal from its impurities by jigging consists of inducing pulsation in water in which the raw coal is suspended. An open-top rectangular box having a perforated bottom, or screen plate, is used. The pulsations are induced through the screen plate by reciprocating members, causing alternate expansion and compaction of the coal water mixture, resulting in stratification of the coal according to the specific gravity of the particles. The lighter coal particles rise to the top, overflow the end of the jig, and are removed as clean product. The denser impurities settle on the screen plate and are withdrawn as *refuse*. Most jigs are of the Baun type, in which the pulsations are generated by air pressure. Plungers and diaphragms are also used as pulsers.

Dense-medium washing

Coal is cleaned by immersing it in a fluid having a density between that of the coal and its impurities. Most dense-medium washers use a suspension of sand or magnetite in water to achieve the desired density. The modern highly automated dense-medium plants are capable of high throughput and sharp separation of the raw coal into a clean product and refuse.

Fine-coal cleaning

The principal fine-coal washers are the wet concentrating table, dense-medium cyclone, hydrocyclone, launder, feldspar jig, hydrotator, and froth-flotation cell. Of these, the concentrating table is the most popular, especially for cleaning bituminous coal. Cyclones are becoming increasingly important for washing both anthracite and bituminous coal, while launders and feldspar jigs are used to a limited extent for cleaning bituminous coal. The hydrotator is used extensively for anthracite.

Wet concentrating table

This is essentially a continuous mechanised form of the classical miner's pan. It consists of a transversely and longitudinally-tilted rectangular or rhombohedral table.

Dense-medium cyclone

Fig. 18.1 shows the essential features of a dense-medium cyclone. The mixture of medium and raw coal enters tangentially near the top of the cylindrical section, producing free-vortex flow. The refuse moves downward along the wall and is discharged through the underflow orifice. The cleaned coal moves radically toward the cyclone axis, passes through the vortex finder to the overflow chamber, and is discharged from a tangential outlet.

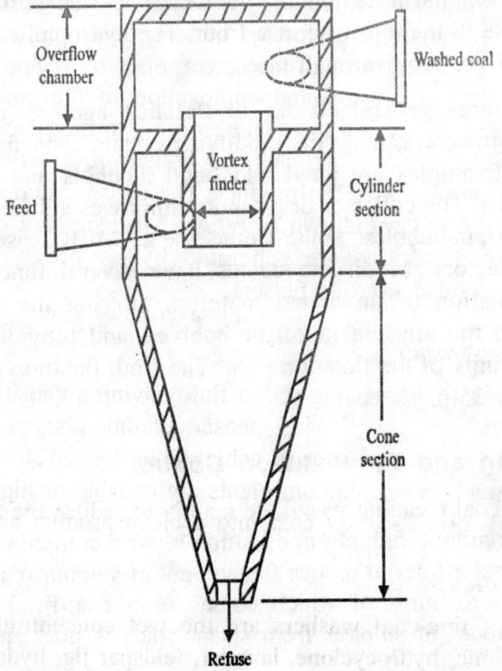

Fig. 18.1. Dense-medium cyclone for cleaning coal.

The dense medium used exclusively is pulverised magnetite suspended in water. The magnetite is recovered magnetically for re-use. Cyclones are available in several sizes, 20-inch and 24-inch diameters being most common. A 20-inch unit will clean about 50 tons of raw coal per hour, and the 24-inch one about 75 tons per hour.

Hydrocyclone

The hydrocyclone, as the name implies, is a cyclone cleaning device that uses only water as a medium. Its design differs from that of the conventional dense-medium cyclone by having a greater cone angle and a longer vortex finder. Hydrocyclones currently are used to clean coal 0.025 inch and smaller, but can be used for coal as course as 1/4 inch.

Froth flotation

In the froth-flotation process fine coal slurry, to which a small quantity of flotation agents is added, is processed through a flotation cell. In the cell, finely disseminated air bubbles pass upward through the slurry. The hydrophobic coal particles attach themselves to the surface of the air bubbles and rise to the top to form a froth. The hydrophilic impurities sink in the cell and are removed as reuse.

There are three general classes of flotation agents: (i) frothers; (ii) collectors or promoters; and (iii) modifying agents. The frother promotes stable froths. Examples are amyl and butyl alcohols and methylisobutyl carbinol (MIBC). The collector or promoter improves adherence of the coal particles to the air bubbles. Some frothers, e.g., MIBC, also function to a degree as collectors. Modifying agents have several functions, such as depressing flotation of unwanted material, altering the surface of the particles to aid the attachment of air bubbles, and providing the desired acidity or alkalinity of the flotation pulp. The froth flotation process is used to clean minus 35-mesh coal.

Coal Cleaning and Air Pollution Control

Until recently, coal cleaning was done mainly to reduce the amount of ash, and any concomitant reduction of sulphur was coincidental. However, owing to national efforts to reduce the amount of sulphur oxides discharged to the atmosphere, most of which comes from coal-fired electric utility plants, the removal of sulphur from steam coals is now an end in itself.

Sulphur occurs in coal in three forms: organic, pyrite (FeS_2), and sulphate. Sulphur as sulphate occurs up to only a few tenths of a per cent, the bulk occurring in the organic and pyritic forms. Since organic sulphur is part of the coal molecule, no method is now known for removing it economically. Consequently the amount o .rganic sulphur represents the lowest limit to which a coal can be mechanically freed of sulphur.

Since pyrite is nearly four times as dense as coal, it would seem possible to separate all of it from the coal substance. In practice, however, the amount of pyritic sulphur that can be removed varies considerably from one coal to another, depending on the way the pyrite is disseminated throughout the coal.

Where most of the pyrite can be removed, mechanical coal preparation is an economically-attractive process. And because pyritic sulphur comprises about one-half of the sulphur in coal, one of the more serious air pollution problems would be substantially reduced in such cases.

Coal Utilisation

The principal uses of coal may be classified broadly as follows :

(1) Combustion to obtain thermal energy for power generation and process heating, including sintering and calcining.

(2) Carbonisation to convert coal to metallurgical and chemical coke, tars, chemicals, and industrial gases.

(3) Gasification to obtain industrial gases for heating, chemical reduction, and hydrogenation and synthesis reactions.

Raw or processed coal, and fly ash from coal combustion, have several other uses.

Combustion

The greatest percentage of the coal consumed in this country is burned in boilers to generate steam for turboelectric plants in the electric utility industry.

Further improvements in thermal efficiency will be achieved with a new system of power generation that is now being developed. Known as MHD (magnetohydrodynamics), this system is expected to have a thermal efficiency between 50 and 55 per cent if coal is used as a fuel. The principle of MHD is to burn the fuel under conditions that will give a combustion gas temperature of about 4500°F, add an easily ionisable salt like potassium carbonate (seeding) so that the gas will have a relatively high concentration of free ions and electrons, and then pass this conducting gas through a magnetic field and draw off electric energy through electrodes placed in the path of the gas. By analogy, the hot conducting gas moving through a magnetic field behaves like the armature in a conventional turbogenerator. A coal-fired MHD generator would discharge less sulphur and nitrogen oxides to the atmosphere than a conventional plant of equivalent size because the seeding salt will absorb these oxides, which would be recovered during regeneration of the seed.

Combustion equipment

Industrial and electric utility boilers are fired with either stokers, pulverised-coal-fired burners, or cyclone burners, the choice depending on the kind of coal and the amount of steam needed.

Stoker firing: Both fuel and air have the same direction in retort stokers; this is called underfeed burning. The fuel moves across the air direction in chain- or travelling-grate stokers; this is called cross-feed burning. The spreader stoker approximates overfeed burning, the incoming fuel moving toward the air.

Pulverised-coal firing and cyclone firing: Electric-utility and large industrial plants favour pulverised-coal-fired and cyclone-fired furnaces because of their inherent flexibility regarding the kind and quality of coal, their comparatively good availability, their quick response to load changes and their extremely high steam-generating capacity.

Fluidised-bed combustion: Recently, interest has been growing in another mode for burning coal-the fluidised-bed. Although still in the experimental stage, there are clearly seen potential advantages to this burning mode (i) The fuel-bed temperature is low, about 1800°F, which means less formation of nitrogen oxides, and retention of some of the sulphur in the ash of certain coals. (Adding dolomite to the fuel-bed greatly improves sulphur retention.) Equally important is less volatilisation of sodium and potassium in the coal, consequently, there are less deposits on and corrosion of furnace, superheater, and reheater tubes; and (ii) Heat-transfer rates from the fluidised bed to immersed heat-transfer surfaces are relatively high, as much as 100 Btu per hr per sq ft per degree Fahrenheit.

It is believed that successful application of fluidised-bed boilers to electric-utility stations would reduce capital costs about 10 per cent compared with conventional plants of the same capacity. The principle of the fluidised-bed combustor is illustrated schematically in Fig. 18.2.

Boiler types

There are various kinds of industrial and utility boilers, broadly classified as fire-tube and water-tube. In the former, the hot combustion gases pass through tubes, and heat is transferred to water outside the tubes. The most common and least expensive boiler of this type is the horizontal return tubular (HRT) boiler. However, because of the design and construction of fire-tube boilers, there is a definite limitation to their size and the pressure which they can tolerate. Water-tube boilers may be broadly classified as straight-tube and bent-tube types, the latter having several variations in design and being preferred for applications where higher capacities and steam pressures are required. In both types, heat is transferred by radiation or convection to the outside of the tubes, and water flows inside the tubes as the result of thermal circulation, or in the case of certain bent-tube boilers, as the result of forced circulation. A comparatively new version of the forced-circulation, bent-tube boiler for central-station power plants in the "once-through" type. The feedwater passes progressively through the heating, evaporation, and superheater sections; no drum is used for separating the steam from the boiler water as in other boilers, consequently, the ratio of water circulated in the boiler to steam generated is unity.

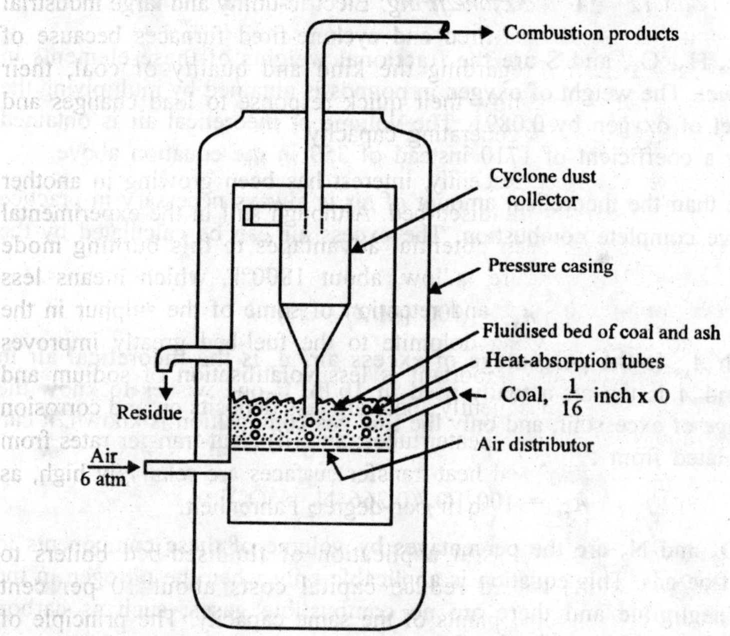

Fig. 18.2. Schematic of pressurised fluidised bed steam generator.

Combustion calculations

The amount of air or oxygen just sufficient to burn the carbon, net hydrogen, and sulphur to carbon dioxide is the theoretical or stoichiometric amount of air or oxygen. The general expression for combustion of a fuel is

$$C_mH_n + [(4m + n)/4]O_2 = mCO_2 + (n/2)H_2O$$

m and *n* being the number of atoms of carbon and net hydrogen, respectively, in the fuel. For example, one mole (an amount equal to the molecular weight of the fuel) of methane (CH_4) requires 2 moles of oxygen for complete combustion to 1 mole of carbon dioxide and 2 moles of water. If air is used, each mole of oxygen is accompanied by 3.76 moles of nitrogen.

The theoretical weight or volume of oxygen or air to burn a given weight of fuel is or primary interest in designing combustion equipment. The volume of theoretical oxygen to burn any fuel can be calculated from the ultimate analysis of the fuel as follows :

$$359\left(\frac{C}{12}+\frac{H_2}{4}-\frac{O_2}{32}+\frac{S}{32}\right)= \text{cu ft of oxygen / lb of fuel}$$

where C, H_2, O_2, and S are the fractional weights of these elements in 1 lb of fuel. The weight of oxygen in pounds is obtained by multiplying the cubic feet of oxygen by 0.0891. The volume of theoretical air is obtained by using a coefficient of 1710 instead of 359 in the equation above.

More than the theoretical amount of air is always necessary in practice to achieve complete combustion. The excess air can be calculated by the equation :

$$A_{xs} = 100\ [(A/A_t - 1)]$$

in which A_{xs} is the percentage of excess air, A_t is the theoretical air in lb/1b, and A is the air actually used in 1b/lb. If one wishes to know the percentage of excess air, and only the flue gas composition is known, it can be calculated from :

$$A_{xs} = 100\ [O_2/(0.266\ N_2 - O_2)]$$

where O_2 and N_2 are the percentages by volume of these components in the dry flue gas. This equation is applicable only when the nitrogen in the fuel is negligible and there are no combustible gases, such as carbon monoxide or hydrogen, in the flue gas.

Sintering and calcining

Sintering is a process of agglomerating finely sized fusible materials to obtain a product suitable in size for a particular use. Typical applications of sintering are the production of lightweight aggregate for use in concrete, and of sintered ore for use in blast furnaces. The Dwight Lloyd-type machine, which is widely used to sinter fine materials, is shown schematically in Fig. 18.3. The mixture of coal (or coke) and the material to be sintered, which may be pelletised ore or fly ash from power plants, is fed to the grate and ignited by a gas- or oil-fired hood. Air is drawn through the charge by induced draft, and ignition and burning occur from the top to the bottom of the bed. The quality of the sinter depends mainly on the composition of the mixture, the thickness of the bed, the speed of the grate, and the amount of air drawn through the bed. Usually, the windbox is compartmented to control the amount of air flowing through the bed at certain points.

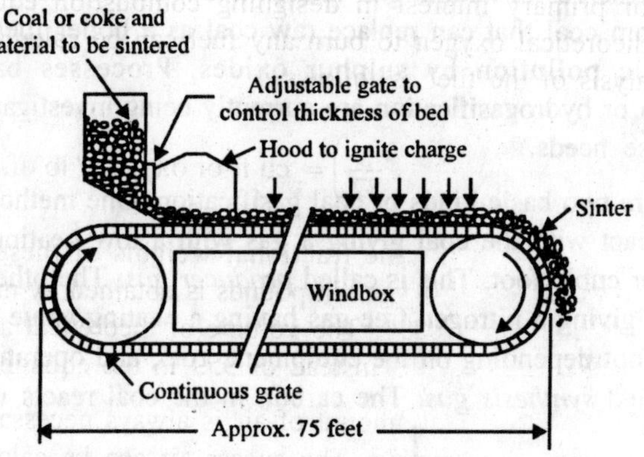

Fig. 18.3. Dwight Lloyd stoker for manufacturing sintered products.

Lightweight aggregate for concrete is also produced by burning pelletised refuse coal from coal-preparation plants on conventional travelling grates.

A significant amount of coal is consumed annually for calcining to make cement, lime, gypsum, alumina, and magnesia, and for firing ceramic products. Calcining is usually done in rotary kilns, and ceramics are fired in tunnel kilns and muffles.

Considerable attention is being given to the production of lightweight aggregate from pulverised fuel ash (PFA), also called fly ash. Several different processes have been devised, but they all depend on sintering. The PFA, to which carbon in some form is added to bring the total amount up to 5 per cent, is usually pelleted with water to give pellets about the size of marbles. The pellets are sintered by heating them in a furnace or kiln to about 2200°F. The processes yield a hard, cellular, lightweight material having a bulk density of 50 pounds per cubic foot.

Coal gasification

Gasification as used here means reacting coal with air, oxygen, steam, carbon dioxide, or a mixture of these, in a manner that yields a gaseous product suitable for use either as a source of energy, or as a raw material for the synthesis of industrial chemicals, liquid fuels, or other gaseous fuels. Some of the important synthesis products are synthetic pipeline gas (methane), hydrogen, ammonia, alcohols, olefins, and distillable oils. Still in the experimental stage is hydrogasification, which involves reacting coal with hydrogen at high pressure and temperature to obtain a methane rich gas. Currently in our country there is an urgent need to supplement the dwindling reserves of natural gas. Also, there is serious need for a sulphur-free gas from coal that can replace raw coal as a boiler fuel, thus abating atmospheric pollution by sulphur oxides. Processes based on coal gasification or hydrogasification are currently being investigated in India to satisfy these needs.

There are two basic kinds of coal gasification. One method uses air and steam to react with the coal giving a gas with a low heating value, about 150 Btu per cubic foot. This is called *producer gas.* The other uses oxygen and steam, giving a nitrogen-free gas having a heating value up to 500 Btu per cubic foot depending on the equipment, coal, and operating conditions. This is called *synthesis gas*. The carbon in the coal reacts with the steam as follows :

$$C + H_2O \rightarrow CO + H_2^*$$

* Under certain conditions some methanc is also formed during gasification.

The oxygen added with the coal and steam burns some of the coal to attain the high temperature needed for the reaction to occur at a reasonable rate. Coal gasification can be conveniently divided into three categories:

(1) Lump coal is supported on a grate or by other means, and the flow of gas and coal may be countercurrent or concurrent. The former is the most common process. This is called *fixed-bed gasification*.

(2) Crushed or fine coal is fluidised by the gasifying medium, giving an expanded fuel-bed that can best be visualised as boiling or jiggling. This is called *fluidised-bed gasification.*

(3) Fuel coal is suspended in the gasifying medium, the particles moving with the gas, either linearly or in a vortex pattern. This is called *suspension* or *entrainment gasification.* These processes are shown schematically in Fig. 18.4.

Fixed-bed and fluidised-bed systems cannot be operated with strongly coking coals, which go through a plastic stage during heating and agglomerate into a mass that impedes the flow of gas. However, such coals can be pretreated to lessen their coking tendency, but this adds to the cost of the product gas. Or, under certain circumstances, agglomeration of coking coals in the fuel bed can be prevented by mechanical stirrers, thereby keeping the bed open to the flow of gases. Any rank of coal can be used in entrained gasification systems.

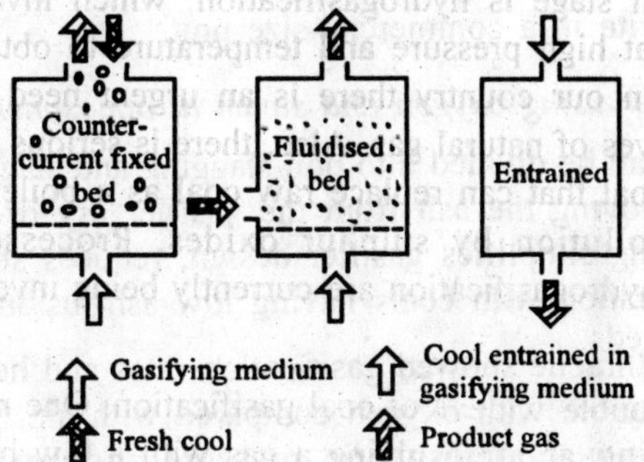

Fig. 18.4. Coal gasifaction systems.

There exist, at least as theoretical schemes, over a hundred different forms of gasifiers and gas producers, but only a few of those that have been built and tested and described here.

In the first technical application of a second-generation coal-gasification process, in which the resulting synthesis gas is used as a chemical feedstock, acetic acid is produced from coal (Fig. 18.5).

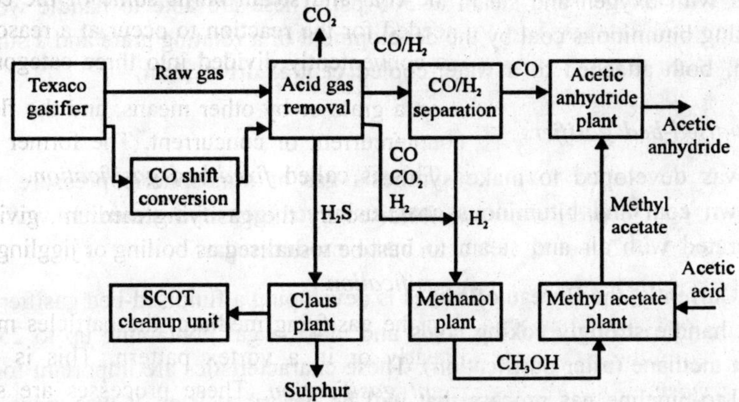

Fig. 18.5. Texaco coal gasification process.

Fixed-bed gasifier using air and steam at atmospheric pressure

The conventional gas producer operates at atmospheric pressure and gasifies coke, anthracite coal, and non-coking or weakly coking bituminous coal, using air and steam for the gasifying medium. The optimum fuel size is 1/2 to 2 in., and the ash-fusion temperature should exceed 2100°F. With coal, the heating value of the gas may be 170 Btu per cubic foot, and the gasification rate up to 50 pounds per hour per square feet of grate area. It can be used for an investigation to determine the technical feasibility of gasifying anthracite in a commercial-size unit.

Fixed-bed gasifier using oxygen and steam at atmospheric pressure

These gasifiers can be divided into non-slagging and slagging, which mean, respectively, removing the ash from the gasifier dry or as a molten slag. Operating slagging simplifies gasifier design, reduces steam consumption, and permits operation with coals having low ash-fusion temperatures.

Tests with anthracite showed gasification rates and heating value of the gas are nearly double with oxygen compared with air. Slagging fixed-bed gasifiers operating at atmospheric pressure and having high specific throughputs, are used in Hungary to make synthesis gas. This gasifier is essentially a small balst furnace with tuyeres at three levels. At the lowest row of tuyeres, some of the make gas is burned in a mixture of steam and oxygen to produce a highly-heated blast.

Fixed-bed gasifiers using oxygen or air at high pressure

The Lurgi gasifier developed in Germany, and it was being used to furnish town gas having a heating value of 450 Btu per cu ft, by gasifying brown coal with oxygen and steam at 300 psi. These are able to handle weakly coking bituminous coal by the development of a rotating grate and a stirring arm, both attached to a water-cooled central drive shaft.

Fluidised-bed gasifiers

It was developed to make synthesis gas at atmospheric pressure from brown coal and bituminous coal, using oxygen and steam. It was also operated with air and steam to make producer gas.

Currently, the Bureau of Mines is developing a fluidised-bed gasifier that can handle strongly coking coals and make a gas containing up to 25 per cent methane (after purification). These characteristics are important for the coal-to-pipeline gas process that will be described later. It was mentioned earlier that strongly coking coals must be pretreated before they can be used in fixed-bed or fluidised-bed gasifiers. The Bureau of Mines gasifiers (Fig. 18.6). integrates three steps in the gasification process: pretreatment of freely falling coal particles with oxygen and steam in the upper section to destroy their coking properties; devolatilisation in the expanded section;

and gasification in the lower section. Ash is removed from the bottom. The unit operates at 600 psig with a fluid-bed temperature of about 1800°F.

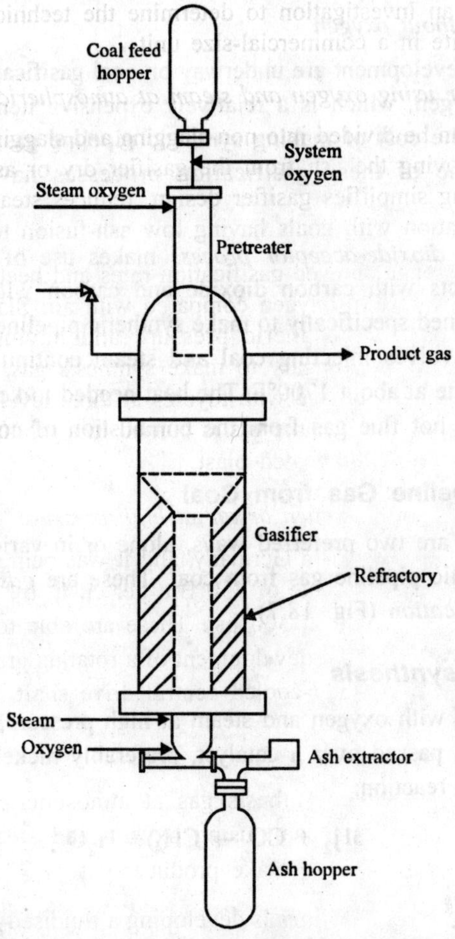

Fig. 18.6. High-pressure fluidised-bed gasifier.

Suspension gasifiers

The suspension gasifier can handle any rank of pulverised coal; produce a gas with no tar and very little methane, which is important for many synthesis processes; achieve high production rates per unit volume of reaction space; and operate either non-slagging or slagging. These advantages must be weighed against the need to recover the heat in the high-temperature gas, and the relatively low carbon conversion, 85 to 90 per cent, compared with 100 per cent in other gasification systems.

Gasification without oxygen

Research and development are underway on coal gasification processes that do not use oxygen, which is a relatively expensive item. Such processes would reduce the cost of making synthetic pipeline gas, synthesis gas, or hydrogen. None of these gasification processes have yet come into industrial use.

The *carbon dioxide-acceptor process* makes use of the heat evolved when lime reacts with carbon dioxide and carbon with hydrogen. This process is designed specifically to make synthetic pipeline gas. The *Molten-salt process*, involves injecting coal and steam continuously into molten sodium carbonate at about 1700°F. The heat needed to keep the salt molten is furnished by hot flue gas from the combustion of coal.

Synthetic Pipeline Gas from Coal

Basically, there are two preferred ways, alone or in various combinations, to make synthetic pipeline gas from coal. These are *gasification-synthesis* and *hydrogasification* (Fig. 18.7).

Gasification-synthesis

Coal is gasified with oxygen and steam at high pressure, and the resulting synthesis gas is passed over a catalyst, preferably nickel, converting it to methane by the reaction:

$$3H_2 + CO \rightarrow CH_4 + H_2O$$

Synthane process

Synthane process illustrates the conversion of coal to synthetic pipeline gas via the gasification-methanation route. The unique feature of this process is the comparatively high concentration of methane in the synthesis gas, up to 25 per cent carbon dioxide-free basis. The advantages of this are: (i) because of the exothermicity of methane formation; (70 Btu per cu ft of synthesis gas converted) less process oxygen is needed; and (ii) a lower capital cost for purification, shifting, and methanation that if the feed gas contained no methane. For example, without methane in the purified feed gas the methanator must convert 400 cubic feet of synthesis gas to make

100 cubic feet of methane. With 25 per cent methane in the feed gas, the methanator would need to convert only 227 cu ft of synthesis gas—a volume reduction of 42.5 per cent.

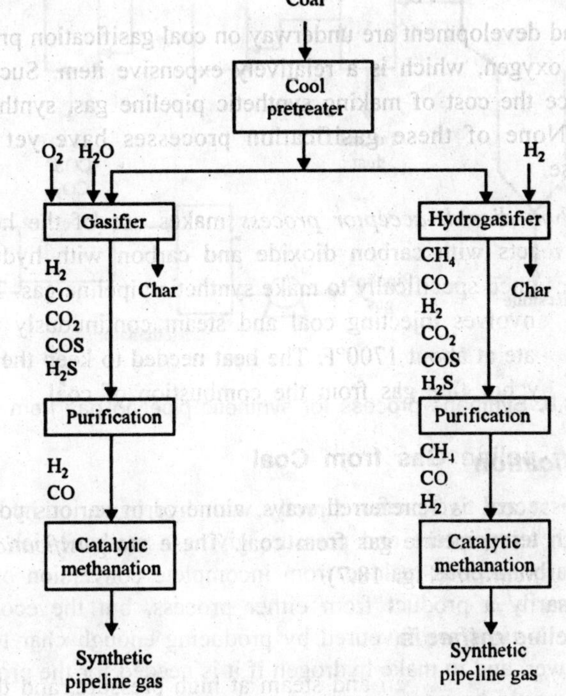

Fig. 18.7. Processes for producing synthetic pipeline gas from coal.

The Synthane process is shown schematically in Fig. 18.8. The high-pressure (600 psig) fluidised-bed gasifier described earlier and shown in Fig. 18.3 is used for the experimental studies. The shift converter adjusts the hydrogen-carbon monoxide ratio to 3:1. The raw gas is purified with a hot potassium carbonate scrubber and activated carbon, so that the gas fed to the methanator contains no more carbon dioxide than 2 volume per cent and no more than 1 ppm of sulphur compounds. An alloy of nickel and aluminium, Raney nickel, is the methanation catalyst.

Gas from any of the high-pressure gasifiers described earlier, with or without oxygen, are feasible sources of gas for the front end of the process shown in Fig. 18.8, but the economics are favoured by methane in the feed gas.

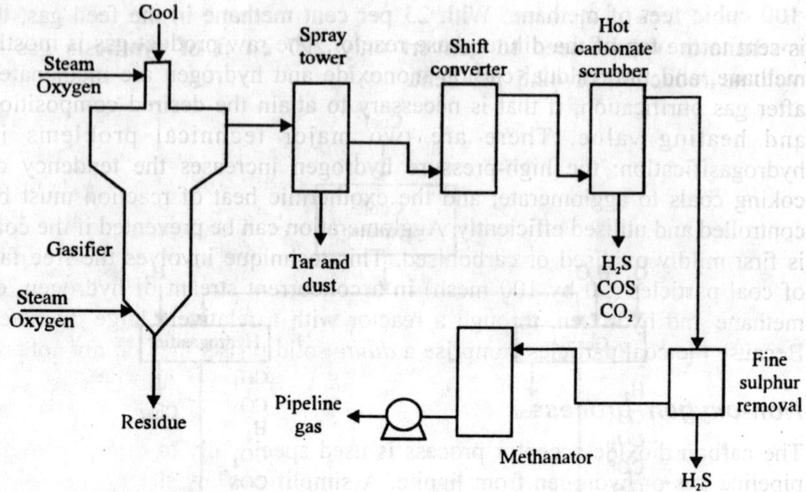

Fig. 18.8. Synthane process for synthetic pipeline gas from coal.

Hydrogasification

In this process coal is reacted directly with hydrogne, or hydrogen and steam, at high temperature and pressure giving a methane-rich gas. Char, which is a carbonaceous residue from incomplete conversion of the coal, is not necessarily a product from either process, but the economics of synthetic pipeline gas are favoured by producing enough char to burn for steam and power, and to make hydrogen if it is needed for the process. Coal pretreatment would be needed only if agglomeration in the reactor were a problem. Methanation of the gas from a hydrogasifier would be needed only if it contained an excessive amount of carbon monoxide and hydrogen. Another step, *shift conversion*, would be needed if the hydrogen-carbon monoxide ratio of the gas going to the methanator were not 3:1.

Coal pyrolysis

Coal pyrolysis, that is, destructive distillation by heat in the absence of oxygen, also forms methane but the yield is too low for an industrial process for pipeline gas only. However, an experimental pyrolysis process for making several coproducts, of which methane is one. Both the Bureau of Mines and IGT are conducting research and development on the hydrogasification route to synthetic pipeline gas. The hydrane process is shown in Fig. 18.9. In the upper reactor, the hydrogasifier, fine coal falls in dilute phase concurrently with a mixture of methane and hydrogen. In the lower reactor the devolatised coal, or char, is fluidised with hot hydrogen. The gas from the fluidised-bed, consisting mainly of methane and hydrogen, is sent to the top of the dilute-phase reactor. The raw product gas is mostly methane, and the residual carbon monoxide and hydrogen are methanated after gas purification, if that is necessary to attain the desired composition and heating value. There are two major technical problems in hydrogasification: the high-pressure hydrogen increases the tendency of coking coals to agglomerate; and the exothermic heat of reaction must be controlled and utilised efficiently. Agglomeration can be prevented if the coal is first mildly oxidised or carbonised. This technique involves the free fall of coal particles (50 by 100 mesh) in a concurrent stream of hydrogen, or methane and hydrogen, through a reactor with a relatively large diameter. Because the coal particles comprise a *dilute* solid phase, they do not cohere.

Non-oxygen process

The carbon dioxide-acceptor process is used specifically to make synthetic pipeline gas or hydrogen from lignite. A simplified flow sheet is shown in Fig. 18.10. Preheated lignite is lock-hoppered into the *fluidised-bed devolatiliser*. Other streams entering the devolatiliser are synthesis gas from the gasifier and calcined dolomite (CaO.MgO) from the regenerator. The calcined dolomite is the core of the process because it reacts with carbon dioxide and liberates heat for devolatilisation and gasification, thereby obviating the need for process oxygen.

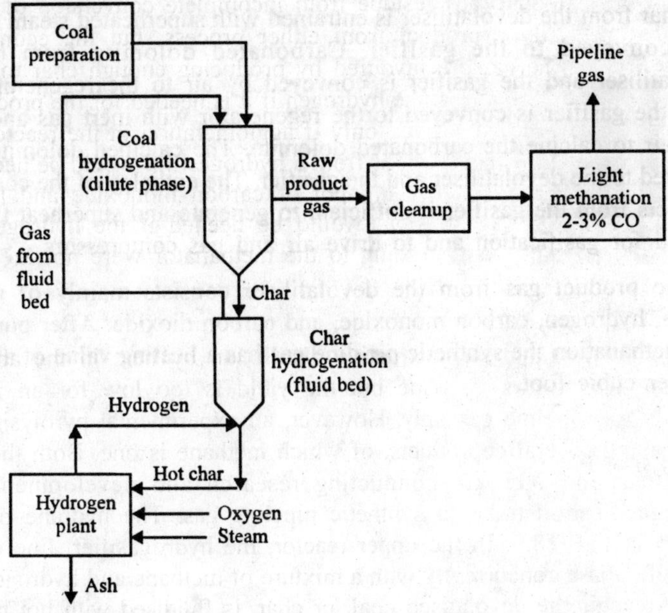

Fig. 18.9. Hydrogasification process for synthetic pipeline gas from coal.

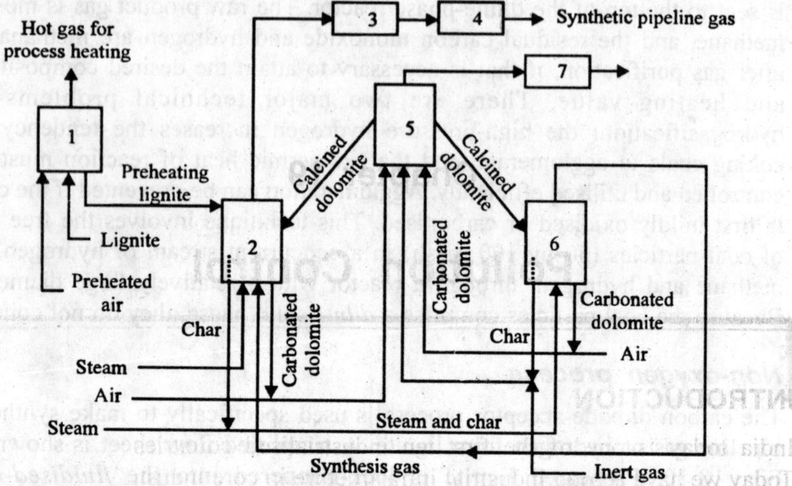

Fig. 18.10. Carbon dioxide-acceptor process for synthetic pipeline gas from lignite. (1) Fluid-bed preheater, 570°F; (2) Fluidised-bed devolatiliser, 1,500°F, 290 psia; (3) Gas cooling, dust removal, and purification; (4) Methanator, 680°F, 225 psia; (5) Regenerator for calcining dolomite, 1,940°F; (6) Fluidised-bed gasifier, 1,575°F; (7) Steam generator.

Char from the devolatiliser is entrained with superheated steam (1200°F) and conveyed to the gasifier. Carbonated dolomite from both the devolatiliser and the gasifier is conveyed by air to the regenerator. Char from the gasifier is conveyed to the regenerator with inert gas and burned with air to calcine the carbonated dolomite. The calcined dolomite then is recycled to the devolatiliser and the gasifier. The enthalpy of the combustion products from the gasifier is sufficient to generate and superheat the steam needed for gasification and to drive air and gas compressors.

The product gas from the devolatiliser consists mainly of methane, ethane, hydrogen, carbon monoxide, and carbon dioxide. After purification and methanation the synthetic pipeline gas has a heating value of about 950 Btu per cubic foot.

Chapter 19

Pollution Control

INTRODUCTION

India today is one of the first ten industrialised countries of the world. Today we have a good industrial infrastructure in core industries like metals, chemicals, fertilisers, petroleum, food etc. What has come out of these? Pesticides, detergents, plastics, solvents, fuels, paints, dyes, food additives, etc. are some examples of products from these industries among others. Besides these useful products, there are a number of industrial effluents and emissions particularly poisonous gases in the atmosphere. Mining activities also added to this problem particularly as solid waste. A serious concern has grown up all over the world about the rivers turning murky, fish rotting on sea shores, tree withering, cities chocking with foul air, toxic chemicals being cycled into food stuffs and disease epidemics appearing so frequently. Due to advances made in nuclear sciences and increasing use of atomic energy, there has also been an increase in radioactivity in the biosphere. Thus, pollution is a necessary evil of all developmental processes.

Pollution control is recognised as an important aspect of chemical engineering. Due to lack of development of a culture of pollution control, a heavy backlog of gaseous, liquid and solid pollution in our country has resulted. Thus pollution control in our country is a recent environmental concern. Pollution is a man-made problem, mainly of effluent countries. The developed countries have been in a mad race to exploit every bit of natural resource to convert them into goods for their comfort and to export them to needy developing world. In doing so, the industrialised countries dump lot of materials in their environment which becomes polluted. In a way, pollution in fact is being forced upon human beings.

Pollution is an undesirable change in the physical, chemical or biological characteristics of air, water and soil that may harmfully affect the life or create a potential health hazard of any living organism. Pollution is thus direct or indirect change in any component of the biosphere that is harmful to the living component(s), and in particular undesirable for man, affecting adversely the industrial progress, cultural and natural assets or general environment.

A social awareness of the pollution problem has arisen with a knowledge of its effects and the discharge of untreated effluent is rightly considered to be irresponsible behaviour. The problem has become more urgent on a global scale with the spread of industrialisation so that eventual dilution can no longer be accepted as a universal solution.

An ever present problem in legislation is in setting the permissible levels of emission of individual contaminants. These should be based on a sound knowledge of the effects of the contaminants on the environment and they must be changed from time to time as our knowledge of the effects of the contaminants changes. It must also be remembered that, without some pollution, modern society as we know it would not be able to exist. Social, political and economic pressures dictate that some pollution should be accepted. In consequence, a compromise is made—emissions are not permitted to exceed levels where they are considered to be harmful and it is required that the best practicable means are used to minimise them.

The cost of pollution abatement in the chemical industry is becoming an increasingly important factor as the trend continues towards tighter controls on emission levels. This has stimulated research and development activity in the field.

ENVIRONMENTAL POLLUTANTS

Any substance which causes pollution is called a pollutant. A pollutant may thus include any chemical or geochemical (dust, sediment, grit, etc.) substance, biotic component or its product, or physical factor (heat) that is released intentionally by man into the environment in such a concentration that may have adverse harmful or unpleasant effects on human health. A pollutant has also been defined as "any solid, liquid or gaseous substance present in such concentration as may be or tend to be injurious to the environment". Pollutants are the residues of things we make, use and throw away. There are many sources of such pollutants. The lakes and rivers are polluted by wastes from chemical and other factories, and the air by gases of automobile exhausts, industries, thermal power plants, etc.

The various principal pollutants which pollute air, water, land are as follows :

(1) Deposited matter—Soot, smoke, tar, dust, grit etc.

(2) Gases—Oxides of nitrogen (NO, NO_2), sulphur (SO_2), carbon monoxide, halogens, (chlorine, bromine, iodine).

(3) Acids droplets—Sulphuric acid, nitric acid etc.

(4) Fluorides

(5) Metals—Mercury, lead, iron, zinc, nickel, tin, cadmium, chromium etc.

(6) Agrochemicals—Biocides (pesticides, herbicides, fungicides, nematicides bactericides, weedicides, etc.) and fertilisers.

(7) Complex organic substances—Benzene, ether, acetic acid, benzpyrenes etc.

(8) Photochemical oxidants—Photochemical smog, ozone, peroxyacetyl nitrate (PAN), peroxybenzoil nitrate (PB_zN), nitrogen oxides, aldehydes, ethylene, etc.

(9) Solid wastes.

(10) Radioactive waste.

(11) Noise.

(12) Odour.

Kinds of Pollutants

On the basis of the kind of pollutant involved, we may have sulphur dioxide pollution, fluoride pollution, carbon monoxide pollution, smoke pollution, lead pollution, mercury pollution, solid waste pollution, radioactive pollution, noise pollution, etc. Of the variety of pollutants, the following two basic types of pollutants non-degradable and biodegradable are recognised.

Non-degradable pollutants

These are the materials and poisonous substances like aluminium cans, mercuric salts, long-chain phenolics, DDT etc. that either do not degrade or degrade only very slowly in nature. They are cycled in ecosystem naturally. They not only accumulate but are often biologically magnified with their subsequent movement in food chains and biogeochemical cycles.

Biodegradable pollutants

They are the domestic wastes that can be rapidly decomposed under natural conditions. They may create problems when they accumulate (i.e., their input into the environment exceeds their decomposition).

Various types of pollution are classified in different ways on the basis of the types of environment being polluted, we may recognise air pollution, water pollution, land (soil) pollution, marine pollution etc.

AIR POLLUTION

Air pollution may be defined as any atmospheric condition in which certain substances are present in such concentration that they can produce undesirable effects on man and his environment. These substances include gases (sulphur oxides, nitrogen oxides, carbon monoxides, hydrocarbons, etc.), particulate matter (smoke, dust, fumes, aerosols), radioactive materials and many others. Most of these substances are naturally present in the atmosphere in low (background) concentrations and are usually considered to be harmless.

Air Quality

The atmosphere is an insulating blanket around the earth. It is source of essential gases, maintains a narrow difference of day and night temperatures and provides a medium for long-distance radio-communication. It also acts as shield around the earth against lethal uv radiations and meteors. Without atmosphere, there will be no lightening, no wind, no clouds, no rains, no snow and no fire. Normal composition of clean air at or near sea is as follows:

Gases	*Per cent (by volume)*
Nitrogen	78.084
Oxygen	20.9476
Argon	0.934
Carbon dioxide	0.0314
Methane	0.0002
Hydrogen	0.00005
Other gases	Minute

Although, there are several parameters to judge air quality, generally three—SO_2, NO_x and SPM (suspended particulate matter) are used which give a fair idea of pollutional load carried by the air.

Unfortulately, the concentration levels of different gases show variations due to pollutions. Moreover, there are also added a number of harmful gases to atmosphere. These adversely affect air quality and make it unfit for living beings.

Industrial air pollution makes an immediate impact on public opinion since it is generally visible and obnoxious and its effects on human health, vegetation and the general environment are well known.

Sources of Air Pollution

Air pollution results from gaseous emissions from mainly industry, thermal power stations, automobiles, domestic combustion, etc.

Industrial chimney wastes

There are a number of industries which are source of air pollution. Petroleum refineries are the major source of gaseous pollutants. The chief gases are SO_2 and NO_x. Mathura-based petroleum refinery is posing threat to Taj Mahal in Agra and other monuments at Fatehpur Sikri complex. Cement factories emit plenty of dust, which is potential health hazard. Stone crushers and hot mix plants also create a menace. The SPM levels in such areas of stone crushing are more than five times the industrial safety limits. There are many food and fertilisers industries which emit gaseous pollutants. There are also chemical manufacturing industries which emit acid vapours in air.

Thermal power stations

There are a number of thermal power stations and super thermal power stations in the country. The coal consumption of thermal plants is several millions tons. The chief pollutants are fly ash, SO_2 and other gases and hydrocarbons. Table 19.1 shows various gaseous pollutants from a 200 MW thermal power plant.

Table 19.1. Gaseous pollutants from a 200 MW thermal power plant.

Components	*Emission factor kg/ton of coal*	*Emitted quantity (tons a day)*
Aldehydes	0.0025	0.0035
Carbon monoxide	0.25	0.35
Hydrocarbons	0.10	0.14
NO_x	10.00	14.00
Oxides of sulphur (0.5% S)	19[s]	13.30
Particulate matter	8[a]	369.60
(33% ash) Ash	2[a]	92.40

[a] Ash content in coal in per cent.

[s] Sulphur content in coal in per cent.

Automobiles

The toxic vehicular exhausts are a source of considerable air pollution, next only to thermal power plants. The ever increasing vehicular traffic density, posed continued threat to the ambient air quality. In the major metropolitan cities, vehicular exhaust accounts for 70% of all CO, 50% of all hydrocarbons, 30–40% of all oxides and 30% of all SPM. The sources of emission in automobiles are:

(1) Exhaust system
(2) Fuel tank and carburettor
(3) Crankcase

The exhaust produces many air pollutants including unburnt hydrocarbons, CO, NO_x and lead oxides. There are also traces of aldehydes, esters, ethers, peroxides and ketones which are chemically active and combine to form smog in presence of light. Evaporation from fuel tank goes on constantly due to volatile nature of petrol, causing emission of hydrocarbons. The evaporation through carburettor occurs when engine is stopped and heat builds up and as much as 12 to 40 ml of fuel is lost during each long stop causing emission of hydrocarbons. Some gas vapour escapes between walls and the piston, which enters the crankcase and then discharges into the atmosphere. This accounts for 25% of the total hydrocarbon emissions of an engine.

Air Pollutants

From the different sources of air pollution, a variety of pollutants are released into atmosphere. The principal air pollutants emitted from these different sources are as follows:

(1) *Carbon compounds:* These are mainly CO_2 and CO, the former released by complete combustion of fossil fuels and the latter by automobile exhausts.
(2) *Sulphur compounds:* These include SO_2, H_2S and H_2SO_4, mostly released by fossil fuel (coal etc.) based power generating plants (thermal plants) and industrial units as refineries.
(3) *Nitrogen oxides:* These include chiefly NO, NO_2, HNO_3, mostly released by automobiles, power plants and industries.
(4) *Ozone:* Its level may rise in atmosphere due to human activities.
(5) *Fluorocarbons:* These come from industries, insecticides spray, etc.
(6) *Hydrocarbons:* These are chiefly benzene, benzpyrene, etc., which are mostly discharged by automobiles and industries.

(7) *Metals:* These include chiefly lead, nickel, arsenic, beryllium, tin, vanadium, titanium, cadmium etc., present in air as solid particles or liquid droplets or gases. They are produced mostly by metallurgical processes, automobiles, seaspray etc.

(8) *Photochemical products:* These are the photochemical smog, PAN, PB_zN, etc., released mostly by automobiles.

(9) *Particulate matter:* These are fly ash, dust, grit and other suspended particulate matter (SPM) released from power plants and industries (stone crushers etc). There are also bacterial cells, fungel spores and pollens in air as biological particulate pollutants.

(10) *Toxicants other than heavy metals:* These are complex chemical substances released during manufacture of other goods.

Acid Rains—An Invisible Threat

The oxides of sulphur and nitrogen are important gaseous pollutants of air. These oxides are produced mainly by combustion of fossil fuels, smelters, power plants, automobile exhausts, domestic fires etc. These oxides are swept up into the atmosphere and can travel thousands of kilometres. The longer they stay in the atmosphere, the more likely they are to be oxidised into acids. Sulphuric acid and nitric acid are the two main acids, which then dissolve in the water in the atmosphere and fall to the ground as acid rain or may remain in atmosphere in clouds and fogs.

Acidification of environment is a man-made phenomenon. The acid rain is infact a cocktail of H_2SO_4 and HNO_3 and the ratio of the two may vary depending on the relative quantities of oxides of sulphur and nitrogen emitted. On an average 60–70% of the acidity is ascribed to H_2SO_4 and 30–40% to HNO_3. The acid rain problem has dramatically increased due to industrialisation. Burning of fossil fuels for power generation contributes to almost 60-70% of total SO_2 emitted globally. Emission of NO_x from anthropogenic sources ranges between 20–90 million tons annually over the globe. Acid rains have assumed global ecological problem, because oxides travel a long distance and during their journey in atmosphere they may undergo physical and chemical transformations to produce more hazardous products.

Acid rains create complex problems and their impacts are far reaching. They increase soil acidity, thus affecting land flora and fauna; cause acidification of lakes and streams thus affecting aquatic life, affects crop productivity and human health. Besides these they also corrodes buildings, monuments, statues, bridges, fences, railings etc. British parliament building also suffered damage due to H_2SO_4 rains. Due to acidity, levels of heavy metals as aluminium, manganese, zinc, cadmium, lead and copper in water increases beyond the safe limits.

PREVENTION AND CONTROL OF AIR POLLUTION

Steps are to be taken to control pollution at source (prevention) as well as after the release of pollutants in the atmosphere. Ideally, the engineering designer's strategy with regard to air pollution control should be to :

(1) Prevent the formation of the pollutant.

(2) Reduce emission by the removal of the pollutant from the process gas stream.

(3) Ensure efficient dispersal in the atmosphère.

Vehicular Pollution

Following are some of the measures to be adopted for control of air pollution by motor vehicles.

(1) To check pollutant emission from vehicular exhaust.

This can be achieved by the following practices:

(a) Using new proportion of gasoline and air.

(b) More exact timing of fuel feeding.

(c) Using gas additives to improve combustion.

(d) By injecting air into the exhaust to convert exhaust compounds to less toxic materials.

(e) Updating of engine design and/or install abatement equipment (device) to improve combustion with the existing engine design.

Carbon monoxide results from low air content of the fuel mixture, whereas NO_x production is promoted by high combustion temperatures. Hydrocarbons follows more or less the pattern of CO. The complete elimination of these three pollutants may be achieved by either updating the present design of engines or by making appropriate changes in devices for improving combustion.

(2) To control evaporation from fuel tank and carburettor.

This can be done by following methods:

(a) Collection of vapours with activated charcoal when the engine is turned off and its ignition when the engine is started.

(b) Subjecting the gasoline in the tank to slight pressure to prevent the gas from evaporation.

(c) Developing low-volatile gasoline that does not evaporate easily.

(3) Use of filters.

Some gas vapours escape between walls and the piston which enter the crankcase and then discharge into atmosphere. Hydrocarbons (about 25%) are released in this way. Thus use of filters that capture and recycle these escaped gases in the engine should help control the emission of these hydrocarbons.

(4) Control through law.

There are to be enforced some standards throuh Motor Vehicles Act and other Acts for design of engines etc.

Industrial Pollution

To check air pollution by industrial and power plant chimney wastes, designer must device measures for removal of the particulate matter and gaseous pollutants from the wastes. Before a pollution control system can be designed or assessed, the nature and quantity of the effluent must be determined. Normally this is done by withdrawing a representative sample from the atmosphere or process and analysing for chemical composition and/or particulate size distribution and concentration.

Particulates

The particle size distribution of a pollutant is important not only in assessing its potential as a hazard (the respirable size range is 0.3-5 μm) but in specifying the correct gas cleaning equipment since this must be used in conjunction with the grade efficiency curve to relate the outlet to the inlet pollutant concentrations.

With regularly shaped particles, it is easy to describe their size—the obvious examples is a sphere in which case the diameter is an adequate description.

However, when the shape is irregular the problem is extremely difficult and numerous definitions are used. The most widely used of these is probably the Stokes diameter—that of a sphere of the same density which settles in a fluid with the same speed. Most methods are related to the technique of analysis used. However, it is sound practice to examine a sample of the particulates by microscope as a preliminary step to a complete size distribution analysis.

Where the particulates are precipitated from the gas stream for subsequent analysis, further size classification can occur. Methods which have been introduced to obviate this include electrostatic precipitation. Notable amongst electrostatic precipitators are the Rochester instrument, Lovelace E.S.P. design, and pulsed voltage precipitator. Thermophoresis has also been used as a precipitation mechanism, but perhaps the simplest and most direct method is to draw the air through an absolute master filter. Membrane type filters which are manufactured with extremely small pore sizes are widely used for this purpose.

Particle size analysis

A very wide range of methods of analysis is now available. Much of the equipment can be obtained commercially and is automatic, requiring very little operation. The more important methods are listed in Table 19.2 in which the mechanisms of operation are shown together with the size range over which they are applicable.

The principle of operation of particle counting instruments is shown in Fig. 19.1. A beam light is focused onto a sensing zone through which the particles are passed individually in a sheath of clear air. The resultant scattered light pulse, which is a function of particle size, is picked up by a photomultiplier tube or solid state detector and fed to a pulse height analyser.

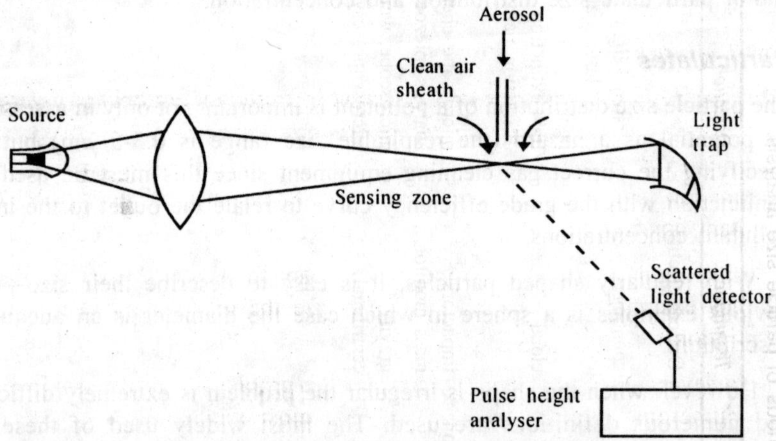

Fig. 19.1. Light scattering particle counter.

The lower limit on particle size which can be measured is imposed by the signal to noise ratio which can be obtained in practice. Use of the laser cavity system in which the particle is fed through the cavity of the laser itself and in which a very high intensity of incident, and hence scattered, beams is achieved has led to an extension of the lower limit on particle size

Table 19.2. Methods of particle size analysis.

Mechanism	*Technique*	*Detection*	*Size range (μm)*	*Resolution*	*Application*
Microscopy	Optical electron		> 1	high	general
			0.005–5.0	high	general-fine aerosols
Optical	counting	scattered light	0.3–15	good	monitoring clean rooms, research
		cavity laser-scattered light	> 0.1	good	monitoring clean,
	cloud	light extinction	> 0.01	none	research concentration
		diffraction pattern	2–100	mean and spread	meter mist analysis
Inertial classification	impactors	microscope, chemical	0.5–50	fair	dust sampling
	mass monitor	analysis, piezoelectric crystal	> 0.01	none	concentration metre
Mobility analysis	gravity	microscope, weight	> 1	fair	dust sampling
	centrifugal	microscope, crystal	0.1–10	good	dust analysis and classification, research
	electrostatic	electrometer	0.006–0.50	good	atmos. pollution studies research
Diffusion	diffusion battery	condensation nuclei counter	0.05–0.50	fair	atmos. pollution studies research
Condensation	nuclei counter	scattered light	> 0.005	none	atmos. pollution studies research
Sedimentation in liquid	gravity	gravimetric, light and	> 2	high	dispersion in liquid
	centrifugal	x-ray attenuation	> 0.2	high	dispersion in liquid
Coulter chamber	resistance change	current flow	> 1.0 μm	high	dispersion in liquid

from about 0.3 to 0.1 μm. Counting methods which employ hydrodynamic particle focusing are restricted by particle coincidence errors for use in relatively low concentration situations. A recent development which allows *in situ* operation at higher concentrations is optical focusing in which a very small sensing volume is used.

Gaseous contaminants

Representative samples of gaseous contaminants are easier to obtain than particulates since inertial classification does not take place. However, precautions may have to be taken to ensure that selective reaction does not occur with components of the sampling system or other gaseous components. Removal of the required contaminants may be achieved by the use of condensation traps, or by gas washers in which chemically selective absorbents are used or by adsorption on, for example, activated carbon, silica gel or molecular sieves. The sample may be removed for total volumetric analysis in the laboratory for it may be passed continuously to a monitor. Recently, there have been a number of developments in the application of remote monitoring.

The methods of analysis can be classified as chemical or physical. In the former the gas is trapped in a suitable medium and analysed chemically. The latter involves the direct measurement of some physical property, perhaps following chromatographic separation. Methods of chemical analysis used include colorimetric, acidimetric and particularly Coulometric methods. Of the physical methods, chemiluminescence and flame photometry, fluorescence and adsorption spectroscopy are all widely used. Both gas chromatography with flame ionisation detection and spectroscopic techniques are used in commercially available instruments for hydrocarbon analysis.

Control of emission of particulates

The selection of equipment for a specific application is dependent on a wide range of factors including the permissible pressure drop, dust loading, gas throughput and temperature, all of which are imposed by the process conditions. An extremely important parameter, however, is the size distribution of the particulates to be removed. The equipment is best characterised with regard to this parameter by its grade efficiency which is a graph of fractional collection efficiency drawn against particle size. There is a wide range of equipment available to the engineer and this is very simply classified in Table 19.3.

The mode of operation of these units is best considered by recourse to the particle removal mechanisms involved.

Table 19.3. Methods of gas/solid separation.

Equipment	*Main collection mechanisms*	*Normal applicable in particle size range (μm)*
Settling chambers	Gravity settling	> 100
Inertial separators	Inertia	> 50
Cyclones	Centrifugal settling	> 5
Scrubbers	Inertia	> 3
Venturi scrubbers	Inertia	> 0.3
Fabric filters	Sieving	> 0.1
Electrostatic precipitators	Electrostatic migration	> 0.001
Fibrous filtration	Inertial interception,	> 0.5
	Diffusion, eletrostatic precipitation	> 0.5

Gravity settlers

Typical gravity settlers (Fig. 19.2) and inertial collectors (Fig. 19.3) are used to collect fairly coarse dusts and may be used as pre-filters. The gravity settler, in particular, works only for large particles the limiting factor being their settling speed in air. That is only particles in excess of 50 μm can be reasonably removed on an industrial scale using gravity separation. The units are, however, widely used as prefilters and their efficiency is enhanced by the inclusion of horizontally placed trays which reduce the sedimentation height and allow greater capture in the same residence time.

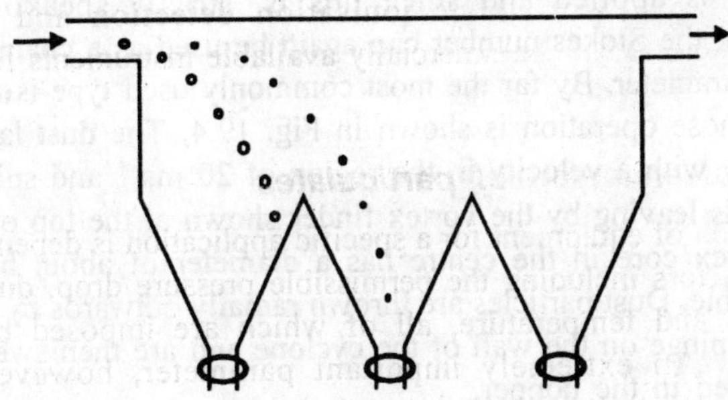

Fig. 19.2. Gravity settler.

Inertial collectors

In inertial collectors (Fig. 19.3) an object is placed in the path of the gas. While the gas passes round the target, particles with sufficiently high inertia impinge on it and are removed from the stream. The inertial mechanism increases in significance with particle size, density and velocity and is reduced as the target size is increased, i.e. with decreasing gas streamline tortuosity. From Fig. 19.3 it can be seen that in the case of inertial separators, an attempt is made to achieve a high gas tortuosity but a pressure drop is incurred as a result of this.

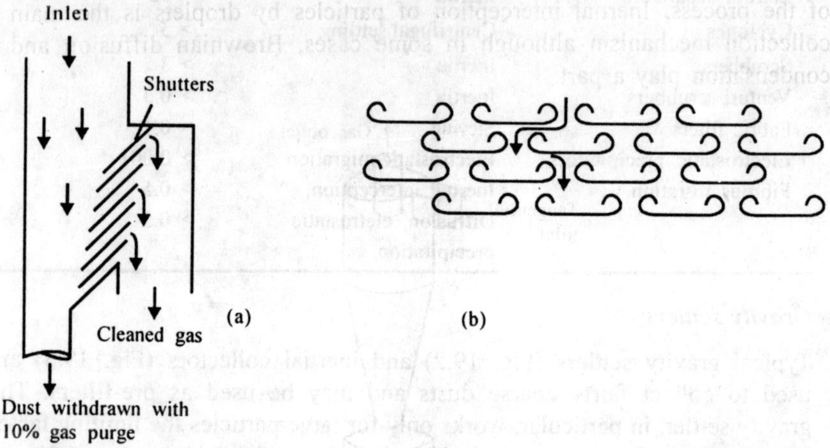

Fig. 19.3. Typical inertial separators: (a) shutter collector; (b) curtain collector.

Cyclones

Cyclones are also primarily used as pre-collectors. In these units centrifugal separation is applied and since this is, strictly speaking, an inertial mechanism, the Stokes number can again be used as a characterisation and scale-up parameter. By far the most commonly used type is the return flow cyclone whose operation is shown in Fig. 19.4. The dust laden gas enters tangentially with a velocity in the region of 20 ms^{-1} and spins downwards and inwards leaving by the vortex finder shown at the top of the unit. The strong vortex core in the centre has a diameter of about half that of the outlet thimble. Dust particles are thrown radially outwards by the centrifugal force to impinge on the wall of the cyclone and are then swept downwards and collected in the hopper.

Wet scrubbers

Wet scrubbers are designed to contact a liquid with a gas and the dust which the gas may contain. A large number of different designs of contacting equipment have been developed and the technique has recently required much attention. Scrubbers offer a number of advantages, the most obvious ones being that they can be used to remove gaseous as well as particulate pollutants; the gas stream is cooled so that there is no fire or explosion hazard if the correct liquid is used and they are normally small compared with dry collectors. Use of a liquid, however, results in economic disadvantages and the gas stream will leave saturated. The creation of a large liquid-gas surface area requires the use of energy so that in general, the greater the energy consumption, the greater is the collection efficiency of the process. Inertial interception of particles by droplets is the main collection mechanism although in some cases, Brownian diffusion and condensation play a part.

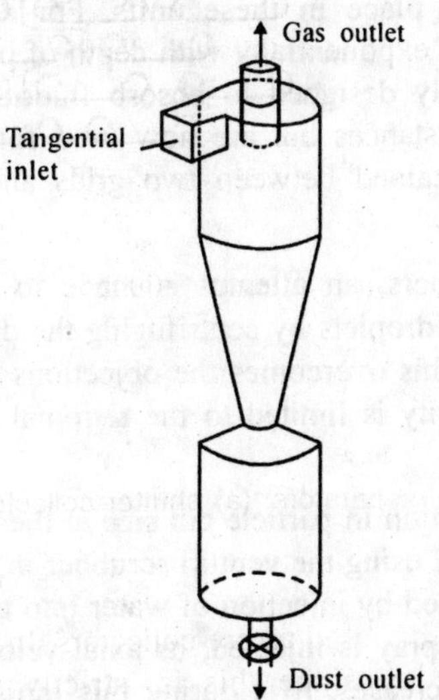

Fig. 19.4. Cyclone separator.

The efficiency of these devices increases with particle size and density and small target sizes and high relative velocities are required from the design and operation point of view. Unfortunately, small droplets accelerate to the gas velocity very quickly, so that a compromise is often made between the two latter requirements. Some of the more common types of scrubbers are described below; in each case a different technique is used to achieve a high relative velocity between the gas and the small droplets.

In the wet impingement scrubber, water flows across an orifice plate through which the gas is passed (Fig. 19.5a). The main mechanism of collection is inertial impaction from the gas stream with the water or the plate located above the orifice. In the self induced spray type (Fig. 19.5b), droplets are sheared from the water surface and inertial collection takes place as they are accelerated by the gas stream. Simple water recirculation takes place and the low water consumption with these units is an advantage.

Packed towers are very widely used in gas absorption but particulates are also removed in the process. The liquid flows downwards through the packing and is contacted with the upward moving gas (Fig. 19.5c). A wide range of packings is available and the liquid flow for a given gas rate is critical. It must exceed the wetting point at which the flow of water is sufficient to wet the packing surface, but must be less than the flooding point at which water down-flow is prevented by the upward motion of the gas. It is most likely that both the droplet collection and impingement referred to above take place in these units. For a given particle size, penetration must fall off exponentially with depth of packing. Fluidised-bed scrubbers were originally designed to absorb fluorides from process gas that contained tarry substances but are now used for particulate removal. The fluidised-bed is retained between two grids and the action ensures continuous cleaning.

In centrifugal scrubbers, an attempt is made to increase the relative velocity of particles and droplets by centrifuging the droplets in an outward direction (Fig. 19.5d). This overcomes the objections to the spray tower in which the relative velocity is limited to the terminal settling speed of the droplets under gravity.

A considerable reduction in particle cut size at the expense of increased pressure drop is achieved using the venturi scrubber in which a high relative droplet velocity is obtained by injection of water into the throat of a venturi (Fig. 19.5e). When the spray is injected, its axial velocity resolute is close to zero but it rapidly increases. It is during this initial period that a high Stokes number is achieved even for submicron particles and a high efficiency results.

Recent developments which have been made in an attempt to reduce power requirements include the use of electrostatic enhancement. In some devices, the incoming gas is ionised by a corona discharge so that the particles become charged, usually negatively, and are attracted to droplets or to an irrigated surface. In other devices the droplets are charged. The pressure drop required to collect submicron particles is significantly reduced and a few such devices are now available commercially. Another development is the use of 'flux force condensation' scrubbing. In process in which sufficient steam is available condensation contributes significantly towards collection. The 'particle size', and hence its inertia, is increased by condensate and both diffusphoresis and thermophoresis assist deposition during the condensation process.

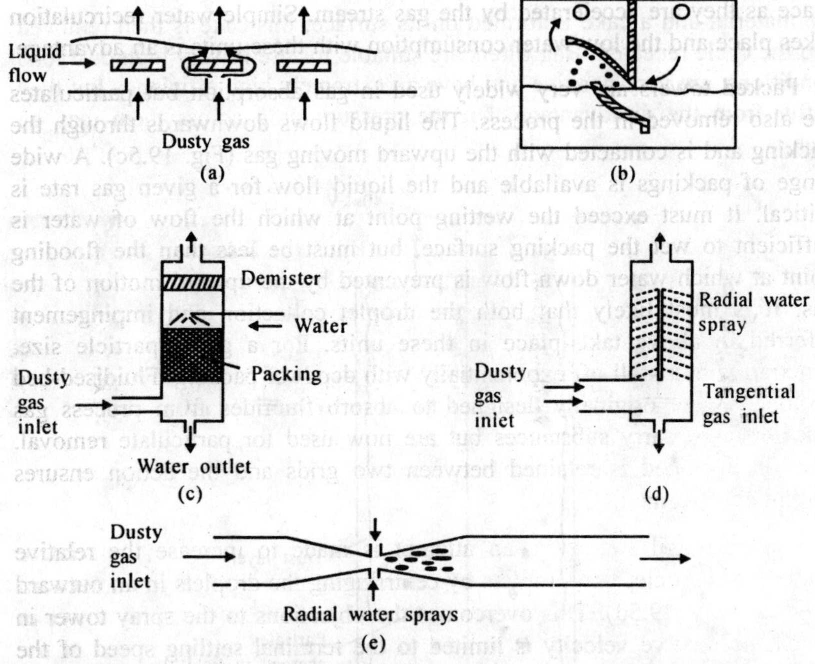

Fig. 19.5. Principles of operation of typical scrubbers: (a) jet impingement; (b) self-induced scrubber; (c) packed tower; (d) centrifugal scrubber; (e) venturi scrubber.

Electrostatic precipitator (ESP)

Electrostatic precipitators are used where collection of fine particles at a high efficiency coupled with a low pressure drop is necessary. The principle of operation is shown in Fig. 19.6 in which dust laden gas is shown to enter a number of tubes or pass between parallel plates. The particulates are charged by corona discharge at the entrance and are subsequently deposited on the earthed plates or tube walls. The walls are 'rapped' periodically to remove the accumulated dust layer. These units are normally used in large installations such as power stations.

Bag filters

Bag filters comprise the simplest and probably the most common class of dust cleaners. Here a cloth or felt filter material which is impervious to the particles is used. Initially, particles are collected within the depth of the media by collision with fibres to which they adhere (the pore size being much greater than the particle size) but eventually the dust builds up within the material and a cake is formed on its surface; this cake is then removed and the cycle repeated. Bag filters are suitable for use in very high dust load conditions, are very positive and have an extremely high efficiency but they suffer from the disadvantage that the pressure drop across them may be high.

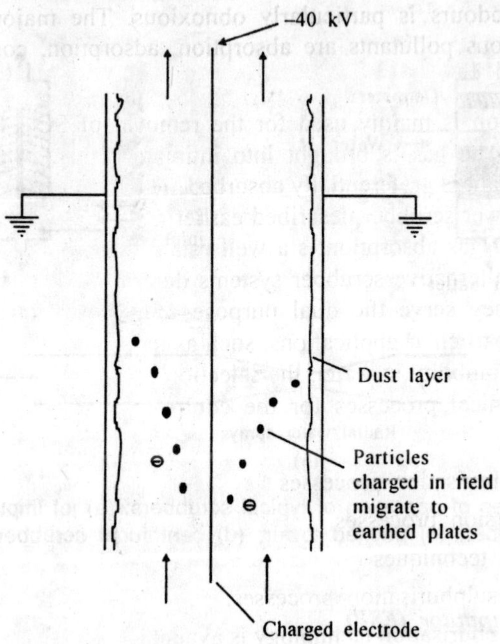

Fig. 19.6. Principle of electrostatic precipitation.

Fibrous filters

Fibrous filters are used to create very clean conditions, such as in clean room installations, to provide a sterile environment for the removal of possible radioactive dusts from nuclear exhausts and in air-conditioning systems. They are used in situations in which the dust load is very low but they can be extremely efficient while offering a very low pressure drop.

Control of emission of gaseous contaminants

The major primary gaseous pollutants arising from industrial sources are those containing sulphur (SO_2, SO_3, H_2S), nitrogen (NO_x, NH_3), halogen compounds, hydrocarbons and carbon monoxide and dioxide. The most important to these from the point of view of volume produced are the products of combustion. CO_2 is not normally considered a pollutant from the industrial arrestment point of view and CO is released in minimum concentrations in the interests of efficiency. SO_2, however, is released in large quantities and presents one of the most difficult problems. The effects of cynogerism in which two relatively inoffensive emissions may combine in the atmosphere to cause an offensive haze must be considered and the production of odours is particularly obnoxious. The major methods of control of gaseous pollutants are absorption, adsorption, combustion and condensation.

Gas absorption is mainly used for the removal of SO_2, HF, Cl_2, HCl, H_2S and NO_x. The gas is brought into intimate contact with a liquid in which the pollutant is preferentially absorbed. The most common equipment is the packed tower scrubber described earlier and the design and operation of such units for gas absorption is a well established chemical engineering operation. The alternative scrubber systems described earlier are also used frequently as they serve the dual purpose of removing both gases and particulates. In particular applications, such as in the removal of SO_2 which has a very low solubility in water, the selection of a suitable solvent is most important. Technical processes for the control of SO_2 emissions can be classified as:

(1) Fuel desulphurisation processes

(2) Fuel conversion processes

(3) Combustion techniques

(4) Flue gas desulphurisation processes.

Fuel oil desulphurisation technology is available and its application could significantly reduce the pollution problem. However, coal cleaning processes are not capable of removing sufficient sulphur from the feedstock to be used on their own. Fuel conversion processes are not yet feasible on economic grounds. The principle of the flue gas desulphurisation system is shown schematically in Fig. 19.7.

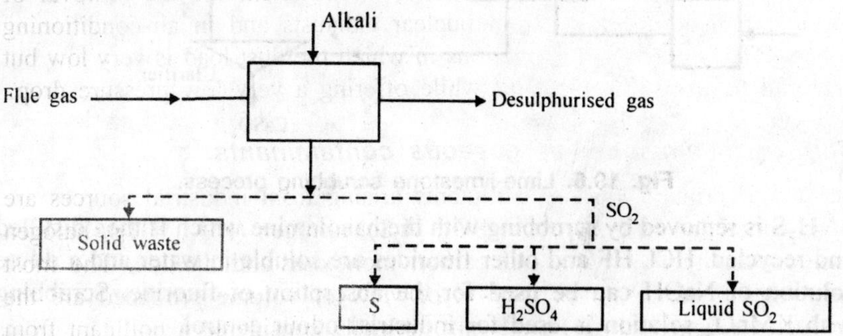

Fig. 19.7. Principle of flue gas desulphurisation.

The most popular process in the U.S., lime-limestone scrubbing, is shown in Fig. 19.8. Particulates are first removed and the flue gas passed to a scrubber where it is contacted with a 5–6% lime or limestone slurry. The following reactions take place:

$$CaCO_3 + SO_2 + \tfrac{1}{2}H_2O \longrightarrow CaSO_3.\ \tfrac{1}{2}H_2O + CO_2$$

or

$$CaO + SO_2 + \tfrac{1}{2}H_2O \longrightarrow CaSO_3.\ \tfrac{1}{2}H_2O$$

Other processes include MgO scrubbing in which the MgO is regenerated by calcination of the $MgSO_3$ and $MgSO_4$ precipitates and the Wellman-Lord process in which the SO_2 is again recovered as a by-product. Catalytic oxidation of the SO_2 to SO_3 using a V_2O_5 catalyst at about 450°C has been used but the dilute H_2SO_4 product has limited commercial value. Other control methods include simple packed tower scrubbing with dimethylaniline from which the SO_2 is distilled off.

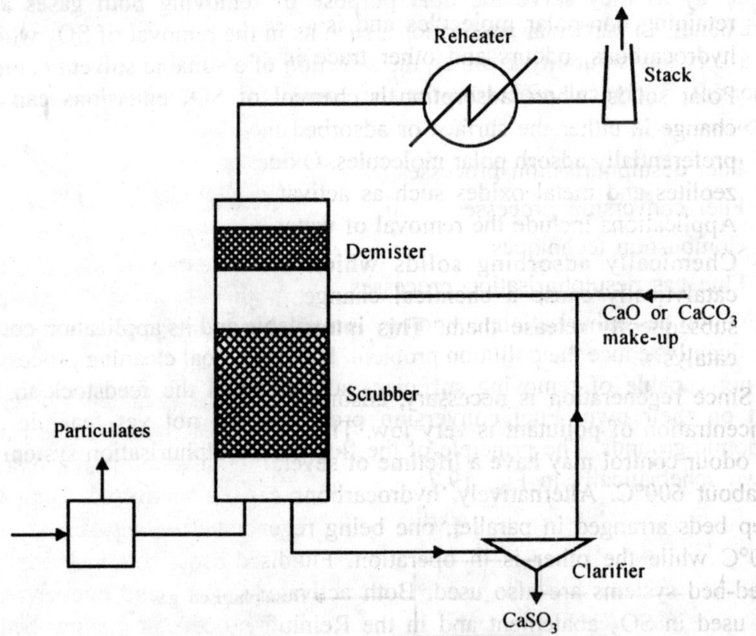

Fig. 19.8. Lime-limestone scrubbing process.

H_2S is removed by scrubbing with diethanolamine which is then distilled and recycled. HCl, HF and other fluorides are soluble in water and a dilute solution of NaOH can be used for the absorption of fluorine. Scrubbing with $KMnO_4$ solution is used for industrial odour control.

98% of NO_x emission from stationary sources is a result of combustion so this can be minimised by control of the combustion conditions. Operation at close to stoichiometric conditions using two stage combustion is effective as a operation at low combustion temperatures, which has been achieved using pressurised fluidisation, with the addition of limestone. NO_x emission from nitric acid plants is controlled by catalytic reduction, adsorption on synthetic zeolites and wet scrubbing. HNO_3 (30%) and hydrogen peroxide are suitable absorbents.

Absorption of pollutants as a method of abatement is well established. The gas is passed through a bed of solid adsorbent on the surface of which the pollutant molecules are preferentially adsorbed and retained by either physical or chemical forces. When the surface is saturated, regeneration of the adsorbent is necessary. Adsorbents can be classified as follows:

(1) Non-polar solids where adsorption is mainly physical. The most important of these is activated carbon which is very effective in retaining non-polar molecules and is widely used for the removal of hydrocarbons, odours and other trace impurities.

(2) Polar solids, where adsorption is chemical but in which no chemical change in either the surface or adsorbed molecules takes place. These preferentially adsorb polar molecules. Oxides of silica such as synthetic zeolites and metal oxides such as activated alumina are widely used. Applications include the removal of water vapour, NH_3, H_2S and SO_2.

(3) Chemically adsorbing solids which either catalytically or non-catalytically cause a chemical change in the adsorbed species and subsequently release them. This introduces the enormous subject of catalysis.

Since regeneration is necessary, adsorption is most suitable where the concentration of pollutant is very low. Thin beds of activated carbon used for odour control may have a lifetime of several months before reactivation at about 600°C. Alternatively, hydrocarbons can be recovered using two deep beds arranged in parallel, one being regenerated with steam at 100–150°C while the other is in operation. Fluidised-beds, rotating-beds and fixed-bed systems are also used. Both activated carbon and copper oxide are used in SO_2 abatement and in the Reinluft process, a moving bed of activated carbon is used as a contact catalyst in the removal of sulphur containing components.

Combustion is a well established process for the control of oxidisable pollutants particularly hydrocarbons, the presence of which in the atmosphere may lead to the formation of photochemical smog. Other applications include the control of odours, smoke, total hydrocarbons and carbon dioxide. Both thermal and catalytic after burners and flares are used.

WATER POLLUTION

Water pollution is defined as "the addition of any substance to water or changing of water's physical and chemical characteristics in any way which interferes with its use for legitimate purposes". Normally water is never pure in a chemical sense. It contains impurities of various kinds—dissolved as well as suspended. These include dissolved gases (H_2S, CO_2, NH_3, N_2), dissolved minerals (Ca, Mg, Na salts), suspended matter (clay, silt, sand) and even microbes. These are natural impurities derived from atmosphere, catchment areas and the soil. They are in very low amounts and normally do not pollute water and it is potable. Polluted waters, however, are turbid, unpleasant, bad smelling, unfit for drinking, bath and washing or other purposes. They are harmful and are vehicles of many diseases as cholera, dysentry, typhoid, etc.

The major pollutants in industrial waste-waters include suspended solids and BOD (Biological Oxygen Demand) or COD (Chemical Oxygen Demand) which are measures of the concentration of oxygen consuming materials. The level of specific toxic materials is of course strictly controlled as are the pH, colour, odour etc. Clearly the quantity of flow and subsequent dilution are also important.

Sources of Pollution and the Pollutants

The chief sources of water pollution are as follows :

(1) Sewage and other waste

(2) Industrial effluents

(3) Agricultural discharges

(4) Industrial wastes from chemical industries, fossil fuel plants (thermal power plants) and nuclear power plants.

Each of these sources of pollution carries a variety of pollutants that enter our water bodies. Following are the sources of water pollution and kind of the pollutants carried by them.

Ground Water Pollution

Ground water is threatened with pollution from seepage pits, refuse dumps, septic tanks, braynard manures, transport accidents and different pollutants. Important sources of ground water pollution are sewage and other wastes otherwise. Raw sewage is dumped in shallow soakpits. This gives birth to cholera, hepatitis, dysentry etc., especially in areas with high water table. The industries of woollens, bicycles contribute high amounts of Ni, Fe, Cu, Cr and cyanides to ground waters.

Marine Pollution

All the what is carried by rivers ultimately ends up in the seas. On their way to sea, rivers receive huge amounts of sewage, garbage, agricultural discharge, biocides, including heavy metals. These all are added to sea. Besides these discharge of oils and petroleum products and dumping of radionuclides waste into sea also cause marine pollution. Huge quantity of plastic is being added to sea and oceans. The pollutants in sea may become dispersed by turbulence and ocean currents or concentrated in the food chain. They may sediment at the bottom by processes like adsorption, precipitation and accumulation. Bioaccumulation in food chain may result into loss of species diversity.

In marine water the most serious pollutant is oil, particularly when afloat on sea. An spill in oil or petroleum product due to accidents or a deliberate discharge of oil polluted waste brings about pollution.

Mercury Pollution

Mercury enters water naturally as well as through industrial effluents. It is a potent hazardous substance. Both, inorganic and organic forms are highly poisonous. Methyl mercury gives off vapours. Mercury was responsible for the Minamata epidemic that caused several deaths in Japan and Sweden.

From the effluents, mercury compounds enter the water body and at their bottom these are metabolically converted into methyl mercury compounds by anaerobic microbes. Methyl mercury is highly persistent, soluble in lipids and after being taken by animals enters into food chain and accumulates in fatty tissues. Fish may accumulate the methyl mercury ions directly. There may be nearly 3000 times more mercury in fish than in water.

Mercury poisoning is caused due to inactivation of several sulphydral enzymes by replacement of hydrogen atoms in sulphydral groups. The antidote, BAL (dimercaptol) is used for mercury poisoning.

Lead Pollution

Lead poisoning is common in adults. The chief source of lead to water are the effluents of lead and lead processing industries. Lead toys may be chewed by children. Painters also have a risk of lead consumption. In some plastic pipes lead is used as stabiliser. The water may become contaminated in these pipes. Lead is also used in insecticides, food, beverages, ointments and medicinal concoctions for flavouring and sweetening.

Lead pollution causes damage to liver and kidney, reduction in haemoglobin formation, mental retardation and abnormalities in fertility and pregnancy. Chronic lead poisoning may cause three general disease syndromes:

(1) Gastrointestinal disorders.

(2) Neuromuscular effects (leadlapsy)—weakness, fatigue, muscular atrophy.

(3) Central nervous system effects or CNS syndrome—that may result in coma and death.

Lead poisoning also causes constipation, abdominal pain, etc.

Fluoride Pollution

Fluoride is also regularly present in water and soil besides air. In nature it is found as fluoride. The crop plants grown in high-fluoride soils in agricultural, non-industrial areas had a fluoride content as high as 300 ppm.

Yamuna—An Open Sewer

A dip in the Yamuna in Delhi no longer cleans the body or uplifts the soul; it only saps the body and sags the soul. A pure holy river has been turned into an open sewer. The river has been affected by the dumping of toxic effluents by various industries of Delhi. The Yamuna water near Delhi is not fit even for irrigation, what to say of drinking, bathing, swimming, fisheries, industrial cooling etc.

Ganga—The Polluted Purifier

Ganga along with its tributaries is the largest and a very important river basin of the country. It has been a symbol of purity but today it is grossly polluted, that is in utter disregard to its sanctity.

Essentially there are three types of pollutants—silt, biological and chemical. Sedimentation may be reduced by rehabilitation of catchment area through tree plantation. The present emphasis is on biological and chemical pollution.

Biological Pollution

There are three main sources :

(1) Urban liquid and solid waste (largest amount of pollution).

(2) Dead bodies of animals and humans.

(3) Wallowing of cattle.

The urban waste is mostly in the form of sheer faecal matter that also goes into the river as surface run off. Since the river is used for mass bathing at Rishikesh, Hardwar and Allahabad, there is problem of water-borne diseases.

Chemical Pollution

There are various chemical industries in India. But only very few have effluent treatment plants in operation. The control of urban waste is indeed a difficult task since most cities and town are on the bank of river which do not have proper sewer and treatment plant facilities. Therefore, first phase in control strategy should include diverting waste-water so as not to flow in the river and then put the same to beneficial use. Among other things the programme must include :

(1) Renovation of sewage pumps, treatment plants and augmentive digester capacity.

(2) Renovation of trunk sewer and other cognate elements.

(3) Extended sewerage in unsewered areas.

(4) Bring waste from unsewered areas to treatment plants.

(5) Provide pumping stations and intercepting sewer and lead to existing treatment.

(6) Install new treatment plants.

Wherever possible, sewage/sullage needs to be recycled for harvesting energy, feed for poultry and animal husbandry, manure and raising aquaculture and supply irrigation water.

The foregoing methods aim the pollution control at point sources. Also there is distributive pollution that reduces self-purifying capability of the river. Point sources are easy to control, distributive pollution cannot be controlled by the orthodox sewage systems. There are plenty of rains during monsoon. The extra water is not only a problem but also goes to sea without any use. There is a need to increase seewage and storing of the surplus monsoon water somewhere. One way is to create big reservoirs at ground level or in underground cavities. This surplus water may be used for different purposes. It also needs to pump water in rabi season and use in kharif. The monsoon water will sink and recharge underground systems.

PREVENTION AND CONTROL OF WATER POLLUTION

Biodegradable pollutants alone are not responsible for water pollution, though these indicate level of pollution (through BOD values). Besides these, a substantial pollution load is contributed by non-degradable or slow degrading pollutants, such as heavy metals, mineral oils, biocides, plastic materials etc. that are dumped into water. For biodegradable pollutants, pollution may be controlled at source by their treatment for reuse and recycling. The non-degradable toxic substances can be removed from water by suitable methods.

In addition to these methods, some standards, conditions and requirements are to be legally enforced by the government through various Acts. The various ways/techniques suggested for control of water pollution is given as under.

Stabilisation of the Ecosystem

This is the most scientific way to control water pollution. The basic principles involved are the reduction in waste input (thus control at source), harvesting and removal of biomass, trapping of nutrients, fish management and aeration. Various methods may be used (biological as well as physical) to restore species diversity and ecological balance in the water body to prevent pollution.

Reutilisation and Recycling of Waste

Various kinds of wastes which include industrial effluents (as paper pulp or other industrial chemicals), sewage/sullage of municipal and other systems and thermal pollutants (waste-water etc.) may be recycled to beneficial use. For instance, urban waste (sewage/sullage) may be recycled to generate cheaper fuel gas and electricity.

Removal of Pollutants

Various pollutants (radioactive, chemical, biological) present in water body can be removed by appropriate methods such as adsorption, electrodialysis, ion-exchange, reverse-osmosis etc. Reverse-osmosis is based on the removal of salts and other substances by forcing the water through a semi-permeable membrane under a pressure exceeding the osmotic pressure. Due to this, flow occurs in reverse direction. For this, we use a power membrane that attracts the solvent and repulses the solute. Reverse-osmosis is commonly used to desalinate the brackish water and can also be used for purifying water from sewage.

Fibrous filters are used to create very clean conditions, such as in clean room installations, to provide a sterile environment for the removal for the possible radio-active dusts from nuclear exhausts and in air conditioning systems. They are used in situations in which the dust load is very low but they can be extremely efficient while offering a very low pressure drop.

Industrial Waste-water Treatment

The techniques applied in industrial waste-water treatment are summarised in Table 19.4 where they are generally classified as primary, secondary and tertiary. Primary treatment normally involves the removal of settleable material where this is present. Secondary, or biological, treatment is often sufficient on its own but with tighter effluent control levels tertiary treatment or 'polishing' is being widely applied. These physical chemical methods are also needed for the treatment of non-biodegradable materials and for specific toxics. Obviously the most suitable effluent control process for any situation depends on the specific nature of the waste materials. For example considerable treatment which may be termed primary is required in the cleaning of refinery waste-waters. The broad classification of 'biological' and 'physical-chemical' processes is, however, convenient and the subject is dealt with under these headings below:

Table 19.4. Waste-water treatment processes.

Primary	*Secondary*	*Tertiary*
Sedimentation	Biological	Physical-chemical
	suspended growth	coagulation and filtration
	film	adsorption
	others (lagoon,	chemical oxidation
	anaerobic etc.)	membrane filtration (ultrafiltration,
		osmosis, electrodialysis)
		ion exchange
		solvent extraction

Biological Treatment

This is the most common waste-water treatment process. A concentrated mass of micro-organisms is used to break down organic matter into stabilised wastes. These organisms can be classified as aerobic, which require molecular oxygen for synthesis; and anaerobic, which function without molecular oxygen and derive energy from organic compounds in the waste. Facultative organisms may operate in either environment. The most widely applied industrial techniques employ aerobic micro-organisms and these techniques can be classified according to the method of accumulating and storing the massed organisms. In the suspended growth method (activated sludge), the organisms are stored in the reacting liquid as a flocculated mass and in the film methods, of which the trickle filter is the most common, the biological mass is fixed as a film to an inert solid material over which the waste-water is passed.

Suspended growth method

A detailed and comprehensive review of this method, is shown in Fig. 19.9. Biological treatment takes place in the aeration tank where the waste is mixed with the flocculated biological sludge and oxygen. Conventionally, air but also pure oxygen and enriched air, are applied to the tank by surface and diffused aeration or by using static mixers. Sludge separation is normally achieved by gravity sedimentation.

The biological culture is in the form of large flocs comprised of a very large number of active bacterial cells each of which is about 1-2 μm diameter. A well flocculated system is necessary for efficient separation of the sludge from the treated water. The mechanisms of cell synthesis and oxidation may be depicted as follows.

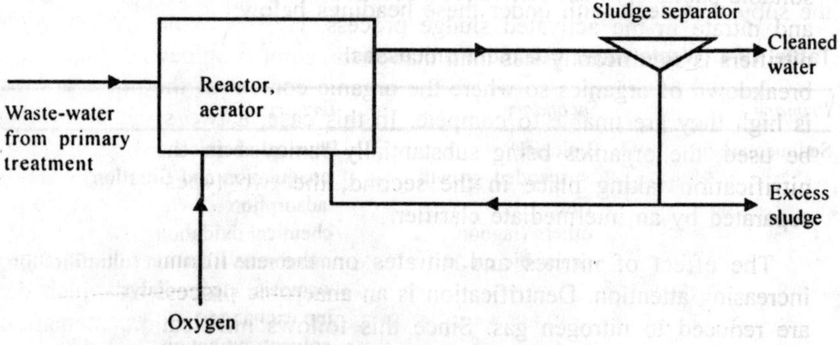

Fig. 19.9. Activated sludge process.

The degradable organic matter in the influent is used as food by the micro-organisms, new cells are formed, the organic waste is rendered inert and CO_2 and water are produced. The rate increases rapidly during the initial stages of the process when the food supply is unlimited and bacteria multiply rapidly. When the food supply runs short, bacterial reproduction ceases and eventually an endogenous phase is reached in which the mass uses up its stored food supply to remain alive. The net increase in biological mass is equal to that produced by food consumption less that lost by endogenous respiration. Clearly this is a very much simplified picture of an extremely complex situation.

The biochemical population can be specifically adapted to particular pollutants. However, in the majority of cases, a wide range of organics must be dealt with and mixed cultures are used. This also adds flexibility to the process since an increase in load of a specific organic will result in an increase in the culture which feeds on it and hence a more rapid removal rate of that material.

Some influent streams may be hydrolysed using caustic soda, carefully controlling the pH level, to render them more easily biodegradable. Excessively high organic loads can be reduced by steam or solvent stripping prior to biological treatment, a technique which is commonly used in the treatment of sour waters in petroleum refineries. Potentially inhibitory and toxic materials can also be removed using this method. Removal of solids prior to treatment and operation at high sludge concentrations (upto 10,000 mg/l) are also considered to enhance performance.

It is frequently required to remove the nutrients, nitrogen and phosphorus, from a waste stream. Nitrification is accomplished using suitable bacteria which can convert reduced nitrogen (as NH_3) to the nitrite and nitrate in the activated sludge process. However, the growth rate of nitrifiers is significantly less than that of the competing bacteria used in the breakdown of organics so where the organic content of the influent stream is high they are unable to compete. In this case, a two stage process can be used, the organics being substantially removed in the first stage and nitrification taking place in the second; the two types of bacteria are separated by an intermediate clarifier.

The effect of nitrites and nitrates on the environment is attracting increasing attention. Dentrification is an anaerobic process by which these are reduced to nitrogen gas. Since this follows nitrification, methanol is often added as a source of energy and carbon but this is expensive and careful control is needed. Soluble phosphorus is removed by the direct addition of lime, aluminium salts and ferric-iron salts.

The addition of powdered activated carbon to the aeration tanks results in the removal of non-biodegradable matter by adsorption. This adds a degree of tertiary treatment and it has been suggested that it adds resistance to shock loading and acts as a toxic sink but the economics of activated carbon and reuse are extremely important.

The treatment and disposal of sludge waste is a major problem. The flow can be as high as 2% of the effluent rate in the activated sludge process and handling and treatment are costly. Concentration by gravity thickening or sometimes by flotation reduces the volume flow rate. Anaerobic digestion is often used to stabilise the sludge into a less offensive material. This is followed by conditioning and finally dewatering.

Anaerobic digestion is often used to stabilise the sludge. In the absence of organic feed, endogenous respiration takes place, the culture consumes itself and gases, primarily methane, are released. It has been found that if the sludge is heated to 24–32°C the rate is increased and the residence time in the digester tank can be reduced from 60 to 30 days. The process is sensitive to the presence of contaminants so some pre-treatment may be necessary.

The material is in a highly flocculated state and further chemical or physical conditioning is necessary before the dewatering stage. Heat treatment under pressure at 18–205°C for a period of 15–45 minutes breaks-down the gel structure. Alternatively a mixture of sludge and air may be heated. Pressure filtration and rotary vacuum filtration are used for dewatering; an alternative is centrifugal separation. Despite much research activity, the gulf between filtration theory and practice is still too wide and the filterability of a sludge is best determined by pilot plant tests. Work on slurry characterisation using the compression permeability cell and capillary suction methods is promising.

The sludge may finally be used for agricultural purposes, composed by drying, or burnt to an ash using open hearth fluidised-bed incinerators.

Fixed film method

The best known fixed film method is the trickle filter in which the waste-water is trickled over a packed-bed in which the medium is covered with biological slime. Oxygen and biological matter diffuse into the slime, where oxidation and synthesis take place and the products appear in the filter effluent. Removal rates are normally low for soluble industrial wastes so the method is generally used only for pre-treatment.

In rotating biological contactors, which are now widely used, the film is attached to the surface of large plastic discs which are rotated slowly with the bottom 1/2–1/4 submerged in the waste-water. Aeration is achieved by exposing the film of waste-water to the air and by air entrainment as the disc re-enters the tank. Since the rate process is mass transfer controlled, its advantages over the trickle filter are obvious. It is also claimed to have stable operation and low maintenance and power costs.

However, with high influent organic concentrations, many discs are required and the capital cost may be high. The system is readily adapted to situations where nitrification and dentrification are needed because the bacteria can be easily separated.

In the biological fluidised-bed process, the waste-water fluidised particulate media, usually activated carbon or sand, to which the film is attached. The mass transfer rate is again high and the agitation prevents slugging.

Physical and chemical methods

Solids separation

Methods of particulate separation include flocculation or coagulation which may be required as a first step. Gravity separation and thickening are often used. Recent development include rapid deep-bed granular filtration. Here, the water is passed through a granular-bed, usually of sand, which is about 1 m in depth and suspended particles, which are much smaller than the pores in the bed, migrate to the media surface where they are retained. A range of washing procedures are used to regenerate the media, including fluidisation and air scouring. The use of graded media in which the grains become progressively finer in the direction of flow results in a higher bed capacity before regeneration and the use of a range of materials of varying density which settle with the largest grains at the top of the bed after washing and up-flow filtration are attracting much interest in this respect. Continuous media recycling-washing processes have also been developed. Precoat filters which use a thin particulate bed laid on cloth media are also widely used. The rotary vacuum filter is probably the most popular of these.

Adsorption

Adsorption, primarily on activated carbon, is being increasingly used for the removal of toxic materials, refractory organics and colour and typically the BOD content may be reduced to 3–10 mg/l. The system consists of carbon handling and storage facilities, reactivation facilities and the adsorption equipment. Both multiple-hearth and rotary-kiln furnaces are used for reactivation. In downflow fixed-bed adsorbers, the water passes down through the bed in which organics are adsorbed. The beds which will also remove some particulate material may be arranged in parallel or in series, depending on process requirements and one bed at a time is removed for reactivation. In packed moving bed absorbers, the fluid passes upwards and the bed downwards, portions being removed periodically from the bottom; upflow expanded bed systems are also used.

In general, molecules of low solubility and polarity tend to be preferentially adsorbed, branched chain compounds are preferred to straight chain compounds and adsorptivity increases with molecular weight for members of a homologous series until the molecules physically block the pores. pH can strongly effect selectivity. Many classes of organics, particularly the oxygenated organics, are not readily adsorbed by activated carbon and much work remains to be done on multi-component systems and on process optimisation.

Chemical oxidation

Chemical oxidation can be used for the oxidation of non-biodegradable organics; the most common agents are chlorine, ozone and hydrogen peroxide.

The use of hydrogen peroxide is particularly effective with sulphur compounds and has been used in the treatment of wastes containing free chlorine, phenols and photographic chemicals. When mixed with formaldehyde, it is used in the treatment of plating wastes. In common with ozone it is a source of dissolved oxygen but the cost of the process has restricted its use.

A widespread example of chemical precipitation followed by solids separation is the lime-soda ash water softening process in which the bivalent cations which are the cause of hardness in water are precipitated on the addition of slaked lime. Many other ions may be precipitated by adjusting the pH and the method is useful in effluent control.

Membrane processes

Ultrafiltration, reverse osmosis and electrodialysis have recently been considered as tertiary effluent control and product recovery methods.

In ultrafiltration the effluent is passed across a semi-permeable membrane. Water passes through the membrane while submicron particles and large molecules are retained in the effluent and concentrated by recycling. The degree of separation and water flow rate depend largely on the pore size of the membrane which is available in a range of materials such as cellulose acetate, triacetate and combinations of these. Membranes based on polyamides can be used over a wide pH range and inorganic membranes can be used in the concentration of oily and latex waste-waters. The membrane is supported on a porous medium for strength and the system arranged geometrically to expose a large filtration surface per unit volume. Configurations used include tubes (1–2.5 cm dia.), plate and frame arrangements and spiral wound modules and hollow fibres. Relatively low pressure differentials (1–10 atmos) are used. These systems are primarily applied where product recovery is desirable such as in the recovery of electrophoretic paint, protein, starch and enzymes and in the recovery of latex concentrated product streams of 20–40% solids can be obtained. They are also used in oil separation and in the treatment of laundry wastes and in PVC recovery.

Reverse osmosis is a high pressure membrane separation process (20–100 atm) which can be used to reject dissolved inorganic salts. It is used industrially as a roughing demineraliser for water with a high total dissolved solids content. Pretreatment is often advisable to prevent membrane fouling. It is used in both product recovery and in water reuse schemes but its major impact is in desalination systems.

Electrodialysis is currently used in water desalination but because of its action in concentrating ionic materials, its potential as a waste-water treatment process is receiving attention and its use in cooling water systems is particularly interesting.

Ion exchange

Ion exchange is used for selective ion removal and is extensively applied in water treatment processes and finds some application in the recovery of specific materials from waste-waters. As with adsorption processes regeneration of the media is necessary. A continuous process incorporating removal of ion exchange resin from the primary tower and its regeneration with hot water is also carried out. It can be used to partially desalinate brackish water in the recovery of chromate and chromic acid from waste-water and in ammonia and dye removal.

Solvent extraction

In this conventional chemical engineering operation the effluent is contacted with a solvent in which organic waste is more soluble. The waste is then separated from the solvent which is recycled. Chemical extractants which are dissolved in a carrier solvent are used to effect the separation of both organic and inorganic chemicals from waste-waters that are difficult to separate using conventional solvents.

SOLID WASTE POLLUTION

The solid waste includes glass containers as bottles, crockeries, plastic containers, polythene and other packing materials that are used and then thrown away as garbage. These pile up at public places and cause obstruction in daily life. Besides these there are also other used things like automobile spares, machines, cycle parts etc. that are thrown as junk. The wastes from building material (during construction and demolition), sludge, dead animal skeletons, heaps of crop residues also contribute to solid waste.

To solve these problems technologies have been developed to recycle most of the solid waste items. Thus paper cans, newspapers and other waste paper may be easily recycled. Metallic components of vehicle spares may be recycled by cheap methods. However, there is problem of recycle of heavy metals, plastics, nylon, polythene, etc.

To control solid waste pollution, following methods have been suggested:

(1) Recycling.

(2) Burning of waste and utilising heat to warm residential units, generation of electricity.

(3) Composting of organic waste for preparation of manures and biogas.

RADIOACTIVE POLLUTION

Discharge of potentially noxious materials into the environment involves some risk, which may or may not be measurable. Within very broad limits research in nuclear hazards enables us to forecast the effects of exposure of large groups of people, for extended periods, to low doses of radiation.

Among the hazardous substances, radioactive substances are most toxic. As compared to organic poisons, injurious effects of radio-nuclides are exceedingly high. For instance, radium is 25,000 times more lethal than arsenic. Nuclear war materials, test explosions, mad rush for power plants and radioisotope use in medicine, industry and research are the main sources of radioactive pollution that could threaten our environmental security.

Sources of Wastes or Nuclear Pollutants

Some of the sources which can cause radioactive pollution and radioactive wastes are listed below:

(1) Uranium mining and milling
(2) Processing of uranium oxide
(3) Fuel fabrication
(4) Reactor wastes
(5) Spent fuel processing
(6) Research and development
(7) Hospitals and biological laboratories
(8) Industrial application

Disposal Principles

If wastes are truly confined, in the sense that in no credible circumstances could they be liberated into the environment, then the only additional requirement is "perceptual custody" to ensure that the confinement is never broken. This is easier said than done. Few private firms go back for 100 years, political regimes have seldom lasted for as long as 500 years, and there are few civilisations that have survived for 2000 years. What then can be done to perpetual custody of wastes containing for example, plutonium with a half-life of 24,000 years?

Dilution and dispersion is the traditional method than men have always used for dealing with their wastes. It depends upon the capacity of the environment to dilute or detoxify the wastes to a level that is innocuous to man and to organisms of interest to man. Safety in discharge to the environment depends upon three factors:

(1) Dispersion by such means as atmospheric dilution, mixing into big bodies of water, or spreading through large volumes of soil.
(2) Fixation of radionuclides on soil minerals and organic detritus.
(3) Decay of radionuclides, dispersed or fixed, before they are able to affect man.

Disposal Practices

Gases

Radioactive gases arise mainly in reactors, spent fuel processing isotope production and research and development facilities. The general principles are the same for all procedures that depend upon dispersion into the atmosphere.

The maximum permissible emission rate—or in some cases the MPC at the stack mouth—is given in the regulations governing the plant or laboratory. It is then the responsibility of the operator to ensure that emissions are kept as for below the permissible level as may be practicable. Numerous methods are available :

(1) Filtration

(2) Electrostatic precipitators

(3) Incinerator off-gas

Liquids

(1) Storage

(2) Evaporation

(3) Flocculation and precipitation

(4) Ion exchange

Solids

(1) Conditioning

(2) Ground disposal

NOISE POLLUTION

The human ear is constantly being assailed by man-made sounds from all sides and there remain few places in populous areas where relative quiet prevails.

Aeroplanes, trains, cars and numatic drills and radio and television sets, all produce noise, the most dangerous pollutant of man's environment. Noise has become a permanent part of our lives these days because of the development of machinery, industry and technology. Noise harms the body and mind. It not only causes irritation or annoyance but is constricts arteries, increases the flow of adrenaline and forces the heart to work faster.

The word noise is usually defined as unwanted or unpleasant sound that causes discomfort. Noise is also defined as "wrong sound, in the wrong place at the wrong time". Noise pollution means "the unwanted sound dumped into the atmosphere leading to health hazards". Formerly noise was limited only to the industry. This too was not much as there were only few industries. These days there has been rapid industrial growth. Moreover, there has been population explosion, due to which there is heavy traffic, urban crowd and electric equipment (luxury items and entertainment). All these have added to the noise nuisence in environment. In our country, besides these the two other factors are the religious and social functions which increase the gravity of situation.

Sources of Noise

The main contributors to noise are factories and industries, transportation (air, rail and road) and community and religious activities. There should not been exaggeration to say that we Indians are a noisy people and every sentiment and occasion is manifested in a noisy manner—be it a religious occasion, elections or a family celebration.

The chief man-made sources in urban areas are automobiles, factories, industries, trains, aeroplanes. Noise makers are horns, sirens, lawn mover, musical instruments, TV, radio, transistor, telephone, dogs, loudspeakers, washing machines, vaccum cleaner, food mixers, pressure cookers, fans, air conditioners, coolers, etc. Ever since the industrial revolution, there has been doubling every 10 years of environmental noise.

Effects of Noise Pollution

The different effects are categorised as :

(1) Auditory effects (affecting hearing faculty).

(2) Non-auditory effects (other than auditory ones).

CONTROL OF NOISE

There are following ways to control and reduce the noise menace.

Source Control

The noise can be reduced and controlled by the following ways:

(1) Designing and fabricating silencing devices in aircraft engines, automobiles, industrial machines and home appliances.

(2) By segregating the noisy machines.

There could be developed gadgets to control noise at source.

Transmission Control

This can be achieved by covering the room walls with sound absorbers as acoustic tiles and construction of enclosures around industrial machinery.

Protect Exposed Person

The workers exposed to noise can be provided with wearing devices as ear plugs and ear muffs.

Create Vegetation Cover

Plants absorb and dissipate sound energy and thus act as buffer zone. Trees should be planted along with high ways, streets and other places. Ashok, Neem, Tamarind etc. are good for this purpose.

Noise Pollution Control Through Law

Silence zones must be created near schools, hospitals and indiscriminate use of loudspeakers at public places may be done by laws. Adequate restrictions must be put on unnecessary use of horns and vehicles plying without silencers.

Public must be made aware and educated about noise nuisance through adequate news media, lectures and other programmes. The movement against noise pollution is very weak in India. The main reason being that most of us do not consider noise as a pollution but as a part of routine life.

Chapter 20

Green Technologies through Chemical Engineering

INTRODUCTION

'Green chemistry is the design, development and implementation of chemical products or processes to reduce or eliminate the use and generation of hazardous substances'. Green chemistry includes concepts such as waste minimisation, solvent selection, atom utilisation, E factor, intensive processing and alternative synthetic routes from sustainable resources. The area of green chemistry has been introduced as a way of using basic science to address environmental issues in an economically profitable manner. In its essence, green chemistry is a science based non-regulatory and economically driven approach to achieving the goals of environmental protection and sustainable development. The approach has been utilised in significant number of countries around the world and the result in these countries have been extremely positive in terms of protection to human health and the environment as well as significant economic benefit to the industrial interests involved. Green chemistry is a scientifically based approach, it addresses problems at the most fundamental level, the molecular level. Through these techniques and methodologies, it can be an effective way to ensure that substances, materials, products and the manner in which they are manufactured are as environmentally benign as possible.

Here, in the Green chemistry, the scope of the process does not include approaches such as waste treatment, waste control or remediation etc., even though these elements are recognised as important, but separate elements of an environmental protection programme. While many countries have those elements capable of reacting to environmental problems once they are formed, there is need for the introduction and focus initiatives that design products and processes such that these environmental problems never occur. This is the focus of Green chemistry. Green chemistry is carrying out chemical activities, including chemical design, manufacturing, usage and disposal, such that hazardous substances will not be used and generated. This means that not only the molecular structures of a final product can be designed to be nonhazardous but also each of the transformations along the way to manufacture of a product are designed so that they does not use or generate hazardous substances. The goal of green chemistry is to reduce the hazards associated with products and processes that are essential not only to maintain the quality of the life achieved by society through chemistry, but also to further advance the technological achievement of chemistry and to do so in a sustainable manner.

PRINCIPLES OF GREEN CHEMISTRY

Green chemistry is the utilisation of a set of principles that reduces or eliminates the use or generation of hazardous substances in the design, manufacturing and application of chemical products. The main principles of Green chemistry may be expressed as follows:

Prevention

'Prevention is better than cure'. It is better to prevent waste than to treat or clean up waste after it has been created. Wealth is converted into waste in this way and wealth is needed to create the waste. Treatability costs are raising steadily, pollution control are becoming stringent making waste treatment almost an unviable approach. However, regulatory requirements necessitate us to do the same by draining the profits.

Atom Economy

Synthetic methods should be designed to maximise the incorporation of all materials used in the process into the final product.

Less Hazardous Chemical Synthesis

Wherever practicable, synthetic methods should be designed to use and substances generate that possess little or no toxicity to human health and the environment.

Designing Safer Chemicals

Chemical products should be designed to effect their desired function while minimising their toxicity.

Safer Solvents and Auxiliaries

The use of auxiliary substances (e.g., solvents, separation agents, etc.) should be avoided wherever possible and innocuous when used. The discovery of next-generation solvents such as supercritical fluids and ionic liquids or the development of solventless systems provide a future alternative to the ubiquitous use of organic and halogenated solvents that are of great concern to human health and the environment.

Design for Energy Efficiency

Energy requirements of chemical processes should be recognised for their environmental and economical impact and should be minimised. If possible, synthetic methods should be conducted at ambient temperature and pressure.

The making and breaking of chemical bonds is intrinsically an energetic process and the material transformation and the basis for energy efficiency must be designed at the molecular level. New methods for improved fundamental design of reaction conditions and pathways are needed to reduce energy intensiveness.

Use of Renewable Feedstocks

Raw materials or feedstocks should be renewable, rather than depleting whenever technically and economically practicable.

Reduce Derivatives

Unnecessary derivatisation (use of blocking groups, protection/deprotection, and temporary modification of physical/chemical processes) should be minimised or avoided if possible, because such steps require additional reagents and can generate waste.

Catalysis

Catalytic reagents (as selective as possible) are superior to stoichiometric reagents.

Design for Degradation

Chemical products should be designed so that at the end of their function they break down into innocuous degradation products and do not persist in the environment.

Real-time Analysis for Pollution Prevention

Analytical methodologies need to be further developed to allow for real-time, in-process monitoring and control prior to the formation of hazardous substances.

Inherently Safer Chemistry for Accident Prevention

Substances and the form of substance used in a chemical process should be chosen to minimise the potential for chemical accidents, including releases, explosions, and fires.

EXAMPLES OF GREEN CHEMISTRY

Alternative Feedstock

The use of alternative feedstocks and starting materials is one of the areas of green chemistry that has seen a great deal of activity in the recent years. Many researchers suggest the use of alternate feedstocks such as carbohydrates, lignin and cellulosic materials in the production of highly functionalised chemicals through atom economical and catalytic reaction pathways. All these starting materials are innocuous to human health and the environment and have the added benefit of providing a way to utilise biomass, that otherwise would be disposed off as solid waste.

Recently many researchers at the Texas University developed a class of technologies that convert waste biomass into animal feed, industrial chemicals and fuels. Firstly, the waste biomass is treated with lime. Lime-treated agricultural residues such as straw, stover etc. may be used as ruminant animal feed. The lime treated biomass can be converted into various chemical products by being fed to a large anaerobic fermentator in which micro-organisms convert the biomass into volatile fatty acid (VFA) salts such as calcium acetate, propionate and butyrate. These salts are concentrated and may be converted into chemicals or fuels via three routes. In the first route, the VFA salts are acidified, releasing acetic, propionic and butyric acids. In the second route, the VFA salts are thermally converted to ketones such as acetone, methyl ethyl ketones and diethyl ketones. In the third route, the ketones may be hydrogenated to their corresponding alcohols such as isopropanol, isobutanol and isopentanol.

Alternative Reagents

Stoichiometric reagents are being developed by using green chemistry principles. Non-phosgene isocyanate synthesis can be sited as a good example for this.

Polyurethanes are important polymers that are widely used for a variety of commercial applications. The environmental difficulty with polyurethanes is that they have been traditionally manufactured through the usage of phosgene, which is an extremely toxic gas. A research team from Monsanto Company developed a method of synthesising polyurethanes and their isocyanate precursors that totally eliminates the use of phosgene.

Synthesis of polyurethanes using phosgene.

$$\underset{\text{Amine}}{RNH_2} + \underset{\text{Phosgene}}{COCl_2} \longrightarrow \underset{\text{Isocyanate}}{RNCO} + 2HCl \xrightarrow{R'OH} \underset{\text{Urethane}}{RNHCO_2R^1}$$

Alternative synthesis of polyurethanes without using phosgene.

$$\underset{\text{Amine}}{RNH_2} + \underset{\text{Carbon dioxide}}{CO_2} \longrightarrow \underset{\text{Isocyanate}}{RNCO} + H_2O \xrightarrow{R'OH} \underset{\text{Urethane}}{RNHCO_2R^1}$$

Catalysis and Biocatalysis

In addition to reagent used in stoichiometric quantities, catalystic and biocatalytic reagents are also being developed by using green chemistry principles.

Following are the some of the major advantages of biocatalysis:

(1) Often run in water with no need for organic solvents.
(2) Often run at or near room temperature at atmospheric pressure.
(3) No toxic metal ions.
(4) No carcinogens.
(5) No noxious emissions, just carbon dioxide.
(6) No toxic wastes; washes can be used as animal feed or composted.
(7) Often use organic wastes as starting materials.
(8) Higher selectivities. It is often possible to react only one of two similar sites.
(9) Chiral compounds can be produced.
(10) No protecting groups are needed in the reaction.
(11) Less energy needed.

Alternative Solvents

A significant amount of research have been devoted recently in the pursuit of alternative solvents, which are more environmentally benign. There have been a number of investigations of using water as a solvent, in place of traditional organic solvents as the medium for carrying out synthetic transformations. *R.Breslow* has researched the Diels-Alder reaction in water and found that the rate of reaction and selectivity can be increased in aqeuous system. He has also investigated the use of Cyclodextrin to facilitate the use of water as a reaction medium and has carried out transformations such as catalytic Cytochrome P-450 biomimetic reaction.

Reduction in Energy Usage

Energy generation and consumption has long been known to include a major environmental impact. Actually, the energy requirements should be recognised for their environmental and economic impacts and should be minimised.

A good example in this category is the use of microwave energy. The use of microwave energy is a technique that is being utilised in order to effect chemical transformations rapidly, and often in the solid state, that have classically been conducted in liquid solutions. In many cases the microwave techniques have shown distinct advantages in not requiring prolonged heating in order to carry out a reaction. The applications of microwave energy as an alternative to conventional heating needs to be tried wherever microwave absorbing i.e., polar materials are present.

These include not only inorganic compounds, but also water, alcohols, amides and many other oxygen containing compounds, but not hydrocarbons, such as hexane and toluene. Microwaves should be well suited for continuous tubular reactors, where they might reduce the amount of energy needed, reduce costs and increase the output of a plant.

MANUFACTURING OF CHEMICAL PRODUCTS

Chemical products can be manufactured using a wide variety of synthesis routes. The designer of a chemical process must choose from alternative raw materials, solvents, reaction pathways, and reaction conditions, and these design choices can have a significant impact on the overall environmental performance of a chemical process. Ideal chemical reactions would have attributes such as:

(1) Simplicity.

(2) Safety.

(3) High yield and selectivity.

(4) Energy efficiency.

(5) Use of renewable and recyclable reagents and raw materials.

In general, chemical reactions cannot achieve all of these goals simultaneously and it is the task of chemists and chemical engineers to identify pathways that optimise the balance of desirable attributes.

Identification of environmentally preferable pathways requires creative advances in chemistry as well as process design. Because the number of choices in selecting reaction pathways is so large and the implications of those choices are so complex, systematic, quantitative design tools for identifying green chemistries are not available. Nevertheless, an extensive body of knowledge concerning green chemistry exists and some qualitative and quantitative design tools are emerging.

Green chemistry, defined as the design of chemical products and processes that reduce or eliminate the use and generation of hazardous substances, is presented in two basic parts–one qualitative and the other quantitative. The first part describes qualitative principles to be used in developing alternatives–alternative solvents, alternative reactants, alternative chemistries–that may lead to environmental improvements.

GREEN CHEMISTRY METHODOLOGIES

The design of a chemical manufacturing process involves feestock selection, selection of solvents, catalysts and other materials, and selection of reaction pathways. This section describes some of the alternatives that are available in making these design decisions and suggests a set of principles that process designers can use to identify alternatives. Specifically, the following issues are addressed:

(1) Alternative feedstocks.

(2) Green solvents.

(3) Synthesis pathways.

Feedstocks

The synthesis and manufacture of any chemical substance begins with the selection of a starting material from which the final product will be synthesised. In many cases, the selection of a starting material can be the most significant factor in determining the impact of a chemical manufacturing process on the environment. There are a number of criteria that can be used in evaluating the potential environmental impacts of materials. For now, note that criteria that may be important in evaluating the environmental performance of a material include its persistence in the environment, its bioaccumulation, potential, its ecotoxicity and human toxicity.

The scarcity of the material and whether it is a renewable or non-renewable resource may also be considered. In addition to considering the persistence, bioaccumulation, and toxicity of materials, the designer should be cognizant of the environmental impacts associated with creating the feedstock material. If a material does not pose any hazard to human health or the environment, for example, but the retrieval or isolation of the substance causes significant risk, then this should be taken into account in the selection. Setting aside, for the moment, the issue of what criteria will be used to evaluate environmental performance of materials, the question remains—How do we identify alternative raw materials that might improve environmental performance ?

Innocuous

The selection of starting materials should start with an evaluation of the materials themselves, to determine if they possess any hazardous properties. Inherent to this analysis is determining whether the process or reaction step requiring the hazardous material is necessary, or whether the final target compound could be obtained from an alternative pathway that uses a less hazardous material.

Generates less waste

An important consideration associated with the use of a particular raw material is whether it is responsible for the generation of the more or less waste than other materials. The amount of waste either generated or eliminated, however, cannot be the only consideration. The type of waste generated must also be assessed. Just as all chemical products are not equal in terms of their hazard, neither are chemical waste streams. Waste streams therefore must also be assessed for any hazardous properties that they possess.

Selective

Utilising a raw material or reaction pathway that is more selective means that more of the starting material will be converted into the desired product. High product selectivity does not always translate into high product yield (and less waste generated), however. Both high selectivity and high conversion must be achieved for a synthetic transformation to generate little or no waste. Using highly selective reagents can means, however, that separation, isolation and purification of the product will be significantly less difficult. Since a substantial portion of the burden to the environment that chemical manufacturing processes incur result from separation and purification processes, highly selective materials and reaction pathways are very desirable.

Efficient

Reaction efficiency, much like product selectivity, has long been a goal of synthetic design, and even prior to the advent of green chemistry principles, has offered benefits. When the overall yield of a process is increased by 10 or 20 per cent, less material ends up in waste streams and more is converted into product. However, yield and selectivity are not entirely adequate as a measure of reaction efficiency.

Solvents

The use of solvents in the chemical industry and the chemical related industries is ubiquitous. According to Toxic Release Inventory (TRI) data, of the chemicals and chemical categories tracked by the programme in 2000, 5 of the top 10 chemicals released or disposed were solvents and include methanol, toluene, xylene, methyl ethyl ketone, and dichloromethane. The total quantity of these chemicals released or disposed was over 687 million pounds, which accounts for 27 per cent of the total quantity of TRI chemicals released and disposed in that year. With increasing regulatory pressure focusing on solvents, there is significant attention being paid to the use of alternative to traditional solvents. General guiding principles in the selection of solvents are given below.

Less hazardous

Solvents have been developed with an eye toward safety since they are used in such large volumes. The earliest and most obvious hazards that were addressed in the design of solvent molecules were their ability to explode or ignite. With the greater understanding of the health and environmental effects that could be caused by a large number of solvents, new solvents are being scrutinised for other hazards as well.

Human health

Solvents are of particular concern because the likelihood of significant levels of exposure is high. Many solvents, by their nature, have high vapour pressure and, in combination with the volumes that are often used, can result in significant exposures. Halogenated solvents such as carbon tetrachloride, perchloroethylene, and chloroform have been implicated as potential and/or suspect carcinogens while other classes of solvents have demonstrated neurotoxicological effects. However, the direct toxicity to human is only one aspect of the total hazards that solvents possess. There are a number of environmental implications of the use of large volumes of solvents.

Environment (local and global)

The use of solvents has caused both global and local environmental concerns. At the global level, the role of chlorofluorocarbons (CFCs) in stratospheric ozone depletion has led to a global phase-out of the substances from virtually all uses. Other solvents have been found to posses significant global warming potential and are thought to contribute to the overall green house gas loading in the environment. At a more local level, the use of certain volatile organic compound (VOCs) as solvents and in other applications has generated concern about their ability to elevate air pollution levels. Some of the main alternatives to traditional solvents include supercritical fluids, aqueous applications, polymerised/immobilised solvents, ionic liquids, solvent-less systems and reduced hazard organic solvents.

Synthesis Pathways

Identifying chemical synthesis pathways that may lead to superior environmental performance is complex and relies on extensive knowledge of synthetic organic chemistry. It is beyond the scope of this chapter to provide a detailed review of this rapidly advancing field, but it is important for a process engineer to be able to identify classes of chemical reactions that have the potential for improvement. Addition reactions ($A + B \longrightarrow AB$), substitution reactions ($AB + C \longrightarrow AC + B$), and elimination reactions ($AB \longrightarrow A + B$), for example, can have different degrees of impact on human health and the environment. Addition reaction incorporate the starting materials into the final product and, therefore, do not produce waste that needs to be treated, disposed of, or otherwise dealt with. Substitution reactions, on the other hand, necessarily generate stoichiometric quantities of substances as by-products and waste. Elimination reactions do not require input of materials during the course of the reaction other than the initial input of a starting material, but they do generate stoichiometric quantities of substances that are not part of the final target molecules. This guidance is qualitative.

A semi-quantitative tool that a process engineer or chemist can use in evaluating synthetic pathways is the concept of atom efficiency. The atom efficiency characterises the fraction of starting materials that are incorporated into desired products and is best illustrated through an example. Friedel Crafts acylations have atom efficiencies that are relatively low. The process typically involves the substitution reaction of an acid chloride with an aromatic substrate. The reaction is frequently accomplished using an aluminium chloride catalyst. The product forms a complex with the catalyst, requiring a water wash, resulting in the formation of hydrochloric acid and salt wastes. The overall reaction is shown in Fig. 20.1.

$$C_6H_6 + RCOCl + AlCl_3 \rightarrow C_6H_5COR$$

$$3HCl + Al(OH)_3(aq) \xrightarrow{3NaOH} 3NaCl$$

Fig. 20.1. Friedel Crafts acylation generates a relatively large amount of waste material, even if the reaction is carried out at 100 per cent yield and 100 per cent selectivity, because of the dissipative use of the aluminium chloride catalysts.

Wastes can sometimes be reduced by simplifying synthesis pathways. Few brief examples—Friedel Crafts reactions, partial oxidations, and reducing the number of steps in synthesis pathways—only scratch the surface of the rich variety of work in synthetic organic chemistry that can improve the performance of chemical manufacturing. In examining reaction pathways, the process engineer should employ tools, such as calculating atom and mass efficiencies, that help determine the magnitude of environmental improvements that may be possible. Additional methods for assessing reaction pathways that may lead to critical environmental improvements.

WIDENING THE SCOPE OF GREEN TECHNOLOGIES THROUGH CHEMICAL ENGINEERING

Green chemistry has become the darling of practitioners and learners alike. Chemical engineers will be at the forefront of developing green technologies. Heterogeneous catalysis, solventless synthesis, use of microwaves, ultrasound, etc. have been attempted by many researchers all over the world to suggest how atom economical processes are also the most cost effective.

Sustainable development will then be a reality. The principles of selectivity engineering can be applied to a variety of important reactions in pharmaceutical and fine chemical industries with reference to environmental issues and a few examples, covering phase transfer catalysis and solid acid catalysis, are cited to illustrate these aspects. The process engineering and reactor engineering aspects are also briefly outlined.

In pharmaceutical and fine chemical industries as also speciality, dyestuff, agrochemical and fragrance industries, multi-step and multi-phase batch reactions are frequently encountered, which involve several cumbersome and expensive separation/purification stages that result into poor atom economy. Pharmaceutical and drug industry typically produce 80–100 kg of by-product or waste per kg of desired product. A challenging aspect of multi-stage synthesis, particularly of bioactive compounds, is that the complexity of the process goes on increasing with each additional stage. This affects the desired selectivity because the number of functional groups and the number of chemically reactive centres in the molecule increase, steric hindrances influence, and thermal stability of compound decrease. Furthermore, recovery and reuse of solvents and/or reactants add to costs.

Sources of Pollution

The basic building blocks in multi-step synthesis are mostly aromatic compounds with incorporation of such groups as halo, $-NH_2$, $-OH$, $-NO_2$, $-SO_3H$, alkyl, acryl, aryl etc. which change the reactivity of the cyclic compound and sometimes the colour and odour. Multi-step synthesis is typically done via routes which do not take into account the atom economy or the amount and nature of co-products or by-products. Catalysis, if any, is limited to homogeneous catalysis involving highly corrosive substances posing disposal problems. Use of solvents is also very common. Low boiling, halogen containing or hazardous inflammable solvents are used.

Some of the important precursors used in these industries are: aniline, acetanilide, benzyl chloride, benzoic acid, monochlorobenzene, chloronitrobenzenes, cresols, cresylic acid, cumene, cyclohexane, cyclohexanone, dichlorobenzenes, dihydroxybenzenes, dimethyl sulphate, ethylbenzene, bisphenol-A, alpha-methylstyrene, nitrobenzene, nonyl-phenol, xylenes, phenol, phthalic anhydride, pyridine, picolines, salicylic acid etc. Most of these are hazardous to handle and wasted during separation or washings. Irrespective of the best solutions to reduce waste at source, it may not be totally avoided and in such cases, it should provide an impetus to convert the waste into an asset. Plant engineers should look for integrated facility to convert liabilities into assets.

The environmental challenges faced by these industries during the 21st century will be:

(1) Development of theoretically zero waste (or minimum waste) processes.
(2) Minimisation of hazardous products and greenhouse gases.
(3) Replacement of corrosive and non-reusable catalysts.
(4) Evolution of sustainable systems; zero energy input processes.
(5) Development of single-pot or cascade engineered processes.
(6) Reduction in number of steps.
(7) Minimum use of solvents, environmentally benign solvents or solvent-free synthesis.
(8) Design of electrically engineered systems (catalysts, different forms of energies, ultrasound; microwave radiation).
(9) Replacement of petroleum feedstock by renewable resources.

Selectivity Engineering Principle

From the process economics and environmental viewpoints, the intensification of reaction rates and enhancement of selectivities can be achieved through selectivity engineering principle which adopts the following strategies:

(1) In a biphasic reaction, addition of a third solid phase (porous or non-porous and/or catalytic or non-catalytic) with particle size smaller than diffusion film thickness. The triphasic system can lead to an enhancement in rate by orders of magnitude with reference to the original gas-liquid, and liquid-liquid systems (micro-phase catalysis).
(2) Converting a liquid-liquid reaction into liquid-liquid-liquid or liquid-liquid-solid phase transfer catalysis.
(3) Heterogenising a homogeneous catalysis.
(4) Cascade engineering or single pot synthesis.
(5) Electrically engineered composite catalysts.
(6) Deliberate incorporation of mass transfer and/heat transfers resistance.
(7) On a larger scale, integration of different plants on the same site to achieve environmental goals.

Process hazards

Industrial accidents are reported to have occurred mostly in batch processes involving hazardous reactants, solvents and exothermic reactions.

Nitrations, oxidations, alkylations, polymerisations, halogenations and hydrogenations are important unit processes, the risk being in that order approximately. Reagents such as phosgene, dimethyl sulphate, formaldehyde/hydrogen chloride (for chloromethylation), sodium azide, hydrogen fluoride, hydrogen cyanide, methyl isocyanate, bromine and for that matter chlorine are hazardous to handle and their use calls for special precautions. Some of these reagents can be replaced by catalytic options such as carbonylation instead of phosgenation, solid superacids for HF etc.

Reactors and process engineering

Innovation in processes can be achieved through reactor design and making the processes continuous. In contrast to the traditional stirred tank, several other combinations appear promising in pharmaceutical and fine chemical industries. The range of options available include:

(1) Fixed bed of catalysts with film flow or flooded, loop reactors.

(2) Fluidised beds.

(3) Moving beds.

(4) Multi-stage suspended beds.

(5) Pipeline reactors with static mixers.

(6) Distillation column reactors for equilibrium limited reactions.

(7) Catalysts in tea-bags.

(8) Catalyst-coated structured packings.

(9) Honey-comb catalysts.

The conventional stirred tank reactor can be modified in several ways to improve mass transfer and heat transfer rates, via new designs of energy saving impellers, leading to drastic reduction in batch times, favourable mixing environments for microbes and enzymes without affecting their activity and selectivity. Spargers have been efficiently designed to aid mass transfer rates and reactors and separators have been combined (Combo-reactors); for instance, reactive distillation, membrane reactors for overcoming thermo-dynamic barriers on conversions and yields.

The role of mixing and micro-mixing are not properly understood in several reactions which are typically conducted in laboratory in round bottom flasks without proper suspension of catalyst or dispersion of immiscible liquids in multi-phase reactions. The data so collected are invariably inadequate for scale-up and lead to lower yields or different product distribution altogether on plant scale.

In fact, a multi-pronged approach is essential in process development with a proper flow-sheet and optimisation of a number of variables simultaneously starting from a micro-reactor, bench-scale prototype, pilot plant, and commercial reactor. Such an optimisation must take into account reaction chemistry, kinetics, molecular modelling, hydrodynamics, safety, hazard and process control. The role of computers for control and optimisation and computational fluid dynamics will be very critical.

For making pharmaceutical and speciality chemicals in a multi-product plant, retro-synthesis, chemical tree building and batch scheduling will also be the key elements, where experience gathered in handling hazardous but atom efficient routes will come handy.

Eco-friendly solvents

The use of solvents has to be properly understood and a variety of eco-friendly solvents including water and supercritical carbon dioxide should be considered. The role of micro-emulsions with water is also a beneficial way of conducting reactions. Various eco-freindly solvents are aqueous blends, oxygenates, terpenes, *N*-methyl-2-pyrrolidone, propylene glycol, methyl isopropyl ketone and methyl amyl ketone, ethyl acetate, *n*-butyl, isopropyl and *n*-propyl acetate, hybrid aqueous-non aqueous solvents, terpene *d*-limonene (citrus-peel derived), tropical derived esters, aliphatic and cycloaliphatic hydrocarbons, isopropyl alcohol, butyl acrylate, 3-ethoxy propionate, butyl propionate, pentyl propionate, dimethyl glutarate, dimethyl succinate, dimethyl adipate and methylal.

Continuous processes

Continuous processes are inherently safer not only due to low inventory of reactants, but also due to improvements in selectivity. Stirred reactors and tubular reactors of small capacity provide high heat transfer rates, withstand high pressures, can be stopped easily and in some cases can be operated adiabatically without leading to run-away situations by introduction of pre-cooled reactants. In the manufacture of intermediates for Vitamin A, a nozzle reactor has been used adiabatically.

Limitations of thermodynamic equilibrium have been overcome in multi-stage sparged reactors, e.g., in the manufacture of phthalimide from molten phthalic anhydride and NH_3. The production of hazardous peroxy esters from the corresponding acid chloride and hydroperoxide is much safer in a continuous mode as also the production of 4,4′-bipyridyl from pyridine.

Miniature technology has been exploited by using micro-reactors, which are 3-D structures imbedded like microchip circuits. The principle advantages of micro-channel reactors are enhanced heat and mass transfer rates, suitability for highly exothermic and hazardous reactions with mass transfer limitations leading to unstable products.

Zero discharge processes

Replacement of a large number of liquid acids by solid acid catalyst in alkylation, acylation, isomerisation, esterification, nitration, hydration has been now widely studied leading to zero discharge processes. Reduction of nitro compounds with hydrogen rather than iron/acid, and epoxidations with hydrogen peroxide or hydroperoxides, rather than using chlorohydrin or peracetic acid, lead to practically zero discharge.

m-phenoxybenzaldehyde can be made using a safer and cleaner process based on *m*-phenoxytoluene, which in turn is obtained by vapour phase dehydration of *m*-cresol and phenol, instead of the polluting process based on bromination of benzaldehyde followed by Ullmann reaction with phenol in the presence of copper catalyst. 2,4-dihyroxyacetophenone can be synthesised from resorcinol and acetic acid in diisopropyl ether over Amberlyst-15. The synthesis of paracetamol is typically done starting from *p*-chloronitrobenzene or phenol in 3–4 stages involving high raw material cost and effluent costs. Instead, the oximation of *p*-hydroxyacetophenone followed by rearrangement by the Celanese process looks attractive.

The Friedel-Crafts acrylation of aromatics is the method of choice in today's organic chemistry for synthesising aromatic ketones as reactive intermediates for the production of fine chemicals. The conventional method of preparation of these aromatic ketones is the homogeneous Friedel-Crafts acrylation of aromatic hydrocarbons with carboxylic acid derivatives using Lewis acids ($AlCl_3$, $FeCl_3$, BF_3, $ZnCl_2$, $TiCl_4$, $ZrCl_4$) or Bronsted acid (polyphosphoric acid, HF). For this purpose, stoichiometric-to-excess amounts of the catalyst are required for the reaction to proceed.

On the industrial scale, the use of metal halide type of acids, which are preferred catalyst, creates work-up and effluent problems. Indeed, during the work-up of acrylation mixtures, catalysts are destroyed which produce relatively large amounts of hydrochloric acid in the off-gas or in the effluent. This hydrochloric acid, which is to be disposed off, originates both from the catalyst and from acryl chloride employed for the acrylation. In addition, to this disposal as a considerable environmental problem, the corrosion problem, due to the hydrochloric acid, must be solved. Pressure from legislative and environmental bodies together with a growing awareness within the chemical industry led to a search for new eco-freindly processes to replace unacceptable outdated reactions. Therefore, a process that could be environmentally friendly and inexpensive, with respect to the disadvantage indicated, is clearly desirable. Solid acids that give the desired level of activity, but which can be easily removed from the reaction mixture with no residual inorganic contamination of the organic products, offer behaviour advantage over existing methods. A 100 per cent atom economical process for the preparation of cyclohexyl esters from the carboxylic acids and cyclohexene over ion exchange resin catalysts where all atoms of the reactants are utilised without any by-product formation is reported in Table 20.1.

Table 20.1. Effect of different substrates.

Acid	*% Conversion*
Phenylacetic acid	74
Benzoic acid	39
Butyric acid	11
Iso-butyric acid	10
Anthranilic acid	15
Acetic acid	30
Propionic acid	21
Heptanoic acid	11

Reaction conditions: acid: cyclohexene; 1:3 mol, time: 6 hours; catalyst: Amberlyst-15; solvent: cyclohexane; catalyst loading: 0.015 g/cm^3; temperature 100°C; speed: 1,000 rpm.

The preparation of perfumery esters such as cyclohexyl-phenyl acetate, cyclohexyl acetate, cyclohexyl anthranilate, cyclohexyl benzoate, cyclohexyl butyrate, cyclohexyl isobutyrate, cyclohexyl heptoate, and cyclohexyl propionate is covered. A complete theoretical and experimental analysis is presented for the model studies with phenylacetic acid and cyclohexane.

Esterification

For the esterification of phthalic anhydride with various alcohols such as isoamyl alcohol, *n*-butanol and 2-ethylhexanol, a large number of catalysts have been reported; for instance, H_2SO_4, *p*-toluenesulphonic acid (PTSA), methanesulphonic acid (MSA), hydrochloric acid and phosphoric acid, which are all liquid-phase catalysts.

Dialkyl phthalates have been prepared from the reaction of phthalic anhydride with alkanols by using various solid acids. Esters of salicylic acid such as methyl salicylate and isoamyl salicylate are known to be very good perfumery products used in industry. Methyl salicylate can be prepared from methanol and salicylic acid by using ion-exchange resins as the catalyst.

Ultrasound-promoted acylation of yara-yara in the presence of solid-acid catalysts

Acylation of 2-methoxynaphthalene with acetic anhydride can be carried out by using different solid acid catalysts such as zeolites, acid activated clays, ion-exchange resins, sulphated zirconia and trifilic acid supported zeolites under irradiation by sonic waves. The product obtained is 2-acetyl-1-methoxy naphthalene.

Acrylation of anisole gives products, which are extremely useful in fine chemical industry. Acrylation of anisole with acetic anhydride is carried out by using trifilic acid supported on mesoporous molecular sieve (HMS), under irradiation by sonic waves. The final product obtained is 4-acetyl-1-methoxy benzene and the selectivity towards *para*-product is more than 95 per cent.

Thus Green chemistry has demonstrated its effectiveness in achieving its mission of meeting environmental and economical goals simultaneously. In order to realise the potential of Green chemistry science, technology and commerce, every chemist has to develop products, processes and services in a sustainable manner to improve the quality of life, the natural environment and industry competitiveness. Green chemistry issues are here to stay. The most successful chemical companies of the future will be those who exploit its opportunities to their competitive advantages, and the most successful chemists of the future will be those who use Green chemistry concepts in R & D, innovation and education.

References

Adams J.A., and **Rogers D.F.**, *Industrial Chemistry*, McGraw-Hill, New York.

Backhurst J.R. and **Parker T.H.**, *Process Plant Design*, Elsevier, New York.

Baum B. and **Parker C.H.**, *Solid Waste Disposal Recycle and Pyrolysis*, Reinhold, New York.

Bauman H.C., *Fundamentals of Cost Engineering in the Chemical Industry*, Reinhold, USA.

Berenson C., *Administration of the Chemical Enterprise*, Wiley-Interscience, New York.

Boynton R.S., *Chemistry and Technology of Paints*, John-Willey, USA.

Budnikov P.O., *Chemistry of Food Processing*, Prentice-Hall, UK.

Burke J.E., *Chemistry and Technology of Dyes*, MacMillan-Pergamon, New York.

Camp, T.R., *Water and Its Impurities*, Reinhold, New York.

Chambers R.D., *Fluorine in Organic Chemistry*, Wiley-Interscience, New York.

Chanlett E.T., *Environmental Protection*, McGraw-Hill, New York.

Clauser H.R., *Encyclopaedia of Engineering Materials and Processes*, Reinhold, USA.

Conover M.S., *Pesticides and Insecticides*, Wiley-Interscience, New York.

Considine D. M., *Encyclopaedia of Instrumentation and Control*, McGraw-Hill, New York.

Cooper H.B.H. and **Rossano A.T.**, *Source Testing for Air Pollution Control*, McGraw-Hill, New York.

Copper R.L. and **Wheatstone K.C.**, *Chemistry of Sugar and Starch*, Wiley-Interscience, New York.

Coughnour D.R. and **Koppel L.B.**, *Process Systems Analysis and Control*, McGraw-Hill, New York.

DeGarmo E.P., *Materials and Processes in Manufacturing*, John Willey-Interscience, New York.

Dickinson D., *Practical Waste Treatment and Disposal*, Wiley-Interscience, New York.

Doremus R.H., *Chemistry of Coal and Coal Chemicals*, Wiley-Interscience, New York.

Eckenfelder W.W. and **O'Connor D.J.**, *Biological Waste Treatment*, Pergamon, UK.

Evans L., *Selecting Engineering Materials for Chemical Plants*, Wiley-Interscience, New York.

Ewing G.W., *Instrumental Methods of Chemical Analysis*, McGraw-Hill, New York.

Fawcett H.H. and **Wood W.S.**, *Safety and Accident Prevention: Chemical Operation*, Wiley-Interscience, New York.

Francis A.W., *Liquid-Liquid Extraction*, Wiley-Interscience, New York.

Hamilton A. and **Hardy H.L.**, *Industrial Toxicology*, McGraw-Hill, New York.

Hampel, *Electrochemical Encvclopaedia*, Reinhold, UK.

Harvey B.G., *Nuclear Physics and Chemistry*, Prentice-Hall, UK.

Hove J. and **Riley W.C.**, *Ceramics for Advanced Technologies*, Wiley-Interscience, New York.

Jackson F.R., *Recycling and Reclaiming of Municipal Solid Wastes*, Noyes, UK.

Jacobs M.B., *Chemical Analysis of Air Pollutants*, Wiley-Interscience, New York.

James A.N., *Industrial Chemistry*, Noyes, UK.

Jolles Z.E., *Chemistry of Polymerisation*, Noyes, UK.

Jurah J.M., *Chemistry of Petroleum*, McGraw-Hill, New York.

Kathera R.L. *Chemistry of Synthetic Fibres,* Harwood Academic Publishers, New York.

Kent J.A., *Riegel's Industrial Chemistry*, Reinhold, USA.

Kingzett, *Chemical Encyclopaedia*, Van Nostrand, USA.

Kirk-Ohmer, *Encyclopaedia of Chemical Technology*, Wiley-Interscience, New York.

Kit B. and **Evered D.S.**, *Chemistry and Technology of Plastic and Rubber*, MacMillan, New York.

Kriegel W.W. and **Palmour H.**, *Mechanical Properties of Engineering Ceramics*, Wiley-Interscience, New York.

Laitinen H.A. and **W.E. Harris**, *Chemical Analysis*, McGraw-Hill, New York.

Landau R., *Chemical Plant Design*, Reinhold, USA.

Levenspiel O., *Chemical Reaction Engineering*, Wiley-Intersicence, New York.

Luyben W.L., *Process Modelling, Simulation and Control for Chemical Engineers*, McGraw-Hill, New York.

Mantell C.L., *Chemistry of Wood*, Wiley-Interscience, New York.

McCoy J.W., *The Chemical Treatment of Cooling Water*, McGraw-Hill, New York.

Miller S.A., *Acetylene: Its Properties, Manufacture and Uses*, Academic, UK.

Morton M., *Rubber Technology*, Van Nostrand, Reinhold, USA.

Nagiev M.F., *The Theory of Recycle Processes in Chemical Engineering*, MacMillan-Pergamon, UK.

Perry R.H., *Chemical Engineers' Handbook*, McGraw-Hill, New York.

R. Norris Shreve and **Joseph A. Brink**, *Chemical Process Industries*, McGraw-Hill, New York.

Ray W.H., *Process Optimisation*, Wiley-Interscience, New York.

Rhodes T. and **Carroll G.**, *Industrial Instruments for Measurements and Control*, McGraw-Hill, New York.

Rudd D., *Process Synthesis,* Prentice-Hall, New York

Rudge A.J., *Manufacture and Use of Fluorine and its Compounds*, Oxford, USA.

Sauchelli V., *Chemistry and Technology of Fertilisers*, Reinhold, UK.

Sconce J.S., *Leather Technology*, Reinhold, UK.

Sherwood T.K., *Process Design*, Wiley-Interscience, New York.

Singer F.S., *Agro-Chemical Industries*, Chemical Publishing, UK.

Siting M., *Catalyst Manufacture: Recovery and Use*, Noyes, UK.

Siting M., *Chemistry of Fermentation*, Noyes, UK.

Stephenson R.N., *Organic Chemical Process Encyclopaedia*, Noyes, UK.

Stephenson R.N., *Introduction to Chemical Process Industries*, Reinhold, USA.

Swern W., *Environment Engineering*, Reinhold Publishing Company, USA.

Urbanski T., *Chemistry and Technology of Soaps and Detergents*, Pergamon, UK.

Urdang George, *Essential Oils*, Van Nostrand, New York.

Van Vlack L.H., *Physical Ceramics for Engineers*, Addison-Wesley, UK.

Van Wazer J.R., *Phosphorus and its Compounds*, Wiley-Interscience, New York.

Weiss S., *Hydrofluoric Acid Manufacture*, Noyes, UK.

Yaffe L., *Nuclear Chemistry*, McGraw-Hill, New York.

Yoshiro Masada, *Analysis of Essential oils by Gas Chromatography and Mass Spectrometry*, John Wiley and Sons, Inc., New York, London.

Index